AF594929

Molecular and Clinical Advances in Understanding Early Embryo Development

Molecular and Clinical Advances in Understanding Early Embryo Development

Editor

Lon J. van Winkle

MDPI • Basel • Beijing • Wuhan • Barcelona • Belgrade • Manchester • Tokyo • Cluj • Tianjin

Editor
Lon J. van Winkle
Department of Medical
Humanities
Rocky Vista University
Parker
United States

Editorial Office
MDPI
St. Alban-Anlage 66
4052 Basel, Switzerland

This is a reprint of articles from the Special Issue published online in the open access journal *Cells* (ISSN 2073-4409) (available at: www.mdpi.com/journal/cells/special_issues/Embryo-Development).

For citation purposes, cite each article independently as indicated on the article page online and as indicated below:

LastName, A.A.; LastName, B.B.; LastName, C.C. Article Title. *Journal Name* **Year**, *Volume Number*, Page Range.

ISBN 978-3-0365-7493-6 (Hbk)
ISBN 978-3-0365-7492-9 (PDF)

Contents

About the Editor

Lon J. van Winkle

Lon J. Van Winkle earned his Ph.D. in Biochemistry from Wayne State University in 1975. He joined the faculty at several community colleges near Detroit in 1974 and became a full-time faculty member in Natural Sciences at the Dearborn campus of the University of Michigan in 1977. He then took a position at Midwestern University in 1979 where he remained until he retired as the Professor and Chair of Biochemistry in August 2015. Dr. Van Winkle is currently a Professor of Medical Humanities at Rocky Vista University and continues to publish regularly in the scientific literature on both early embryo development and medical science education. His work has been supported by grants from the Illinois Board of Higher Education, the Illinois Regenerative Medicine Institute, and the National Institutes of Health, which have helped him produce over 100 peer-reviewed publications and 5 books concerning biomembrane transport, embryo development, stem cells, emerging talents in pharmacology, and interprofessional education. He is a member of Sigma Xi, the American Society for Biochemistry and Molecular Biology, the American Physiological Society, the Society for the Study of Reproduction, the International Association of Medical Science Educators, and the Gold Humanism Honor Society.

Preface to "Molecular and Clinical Advances in Understanding Early Embryo Development"

This reprint considers a wide variety of topics concerning molecular and clinical advances in understanding early mammalian embryo development. The authors of various chapters of this reprint examine oocyte biology, preimplantation embryo development, blastocyst implantation in the uterus, and second cellular lineage differentiation. In addition, epigenetic contributions to reproduction, which may occur throughout pregnancy and in resultant offspring, are also considered. Environmentally induced epigenetic changes in germ cells can render mammalian offspring less healthy, and these changes can be difficult to reverse. Included in these changes are those caused by assisted reproductive technologies (ARTs). It remains to be determined whether the healthiest preimplantation embryos develop in vitro in conditions that mimic the physiological environment in vivo or whether simpler conditions can also foster this development in vitro. In this regard, a number of advances as well as challenges in ART are considered by the authors of the chapters included in this reprint.

Lon J. van Winkle
Editor

MDPI

Editorial

Molecular and Clinical Advances in Understanding Early Embryo Development

Lon J. Van Winkle [1,2]

[1] Department of Medical Humanities, Rocky Vista University, Parker, CO 80122, USA; lvanwinkle@rvu.edu or lvanwi@midwestern.edu
[2] Department of Biochemistry, Midwestern University, Downers Grove, IL 60515, USA

Citation: Van Winkle, L.J. Molecular and Clinical Advances in Understanding Early Embryo Development. *Cells* **2023**, *12*, 1171. https://doi.org/10.3390/cells12081171

Received: 10 April 2023
Accepted: 14 April 2023
Published: 16 April 2023

1. Introduction

The articles in this Special Issue address a wide variety of topics concerning molecular and clinical advances in understanding early embryo development. For convenience, the papers presented in this editorial address topics from the examination of oocyte biology to the emergence of second cellular lineages in association with and following mammalian embryo implantation in the uterus. In addition, epigenetic contributions to reproduction, which may occur throughout pregnancy and in resultant offspring, are also considered.

2. Early Development

2.1. Oocytes

Four papers in this Special Issue consider molecular, clinical and even evolutionary advances in understanding oocyte development. A possible metabolic function of the Na^+-dependent amino acid transport system $B^{0,+}$ in porcine oocytes is the provision of amino acids for protein synthesis and leucine to initiate mTOR1 signaling [1]. This also appears to be the case in Xenopus oocytes, where $B^{0,+}$ disappears prior to egg deposition in fresh water. This transporter appears not to be present in the oocytes of mammalian species exhibiting invasive embryo implantation in the uterus. Instead, $B^{0,+}$ first appears in rat, mouse, and likely human embryos at the blastocyst stage (see description of function below). Contrastingly, $B^{0,+}$ is highly expressed in sea urchin oocytes to take up amino acids from sea water, but only after they are fertilized [1]. These differences in the timing of $B^{0,+}$ expression are an important mechanism of evolution known as heterochrony [1].

In mouse oocytes, there appear to be other needs for amino acid transport [2], which may also be the case for human oocytes [1]. For example, proline transport into mouse oocytes by the Na^+-independent transporters, PAT1 and PAT2, decreases mitochondrial activity and the production of reactive oxygen species (ROS), and thus improves subsequent early embryo development after in vitro fertilization [2]. In humans, and likely other species, fertility may be altered by mutations in mitochondrial DNA in association with ovarian aging, possibly due to ROS production [3]. Interestingly, the replacement of damaged mitochondria by healthier counterparts in human oocytes may soon improve assisted reproductive technologies (ART) [4].

2.2. Preimplantation Development

Remarkably, our knowledge of the importance of amino acid transport, signaling, and metabolism to preimplantation development continues to develop. For example, in a paper of this Special Issue, Treleaven et al. report how proline transport into preimplantation mouse embryos fosters their development but in a stage-specific manner [5]. In other words, proline improves the preimplantation development of mouse zygotes, and the B^0 amino acid transporter appears to be mainly responsible for this uptake on day 4 but not on other days of development [5]. Moreover, proline fosters development in a growth-factor-like manner, in part because proline can substitute for the apparent paracrine signaling among these embryos grown at a high density [1,6]. One or more proline transporter also seems to function as a transceptor to activate mTOR1, Akt, and ERK signaling in preimplantation embryos [1].

Similarly, glycine fosters the preimplantation development of mouse embryos in a hypertonic oviductal fluid-like medium, but it also antagonizes the positive effects of proline, possibly by inhibiting proline transport [1]. Of the amino acids, proline seems to be most beneficial for preimplantation development in vitro. Therefore, the most effective way to culture healthy mouse and possibly human embryos is to supply only proline in these media [5,6]. Although complex, conditions in vivo can also be highly suitable for fostering the development of the healthiest preimplantation embryos under physiologically normal conditions [1,7]. Hence, it still needs to be determined whether the latter conditions can be mimicked in vitro in ways that do not cause unwanted effects on offspring at birth or later in adult life [1].

As discussed by Chen et al. in this Special Issue [8], the detrimental effects of in vitro culture are significant for porcine embryos where only about 40% of fertilized porcine oocytes develop into blastocysts. While such cultures are considerably more successful in mouse and human embryos, children conceived through ART have much greater risks of developing metabolic syndrome and associated conditions, such as higher body fat, greater fasting blood glucose, hypertension, and cardiovascular disease, in comparison to children conceived under physiological conditions in vivo [7]. Supporting the notion that mimicking in vivo conditions can improve development in vitro, in this Special Issue, Saadeldin et al. report that extracellular vesicles from the endometrium can increase the frequency of attachment of porcine blastocysts [9]. Conversely, a model to mimic type 2 diabetes by culturing early rabbit embryos in a medium containing high levels of glucose and insulin showed extensive and unwanted initiation of a number of gene expressions in the resultant inner cell mass (ICM) and trophectoderm (TE) of blastocysts [10]. Hence, attempts to mimic physiological conditions in vitro can be complex, considering the possible variability of these conditions in prospective mothers.

2.3. Implantation

Signaling and metabolic changes are also involved in initiating implantation and maintaining pregnancy in vivo. The uptake of leucine via amino acid transport system $B^{0,+}$ initiates mTOR1 signaling in the TE of mouse and likely blastocysts of all mammalian species exhibiting invasive implantation [1]. This signaling leads to development of the trophoblast motility needed by blastocysts to invade the uterine epithelium. For unknown reasons, system $B^{0,+}$ then becomes relatively inactive in blastocysts in utero probably due to the action of extracellular histones. However, it likely must be reactivated at the time of uterine penetration in order to help deprive T-cells of another $B^{0,+}$ preferred substrate, tryptophan, thus preventing the rejection of blastocysts in implantation chambers [1]. At the same time, ICM cells are maintained in a pluripotent state, partly due to the trimethylation of lysine 4 in histone H3 to form H3K4me3 [1]. However, these ICM cells will soon begin to differentiate into a number of tissues including, eventually, those in the post-implantation embryo and fetus.

2.4. Second Lineage Differentiation

Following differentiation of the ICM and TE from the morula, and in association with implantation, the TE begins to form extraembryonic tissues, such as the placenta, while the ICM gives rise to the primitive endoderm (hypoblast) and epiblast (EPI). Thus, second lineage differentiation begins, followed by endoderm, mesoderm, and ectoderm development from EPI cells. The yolk sac develops from hypoblast cells, whereas tissues of the embryo and fetus arise from the endoderm, mesoderm, and ectoderm.

Several papers in this Special Issue concern these second lineage differentiations. Firstly, Duan et al. show that the silencing of a long terminal repeat element of an endogenous retrovirus, known as MacERV6-LTR1a, postpones the differentiation of TE, EPI, and hypoblast cells in cynomolgus monkey embryos at day seven [11], although the mechanism for this delay is still being explored. In the mouse, NANOG expression is needed for hypoblast formation, but Springer et al. demonstrate that this is not the case for bovine embryos [12]. Hence, species differences clearly exist in the mechanisms of these second lineage differentiations. The delineation of these differences is essential in species such as mouse and pig that are sometimes used as models for human embryo development.

In addition to signaling molecules, epigenetic mechanisms likely play roles in regulating the emergence of these cell lineages in early mammalian embryo development, and the detailed mechanisms of these epigenetic modifications also vary among species. For example, the formation of the H3K4me3 needed to maintain ICM cell pluripotency in mice exclusively depends on threonine metabolism in mice, but in humans, H3K4 methylation likely relies on serine metabolism [1]. Further complicating this regulation are differences in chromatin structure in different cell lineages. As reported by Quan et al. in this Special Issue, domains have been identified as CpG-rich (forests) and CpG-poor (prairies) in chromatin [13]. In both early human and early mouse embryos, the ectoderm cell lineages show the weakest domain segregation, whereas the endoderm cell lineages display the strongest domain segregation in germ layers [13]. The significance of these chromatin domain segregations for epigenetic contributions to cell development is yet to be investigated.

3. Epigenetics in Human Reproduction

The importance of epigenetics to human reproduction cannot be over emphasized. For example, in this Special Issue, Wen et al. report that the loss of *H19/IGF2* epigenetic imprinting in the decidual microenvironment of early human pregnancy is associated with recurrent spontaneous abortions [14]. More broadly regarding epigenetics and pregnancy outcome, diet-induced obesity in mice is associated with numerous methylation changes in DNA in genes associated with type 2 diabetes mellitus [15]. Additionally, offspring of such mice are also likely to be obese [16], probably due to the transmission of these epigenetic changes to offspring by their mothers. Importantly, a study in this Special Issue—to investigate whether a pre-pregnancy lifestyle intervention in obese women reduces the risk of obesity and cardiometabolic disease of their offspring—showed no effect of an intervention [16]. Moreover, these authors summarize numerous studies on humans and other animals that demonstrate little or no effect of such interventions on the health of offspring (Table 3 of [16]). Hence, the epigenetic changes associated with obesity and related conditions seem resistant to modification before transmission of the pertinent epigenetically modified genes to offspring via maternal germ cells.

4. Conclusions

- Amino acid transport and signaling in oocytes influence their mitochondrial metabolism, ROS production, and health; therefore, the replacement of damaged mitochondria in oocytes may soon improve ART in humans.
- During the preimplantation period, amino acid transport and signaling also foster more normal embryo development.
- It remains to be determined whether the healthiest preimplantation embryos develop in vitro in conditions that mimic the physiological environment in vivo or whether simpler conditions can also foster this development in vitro.
- Amino acid transport, metabolism, and signaling are also needed in blastocysts to maintain their pluripotent ICM cells and to foster the trophoblast invasion of the uterine epithelium during implantation in the uterus.
- The details of signaling needed to promote second cellular lineage differentiation in peri-implantation embryos varies among mammalian species used as models for human embryo development.
- Environmentally induced epigenetic changes in germ cells can render mammalian offspring less healthy, and these changes can be difficult to reverse.

Conflicts of Interest: The author declares no conflict of interest.

References

1. Van Winkle, L.J. Amino acid transport and metabolism regulate early embryo development: Species differences, clinical significance, and evolutionary implications. *Cells* **2021**, *10*, 3154. [CrossRef] [PubMed]
2. Treleaven, T.; Hardy, M.L.; Guttman-Jones, M.; Morris, M.B.; Day, M.L. In vitro fertilization of mouse oocytes in L-Proline and L-Pipecolic acid improves subsequent development. *Cells* **2021**, *10*, 1352. [CrossRef] [PubMed]
3. Podolak, A.; Woclawek-Potocka, I.; Lukaszuk, K. The Role of Mitochondria in Human Fertility and Early Embryo Development: What Can We Learn for Clinical Application of Assessing and Improving Mitochondrial DNA? *Cells* **2022**, *11*, 797. [CrossRef] [PubMed]
4. Palmerini, M.G.; Antonouli, S.; Macchiarelli, G.; Cecconi, S.; Bianchi, S.; Khalili, M.A.; Nottola, S.A. Ultrastructural evaluation of the human oocyte at the germinal vesicle stage during the application of assisted reproductive technologies. *Cells* **2022**, *11*, 1636. [CrossRef] [PubMed]
5. Treleaven, T.; Zada, M.; Nagarajah, R.; Bailey, C.G.; Rasko, J.E.; Morris, M.B.; Day, M.L. Stage-Specific L-Proline Uptake by Amino Acid Transporter Slc6a19/B0AT1 Is Required for Optimal Preimplantation Embryo Development in Mice. *Cells* **2023**, *12*, 18. [CrossRef] [PubMed]
6. Morris, M.B.; Ozsoy, S.; Zada, M.; Zada, M.; Zamfirescu, R.C.; Todorova, M.G.; Day, M.L. Selected amino acids promote mouse pre-implantation embryo development in a growth factor-like manner. *Front. Physiol.* **2020**, *11*, 140. [CrossRef] [PubMed]
7. Van Winkle, L.J. Perspective: One-cell and cleavage-stage mouse embryos thrive in hyperosmotic oviductal fluid through expression of a glycine neurotransmitter transporter and a glycine-gated chloride channel: Clinical and transgenerational implications. *Front. Physiol.* **2020**, *11*, 613840. [CrossRef] [PubMed]
8. Chen, P.R.; Redel, B.K.; Kerns, K.C.; Spate, L.D.; Prather, R.S. Challenges and considerations during in vitro production of porcine embryos. *Cells* **2021**, *10*, 2770. [CrossRef] [PubMed]
9. Saadeldin, I.M.; Tanga, B.M.; Bang, S.; Seo, C.; Koo, O.; Yun, S.H.; Kim, S.I.; Lee, S.; Cho, J. ROCK Inhibitor (Y-27632) Abolishes the Negative Impacts of miR-155 in the Endometrium-Derived Extracellular Vesicles and Supports Embryo Attachment. *Cells* **2022**, *11*, 3178. [CrossRef] [PubMed]
10. Via y Rada, R.; Daniel, N.; Archilla, C.; Frambourg, A.; Jouneau, L.; Jaszczyszyn, Y.; Charpigny, G.; Duranthon, V.; Calderari, S. Identification of the Inner Cell Mass and the Trophectoderm Responses after an In Vitro Exposure to Glucose and Insulin during the Preimplantation Period in the Rabbit Embryo. *Cells* **2022**, *11*, 3766. [CrossRef] [PubMed]
11. Duan, K.; Si, C.Y.; Zhao, S.M.; Ai, Z.Y.; Niu, B.H.; Yin, Y.; Xiang, L.F.; Ding, H.; Zheng, Y. The Long Terminal Repeats of ERV6 Are Activated in Pre-Implantation Embryos of Cynomolgus Monkey. *Cells* **2021**, *10*, 2710. [CrossRef] [PubMed]
12. Springer, C.; Zakhartchenko, V.; Wolf, E.; Simmet, K. Hypoblast formation in bovine embryos does not depend on NANOG. *Cells* **2021**, *10*, 2232. [CrossRef] [PubMed]
13. Quan, H.; Tian, H.; Liu, S.; Xue, Y.; Zhang, Y.; Xie, W.; Gao, Y.Q. Progressive domain segregation in early embryonic development and underlying correlation to genetic and epigenetic changes. *Cells* **2021**, *10*, 2521. [CrossRef] [PubMed]
14. Wen, X.; Zhang, Q.; Zhou, L.; Li, Z.; Wei, X.; Yang, W.; Zhang, J.; Li, H.; Xu, Z.; Cui, X.; et al. Intrachromosomal Looping and Histone K27 Methylation Coordinately Regulates the lncRNA H19-Fetal Mitogen IGF2 Imprinting Cluster in the Decidual Microenvironment of Early Pregnancy. *Cells* **2022**, *11*, 3130. [CrossRef] [PubMed]
15. Multhaup, M.L.; Seldin, M.M.; Jaffe, A.E.; Lei, X.; Kirchner, H.; Mondal, P.; Li, Y.; Rodriguez, V.; Drong, A.; Hussain, M.; et al. Mouse-human experimental epigenetic analysis unmasks dietary targets and genetic liability for diabetic phenotypes. *Cell Metab.* **2015**, *21*, 138–149. [CrossRef] [PubMed]
16. Mintjens, S.; Van Poppel, M.N.; Groen, H.; Hoek, A.; Mol, B.W.; Painter, R.C.; Gemke, R.J.; Roseboom, T.J. The effects of a preconception lifestyle intervention on childhood cardiometabolic health—Follow-up of a randomized controlled trial. *Cells* **2022**, *11*, 41. [CrossRef] [PubMed]

Review

Amino Acid Transport and Metabolism Regulate Early Embryo Development: Species Differences, Clinical Significance, and Evolutionary Implications

Lon J. Van Winkle

Department of Medical Humanities, Rocky Vista University, 8401 S. Chambers Road, Parker, CO 80134, USA; lvanwinkle@rvu.edu or lvanwi@midwestern.edu

Abstract: In this review we discuss the beneficial effects of amino acid transport and metabolism on pre- and peri-implantation embryo development, and we consider how disturbances in these processes lead to undesirable health outcomes in adults. Proline, glutamine, glycine, and methionine transport each foster cleavage-stage development, whereas leucine uptake by blastocysts via transport system $B^{0,+}$ promotes the development of trophoblast motility and the penetration of the uterine epithelium in mammalian species exhibiting invasive implantation. (Amino acid transport systems and transporters, such as $B^{0,+}$, are often oddly named. The reader is urged to focus on the transporters' functions, not their names.) $B^{0,+}$ also accumulates leucine and other amino acids in oocytes of species with noninvasive implantation, thus helping them to produce proteins to support later development. This difference in the timing of the expression of system $B^{0,+}$ is termed heterochrony—a process employed in evolution. Disturbances in leucine uptake via system $B^{0,+}$ in blastocysts appear to alter the subsequent development of embryos, fetuses, and placentae, with undesirable consequences for offspring. These consequences may include greater adiposity, cardiovascular dysfunction, hypertension, neural abnormalities, and altered bone growth in adults. Similarly, alterations in amino acid transport and metabolism in pluripotent cells in the blastocyst inner cell mass likely lead to epigenetic DNA and histone modifications that produce unwanted transgenerational health outcomes. Such outcomes might be avoided if we learn more about the mechanisms of these effects.

Keywords: amino acid transport; amino acid metabolism; embryo development; epigenetic modifications; offspring

Citation: Van Winkle, L.J. Amino Acid Transport and Metabolism Regulate Early Embryo Development: Species Differences, Clinical Significance, and Evolutionary Implications. *Cells* **2021**, *10*, 3154. https://doi.org/10.3390/cells10113154

Academic Editor: Maxime Bouchard

Received: 15 October 2021
Accepted: 11 November 2021
Published: 13 November 2021

1. Introduction

The variety of known functions of amino acid transport and metabolism supporting oocyte and early embryo development are quite remarkable. Amino acids and their metabolites resemble classical signaling molecules, such as hormones and growth factors, as well as the substances needed to regulate histone and DNA epigenetic modifications [1,2]. In addition, some amino acids serve as osmolytes to balance hyperosmotic conditions encountered in the reproductive tract beginning after ovulation [3,4]. Strikingly, many of these amino acid transporters, receptors, and possibly even transceptors normally function primarily in the central nervous system (CNS) of adults.

For example, a CNS glycine neurotransmitter transporter and a glycine-gated chloride channel may work together in vivo to support the cleavage stage development of mouse embryos and probably embryos of other species [3,5]. The full nature of this cooperation remains to be explored. Similarly, other amino acids promote preimplantation development in "a growth factor-like manner," although the mechanisms underlying these phenomena are still being elucidated [2].

In blastocysts, redundant mechanisms exist to insure a continued development in species with invasive implantation [6–9]. Normally, amino acid transport system $B^{0,+}$

must take up leucine to trigger mTOR1 signaling in the trophectoderm during a critical window of time [6], but other mechanisms step in should this one fail [6–9]. This signaling ensures the development of the trophoblast motility needed to invade the uterine epithelium. System $B^{0,+}$ is also expressed in the oocytes, but not the blastocysts, of mammalian species with noninvasive implantation, and even in ovipara oocytes, probably to supply nutrients to these cells [10,11] but also possibly to serve signaling functions. (See below.) These differences in the timing of expression of the $B^{0,+}$ transporter—that is, beginning or exclusively in the oocytes of some species, and during blastocyst formation in others—is termed heterochrony, a process employed in evolution [10].

In the inner cell masses (ICMs) of mammalian blastocysts, and probably the stem cells of other multicellular species including plants, amino acid transport and metabolism support and regulate the maintenance of pluripotency [12]. For example, threonine transport and metabolism in the embryonic stem (ES) of mice and their progenitor (ICM) cells selectively foster the trimethylation of lysine residue 4 in histone H3 (i.e., H3K4me3 formation). H3K4me3 specifically supports ES and ICM cell proliferation and pluripotency in mice, humans, and probably other species [1]. These and other fascinating amino acid actions on oocytes and pre- and peri-implantation embryos are considered in more detail in the following sections and summarized in Table 1.

Table 1. Effect of amino acid transporters/transport systems on early embryo development.

Transporter/System	Preferred Amino Acids	Effect [References]	Mechanism
Cleavage-stage embryos			
Proline-preferring	Proline	Blastocyst formation [2,13]	mTOR1, Akt, ERK signaling
System N	Glutamine	Blastocyst formation [2,4]	Growth factor-like; Osmolyte
System Gly	Glycine	Blastocyst formation [3,4,14,15]	Osmolyte in hypertonic oviductal fluid
System $B^{0,+}$	Branched chain/Benzenoid	Oocyte nutrition in ungulates (e.g., pig) [11]	Amino acid uptake
System L	Bulky side chain	Blastocyst formation [16]	Methionine uptake
System $b^{0,+}$	Arginine	Embryo nutrition [17]	Amino acid uptake/exchange
System b^{+}	Arginine	Embryo nutrition [18]	Arginine uptake
Blastocysts			
System $B^{0,+}$	Branched chain/Benzenoid	Development of trophoblast motility; Suppression of invading blastocyst rejection [6–8]	Leucine uptake initiates mTOR1 signaling; Tryptophan removal suppresses T-cells
System $b^{0,+}$	Arginine	Development of trophoblast motility [6–8]	mTOR1, nitric oxide, polyamine signaling
System b^{+}	Arginine	Development of trophoblast motility [6–8]	mTOR1, nitric oxide, polyamine signaling
ASCT1/2	Threonine [1]	ICM cell pluripotency [19–23]	Transceptor; [2] Formation of H3K4me3
Lysine-preferring	Lysine [1]	ICM cell proliferation [24,25]	Glutamate formation

[1] Selectivity to be determined. [2] Transporter signaling.

2. Amino Acid Transport and Signaling: Proline-Preferring Systems and Systems N, Gly, and $B^{0,+}$

2.1. Proline-Preferring Systems

In recent studies, Morris and associates showed not only that proline (Pro) fosters preimplantation embryo development, but also that this amino acid improves the subsequent development, when present during in vitro fertilization (IVF), of mouse oocytes [2,13]. When present during IVF, Pro increased the blastocyst formation and the number of cells in ICMs, probably owing to a reduction of both mitochondrial activity and the production of reactive oxygen species. However, the presence of Pro during IVF did not raise the number of trophectoderm cells to the level achieved in vivo [13], and that should be a goal for conditions in vitro. These authors provide evidence that Pro uptake by oocytes likely occurs via the Pro-preferring amino acid transport system PROT, a Na^+-dependent transporter, and two Na^+-independent transporters (PAT1 and PAT2). Nevertheless, the ability of excesses of other, often beneficial amino acids to block Pro effects reminds us that a balance of conditions may produce the most desirable outcomes of in vitro culture. The ability of high concentrations of betaine, glycine, and histidine to partially inhibit Pro transport, either competitively or noncompetitively, may account, in part, for their ability to block the positive effects of Pro during IVF [13].

Even more fascinating are the direct effects of Pro on embryos during preimplantation development. Especially when early embryos are cultured under the hyperosmotic conditions of the oviductal fluid [3], Pro fosters blastocyst formation when they are grown at low density (LD), but such is also the case even when the embryos develop at high density (HD) [2]. A stimulation of embryo development occurs after the 2-cell stage. At HD, paracrine signaling among embryos is likely pronounced, whereas this signaling is minimized at LD. Somewhat surprisingly, the HD culture of embryos partially counteracts the negative effects of hyperosmotic media [2], thus providing redundant mechanisms to foster development in such conditions in oviductal fluid in vivo. (See also the consideration of glycine transport below.)

Partially owing to this redundancy to resist hyperosmotic stress, Pro seems to act on preimplantation embryos in a growth-factor-like manner [2]. Moreover, Pro activates mTOR1, Akt, and ERK signaling likely through the action of its transporter(s) as transceptor(s), intracellular signaling owing to Pro or its metabolites, or a combination of these mechanisms. Excesses of glycine, betaine, and leucine seem to counteract the benefits of Pro on development in part by slowing Pro transport. Independently, excess glycine reduces the number of cells per blastocyst at least in an isosmotic medium (i.e., 270 mOsm/kg) [2], so glycine need not act only by inhibiting the action(s) of Pro. Hence, the beneficial effects of various amino acids on preimplantation embryo development likely occur independently of each other in some instances and through interactions with one another in other cases. The elucidation of all of these mechanisms will help us develop clinically useful culture media that minimize the health risks to offspring and future generations. (See below.)

2.2. *System N*

System N preferentially transports glutamine (Gln). However, the evidence of system N's expression in cleavage-stage embryos is circumstantial. We proposed an increase in system N activity at the 4- to 8-cell stage of development owing to an increase in the level of Gln in embryos at that time [19]. Consistent with this interpretation, Gln fosters preimplantation development beginning after the 2-cell stage [2]. The importance of Gln to preimplantation development has been known for decades [26–28], although the mechanism(s) of its action are only now coming to light. While Gln likely functions as an intracellular osmolyte in early embryos under hyperosmotic conditions [4], all we know about other actions of Gln to support preimplantation development is that it does not act through mTOR1 signaling [2]. Beyond the scope of the current discussion, system N is upregulated in the inner cell mass of mouse blastocysts during diapause and is needed there to maintain a diapausing state [29,30]. (See also Section 4.2 below.)

Owing to the growth factor-like nature by which Gln acts on preimplantation embryos, it does not function only as an osmolyte to resist hyperosmolar environments [2,4]. As for Pro, Gln fosters blastocyst formation in hyperosmotic media when early embryos are cultured at LD, although, to a lesser extent, such is also the case even when the embryos develop at HD in vitro [2]. Both paracrine activity and intracellular osmolytes likely contribute to this resistance to hyperosmolarity. Future studies should reveal the elusive mechanisms by which Gln promotes preimplantation embryo development. Despite its possible toxicity, Gln, or a dipeptide form of it, likely needs to be included in the media used to culture early embryos in order to promote the development of the healthiest possible conceptuses and adults [31,32].

2.3. System Gly

We first reported that glycine (Gly) protects cleavage-stage embryos from the detrimental effects of the hyperosmotic oviductal fluid-like medium owing to its uptake as an osmolyte against its total chemical potential and concentration gradients via system Gly [14,15]. Since then, this selective transport of Gly has been studied extensively in early embryos to demonstrate its function as an osmolyte [3,4]. Nevertheless, Gly clearly has other actions, as described in the section on Pro transport above. Some of these effects of Gly occur owing to the expression of a glycine-gated chloride channel in preimplantation conceptuses [5], whereas other mechanisms of Gly action are likely yet to be determined (e.g., interactions between the chloride channel and system Gly, as in the brain). In vivo, these effects of Gly occur in the context of developmental changes in the oviductal fluid.

While somewhat controversial, it seems likely that the oviductal fluid becomes more hyperosmotic as development proceeds from the one-cell to the two-cell stages [3]. Using each of two separate reports, this change can be calculated to range from 18 to 59 mM [33,34], with a total osmolality of mouse oviductal fluid sometimes exceeding 350 mOsmol/kg [3]. Moreover, the increases in Na^+ and Cl^- concentrations in the oviductal fluid [34] could contribute somewhat [14] to a greater Gly accumulation by early embryos as the osmolarity increases [4]. Thus, the physiological conditions in oviductal fluid of about one mM glycine [35] and the hyperosmolality of the fluid interact to foster early embryo development.

Redundant mechanisms exist, however, to resist the otherwise detrimental effects of the hypertonic oviductal fluid, since insulin-like growth factors [36] as well as other embryos nearby [2] foster preimplantation embryo development in hyperosmotic media. Nearby embryos promote development apparently via paracrine signaling [2], that may include the production of extracellular vesicles by the conceptuses [37]. All these mechanisms are likely present in vivo to promote the development of healthy embryos and subsequently adults. An open question remains as to whether early embryos, developing in a hypothetically more physiological hyperosmotic medium containing Gly, are healthier than those developing in a hypotonic medium. That is, do embryos and adults that result from IVF and other assisted reproduction efforts get healthier the more we mimic in vitro the normal physiological conditions in vivo?

2.4. System $B^{0,+}$

Amino acid transport system $B^{0,+}$ was unusual when first characterized in mouse blastocysts because it accepted both cationic and zwitterionic amino acids as good substrates [38]. Subsequently, however, this Na^+-dependent system was found to strongly prefer branched chain amino acids, such as leucine (Leu), and benzenoid substrates like tryptophan (Trp) [8]. The system is pivotal in promoting the implantation and further development of mouse, rat, and human blastocysts, and probably blastocysts of other species exhibiting invasive implantation [6–9,19]. System $B^{0,+}$ fosters this type of implantation in at least two ways [8].

First, system $B^{0,+}$ takes up Leu during a critical time period required to cause mTOR1 signaling in blastocysts [6]. This signaling leads to the development of the trophoblast motility needed to penetrate the uterine epithelium. Meanwhile, the system becomes relatively inactive in blastocysts in utero, but it becomes active again when blastocysts are removed from the uterus [8].

The physiological reasons for these changes in system $B^{0,+}$ activity are likely two-fold. A suppression of system $B^{0,+}$ activity may occur in blastocysts owing to the inhibition of the system by extracellular histones in uterine secretions [9,19,39]. It remains to be determined whether these histones are free in uterine secretions or sequestered in extracellular vesicles produced by uterine epithelial or other cells, e.g., [37,40]. Regardless of their origin, such actions of the histones must somehow be neutralized at the time of implantation [9,19,39]. Following a reversal of this possible histone action and, thus, the reactivation of system $B^{0,+}$ in blastocysts in implantation chambers, the system would help to remove Trp from the surrounding fluid. Trp removal from the chambers likely prevents the immunological rejection of blastocysts by suppressing T-cell proliferation during the initial trophoblast penetration of the uterine epithelium [41,42].

Clinically, imbalances of system $B^{0,+}$-preferred substrates may lead to unwanted developmental consequences. For example, the consumption of an excess of the system $B^{0,+}$'s substrate, isoleucine (Ile), during the pre- and peri-implantation periods of development led mice to deliver pups that were 9% larger than controls on day 19 of pregnancy, but the pups born on day 20 were 9% smaller than controls [43]. Other effects of excess Ile consumption during early development were significantly different fetal and placental growth rates between days 15 and 18 of pregnancy.

Ile supplementation may have inhibited Leu and Trp uptake by blastocysts about 15 h prior to and during their implantation, respectively [39]. Alternatively, since Ile deficiency is an effective activator of trophoblast endocytosis and lysosome production [44], excess Ile consumption during preimplantation embryo development could conceivably reduce the trophectoderm endocytic uptake of uterine fluid proteins and their digestion in lysosome. A decreased lysosomal protein digestion could limit the supply of threonine and lysine to ICM cells [1,19,24]. (See Section 4 below.) Dietary supplementation with the branched chain amino acids, Ile, Leu, and valine, has become part of the solution to some health problems [43]. Hence, the use of such supplements should be cautious in species such as humans, that employ invasive implantation.

More broadly regarding mTOR1 signaling and the development of trophoblast motility, a maternal low protein diet (LPD) during preimplantation development to the blastocyst stage led to a 25% reduction in the Leu concentration in mouse uterine fluid [45]. In a model, the latter authors proposed that the decreased uterine fluid Leu concentration results in less mTOR1 signaling, greater compensatory trophoblast pinocytosis and lysosome biogenesis, and increased histotrophic nutrition beginning at the blastocyst stage [46,47]. Consequently, offspring grow faster, have greater adiposity, and develop cardiovascular dysfunction, hypertension, neural abnormalities, and altered bone growth. This model may apply to all mammalian species, including humans, with invasive trophoblast implantation [46,47].

But what might be the function(s) of system $B^{0,+}$'s expression in species with non-invasive or no placentation? System $B^{0,+}$ is expressed in unfertilized pig oocytes but not blastocysts [11]. Since mTOR1 helps regulate the development of oocytes, cleavage-stage embryos, and blastocysts (e.g., [48]), it is conceivable that system $B^{0,+}$ acts in porcine oocytes via the leucine activation of mTOR1, as well as serving to nourish these cells [10]. Similarly, system $B^{0,+}$ likely helps Xenopus oocytes grow prior to their deposition in fresh water. After deposition, the expression of the system is turned off likely to help retain the amino acids in the oocytes [10]. On the other hand, the activation of system $B^{0,+}$'s expression upon fertilization of sea urchin eggs should help them accumulate nutrients from sea water [10], especially owing to the system's low K_m values for the accumulation of some amino acids [8]. Such changes in the timing of system $B^{0,+}$'s transporter gene expression during the early development of mouse, pig, Xenopus, and sea urchin oocytes and embryos is termed heterochrony, and heterochrony is an important process employed in the evolution of new species [10].

3. Amino Acid Transport and Signaling Sometimes Includes Metabolism: Systems L, $b^{0,+}$, and b^+ or y^+

3.1. System L

System L is present in preimplantation mouse embryos at all stages of their development [17,19,36]. While this system transports branched chain and benzenoid amino acids well, it also likely takes up methionine (Met) [20]. Consequently, the removal of Met from KSOM culture medium causes fewer zygotes to develop into blastocysts [16]. Of the embryos that do develop to the morula and blastocyst stages, their content of histone H3K4me3 was significantly decreased, likely because they did not have enough Met to support one carbon (1C) metabolism and H3K4 methylation [1,16,19,35]. As we shall see, H3K4me3 is needed to maintain pluripotent and proliferating cells in morulae and blastocysts (Section 4.1 below).

3.2. System $b^{0,+}$

We also discovered an unanticipated, Na^+-independent transporter of both cationic and zwitterionic amino acids in mouse blastocysts [49]. While initially it seemed that both types of substrates were equally accepted for transport, arginine (Arg) was soon found to be the highly preferred substrate [17,19,36]. Importantly, this transporter helped to reveal redundant mechanisms to support the development of trophoblast motility in blastocysts [6–8].

Not only Leu, but also Arg, by itself, supports the development of trophoblast motility in blastocysts in vitro through the activation of mTOR1 signaling [6]. Arg is also a precursor to both nitric oxide (NO) and polyamines, and all these substances foster cell motility [7]. In fact, polyamines partially overcome the rapamycin inhibition of mTOR1 and the block to the development of trophoblast motility in mouse blastocysts [8]. Moreover, uterine secretions foster the development of trophoblast motility in mouse blastocysts in vitro, but only when the secretions are obtained at a specific time several hours prior to the time of blastocyst implantation [6]. Since amino acids are likely present continuously in these secretions, we speculate that macromolecular constituents, possibly contained in extracellular vesicles of uterine fluid (e.g., [37]) are responsible for promoting trophoblast motility. This redundancy likely explains why one or the other of system $B^{0,+}$ or system $b^{0,+}$ knockout results in only small, nonlethal effects [9,50]. For other nutritional purposes, several alternative systems for the transport of cationic and zwitterionic amino acids are present in early conceptuses (e.g., system L above for zwitterionic amino acid transport) [17,19,36].

3.3. System b^+ or y^+

In the case of Arg nutrition, at least two cationic amino acid transporters (CATs), CAT1 and CAT2, are expressed throughout the preimplantation development of mouse embryos [19]. In early porcine embryos, one of these transporters (CAT1) is upregulated during the development in vitro vs. the development in vivo [51]. While the latter transport has been attributed to that catalyzed by the well-known amino acid transport system y^+ first identified in adult tissue [18,52–54], a careful characterization of cationic amino acid transport in fertilized mouse eggs, cleavage-stage conceptuses, and blastocysts distinguished this transport from that of system y^+ and termed it b^+ [18]. For example, the K_i (and K_m) values for Arg and lysine (Lys) transport via system y^+ are nearly identical, but these values differ by an order of magnitude or more for transport by one-cell embryos and blastocysts (Table 2).

Table 2. The K_i/K_m values for Arginine and Lysine transport by cationic amino acid transporters (CATs) are nearly identical in adult tissues or during heterologous expression in Xenopus oocytes, but the values are one or two orders of magnitude different for CATs expressed in preimplantation mouse embryos (**, $p < 0.01$).

CAT Expression in:	K_i (K_m) Values (Mean +/− SEM, mM) [1]		
	Arginine		Lysine
Fibroblasts (y^+)	0.041 ± 0.002	n.s.	0.040 ± 0.004
Hepatoma cells (y^+)	0.20 ± 0.04	n.s.	0.14 ± 0.01
Xenopus oocytes (CAT2)	0.19 ± 0.03	n.s.	0.20 ± 0.03
One-cell embryos ($b^+{}_1$)	0.13 ± 0.04	**	1.25 ± 0.18
Blastocysts ($b^+{}_2$)	0.084 ± 0.021	**	8.10 ± 1.00

[1] Data from [18,52–54], ** ($p < 0.01$), n.s. (not significant)

Perhaps the Arg uptake by early embryos is so important that CATs, normally taking up Arg and Lys equally efficiently in adults, are adapted in early embryos to insure Arg is accumulated regardless of the activities of other Arg transporters, such as $b^{0,+}$. Moreover,

CATs' preference for Arg over Lys transport in blastocysts is even more pronounced than in one-cell embryos (i.e., the K_m for Lys uptake is nearly 10-fold higher in blastocysts than in one-cell embryos, while the K_m for Arg remains unchanged, Table 2, $p < 0.01$) [18]. The latter development change in CAT transport activity could help to insure a preferential Arg uptake in blastocysts even when the high-affinity Arg-preferring system $b^{0,+}$ transporter is knocked out [50]. Such Arg transport would likely foster the development of the trophoblast motility needed to penetrate the uterine epithelium [6–8].

Because of the importance of Arg to reproduction and embryo development [6–8,19,35], it has been used as a prenatal dietary supplement in numerous studies. A systematic review and meta-analysis of 47 animal studies including 12 human investigations revealed beneficial effects of dietary Arg supplementation on fetal/birth weight, but only in complicated pregnancies [55]. In another recent review [56], it was emphasized that dietary Arg supplementation for the first seven days of pregnancy in the rat increased litter sizes by 30% while maintaining normal birth weights [57]. Although this Arg effect might help avoid early embryo losses in a variety of species, including humans, one can also argue that the effect might not be beneficial. For example, the average liter size in isonitrogenous control animals in the cited study was 11.4 ± 0.4 pups, which is not significantly different from the normal litter size of 11.0 in Sprague Dawley rats [58]. Thus, it might also be concluded that Arg supplementation saves less healthy embryos otherwise destined to fail in early pregnancy. Unfortunately, offspring growth and other parameters, such as blood pressure and blood glucose levels, were apparently not followed into adulthood in this study [57], so it is not known whether 30% of the experimental pups were less healthy than control pups.

It has also been speculated that the other cationic amino acid, Lys, might be deficient in low-protein diets [24]. Both an increase and a decrease in maternal dietary lysine produced smaller fetuses/birth weights in rats, e.g., [59,60]. Since Lys is not taken up efficiently by systems in early mouse embryos (e.g., Table 2), other possible mechanism(s) for the effects of Lys-deficient diets on early embryo development are discussed in Section 4.2, below.

4. Regulation through Amino Acid Signaling and Metabolism but Apparently Not Involving/Requiring Amino Acid Transporters in the Apical Membrane of the Trophectoderm

4.1. Threonine and Serine in the Inner Cell Mass

Morphologically distinct cell types first appear in preimplantation mammalian blastocysts a few days after fertilization [1,19]. The trophectoderm surrounds the inner cell mass (ICM) and will initiate implantation one to a number of days later depending on the species. The ICM gives rise to all other mammalian tissues, and these cells are used to produce embryonic stem (ES) cells. Consequently, ES cells are used as models for the ICM [1,19].

Murine (mES) and probably bovine embryonic stem cells require Thr to stay pluripotent and proliferate [21,61]. Threonine dehydrogenase (TDH) regulates the conversion of Thr to Gly and acetyl CoA in these ES cells. To control Thr consumption, posttranscriptional and posttranslational effectors as well as gene transcription maintain TDH activity [62]. Glycine produced from Thr in these ES cells is crucial to sustain the specific epigenetic modifications needed to preserve their pluripotency [63,64].

While mES cells maintain an optimal intracellular Thr concentration through Thr uptake from the culture medium and metabolism via TDH, the blastocyst trophectoderm must supply Thr to the ICM. A direct contact between trophoblast and ICM cells might supply amino acids to these pluripotent cells, while ICM cells bordering the blastocoelic fluid inside the trophectoderm could take up amino acids from this fluid [65]. Moreover, the apical membrane of the trophectoderm expresses at least 12 amino acid transport system activities that could help provide amino acids to the ICM [19,36]. However, Thr is not received with a low K_m value by any of these systems [19,36]. Consequently, blastocysts must use a different process to supply Thr to their ICM cells.

In this regard, Thr accumulates in mouse blastocysts as they develop, while most other amino acids do not [1,7,19,66]. Even without the presence of amino acids in vitro, the Thr content of blastocysts increases as the development proceeds. Since Thr is an essential amino acid in animals, the trophectoderm probably generates Thr by hydrolyzing protein in vitro as well as in utero [19]. Trophoblast cells can readily carry out this process via pinocytosis and the subsequent digestion of the protein they take up [67].

Based on studies with mES cells, ICM cells then likely use two Na^+-dependent and at least one Na^+-independent transport systems to take up Thr. These transporters were identified, respectively, as the obligate exchange ASC amino acid transporters, ASCT1 and ASCT2, and a system L exchange amino acid transporter [20]. One or more of these Thr transporters may function for signaling as a transceptor (i.e., a transporter that initiates signaling) [22], as well as by supplying the intracellular Thr metabolized for the specific epigenetic modifications needed to maintain mES cell proliferation and pluripotency.

By promoting cMyc expression, Thr transport initiates mTOR1 signaling in mES cells. This signaling does not proceed, however, upon the disruption of the lipid rafts within which Thr transporter(s) reside [22]. Thr transport may need undamaged rafts, or transport might cause caveolae in rafts to initiate signaling as a transceptor. Thr transport is inhibited by its analogue, 3-hydroxynorvaline (3-HNV), which likely slows mES cell proliferation in this way [20], in addition to slowing Thr metabolism by blocking TDH activity [21]. (See below.) Likely in these same ways, 3-HNV blocks the formation of cavitated blastocysts by pre-compacted morulae [21].

Because Thr transporter(s) seem to function as transceptor(s) [22], we predicted correctly that 3-HNV would block TDH-deficient hES cell proliferation [23]. While it is conceivable that 3-HNV could inhibit the proliferation of hES cells by replacing Thr residues in proteins [68], 3-HNV does not slow the proliferation of other human and mouse cell lines [21], so cell growth is not blocked through 3-HNV incorporation into proteins. An excess Thr rescue of 3-HNV-inhibited hES proliferation supports the hypothesis that the ES cells of all mammalian species require Thr transport itself for signaling, in addition to the signaling provided in most species by Thr catabolism for epigenetic histone modifications [23].

4.1.1. The Glycine Cleavage System (GCS) Is Also Needed to Maintain ES Cells

The GCS is needed to further process Gly generated by most mammalian ES and ICM cells owing to mitochondrial TDH activity (Figure 1). However, human ICM and ES cells produce inactive TDH, and they do not need it to generate Gly [69]. Instead, hES and probably their progenitor cells in the ICM greatly upregulate the expression of enzymes for serine (Ser) synthesis from glycolytic intermediates (Figure 1), whereas such upregulation does not occur in mES or mouse-induced pluripotent stem cells [64]. Gly is then likely generated from Ser by serine hydroxymethyltransferase (SHMT).

The further processing of Gly requires the glycine cleavage system (GCS) to maintain the pluripotency and proliferation of mES, hES, and likely other mammalian ES cells [64,70]. In this regard, the upregulation of GCS expression fosters the formation of induced murine and human pluripotent stem (iPS) cells. The GCS also prevents ES cell senescence by preventing methylglyoxal production (Figure 1) [64,70]. Instead, Gly is used in 1-carbon (1C) metabolism that fosters ES cell pluripotency and proliferation via specific epigenetic histone modifications [1,71–73]. This mitochondrial specialization is supported by glycolysis-generated ATP in ES cells [74]. Somewhat surprisingly, uncoupling protein 2 (UCP2) prevents pyruvate oxidation in mitochondria and, thus, shunts it to lactate, the final step in glycolysis [75,76].

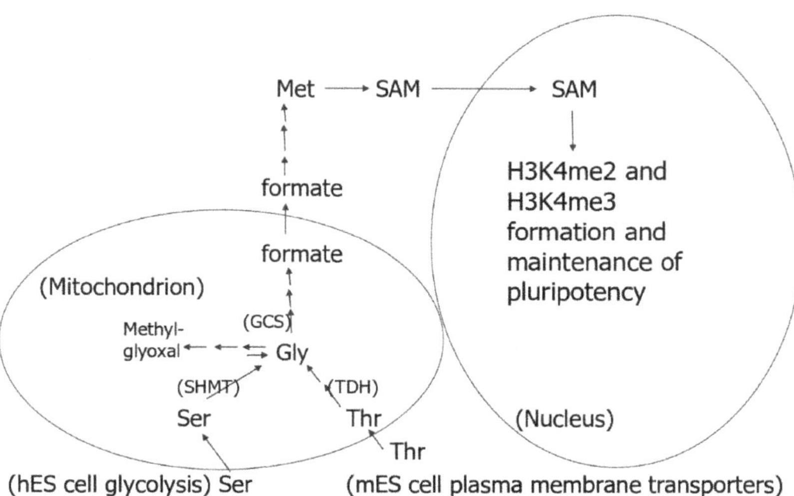

Figure 1. Mouse and probably most other mammalian ES cells require threonine (Thr) for the production of the 1C units needed to methylate histone H3K4. H3K4me3 formation is needed to maintain ES cell proliferation and pluripotency. Since other sources of 1C units cannot substitute Thr, we propose that a subpopulation of perinuclear mitochondria take up Thr, use it to form formate, and then selectively direct the formate to produce the S-adenosylmethionine (SAM) used in the nucleus to methylate H3K4. Since hES cells produce an inactive form of threonine dehydrogenase (TDH), they upregulate the expression of serine (Ser) synthesis enzymes probably to produce 1C units for H3K4me3 formation. GCS, glycine cleavage system; SHMT, Serine hydroxymethyltransferase. (The figure is a modification of those in reference [1]).

4.1.2. mES Cells Require Thr Catabolism for Specific Histone Modifications

Mouse ES cells stop proliferating when Thr is removed from the culture medium [21]. In the absence of Thr, the methylation of histone H3 to form H3K4me3 decreases dramatically, and this decrease is selective, since the methylation of DNA and other histone sites continues normally [63]. Somehow, mES cells take up Thr and direct it selectively to provide 1C units for H3K4 methylation. But how might such selection occur?

Thr catabolism in mES cells is performed by perinuclear mitochondria [77–79]. Thr may be selectively accumulated by a portion of these organelles specialized to do so (Figure 1). These specialized mitochondria might then produce formate from Thr and transport the formate to the cytosol, where it is converted to S-adenosyl methionine (SAM) methyl groups needed for nuclear H3K4 methylation [80,81]. The formate must somehow be directed specifically to H3K4 methylation sites, because the 1C units must originate from Thr (Figure 1) [63].

Another possibility is the selective direction of Thr by its plasma membrane transporter(s) to the portion of the mitochondria helping to form H3K4me3 in mES cells [1,20]. Nevertheless, the mitochondria would need to remain specialized for this purpose. Similarly, and despite being TDH-deficient, some hES cell mitochondria likely specialize in H3K4me3 formation via the GCS. H3K4me3 is required for hES as well as mES cells to proliferate and remain pluripotent [1,71–73].

4.1.3. Why Doesn't TDH Knockout Block Mouse Blastocyst Development?

Surprisingly, TDH knockout embryos appear to develop normally, and knockout adults are not sterile [82]. While both signaling via Thr membrane transport [22] and the subsequent Thr metabolism to form H3K4me3 [1,71–73] support mES cell pluripotency and proliferation, neither is sufficient alone to maintain such mES cell stemness [83]. Compensatory mechanisms to circumvent TDH knockout are, however, easy to envision, and may occur in mES progenitor cells in mouse blastocysts. For example, when Thr is deleted from the mES cell culture medium, the addition of excess Gly (plus pyruvate) to the medium maintains the H3K4me3 formation, which normally requires Thr [63]. Similarly, an increased supply of Gly occurs naturally inside TDH-deficient human iPS and hES cells when they upregulate serine synthesis to generate Gly and become/remain pluripotent (Figure 1) [64]. We propose that TDH knockout mES and their progenitor cells in blastocysts may adapt to mimic TDH-deficient hES cells (Figure 1).

A further understanding of this possible adaptation and regulation is likely essential for us to fully comprehend early embryo development in all mammalian species, since all species except humans express TDH, and even humans likely need specialized mitochondria in their ES and ICM cells. ES progenitor cells in the ICM give rise to all mammalian tissues and organs, so the clinical implications of environmentally altered epigenetic modifications seem to begin with such possible alterations in these cells. Much epigenetic reprogramming occurs during pre- and peri-implantation embryo development, e.g., [84]. Moreover, since Met is central to these epigenetic modifications (Figure 1) and to 1C metabolism in general, it might seem surprising, at first, to learn that a maternal low-protein diet (LPD) causes the Met concentration to increase in mouse uterine fluid on day 4 of pregnancy [45]. However, 1C metabolism might be sluggish owing to a LPD and to dependence on amino acids for 1C metabolism. Such slower 1C metabolism could result in a lower demand for Met in living cells including early embryos.

4.1.4. Clinical Implications of Altered Epigenetic Histone and DNA Modifications

Maternal LPD consumption during early embryo development leads to several changes in trophectoderm behavior, including greater motility during the implantation in the uterus and an increased uptake of extracellular protein via endocytosis [46]. The hydrolysis of more protein releases more amino acids, including Thr, which might be delivered to ICM cells and could be taken up by at least three transporters [1,19,20]. A greater Thr transport into ICM cells may alter both signaling via its transceptor and H3K4me3 formation owing to changes in 1C metabolism. Also due to increased protein hydrolysis, trophoblast motility may increase owing to more Arg and Leu signaling and metabolism in these cells [6].

More efficient placentas also develop in association with a maternal LPD during early embryo development, possibly owing to a more robust trophoblast penetration of the uterine epithelium [46]. Similarly, modified epigenetic histone alterations in the primitive endoderm owing to a maternal LPD leads to the delivery of more nutrients to the embryo by the yolk sac placenta [46]. In addition, changes in dietary protein consumption during early embryo development cause ICM cell lineages to alter their epigenetic DNA modifications. These post-implantation as well as preimplantation alterations in the nutrients supplied to early embryos may affect the epigenetic modification of their DNA and histones, probably including H3K4.

4.1.5. Future Generations Likely Experience Effects from DNA and Histones Modified during the Development of Their Ancestors

A maternal LPD, only during an F_0 pregnancy, causes F_1 rat offspring to produce an F_2 generation with increased blood pressure and abnormal endothelial cells, likely due to transgenerational epigenetic histone and DNA changes [85]. Similarly, older maternal age or hampered placental function lead the F_0 generation to transmit metabolic and cardiorenal changes to F_2 rats, probably owing to epigenetic alterations [86,87]. In addition to the intracellular effects of these DNA and histone modifications, the histones might act extracellularly after their secretion into the uterine fluid by F_0 and F_1 females [9].

Furthermore, the genetic impairment of 1C folate metabolism in F_0 mice suppresses epigenetic methylation, resulting in the altered development of wild-type mice over the following five generations [88]. In the latter case, DNA and histone modifications in F_1 female germlines result in congenital malformations, while growth defects are caused by changes in the uterine environment. This uterine environment likely contains altered histones, because F_1 wild-type females exhibited hypomethylation in their uterine cells.

Histones in uterine fluid may affect blastocyst development and implantation by altering system $B^{0,+}$ activity [9,39]. (See above.) The proposed histone hypomethylation may have contributed to both growth defects and congenital malformations because histones may have extracellular as well as intracellular effects.

Extracellular histones also provide one possible mechanism by which a paternal LPD adversely affects the health of their offspring [89]. Histones, modified owing to a LPD, might be present in the seminal fluid and dead sperm cells [90–92]. As for maternal epigenetic effects on future generations, paternal exposure to unhealthy environments adversely affects early embryo development through epigenetic transgenerational modifications [93,94].

4.2. Conversion of Lys to Glutamate in ICM Cells

While ES cells serve as models for ICM cells, the ES cell environment is not regulated in a physiologically normal way. In contrast, the immediate surroundings of ICM cells are controlled through interactions with the trophectoderm, e.g., [65]. Hence, data concerning the regulation of this environment need to be combined with findings of ES cell metabolism for a further understanding of the ICM cell function.

As for Thr, Lys is not a preferred substrate of any known transport system in the trophectoderm (e.g., Table 2) [19,36]. However, it may be needed in the ICM for glutamate (Glu) production [24]. We suggest that trophoblast cells take up protein by pinocytosis and hydrolyze the protein to generate Lys as well as Thr for the ICM [24]. The transport of Lys across the plasma membrane of ES and their progenitor cells in the ICM has not, to our knowledge, been characterized.

Nevertheless, hES cells remove Lys from their culture medium and release Glu to the medium [24], so the cells must express a transporter(s) to take up Lys. While Glu could conceivably be produced from Gln by extracellular glutaminase, we suggest that it is produced metabolically from Lys in hES cells and then released to the medium. This metabolism of Lys is possible because more Lys is consumed by hES cells than the amount of Glu they produce [24]. But what is the evidence that Lys is metabolized to Glu in hES cells, how does this metabolism matter to ICM cell pluripotency and proliferation, and what may be the clinical consequences of Lys deficiency owing to, say, LPDs?

4.2.1. How Is Lys Converted to Glu in hES Cells?

Somewhat surprisingly, we found a relatively high expression of RNA encoding alpha-aminoadipic semialdehyde synthase (AASS) in hES cells [24]. AASS regulates the Glu synthesis from Lys in mouse and human brain, so such is likely the case for hES cells [95–99]. The resultant Glu is needed for normal brain and ES cell functioning through Glu signaling. Although the Glu-Gln cycle also produces Glu in the brain, it is insufficient for brain health [95–99]. Similarly, system N for Gln uptake is upregulated in mouse ICM cells to help maintain a diapausing blastocyst state [29,30], but diapause is characterized by a relatively slow cell division in blastocysts. Gln transport into ICM cells must decrease dramatically for the proliferation of ICM cells to continue [29,30], and glutaminase activity is likely relatively low in ES and their progenitor cells [24]. Thus, the conversion of Gln to Glu in ICM cells seems not to be a priority and is unlikely to contribute to the pool of Glu needed to support their pluripotency and proliferation. (See below.)

Hence, the pool of Glu produced from Lys seems to be used uniquely for signaling after the release from nerve and ICM cells [24,95–99]. Perhaps the Glu is extruded from mitochondria in the nerve and ICM cells and then preferentially released from the cells for autocrine and paracrine signaling. Contrariwise, instead of the proposed unique production of Glu from Lys, the trophoblast could conceivably supply Glu to ICM cells in a more direct manner. However, the Glu content of mouse blastocysts decreases as they develop [19,66], so there does not seem to be a concerted effort to supply free Glu directly to ICM cells as the development proceeds. So why is Glu needed outside of ES and ICM cells, and why might its presence outside ICM cells be carefully regulated?

4.2.2. Function of Metabolically Produced Glu in ICM Cells

Mouse ES and, likely, ICM cells express mGlu5 metabotropic glutamate receptors, and they require the activation of these receptors to sustain self-renewal [100]. The mES cells produce Glu in vitro, and ICM cells likely do so in vivo, and such endogenous production of Glu fosters cell proliferation and the maintenance of pluripotency. The Glu activation of mGlu5 metabotropic glutamate receptors works to maintain a greater c-Myc expression in mES cells through the interaction with signaling by the leukemia inhibitory factor [101].

In this regard, we found hES cells to express mRNA encoding at least two such Glu receptors—metabotropic glutamate receptor 3 as well as 5 [24]. This mRNA is likely translated to produce glutamate receptor proteins in hES and ICM cells. We propose that endogenous Glu production from Lys in human as well as mouse ICM cells, and the release of this pool of Glu from the cells, help to maintain their pluripotency and proliferation.

4.2.3. What Are the Possible Clinical Consequences of Lys Deficiencies Owing to LPDs?

When Lys is removed from the culture medium, hES and human iPS cells nearly stop proliferating [25]. Similarly, maternal LPDs may deprive ICM cells of Lys, because the concentration of Lys in mouse blastocysts is likely decreased by maternal LPD consumption [45]. Somewhat surprisingly, a maternal LPD increases the concentration of Met in the uterine fluid on day 4 of pregnancy in the mouse [45]. Such an increase in the Met concentration in vivo could lead to decreased blastocyst levels of the cationic amino acids-Arg and Lys—as also observed in mouse blastocysts in vitro in the presence of a physiological concentration of Met [102]. This decrease in the cationic amino acid concentrations in blastocysts occurs owing to the uptake of Met in exchange for (and exodus of) the cationic amino acids via system $b^{0,+}$ [24].

The ability of Lys deprivation to nearly stop hES cell proliferation likely does not result only from the nutritionally essential nature of Lys. Human ES cells, grown without each of several other essential amino acids, continue to divide at almost normal rates [25]. Rather, without Lys, hES cells may not produce the specialized pool of Glu they need to bind metabotropic Glu receptors and, thus, remain pluripotent. (See above.)

Specialized pools of amino acids and their metabolites seem to be an especially important phenomenon in early embryos. For example, Thr is used selectively for H3K4me3 formation in mES and probably ICM cells in blastocysts (Figure 1). Other 1C sources cannot normally substitute Thr. Similarly, system $B^{0,+}$ specifically directs Leu to sites of mTOR1 signaling in the blastocyst trophectoderm in order to foster the development of its motility [6]. Other transporters of Leu cannot substitute this $B^{0,+}$ Leu transport, likely owing to the specialized intracellular Leu pool that $B^{0,+}$ creates.

In contrast, mES cell proliferation does not appear to be as reliant on Glu production from Lys as hES cell proliferation [25]. The removal of Lys from the mES cell culture medium does not slow their proliferation [21], at least when the medium contains Glu [100,101]. However, for mammalian ES progenitor cells in the ICM, we suggest that Lys deprivation, owing to maternal LPDs, adversely alters the ICM production of tissues and organs in newborns and adults [24]. Unwanted effects of maternal LPD consumption during pre- and peri-implantation embryo development include metabolic syndrome and related disorders in adulthood [1,9,19,46,47].

5. Summary

Pro, Gln, and Gly each support preimplantation development apparently as growth factors, and their Na^{+}-dependent transport against total chemical potential gradients allows them to serve as osmolytes to resist the otherwise detrimental effects of the hyperosmolar oviductal fluid. We still need to learn whether we can produce hyperosmotic conditions in vitro that allow oocyte fertilization and the development of the healthiest possible early embryos for transfer to the reproductive tract. Similarly, Leu and Arg transport into blastocysts via systems $B^{0,+}$ and $b^{0,+}$ both likely foster the invasion of the uterine epithelium. If the process is disturbed, however, owing to, say, a maternal LPD, the resultant embryos seem to give rise to adults with increased risk of having metabolic syndrome and related disorders. Finally, mouse and human ICM cells have different requirements for Thr and probably Lys uptake and metabolism [21,25]. Despite these differences, ICM cells of both species need the pertinent metabolic products, H3K4me3 and Glu, to remain pluripotent and proliferate. Disturbances in H3K4me3 and Glu production in ICM cells likely lead to transgenerational metabolic disorders.

6. Conclusions

6.1. Evolutionary Considerations

Amino acid transport system $B^{0,+}$ not only supports the development of trophoblast motility in mammalian species with invasive blastocyst implantation, but it also seems to help nourish oocytes of species that do not penetrate the uterine epithelium after blastocyst attachment. Similarly, it may promote amino acid accumulation in Xenopus oocytes, prior to their deposition in fresh water, and in sea urchin eggs following fertilization. These changes in the timing of $B^{0,+}$ expression are termed heterochrony—a well-known process employed in evolution [10].

Likewise, Thr transport and metabolism support stem cell maintenance in both animals and plants, so this reliance on Thr may be an ancient mechanism to conserve stem cell proliferation in multicellular organisms [12]. Similarly, Glu production from Lys seems to support the maintenance of pluripotency in ICM cells of mammalian blastocysts [24], while, in plants, this Glu production regulates their environmental responses and growth [103]. In the case of animal evolution, it seems likely that the Glu production from Lys was used first for signaling in ICM cells and, only later, for normal brain functioning after the evolution of this organ.

6.2. Several Ways to Foster the Development of Cleavage-Stage Embryos in the Hyperosmotic Oviductal Fluid

Cleavage-stage embryos likely develop in the hyperosmotic oviductal fluid [3]. While the osmolarities of oviductal and uterine secretions change as development proceeds, these secretions remain hyperosmotic at virtually all stages of preimplantation development [3,8]. Consequently, several mechanisms have evolved to help resist the potentially detrimental effects of these hyperosmotic conditions. The mechanisms include amino acid transport into the embryos as osmolytes, signaling by growth factors, such as insulin-like growth factors 1 and 2, and autocrine and paracrine effectors released by nearby embryos and cells of the reproductive tract [2–4,14,15,36]. Nevertheless, these changing hyperosmotic conditions in situ are likely beneficial to the reproductive process, although their possible advantages during embryo development are still to be determined. The elucidation of these benefits seems essential, however, in order to produce the heathiest possible life-long outcomes of assisted reproductive technology.

6.3. Redundant Mechanisms to Insure the Development of Trophoblast Motility and Implantation

Leu uptake by trophoblasts via system $B^{0,+}$ fosters the mTOR1 signaling needed for the development of cell motility and the invasion of the uterine epithelium [6]. Nevertheless, alternate mechanisms, such as Arg transport by system $b^{0,+}$, also serve this purpose [6,8]. Moreover, uterine secretions promote the development of trophoblast motility during a crucial interval several hours prior to implantation, and apparently not because the secretions contain amino acids. (See Section 2.4 above.) Consequently, neither system $B^{0,+}$ nor system $b^{0,+}$ knockout alone is detrimental to early embryo development or the fertility of the knockout mice [9,50]. Blastocyst implantation in the uterus is crucial to mammalian species' survival, so it is not surprising that multiple mechanisms seem to fully support this process. Nevertheless, disturbances in the normal system $B^{0,+}$ functioning, such as maternal consumption of a LPD, apparently lead to greater adiposity, cardiovascular dysfunction, hypertension, neural abnormalities, and altered bone growth in adults [46,47]. In this regard, a culture of mouse blastocysts for only a few hours without amino acids decreases mTOR1 signaling, which is only partially restored by amino acid addition [104].

6.4. Multiple Mechanisms to Maintain Pluripotent ICM Cells

Thr transporters in eukaryotic stem cells may function as transceptors to help maintain their proliferation [1,12]. Similarly, in most animals, TDH-regulated Thr metabolism, to form Gly and subsequently H3K4me3, likely maintains their pluripotent ES progenitor cells in the ICM [1,12]. In TDH-deficient hES progenitor cells, Gly and H3K4me3 are likely formed from Ser synthesized in the cells [64]. The proper maintenance of this H3K4me3 production through 1C metabolism is essential for the optimal development and production of healthy offspring. Gln metabolism can also contribute to ES cell maintenance through α-ketoglutarate production and epigenetic mechanisms [105], while Pro fosters ES cell differentiation [106].

In addition, the conversion of Lys to Glu may be needed to maintain pluripotent stem cells in the mammalian ICM [24]. The removal of Lys from the culture medium blocks hES (but not mES) cell proliferation almost completely [25], whereas the removal of Thr stops mES (but not hES) cell proliferation [21]. Hence, ES cells of these two species appear to have different needs for Lys and Thr uptake and metabolism. Nevertheless, the consumption of both Thr and Lys, and the production of Glu, by the bovine ICM [107], support the notion that most mammalian ICM cells require Thr and Lys to remain pluripotent. Moreover, the bovine trophectoderm produces both Thr and Lys [107], as we predicted in Section 4 above for the mouse and human trophectoderm. Disturbances of any of these processes in the ICM may lead to transgenerational epigenetic modifications with serious health consequences in adults [1,24]. (See Section 4 above.)

Funding: This research received no external funding.

Institutional Review Board Statement: Not applicable.

Informed Consent Statement: Not applicable.

Data Availability Statement: All data are included in the manuscript.

Acknowledgments: We thank Philip Iannaccone for critically reading the manuscript and for the useful discussions.

Conflicts of Interest: The author declares no conflict of interest.

References

1. Van Winkle, L.J.; Ryznar, R. One-carbon metabolism regulates embryonic stem cell fate through epigenetic DNA and histone modifications: Implications for transgenerational metabolic disorders in adults. *Front. Cell Dev. Biol.* **2019**, *7*, 300. [CrossRef]
2. Morris, M.B.; Ozsoy, S.; Zada, M.; Zada, M.; Zamfirescu, R.C.; Todorova, M.G.; Day, M.L. Selected amino acids promote mouse pre-implantation embryo development in a growth factor-like manner. *Front. Physiol.* **2020**, *11*, 140. [CrossRef] [PubMed]
3. Van Winkle, L.J. Perspective: One-cell and cleavage-stage mouse embryos thrive in hyperosmotic oviductal fluid through expression of a glycine neurotransmitter transporter and a glycine-gated chloride channel: Clinical and transgenerational implications. *Front. Physiol.* **2020**, *11*, 1706. [CrossRef] [PubMed]
4. Tscherner, A.K.; Macaulay, A.D.; Ortman, C.S.; Baltz, J.M. Initiation of cell volume regulation and unique cell volume regulatory mechanisms in mammalian oocytes and embryos. *J. Cell. Physiol.* **2021**, *236*, 7117–7133. [CrossRef]
5. Nishizono, H.; Darwish, M.; Endo, T.A.; Uno, K.; Abe, H.; Yasuda, R. Glycine receptor α4 subunit facilitates the early embryonic development in mice. *Reproduction* **2020**, *159*, 41. [CrossRef] [PubMed]
6. González, I.M.; Martin, P.M.; Burdsal, C.; Sloan, J.L.; Mager, S.; Harris, T.; Sutherland, A.E. Leucine and arginine regulate trophoblast motility through mTOR-dependent and independent pathways in the preimplantation mouse embryo. *Dev. Biol.* **2012**, *361*, 286–300. [CrossRef]
7. Martin, P.M.; Sutherland, A.E.; Van Winkle, L.J. Amino acid transport regulates blastocyst implantation. *Biol. Reprod.* **2003**, *69*, 1101–1108. [CrossRef]
8. Van Winkle, L.J.; Tesch, J.K.; Shah, A.; Campione, A.L. System $B^{0,+}$ amino acid transport regulates the penetration stage of blastocyst implantation with possible long-term developmental consequences through adulthood. *Hum. Reprod. Update* **2006**, *12*, 145–157. [CrossRef] [PubMed]

9. Van Winkle, L.J.; Ryznar, R. Can uterine secretion of modified histones alter blastocyst implantation, embryo nutrition, and transgenerational phenotype? *Biomol. Concepts* **2018**, *9*, 176–183. [CrossRef] [PubMed]
10. Van Winkle, L.J. Amino acid transport in developing animal oocytes and early conceptuses. *Biochim. Biophys. Acta (BBA)-Rev. Biomembr.* **1988**, *947*, 173–208. [CrossRef]
11. Prather, R.S.; Peters, M.S.; Van Winkle, L.J. Alanine and leucine transport in unfertilized pig oocytes and early blastocysts. *Mol. Reprod. Dev.* **1993**, *34*, 250–254. [CrossRef] [PubMed]
12. Sahoo, D.P.; Van Winkle, L.J.; Díaz de la Garza, R.I.; Dubrovsky, J.G. Interkingdom comparison of threonine metabolism for stem cell maintenance in plants and animals. *Front. Cell Dev. Biol. Stem Cell Res.* **2021**, in press. [CrossRef]
13. Treleaven, T.; Hardy, M.L.; Guttman-Jones, M.; Morris, M.B.; Day, M.L. In Vitro Fertilization of Mouse Oocytes in L-Proline and L-Pipecolic Acid Improves Subsequent Development. *Cells* **2021**, *10*, 1352. [CrossRef]
14. Van Winkle, L.J.; Haghighat, N.; Campione, A.L.; Gorman, J.M. Glycine transport in mouse eggs ad preimplantation conceptuses. *Biochim. Biophys. Acta (BBA)-Biomembr.* **1988**, *941*, 241–256. [CrossRef]
15. Van Winkle, L.J.; Haghighat, N.; Campione, A.L. Glycine protects preimplantation mouse conceptuses from a detrimental effect on development of the inorganic ions in oviductal fluid. *J. Exp. Zool.* **1990**, *253*, 215–219. [CrossRef]
16. Sun, H.; Kang, J.; Su, J.; Zhang, J.; Zhang, L.; Liu, X.; Zhang, J.; Wang, F.; Lu, Z.; Xing, X.; et al. Methionine adenosyltransferase 2A regulates mouse zygotic genome activation and morula to blastocyst transition. *Biol. Reprod.* **2019**, *100*, 601–617. [CrossRef]
17. Van Winkle, L.J.; Campione, A.L.; Gorman, J.M.; Weimer, B.D. Changes in the activities of amino acid transport systems $b^{0,+}$ and L during development of preimplantation mouse conceptuses. *Biochim. Biophys. Acta (BBA)-Biomembr.* **1990**, *1021*, 77–84. [CrossRef]
18. Van Winkle, L.J.; Campione, A.L. Functional changes in cation-preferring amino acid transport during development of preimplantation mouse conceptuses. *Biochim. Biophys. Acta (BBA)-Biomembr.* **1990**, *1028*, 165–173. [CrossRef]
19. Van Winkle, L.J.; Ryznar, R. Amino acid transporters: Roles for nutrition, signaling and epigenetic modifications in embryonic stem cells and their progenitors. *eLS* **2019**, 1–13. [CrossRef]
20. Formisano, T.M.; Van Winkle, L.J. At least three transporters likely mediate threonine uptake needed for mouse embryonic stem cell proliferation. *Front. Cell Dev. Biol.* **2016**, *4*, 17. [CrossRef] [PubMed]
21. Wang, J.; Alexander, P.; Wu, L.; Hammer, R.; Cleaver, O.; McKnight, S.L. Dependence of mouse embryonic stem cells on threonine catabolism. *Science* **2009**, *325*, 435–439. [CrossRef]
22. Ryu, J.M.; Han, H.J. L-threonine regulates G1/S phase transition of mouse embryonic stem cells via PI3K/Akt, MAPKs, and mTORC pathways. *J. Biol. Chem.* **2011**, *286*, 23667–23678. [CrossRef]
23. Van Winkle, L.J.; Galat, V.; Iannaccone, P.M. Threonine appears to be essential for proliferation of human as well as mouse embryonic stem cells. *Front. Cell Dev. Biol.* **2014**, *2*, 18. [CrossRef]
24. Van Winkle, L.J.; Galat, V.; Iannaccone, P.M. Lysine Deprivation during Maternal Consumption of Low-Protein Diets Could Adversely Affect Early Embryo Development and Health in Adulthood. *Int. J. Environ. Res. Public Health* **2020**, *17*, 5462. [CrossRef]
25. Shiraki, N.; Shiraki, Y.; Tsuyama, T.; Obata, F.; Miura, M.; Nagae, G.; Aburatani, H.; Kume, K.; Endo, F.; Kume, S. Methionine metabolism regulates maintenance and differentiation of human pluripotent stem cells. *Cell Metab.* **2014**, *19*, 780–794. [CrossRef]
26. Chatot, C.L.; Ziomek, C.A.; Bavister, B.D.; Lewis, J.L.; Torres, I. An improved culture medium supports development of random-bred 1-cell mouse embryos in vitro. *Reproduction* **1989**, *86*, 679–688. [CrossRef]
27. Erbach, G.T.; Lawitts, J.A.; Papaioannou, V.E.; Biggers, J.D. Differential growth of the mouse preimplantation embryo in chemically defined media. *Biol. Reprod.* **1994**, *50*, 1027–1033. [CrossRef]
28. Lane, M.; Gardner, D.K. Nonessential amino acids and glutamine decrease the time of the first three cleavage divisions and increase compaction of mouse zygotes in vitro. *J. Assist. Reprod. Genet.* **1997**, *14*, 398–403. [CrossRef]
29. Hussein, A.M.; Wang, Y.; Mathieu, J.; Margaretha, L.; Song, C.; Jones, D.C.; Cavanaugh, C.; Miklas, J.W.; Mahen, E.; Showalter, M.R.; et al. Metabolic control over mTOR-dependent diapause-like state. *Dev. Cell* **2020**, *52*, 236–250. [CrossRef]
30. Van der Weijden, V.A.; Bulut-Karslioglu, A. Molecular Regulation of Paused Pluripotency in Early Mammalian Embryos and Stem Cells. *Front. Cell Dev. Biol.* **2021**, *9*, 708318. [CrossRef]
31. Lane, M.; Hooper, K.; Gardner, D.K. Animal experimentation: Effect of essential amino acids on mouse embryo viability and ammonium production. *J. Assist. Reprod. Genet.* **2001**, *18*, 519–525. [CrossRef]
32. Biggers, J.D.; McGinnis, L.K.; Lawitts, J.A. Enhanced effect of glycyl-L-glutamine on mouse preimplantation embryos in vitro. *Reprod. Biomed. Online* **2004**, *9*, 59–69. [CrossRef]
33. Roblero, L.; Biggers, J.D.; Lechene, C.P. Electron probe microanalysis of the elemental microenvironment of oviducal cleavage stages of the mouse. *J. Reprod. Fert.* **1976**, *46*, 431–434. [CrossRef]
34. Borland, R.M.; Hazra, S.; Biggers, J.D.; Lechene, C.P. The elemental composition of the environments of the gametes and preimplantation embryo during the initiation of pregnancy. *Biol. Reprod.* **1977**, *16*, 147–157. [CrossRef]
35. Leese, H.J.; McKeegan, P.; Sturmey, R.G. Amino acids and the early mammalian embryo: Origin, fate, function and life-long legacy. *Int. J. Environ. Res. Public Health* **2021**, *18*, 9874. [CrossRef]
36. Van Winkle, L.J. Amino acid transport regulation and early embryo development. *Biol. Reprod.* **2001**, *64*, 1–12. [CrossRef]
37. Capra, E.; Lange-Consiglio, A. The Biological Function of Extracellular Vesicles during Fertilization, Early Embryo—Maternal Crosstalk and Their Involvement in Reproduction: Review and Overview. *Biomolecules* **2020**, *10*, 1510. [CrossRef]

38. Van Winkle, L.J.; Christensen, H.N.; Campione, A.L. Na^+-dependent transport of basic, zwitterionic, and bicyclic amino acids by a broad-scope system in mouse blastocysts. *J. Biol. Chem.* **1985**, *260*, 12118–12123. [CrossRef]
39. Van Winkle, L.J. Uterine histone secretion likely fosters early embryo development so efforts to mitigate histone cytotoxicity should be cautious. *Front. Cell Dev. Biol.* **2017**, *5*, 100. [CrossRef]
40. Nangami, G.; Koumangoye, R.; Goodwin, J.S.; Sakwe, A.M.; Marshall, D.; Higginbotham, J.; Ochieng, J. Fetuin-A associates with histones intracellularly and shuttles them to exosomes to promote focal adhesion assembly resulting in rapid adhesion and spreading in breast carcinoma cells. *Exp. Cell Res.* **2014**, *328*, 388–400. [CrossRef]
41. Munn, D.H.; Zhou, M.; Attwood, J.T.; Bondarev, I.; Conway, S.J.; Marshall, B.; Brown, C.; Mellor, A.L. Prevention of allogeneic fetal rejection by tryptophan catabolism. *Science* **1998**, *281*, 1191–1193. [CrossRef]
42. Baban, B.; Chandler, P.; McCool, D.; Marshall, B.; Munn, D.H.; Mellor, A.L. Indoleamine 2, 3-dioxygenase expression is restricted to fetal trophoblast giant cells during murine gestation and is maternal genome specific. *J. Reprod. Immunol.* **2004**, *61*, 67–77. [CrossRef]
43. To, C.Y.; Freeman, M.; Van Winkle, L.J. Consumption of a Branched-Chain Amino Acid (BCAA) during Days 2–10 of Pregnancy Causes Abnormal Fetal and Placental Growth: Implications for BCAA Supplementation in Humans. *Int. J. Environ. Res. Public Health* **2020**, *17*, 2445. [CrossRef]
44. Caetano, L.; Eckert, J.; Johnston, D.; Chatelet, D.; Tumbarello, D.; Smyth, N.; Ingamells, S.; Price, A.; Fleming, T. Blastocyst trophectoderm endocytic activation, a marker of adverse developmental programming. *Reproduction* **2021**, in press. [CrossRef]
45. Eckert, J.J.; Porter, R.; Watkins, A.J.; Burt, E.; Brooks, S.; Leese, H.J.; Humpherson, P.G.; Cameron, I.T.; Fleming, T.P. Metabolic induction and early responses of mouse blastocyst developmental programming following maternal low protein diet affecting life-long health. *PLoS ONE* **2012**, *7*, e52791. [CrossRef]
46. Fleming, T.P.; Watkins, A.J.; Velazquez, M.A.; Mathers, J.C.; Prentice, A.M.; Stephenson, J.; Barker, M.; Saffery, R.; Yajnik, C.S.; Eckert, J.J.; et al. Origins of lifetime health around the time of conception: Causes and consequences. *Lancet* **2018**, *391*, 1842–1852. [CrossRef]
47. Fleming, T.P.; Sun, C.; Denisenko, O.; Caetano, L.; Aljahdali, A.; Gould, J.M.; Khurana, P. Environmental Exposures around Conception: Developmental Pathways Leading to Lifetime Disease Risk. *Int. J. Environ. Res. Public Health* **2021**, *18*, 9380. [CrossRef]
48. Jansova, D.; Koncicka, M.; Tetkova, A.; Cerna, R.; Malik, R.; Del Llano, E.; Kubelka, M.; Susor, A. Regulation of 4E-BP1 activity in the mammalian oocyte. *Cell Cycle* **2017**, *16*, 927–939. [CrossRef]
49. Van Winkle, L.J.; Campione, A.L.; Gorman, J.M. Na^+-independent transport of basic and zwitterionic amino acids in mouse blastocysts by a shared system and by processes which distinguish between these substrates. *J. Biol. Chem.* **1988**, *263*, 3150–3163. [CrossRef]
50. Feliubadaló, L.; Arbonés, M.L.; Mañas, S.; Chillarón, J.; Visa, J.; Rodés, M.; Rousaud, F.; Zorzano, A.; Palacín, M.; Nunes, V. Slc7a9-deficient mice develop cystinuria non-I and cystine urolithiasis. *Hum. Mol. Genet.* **2003**, *12*, 2097–2108. [CrossRef]
51. Redel, B.K.; Tessanne, K.J.; Spate, L.D.; Murphy, C.N.; Prather, R.S. Arginine increases development of in vitro-produced porcine embryos and affects the protein arginine methyltransferase–dimethylarginine dimethylaminohydrolase–nitric oxide axis. *Reprod. Fertil. Dev.* **2015**, *27*, 655–666. [CrossRef]
52. White, M.F.; Gazzola, G.C.; Christensen, H.N. Cationic amino acid transport into cultured animal cells. I. Influx into cultured human fibroblasts. *J. Biol. Chem.* **1982**, *257*, 4443–4449. [CrossRef]
53. White, M.F.; Christensen, H.N. Cationic amino acid transport into cultured animal cells. II. Transport system barely perceptible in ordinary hepatocytes, but active in hepatoma cell lines. *J. Biol. Chem.* **1982**, *257*, 4450–4457. [CrossRef]
54. MacLeod, C.L.; Finley, K.D.; Kakuda, D.K. y^+-type cationic amino acid transport: Expression and regulation of the mCAT genes. *J. Exp. Biol.* **1994**, *196*, 109–121. [CrossRef]
55. Terstappen, F.; Tol, A.J.; Gremmels, H.; Wever, K.E.; Paauw, N.D.; Joles, J.A.; van der Beek, E.M.; Lely, A.T. Prenatal amino acid supplementation to improve fetal growth: A systematic review and meta-analysis. *Nutrients* **2020**, *12*, 2535. [CrossRef]
56. Hussain, T.; Bie Tan, G.M.; Metwally, E.; Yang, H.; Kalhoro, M.S.; Kalhoro, D.H.; Chughtai, M.I.; Yin, Y. Role of Dietary Amino Acids and Nutrient Sensing System in Pregnancy Associated Disorders. *Front. Pharmacol.* **2020**, *11*, 586979. [CrossRef]
57. Zeng, X.; Wang, F.; Fan, X.; Yang, W.; Zhou, B.; Li, P.; Yin, Y.; Wu, G.; Wang, J. Dietary arginine supplementation during early pregnancy enhances embryonic survival in rats. *J. Nutr.* **2008**, *138*, 1421–1425. [CrossRef]
58. Sprague Dawley® SD® Outbred Rats. Available online: envigo.com (accessed on 5 November 2021).
59. Stapleton, P.; Hill, D.C. The effect of maternal dietary lysine and methionine levels on pregnancy and lactation in the rat. *Nutr. Rep. Int.* **1980**, *21*, 231–242.
60. Funk, D.N.; Worthington-Roberts, B.; Fantel, A. Impact of supplemental lysine or tryptophan on pregnancy course and outcome in rats. *Nutr. Res.* **1991**, *11*, 501–512. [CrossRef]
61. Najafzadeh, V.; Henderson, H.; Martinus, R.D.; Oback, B. Bovine blastocyst development depends on threonine catabolism. *bioRxiv* **2018**. [CrossRef]
62. Han, C.; Gu, H.; Wang, J.; Lu, W.; Mei, Y.; Wu, M. Regulation of l-threonine dehydrogenase in somatic cell reprogramming. *Stem Cells* **2013**, *31*, 953–965. [CrossRef]

63. Shyh-Chang, N.; Locasale, J.W.; Lyssiotis, C.A.; Zheng, Y.; Teo, R.Y.; Ratanasirintrawoot, S.; Zhang, J.; Onder, T.; Unternaehrer, J.J.; Zhu, H.; et al. Influence of threonine metabolism on S-adenosylmethionine and histone methylation. *Science* **2013**, *339*, 222–226. [CrossRef]
64. Tian, S.; Feng, J.; Cao, Y.; Shen, S.; Cai, Y.; Yang, D.; Yan, R.; Wang, L.; Zhang, H.; Zhong, X.; et al. Glycine cleavage system determines the fate of pluripotent stem cells via the regulation of senescence and epigenetic modifications. *Life Sci. Alliance* **2019**, *2*, e201900413. [CrossRef]
65. Gardner, R.L. Flow of cells from polar to mural trophectoderm is polarized in the mouse blastocyst. *Hum. Reprod.* **2000**, *15*, 694–701. [CrossRef]
66. Van Winkle, L.J.; Dickinson, H.R. Differences in amino acid content of preimplantation mouse embryos that develop in vitro versus in vivo: In vitro effects of five amino acids that are abundant in oviductal secretions. *Biol. Reprod.* **1995**, *52*, 96–104. [CrossRef]
67. Sun, C.; Velazquez, M.A.; Marfy-Smith, S.; Sheth, B.; Cox, A.; Johnston, D.A.; Smyth, N.; Fleming, T.P. Mouse early extra-embryonic lineages activate compensatory endocytosis in response to poor maternal nutrition. *Development* **2014**, *141*, 1140–1150. [CrossRef]
68. Najafzadeh, V. The Role of Amino Acids and the Threonine-SAM Pathway in the Development of Bovine Inner Cell Mass and Pluripotency. Ph.D. Dissertation, The University of Waikato, Hillcrest, Hamilton, 2018.
69. Edgar, A.J. The human L-threonine 3-dehydrogenase gene is an expressed pseudogene. *BMC Genet.* **2002**, *3*, 1–13. [CrossRef]
70. Kang, P.J.; Zheng, J.; Lee, G.; Son, D.; Kim, I.Y.; Song, G.; Park, G.; You, S. Glycine decarboxylase regulates the maintenance and induction of pluripotency via metabolic control. *Metab. Eng.* **2019**, *53*, 35–47. [CrossRef]
71. Ang, Y.S.; Tsai, S.Y.; Lee, D.F.; Monk, J.; Su, J.; Ratnakumar, K.; Ding, J.; Ge, Y.; Darr, H.; Chang, B.; et al. Wdr5 mediates self-renewal and reprogramming via the embryonic stem cell core transcriptional network. *Cell* **2011**, *145*, 183–197. [CrossRef]
72. Kilberg, M.S.; Terada, N.; Shan, J. Influence of amino acid metabolism on embryonic stem cell function and differentiation. *Adv. Nutr.* **2016**, *7*, 780S–789S. [CrossRef]
73. Chen, G.; Wang, J. A regulatory circuitry locking pluripotent stemness to embryonic stem cell: Interaction between threonine catabolism and histone methylation. In *Seminars in Cancer Biology*; Academic Press: Cambridge, MA, USA, 2019; Volume 57, pp. 72–78.
74. Spyrou, J.; Gardner, D.K.; Harvey, A.J. Metabolism is a key regulator of induced pluripotent stem cell reprogramming. *Stem Cells Int.* **2019**, *2019*, 7360121. [CrossRef] [PubMed]
75. Zhang, J.; Khvorostov, I.; Hong, J.S.; Oktay, Y.; Vergnes, L.; Nuebel, E.; Wahjudi, P.N.; Setoguchi, K.; Wang, G.; Do, A.; et al. UCP2 regulates energy metabolism and differentiation potential of human pluripotent stem cells. *EMBO J.* **2011**, *30*, 4860–4873. [CrossRef] [PubMed]
76. Chakrabarty, R.P.; Chandel, N.S. Mitochondria as signaling organelles control mammalian stem cell fate. *Cell Stem Cell* **2021**, *28*, 394–408. [CrossRef] [PubMed]
77. Wanet, A.; Arnould, T.; Najimi, M.; Renard, P. Connecting mitochondria, metabolism, and stem cell fate. *Stem Cells Dev.* **2015**, *24*, 1957–1971. [CrossRef]
78. Lees, J.G.; Gardner, D.K.; Harvey, A.J. Pluripotent stem cell metabolism and mitochondria: Beyond ATP. *Stem Cells Int.* **2017**, *2017*, 2874283. [CrossRef] [PubMed]
79. Woods, D.C. Mitochondrial heterogeneity: Evaluating mitochondrial subpopulation dynamics in stem cells. *Stem Cells Int.* **2017**, *2017*, 7068567. [CrossRef] [PubMed]
80. Scotti, M.; Stella, L.; Shearer, E.J.; Stover, P.J. Modeling cellular compartmentation in one-carbon metabolism. *Wiley Interdiscip. Rev. Syst. Biol. Med.* **2013**, *5*, 343–365. [CrossRef]
81. Harvey, A.J. Mitochondria in early development: Linking the microenvironment, metabolism and the epigenome. *Reproduction* **2019**, *157*, R159–R179. [CrossRef]
82. Alexander, P.B. Requirement of a High-Flux Metabolic State for Mouse Embryonic Stem Cell Self-Renewal. Ph.D. Dissertation, The University of Texas Southwestern Medical Center at Dallas, Dallas, TX, USA, 2010.
83. Alexander, P.B.; Wang, J.; McKnight, S.L. Targeted killing of a mammalian cell based upon its specialized metabolic state. *Proc. Natl. Acad. Sci. USA* **2011**, *108*, 15828–15833. [CrossRef]
84. Guo, H.; Zhu, P.; Yan, L.; Li, R.; Hu, B.; Lian, Y.; Yan, J.; Ren, X.; Lin, S.; Li, J.; et al. The DNA methylation landscape of human early embryos. *Nature* **2014**, *511*, 606–610. [CrossRef]
85. Torrens, C.; Poston, L.; Hanson, M.A. Transmission of raised blood pressure and endothelial dysfunction to the F2 generation induced by maternal protein restriction in the F0, in the absence of dietary challenge in the F1 generation. *Br. J. Nutr.* **2008**, *100*, 760–766. [CrossRef] [PubMed]
86. Gallo, L.A.; Tran, M.; Moritz, K.M.; Jefferies, A.J.; Wlodek, M.E. Pregnancy in aged rats that were born small: Cardiorenal and metabolic adaptations and second-generation fetal growth. *FASEB J.* **2012**, *26*, 4337–4347. [CrossRef] [PubMed]
87. Master, J.S.; Thouas, G.A.; Harvey, A.J.; Sheedy, J.R.; Hannan, N.J.; Gardner, D.K.; Wlodek, M.E. Low female birth weight and advanced maternal age programme alterations in next-generation blastocyst development. *Reproduction* **2015**, *149*, 497–510. [CrossRef]

88. Padmanabhan, N.; Jia, D.; Geary-Joo, C.; Wu, X.; Ferguson-Smith, A.C.; Fung, E.; Bieda, M.C.; Snyder, F.F.; Gravel, R.A.; Cross, J.C.; et al. Mutation in folate metabolism causes epigenetic instability and transgenerational effects on development. *Cell* **2013**, *155*, 81–93. [CrossRef] [PubMed]
89. Watkins, A.J.; Dias, I.; Tsuro, H.; Allen, D.; Emes, R.D.; Moreton, J.; Wilson, R.; Ingram, R.J.; Sinclair, K.D. Paternal diet programs offspring health through sperm-and seminal plasma-specific pathways in mice. *Proc. Natl. Acad. Sci. USA* **2018**, *115*, 10064–10069. [CrossRef] [PubMed]
90. Fung, K.Y.; Glode, L.M.; Green, S.; Duncan, M.W. A comprehensive characterization of the peptide and protein constituents of human seminal fluid. *Prostate* **2004**, *61*, 171–181. [CrossRef]
91. Miller, D.; Brinkworth, M.; Iles, D. Paternal DNA packaging in spermatozoa: More than the sum of its parts? DNA, histones, protamines and epigenetics. *Reproduction* **2010**, *139*, 287–301. [CrossRef]
92. Miller, D.J. The epic journey of sperm through the female reproductive tract. *Animal* **2018**, *12*, s110–s120. [CrossRef]
93. Clare, C.E.; Brassington, A.H.; Kwong, W.Y.; Sinclair, K.D. One-carbon metabolism: Linking nutritional biochemistry to epigenetic programming of long-term development. *Annu. Rev. Anim. Biosci.* **2019**, *7*, 263–287. [CrossRef] [PubMed]
94. Su, L.; Patti, M.E. Paternal nongenetic intergenerational transmission of metabolic disease risk. *Curr. Diabetes Rep.* **2019**, *19*, 1–9. [CrossRef]
95. Papes, F.; Surpili, M.J.; Langone, F.; Trigo, J.R.; Arruda, P. The essential amino acid lysine acts as precursor of glutamate in the mammalian central nervous system. *FEBS Lett.* **2001**, *488*, 34–38. [CrossRef]
96. Sacksteder, K.A.; Biery, B.J.; Morrell, J.C.; Goodman, B.K.; Geisbrecht, B.V.; Cox, R.P.; Gould, S.J.; Geraghty, M.T. Identification of the α-aminoadipic semialdehyde synthase gene, which is defective in familial hyperlysinemia. *Am. J. Hum. Genet.* **2000**, *66*, 1736–1743. [CrossRef] [PubMed]
97. Crowther, L.M.; Mathis, D.; Poms, M.; Plecko, B. New insights into human lysine degradation pathways with relevance to pyridoxine-dependent epilepsy due to antiquitin deficiency. *J. Inherit. Metab. Dis.* **2019**, *42*, 620–628. [CrossRef] [PubMed]
98. Bell, S.; Maussion, G.; Jefri, M.; Peng, H.; Theroux, J.F.; Silveira, H.; Soubannier, V.; Wu, H.; Hu, P.; Galat, E.; et al. Disruption of GRIN2B impairs differentiation in human neurons. *Stem Cell Rep.* **2018**, *11*, 183–196. [CrossRef] [PubMed]
99. McKenna, M.C. The glutamate-glutamine cycle is not stoichiometric: Fates of glutamate in brain. *J. Neurosci. Res.* **2007**, *85*, 3347–3358. [CrossRef] [PubMed]
100. Cappuccio, I.; Spinsanti, P.; Porcellini, A.; Desiderati, F.; De Vita, T.; Storto, M.; Capobianco, L.; Battaglia, G.; Nicoletti, F.; Melchiorri, D. Endogenous activation of mGlu5 metabotropic glutamate receptors supports self-renewal of cultured mouse embryonic stem cells. *Neuropharmacology* **2005**, *49*, 196–205. [CrossRef]
101. Spinsanti, P.; De Vita, T.; Di Castro, S.; Storto, M.; Formisano, P.; Nicoletti, F.; Melchiorri, D. Endogenously activated mGlu5 metabotropic glutamate receptors sustain the increase in c-Myc expression induced by leukaemia inhibitory factor in cultured mouse embryonic stem cells. *J. Neurochem.* **2006**, *99*, 299–307. [CrossRef]
102. Kaye, P.L.; Schultz, G.A.; Johnson, M.H.; Pratt, H.P.; Church, R.B. Amino acid transport and exchange in preimplantation mouse embryos. *Reproduction* **1982**, *65*, 367–380. [CrossRef] [PubMed]
103. Galili, G. New insights into the regulation and functional significance of lysine metabolism in plants. *Annu. Rev. Plant Biol.* **2002**, *53*, 27–43. [CrossRef] [PubMed]
104. Zamfirescu, R.C.; Day, M.L.; Morris, M.B. mTORC1/2 signaling is downregulated by amino acid-free culture of mouse preimplantation embryos and is only partially restored by amino acid readdition. *Am. J. Physiol. -Cell Physiol.* **2021**, *320*, C30–C44. [CrossRef]
105. Carey, B.W.; Finley, L.W.; Cross, J.R.; Allis, C.D.; Thompson, C.B. Intracellular α-ketoglutarate maintains the pluripotency of embryonic stem cells. *Nature* **2015**, *518*, 413–416. [CrossRef] [PubMed]
106. Comes, S.; Gagliardi, M.; Laprano, N.; Fico, A.; Cimmino, A.; Palamidessi, A.; De Cesare, D.; De Falco, S.; Angelini, C.; Scita, G.; et al. L-Proline induces a mesenchymal-like invasive program in embryonic stem cells by remodeling H3K9 and H3K36 methylation. *Stem Cell Rep.* **2013**, *1*, 307–321. [CrossRef] [PubMed]
107. Gopichandran, N.; Leese, H.J. Metabolic characterization of the bovine blastocyst, inner cell mass, trophectoderm and blastocoel fluid. *Reproduction* **2003**, *126*, 299–308. [CrossRef] [PubMed]

MDPI

Article

In Vitro Fertilisation of Mouse Oocytes in L-Proline and L-Pipecolic Acid Improves Subsequent Development

Tamara Treleaven, Madeleine L.M. Hardy, Michelle Guttman-Jones, Michael B. Morris † and Margot L. Day *,†

Physiology, School of Medical Sciences, Faculty of Medicine and Health, University of Sydney, Sydney, NSW 2006, Australia; tprz7203@uni.sydney.edu.au (T.T.); mhar7073@uni.sydney.edu.au (M.L.M.H.); drmguttmanjones@gmail.com (M.G.-J.); m.morris@sydney.edu.au (M.B.M.)
* Correspondence: margot.day@sydney.edu.au; Tel.: +61-2-9036-3312
† M.B.M. and M.L.D. are joint senior authors.

Abstract: Exposure of oocytes to specific amino acids during in vitro fertilisation (IVF) improves preimplantation embryo development. Embryos fertilised in medium with proline and its homologue pipecolic acid showed increased blastocyst formation and inner cell mass cell numbers compared to embryos fertilised in medium containing no amino acids, betaine, glycine, or histidine. The beneficial effect of proline was prevented by the addition of excess betaine, glycine, and histidine, indicating competitive inhibition of transport-mediated uptake. Expression of transporters of proline in oocytes was investigated by measuring the rate of uptake of radiolabelled proline in the presence of unlabelled amino acids. Three transporters were identified, one that was sodium-dependent, PROT (SLC6A7), and two others that were sodium-independent, PAT1 (SLC36A1) and PAT2 (SLC36A2). Immunofluorescent staining showed localisation of PROT in intracellular vesicles and limited expression in the plasma membrane, while PAT1 and PAT2 were both expressed in the plasma membrane. Proline and pipecolic acid reduced mitochondrial activity and reactive oxygen species in oocytes, and this may be responsible for their beneficial effect. Overall, our results indicate the importance of inclusion of specific amino acids in IVF medium and that consideration should be given to whether the addition of multiple amino acids prevents the action of beneficial amino acids.

Keywords: oocyte; amino acid transporters; proline; pipecolic acid; in vitro fertilisation

Citation: Treleaven, T.; Hardy, M.L.M.; Guttman-Jones, M.; Morris, M.B.; Day, M.L. In Vitro Fertilisation of Mouse Oocytes in L-Proline and L-Pipecolic Acid Improves Subsequent Development. *Cells* **2021**, *10*, 1352. https://doi.org/10.3390/cells10061352

Academic Editor: Lon J. van Winkle

Received: 7 May 2021
Accepted: 26 May 2021
Published: 29 May 2021

1. Introduction

The presence of individual and groups of amino acids in embryo culture medium impacts preimplantation embryo development [1–8]. Much less is known, however, about the impact of individual amino acids on future development when they are exogenously added to oocytes during in vitro fertilisation (IVF). The addition of specific amino acids or groups of amino acids to the IVF medium can improve various aspects of subsequent development, including the percentage of embryos that reach the blastocyst stage and then hatch [9,10], indicating that amino acids can enter the oocyte in sufficient quantities during the period of fertilisation to exert effects on later embryo development.

The amino acid L-proline (Pro) acts in a growth-factor-like manner to stimulate preimplantation development [7]. When zygotes are cultured at low density (1 embryo/100 μL, to eliminate the action of embryo-derived growth factors), the addition of Pro increases both the number of embryos reaching the blastocyst stage and their hatching by stimulating development within the late two-cell to eight-cell stages [7]. The addition of Pro also promotes differentiation and neural lineage commitment in embryonic stem cell (ESC) models of post-implantation development [11,12].

The mechanisms by which Pro stimulates pre- and post-implantation development are only partially understood. In embryos cultured at low density, Pro stimulates nuclear translocation of p-Akt^{S473} and p-ERK1/2$^{T202/Y204}$, and improvement in development is dependent on mTORC1 signalling (rapamycin sensitive) [3]. Activation of these signalling pathways is also involved in the differentiation of ESCs [11,13,14]. Cellular handling of the production of this conditional amino acid is also involved: ESCs limit the endogenous production of Pro via a mechanism involving the amino acid response pathway, to the extent that differentiation is suppressed and ESC self-renewal maintained [15]. The exogenous addition of Pro overwhelms this suppressive mechanism. Pro also promotes changes to the epigenetic landscape of ESCs via enhanced histone methylation (at H3K9 and H3K36). The mechanism may involve direct or indirect inhibition of the Jumonji domain demethylases, which help control methylation status at these sites [16].

Attention has also turned to Pro's unique metabolism as a source of mechanisms controlling developmental fate [13,14,16–18]. Pro is metabolised via the Pro cycle to pyrroline-5-carboxylic acid (P5C) by proline oxidase (POX), and then to glutamate semialdehyde before being converted to glutamate (Glu) and α-ketoglutarate [17,19]. Thus, Pro metabolism can provide a number of intermediates important for cellular function, including the production of ATP via both α-ketoglutarate production and Pro-derived high-energy electrons entering the electron transport system [17].

Pro also has the ability to scavenge intracellular reactive oxygen species (ROS) [20]. Embryos with low metabolism, low glycolytic activity, low amino acid turnover, and high antioxidant levels are hypothesised to be the most viable [21–24], most likely because the embryo expends less energy and has lower production and leakage of superoxide free radicals from mitochondria [21,25,26]. H_2O_2 is another source of ROS, produced by superoxide dismutase conversion of the superoxide radical, especially during the G_2/M phase of cleavage-stage embryos [27–29]. High concentrations of ROS are detrimental to oocyte quality: Oocytes exposed to high levels of ROS in vivo have lower fertilisation rates, produce lower quality embryos, and present increased embryo arrest at the two-cell stage in the mouse [28,30].

The transport of Pro into cells occurs via specific amino acid transporters, specifically neutral amino acid transport systems, and are further characterised by their gene sequence homology in the solute carrier transporter (SLC) family. In the mouse oocyte, a number of amino acid transport systems have been identified (L, $b^{0,+}$, ASC, asc, B^0, Gly, beta), one or more of which may be responsible for the uptake of Pro into the oocyte [31]. Uptake of Pro into cumulus cells surrounding the oocyte also occurs via y+LAT2 (SLC7A6), and thus Pro is transferred to the oocyte via gap junctions in the cumulus oocyte complex [32,33]. mRNA for the Na^+-dependent Pro transporter PROT (SLC6A7) and the Na^+/Cl^--dependent betaine/proline transporter SIT1 (SLC6A20) are present in the mouse oocyte, but it is not known if the transporters themselves are expressed and actively take up Pro [34]. After fertilisation, large amounts of Pro are taken up by SIT1, which is active in the zygote and two-cell stages [35]. Pro uptake in zygotes also involves an unidentified Na^+-dependent betaine-resistant transporter [34,35]. In ESCs, Pro uptake is via the System A Na^+-dependent SNAT2 transporter (SLC38A2) [14]. System A transporters are also the major transport system in the inner cell mass of blastocysts, but it is not known if SNAT2 is the principal contributor [36]. The different amino acid requirements at specific stages of oocyte and embryonic development reflect the dynamic physiology of the preimplantation embryo, including changes in metabolism and amino acid transporter expression.

L-pipecolic acid (PA) is a product of lysine metabolism and a homologue of Pro, having a six-membered piperidine ring rather the five-membered pyrrolidine ring of Pro [20]. Both PA and Pro undergo oxidation in mitochondria and promote cell survival during oxidative stress [37]. It is not known if there is a transporter for PA in oocytes, but in somatic cells PA is transported by the high-affinity Pro transporter PROT [38–40] and SIT1 [41,42]. We hypothesised that because PA is structurally similar to Pro it would be taken up into the oocyte and be either metabolised or act as a ROS scavenger in a similar way to Pro, and thereby improve later embryo development [37,43].

This study investigated the effect of the addition of specific amino acids to IVF medium on subsequent preimplantation embryo development, with a particular focus on Pro and its analogue PA. We also examined the effect of molar excess of specific amino acids on uptake of Pro to determine the transporter(s) responsible for uptake. Finally, we investigated metabolic mechanisms by which Pro and PA may impact oocyte development viability after IVF. Overall, this study shows that Pro and PA can improve preimplantation embryo development when added to IVF medium by reducing mitochondrial activity and ROS in the oocyte.

2. Materials and Methods

2.1. Animals (Mus Musculus)

Quackenbush Swiss (QS) mice (Animal Resource Centre, Perth and Lab Animal Services, University of Sydney) were housed under 12 h light:12 h dark conditions as described previously [44]. Experiments were conducted in accordance with the Australian Code of Practice for the Care and Use of Animals for Scientific Purposes and approved by the University of Sydney Animal Ethics Committee as required by the NSW Animal Research Act (protocols 5583 and 824).

2.2. Isolation of Oocytes and Zygotes Fertilised In Vivo

Superovulation of female mice aged 4–6 weeks was achieved by intraperitoneal injection with 10 IU pregnant mares' serum gonadotropin (PMS; Intervet, Vic. Australia) followed after 48 $\pm$ 2 h by injection with 10 IU human chronic gonadotropin (hCG; Intervet). To obtain zygotes fertilised in vivo, female mice were paired overnight with individually housed QS male mice (2–8 months of age) immediately following hCG injection. Mating was confirmed the following day by the presence of a vaginal plug. Mice were sacrificed by cervical dislocation at either 13–15 h post-hCG to obtain oocytes or 22 h post-hCG to obtain zygotes. Oocytes and zygotes were isolated into HEPES-buffered modified human tubal fluid (Hepes-modHTF) in which the concentration of NaCl was adjusted to 85 mM to give an osmolality of 270 mOsm/kg [7]. Cumulus cells were removed from zygotes and oocytes by treatment with 1 mg/mL hyaluronidase in HEPES-modHTF and then washed 3 times in HEPES-modHTF to remove remaining cumulus cells.

2.3. In Vitro Fertilisation of Oocytes and Culture of Embryos

In vitro fertilisation was performed in modified Whittingham's medium [45,46] which contained (in mM) NaCl (99.3), KCl (2.7), NaH_2PO_4 (0.36), $MgCl_2.6H_2O$ (0.5), $NaHCO_3$ (25), Na pyruvate (0.5), glucose (5.5), $CaCl_2.2H_2O$ (1.8), Na lactate (18.7), and 30 mg/mL BSA. At least 4 cumulus masses containing oocytes were added to 500 µL pre-equilibrated (5% CO_2 at 37 °C) Whittingham's medium, containing the appropriate amino acid(s), overlayed with mineral oil (Sigma-Aldrich, NSW, Australia). Treatment groups were (i) no amino acids (no AA), and (ii) 0.4 mM L-proline (Pro) $\pm$ 5 mM His, Gly, betaine (Bet) or PA (as potential competitive inhibitors of the uptake of Pro). The concentration of Pro used in these experiments was based on the physiological concentration of fluid in the reproductive tract [47] and was shown to improve embryo development [7].

Sperm was isolated from the epididymis of a male QS mouse of proven fertility. An incision was made in the efferent ducts between the testis and the epididymis, and at the inferior end of the caudis epididymis. The dissected tissue was immediately washed, and blood vessels drained in PBS. The epididymides were placed into 500 µL Whittingham's medium at 37 °C and 5% CO_2 and the sperm gently squeezed out of the epididymides using forceps and allowed to capacitate for 1.5 h. At 13.5 h post-hCG, 10^6 sperm were added to each drop containing cumulus masses and incubated together for 3 h, after which fertilisation was determined by the presence of a polar body. Zygotes were washed through at least 3 drops of HEPES mod-HTF in the absence of amino acids and then cultured to the blastocyst stage (120 h) at low density (1 embryo/100 µL) in mod-HTF + 0.3 mg/mL BSA, containing no amino acids, at 37 °C in 5% CO_2 under mineral oil [7]. Embryos were scored for their developmental stage every 24 h. In vivo fertilised zygotes were also cultured under these conditions as a control.

2.4. Measurement of L-[^{3}H]-Pro Uptake in Oocytes

Oocytes were incubated in 20 μL assay medium consisting of 1 μM L-[^{3}H]-Pro (L-[2,3,4,5-^{3}H]-proline; 1 mCi/mL; Perkin-Elmer, Vic, Australia, NET483001MC) in HEPES mod-HTF ± molar excess of individual unlabelled amino acids (5 mM for all amino acids except Pro, which was used at 0.4 mM). Unless otherwise stated, the L-isomer was used for all unlabelled amino acids and purchased from Sigma Aldrich, except L-Lys (Chem-Impex, IL, USA). In experiments using Na^+-free HEPES mod-HTF, the Na^+ was replaced with N-methyl-D-glucamine (NMDG) at 85 mM. At this low concentration NMDG is not known to affect Pro uptake by Na^+-independent transporters.

An initial time course experiment was performed in which oocytes were incubated in L-[^{3}H]-Pro for up to 120 min and samples collected at 20-min intervals to show that the uptake of L-[^{3}H]-Pro remained linear over this time course (data not shown). In all subsequent experiments, oocytes in groups of 4 were incubated for 100 min at 37 °C in each treatment and then washed through at least 3 drops of cold (4 °C) HEPES mod-HTF, aspirated, and placed onto a gridded filter mat (Perkin-Elmer, Vic, Australia, #1450-421). The mat was placed inside a plastic sleeve and 4 mL ULTIMA GOLD scintillation fluid (Perkin-Elmer, Vic, Australia, #6013371) was added. The plastic sleeve was placed in a MicroBeta TriLux Plate Counter (Perkin-Elmer, Vic, Australia) and each sample was counted for 30 min. To determine the rate of L-[^{3}H]-Pro uptake (fmol min^{-1} $oocyte^{-1}$), a standard curve was created for each experiment from standards serially diluted from a stock of 1 μM L-[^{3}H]-Pro in HEPES mod-HTF and fitted by linear regression.

2.5. Immunostaining and Confocal Microscopy

Oocytes were fixed in 4% paraformaldehyde (PFA) for 30 min at room temperature, then washed 3 times in PBS + 1 mg/mL polyvinyl acid (PBS + PVA; Sigma-Aldrich, NSW, Australia), and permeabilised with PBS + PVA + 0.3% Triton X-100 for 30 min at room temperature. The oocytes were blocked by incubation in PBS + PVA + 0.1% Tween-20 + 0.7% BSA for 30 min at room temperature. Antibodies used were rabbit anti-SLC6A7/PROT (GeneTex, CA, USA #GTX51242), rabbit anti-SLC36A1/PAT1 [48], rabbit anti-SLC36A2/PAT2 [48], and Alexa Fluor 488-coupled goat anti-rabbit IgG (Invitrogen, Vic, Australia). Primary antibodies were diluted 1:500, and the secondary antibody 1:200, in PBS + PVA + 0.1% Tween-20 + 0.7% BSA. Samples were incubated in primary antibody for 2 h at room temperature or overnight at 4 °C and then washed 3 times in PBS + PVA + 0.1% Tween-20 + 0.7% BSA. Samples were incubated in the dark in secondary antibody for 1 h at room temperature and then washed 3 times in PBS + PVA + 0.1% Tween-20 + 0.7% BSA before being mounted in 3 μL Vectashield containing 1.5 μg/mL DAPI (Vector Laboratories, CA, USA) to visualise nuclei. Oocytes were imaged using confocal microscopy (LSM Meta 800, Carl Zeiss, Oberkochen, Germany) using 405 nm and 488 nm lasers and a 40× objective. Images were prepared using Fiji by Image J.

For analysis of cell numbers, blastocysts were fixed and permeabilised as described above and mounted in VECTASHIELD containing DAPI. Nuclei were visualised by confocal microscopy using a 405 nm laser. A Z-stack was taken through each blastocyst at 2.5 μm slices to allow for each nucleus to be imaged at least once. Cells in the inner cell mass and trophectoderm were counted manually using Image J. Cell numbers from each blastocyst were averaged.

2.6. Measurement of Mitochondrial Activity and Reactive Oxygen Species

Oocytes were cultured for 3 h in Whittingham's medium containing 0.4 mM Pro ± 5 mM PA Gly, Bet or His. Oocytes were then transferred to Whittingham's medium containing the above amino acid combinations to which 100 nM tetramethylrhodamine methyl ester (TMRM) and 10 μM 2′,7′-dichlorofluorescin diacetate (DCFDA) were added. These were incubated for 30 min at 37 °C in 5% CO_2 to allow uptake of the dyes, then washed twice through 2 drops of Whittingham's medium, containing the amino acid combinations above, to remove excess dye. A 35 mm glass-bottomed petri dish was prepared containing 10 μL drops of medium (under mineral oil) containing each of the treatment conditions ±5 μM carbonyl cyanide-4 (trifluoromethoxy) phenylhydrazone (FCCP) as a positive control for mitochondrial activity, or ±60 μM hydrogen peroxide (H_2O_2) as a positive control for ROS, pre-equilibrated to 37 °C and 5% CO_2. Oocytes were added to drops and imaged using confocal microscopy (Carl Zeiss, Oberkochen, Germany, LSM Meta 800), with the chamber set to 37 °C and 5% CO_2 for live-cell imaging, using a 20× objective and 524 nm and 488 nm lasers. All oocytes were imaged within 4 h of isolation, which was equivalent to the timeline of amino acid treatment in the IVF experiments. Images were analysed using Fiji by Image J to obtain the integrated density of the stain, and the corrected total cell fluorescence was calculated using the following formula: Corrected total cell fluorescence = integrated density—(background fluorescence × mean area) [49].

2.7. Statistical Analyses

Oocyte/embryo culture experiments were performed a minimum of 3 times with at least 10 oocytes/embryos per treatment group. Development in the different treatment groups was compared by calculating the percentage of embryos that developed to the particular cell stage. Differences between groups were determined using 1-way ANOVA and Tukey's or Dunnett's multiple comparisons test as indicated in the figure legends.

L-[^{3}H]-Pro uptake experiments were performed at least 3 times with 4 embryos per treatment group. The uptake of L-[^{3}H]-Pro for each treatment was averaged and the difference between groups compared by 1-way ANOVA and Tukey's multiple comparisons test.

For fluorescence imaging, corrected total cell fluorescence was averaged from 3 independent experiments with a total of 28–71 embryos per treatment, and differences between treatments were determined by one-way ANOVA and Tukey's multiple comparisons test. Data analysis was performed using GraphPad Prism v7.

3. Results

3.1. IVF in the Presence of Pro and PA Improves Subsequent Embryo Development

Oocytes were fertilised for 3 h in the presence or absence of various amino acids, and then subsequently allowed to develop in vitro in the absence of any amino acids. In vitro development was performed at low density (1 embryo/100 μL) and, therefore, in the absence of autocrine support. The proportion of oocytes fertilised was not significantly affected by any amino acid treatment (No AA 45%, Pro 47%, PA 50%, Gly 45%, His 39%, Pro + PA 45%, Pro + Gly 45%, His + Pro 39%). However, addition of 0.4 mM Pro to the medium during IVF improved subsequent development as compared to oocytes fertilised in the absence of amino acids (Figure 1A–D). In particular, the percentage of embryos that had compacted at 72 h and developed to the blastocyst stage at 120 h following IVF in Pro increased compared to oocytes fertilised in the absence of AA. The presence of Pro during IVF also increased the number of ICM cells, with the number reaching levels seen in blastocysts developed from in vivo fertilised oocytes (Figure 1E). IVF in the presence of Pro had no effect on trophectoderm cell number in blastocysts (Figure 1F).

PA is transported by some of the transporters of Pro and may, therefore, compete for and inhibit cellular uptake of Pro by those transporters. Thus, it was hypothesised that the presence of PA in molar excess (5 mM) over Pro (0.4 mM) during IVF would inhibit the effects of Pro. Instead, addition of PA alone to medium during IVF produced results similar to those observed for Pro alone (Figure 1A,E,F), increasing percent development to the compacted and blastocyst stages and increasing ICM cell numbers in blastocysts. The combination of PA with Pro during IVF had no additive or synergistic effects (Figure 1A,E).

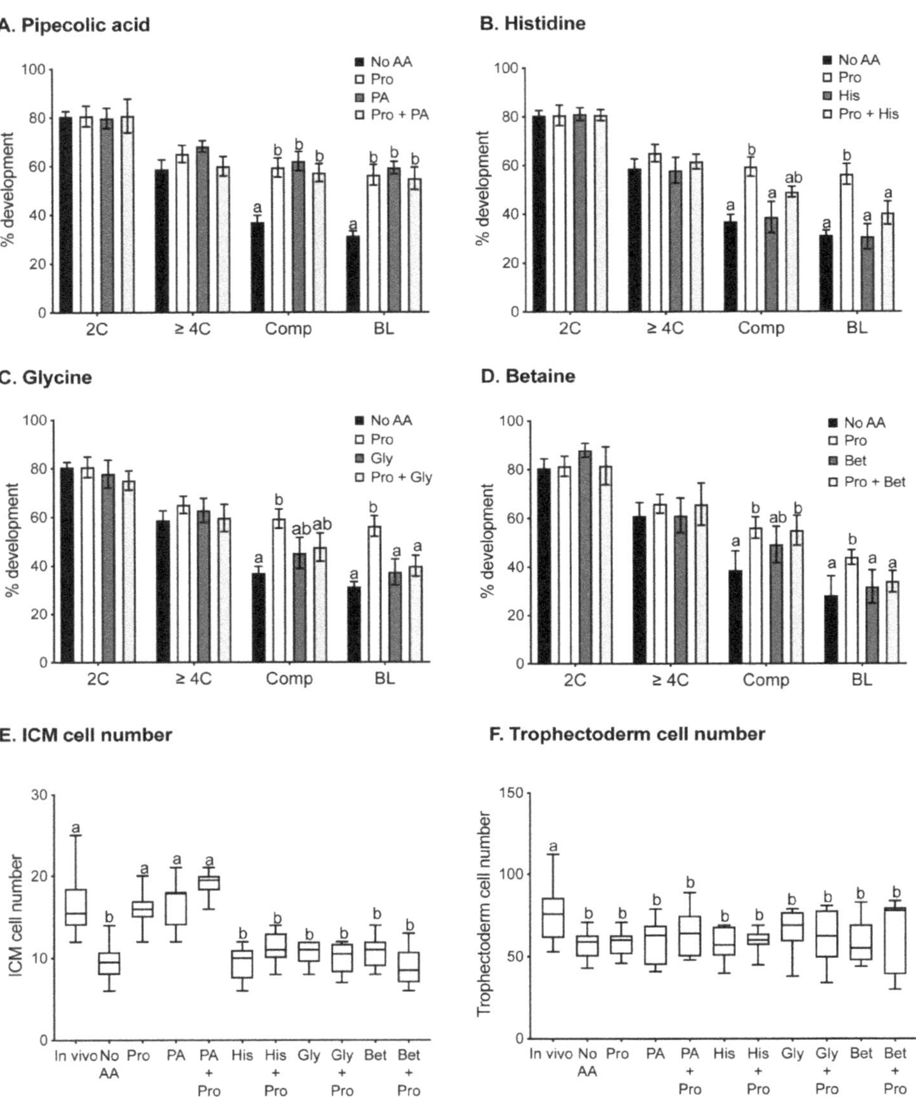

Figure 1. Preimplantation development following IVF in the presence of various amino acids. Fertilisation was performed in the absence of amino acids (No AA) or the presence of 0.4 mM Pro alone or 0.4 mM Pro in the presence of (**A**) 5 mM PA, (**B**) 5 mM His, (**C**) 5 mM Gly, or (**D**) 5 mM Bet. Following fertilisation, zygotes were cultured in mod-HTF in the absence of added amino acids for 120 h, and the percentages of embryos at the 2-cell (24 h), ≥4-cell (48 h), compacted (72 h), and blastocyst (120 h) stages were recorded. Bars represent mean ± SEM obtained from 3 independent experiments with a minimum of 10 zygotes per treatment group in each experiment. Blastocysts were fixed and stained with DAPI, and ICM (**E**) and trophectoderm (**F**) cell numbers were counted. Whiskers indicate highest and lowest values, and the mean is represented by the centre line in the box. Data were analysed using GraphPad Prism using 1-way ANOVA with Tukey's multiple comparisons test (**A–D**) or Dunnett's multiple comparison test (**E**,**F**). Bars sharing the same letter are not significantly different ($p > 0.05$).

3.2. Gly, Bet and His Inhibit Pro-Mediated Improvement in Development after IVF

The effect of other potential competitors of Pro uptake during IVF was also examined. When 5 mM Gly, Bet, or His were added individually to the IVF medium, none had any effect on embryo development compared to IVF undertaken in the absence of added amino acids (Figure 1B–F). However, each of these amino acids inhibited the effects observed with Pro (Figure 1B–E), indicating competition for transport of Pro into the oocyte.

3.3. Transport of Pro into Oocytes Has Na^+-Dependent and Na^+-Independent Components

A number of amino acid transporters are expressed in the mouse oocyte [31]. To determine which of these transporters is responsible for the transport of Pro, we examined the rate of 1 μM L-[^{3}H]-Pro uptake in the presence of a range of unlabelled amino acids in molar excess. Excess Pro itself was used to show the proportion of L-[^{3}H]-Pro uptake that was saturable (i.e., due to a transporter). Common L-amino acids and D-Pro were chosen based on the preferred substrates of known Pro transporters [38]. Trp was chosen as a known inhibitor of specific Na^+-independent transporters [50,51].

We found uptake of Pro involves both Na^+-dependent (Figure 2A) and Na^+-independent (Figure 2B) transport mechanisms. In Na^+-containing medium, Pro, D-Pro, PA, Gly, Ala, Bet, His, and sarcosine (Sar) all inhibited the uptake of L-[^{3}H]-Pro ($p < 0.01$). Lys, Leu, Ser, and Gln did not decrease the L-[^{3}H]-Pro rate of uptake. In Na^+-free medium, D-Pro, Gly, Trp, Bet, Pro, and PA inhibited the L-[^{3}H]-Pro rate of uptake ($p < 0.05$), whereas His and Leu had no effect.

A. Na⁺ present

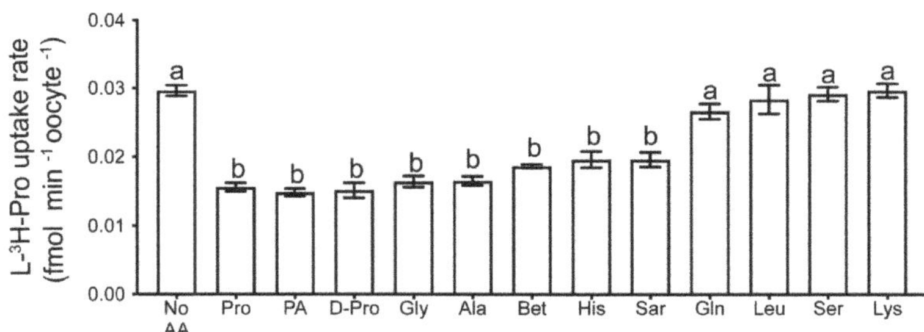

B. Na⁺ free

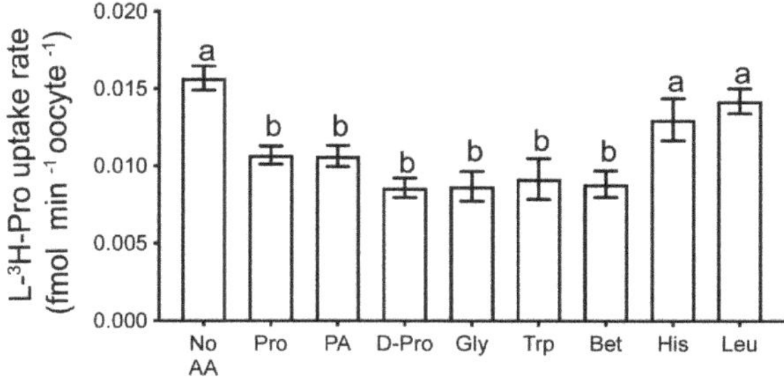

Figure 2. Uptake of L-[^{3}H]-Pro by oocytes in the presence of other amino acids. Rate of uptake of 1 μM L-[^{3}H]Pro in (**A**) mod-HTF containing Na^+ or (**B**) Na^+-free mod-HTF in the presence of unlabelled competing amino acids. The concentration of all competitors was 5 mM, with the exception of Pro (0.4 mM). Each bar represents the mean ± SEM from at least 3 independent experiments, each with treatments performed in triplicates of 4 oocytes. Data were analysed using 1-way ANOVA with Tukey's post hoc test. Bars with different letters are significantly different ($p < 0.05$). Sar: sarcosine.

3.4. PROT and PAT1/2 Are Expressed in Oocytes

We investigated the expression of three Pro transporters with similar amino acid uptake profiles to what we observed in the oocyte. Expression of PROT, PAT1, and PAT2 was examined by immunofluorescent staining. SLC6A7 (PROT) displayed puncta across the cytoplasm as well as some plasma-membrane staining over the location of the metaphase II spindle (Figure 3). This indicates that PROT is involved in both membrane and vesicular transport of Pro. It is likely that expression of PROT in vesicles contributes to Pro accumulation and/or signalling processes important for oocyte competency. For SLC36A1 (PAT1) and SLC36A2 (PAT2), staining was detected in the plasma membrane as well as in sub-cortical puncta (Figure 3). No staining was detected with the IgG control.

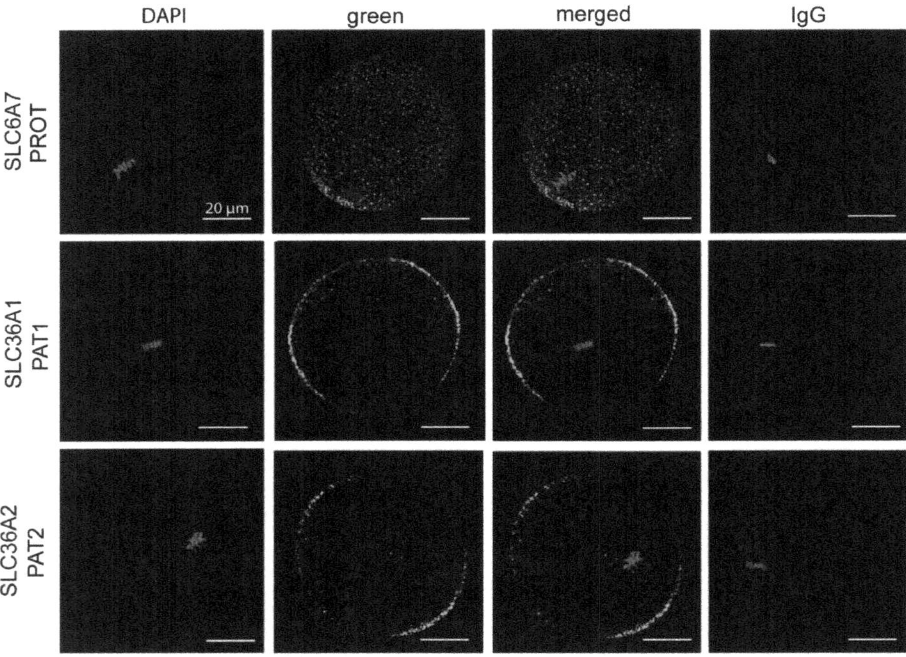

Figure 3. Expression of Pro transporters PROT, PAT1, and PAT2 in mouse oocytes. Oocytes were freshly isolated at 13–15 h post-hCG and exposed to Pro (0.4 mM) for 100 min, fixed, and immunostained for SLC6A7 (PROT), SLC36A1 (PAT1), and SLC36A2 (PAT2) (green). Metaphase chromosomes were counterstained with DAPI. Oocytes were analysed by confocal microscopy using a 40× objective. Images are representative of 6 oocytes taken from 3 independent experiments. A pre-immune IgG equivalent was used as a negative control for the primary antibody. Scale bar = 20 μm for all images.

3.5. Pro and PA Significantly Reduce Mitochondrial Activity in Oocytes

Reduced metabolism has been linked to improved embryo viability [21]. We hypothesised that the presence of Pro or PA during IVF would improve embryo development by decreasing metabolism in the oocyte. Mitochondrial membrane potential is used as a measure of mitochondrial activity and therefore the metabolic activity of cells [21]. Thus, to determine if Pro, PA, or other amino acids were affecting metabolic activity, we examined mitochondrial membrane potential using the potentiometric indicator TMRM in oocytes cultured in medium containing specific amino acids for 3 h. The presence of Pro or PA in the medium reduced TMRM fluorescence compared to the no AA control (Figure 4A,B). The presence of both Pro and PA did not have an additive effect. Gly and His alone had no effect on TMRM fluorescence, while Bet reduced TMRM fluorescence, but not to the same extent as Pro or PA. Gly, Bet, and His all prevented the Pro-induced decrease in TMRM fluorescence. FCCP uncouples mitochondrial oxidative phosphorylation by interfering with the transport of protons across the mitochondrial membrane. As expected, FCCP decreased TMRM fluorescence and mitochondrial activity (Figure 4D).

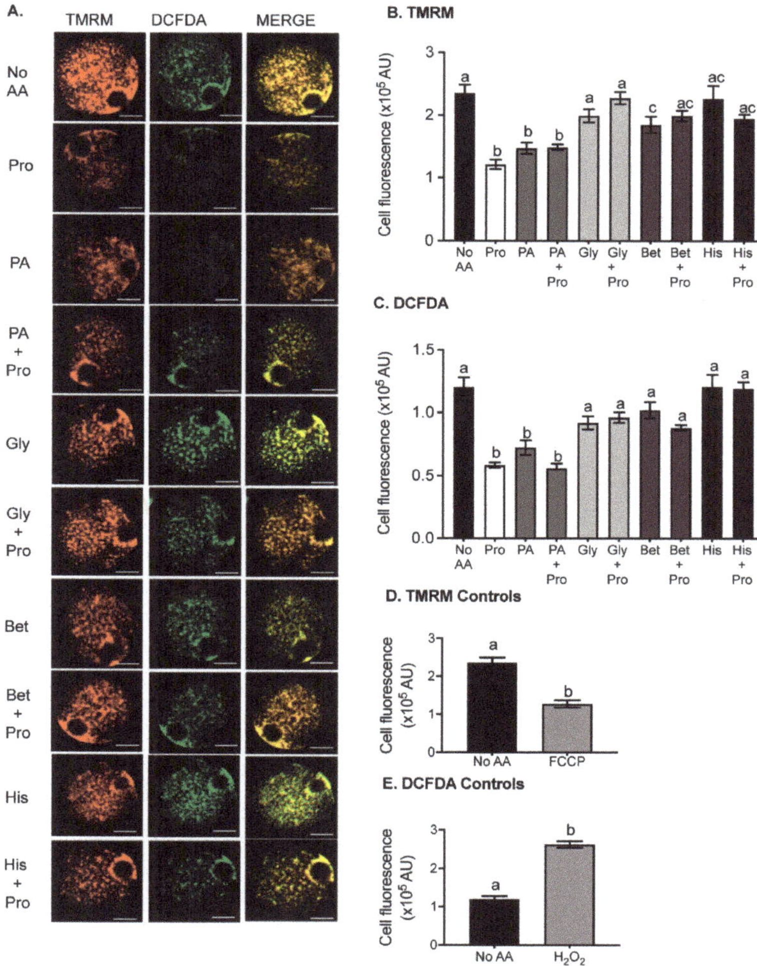

Figure 4. The effect of amino acids on ROS production and mitochondrial activity in oocytes. Oocytes were incubated for 3 h in medium containing either no amino acids (no AA) or 0.4 mM Pro or 0.4 mM PA, Bet, His, or Gly ± 0.4 mM Pro. Oocytes were then loaded with DCFDA for ROS, and TMRM for mitochondrial activity, and imaged by confocal microscopy. (**A**) Representative fluorescent images of oocytes in each condition. Quantified cell fluorescence for (**B**) TMRM and (**C**) DCFDA. Bars represent mean ± SEM for 28–71 oocytes, obtained from at least 3 independent experiments. Control oocytes stained with (**D**) TMRM and (**E**) DCFDA were transferred to a 10 µL medium containing 60 µM H_2O_2 or 5 µM carbonyl cyanide-p-trifluoromethoxyphenylhydrazone (FCCP) before imaging. Scale bar = 20 µm for all images. Data were compared by 1-way ANOVA with Tukey's multiple comparisons test. Bars not sharing the same letter are significantly different ($p < 0.05$).

3.6. Pro and PA Significantly Reduce Reactive Oxygen Species in Oocytes

ROS accumulation is detrimental to fertilisation and embryo development [28,30]. Since Pro and PA can act as ROS scavengers [20,37], we investigated whether they were performing this role in the oocyte. ROS were measured in oocytes using DCFDA, which is oxidized by ROS, increasing fluorescence of the dye [52]. The presence of Pro or PA in the medium reduced DCFDA fluorescence in oocytes reflecting a decrease in ROS in comparison to the no AA control (Figure 4A,C). The presence of Gly, Bet, and His alone did not affect DCFDA fluorescence but prevented the Pro-dependent decrease in fluorescence. The positive control, H_2O_2, caused a large increase in ROS, as expected (Figure 4E).

4. Discussion

In this study, we exposed oocytes to individual amino acids in fertilisation medium to determine if any subsequently improved preimplantation development. In the reproductive tract, oocytes are bathed in oviductal fluid, which contains amino acids and other factors that contribute to normal sperm capacitation, oocyte fertilisation, and preimplantation development. When the oocyte is fertilised outside of its physiological environment, its development potential is significantly impacted by the components of medium [53]. The amino acid composition of commercially available media varies [54].Although it is known that individual amino acids can benefit embryo development [7,55,56], little is known about how the presence of specific amino acids during fertilisation might impact on later embryo development, and this was therefore the focus of this study.

Fertilisation of oocytes was not significantly affected by supplementing fertilisation medium with any of the amino acids tested (Bet, Gly, His, Pro, and PA). Similarly, the addition of 19 common amino acids together does not increase the proportion of oocytes fertilised when IVF is performed on bovine oocytes [57]. This suggests that amino acids do not impact either the sperm–egg interaction or fertilisation processes occurring after sperm–oocyte fusion. However, the presence of Pro or PA during IVF (followed by their removal) resulted in an increase in the percentage of compacted embryos and blastocysts and inner cell mass cell numbers, suggesting improved embryo viability. IVF in the presence of Bet, Gly, or His did not improve embryo development by these criteria, but they did inhibit the improvements mediated by Pro or PA, suggesting that Bet, Gly, and His compete for uptake of Pro by the same amino acid transporter. These results are consistent with a growing body of evidence showing that competitive inhibition of uptake prevents the beneficial effects that selected amino acids have on various stages of the development process [7,8,11,14,34,35].

In somatic cells there are multiple transporters for Pro including SIT1, PAT1 and 2, PROT, SNAT2, B^0AT1 and 2, NTT4, ASCT1, y+LAT2, and GLYT1 [31,34,38,50,58–61]. Of these, SIT1 [34,35], GLYT1 [1,62,63], B^0AT2 [64], PROT [35], and SNAT2 [65] have been described in mouse embryo developmental stages after fertilisation. In oocytes, both Na^+-dependent (~50% of transport) and Na^+-independent (~30% of transport) Pro transport were observed. The Bet/Pro transporter, SIT1, can be eliminated as a candidate for Pro uptake since it is not active until after fertilisation [34]. Similarly, y+LAT2 can be eliminated because it is only expressed in cumulus cells [32,33]. Using competitors for L-[^{3}H]-Pro uptake, it was also possible to eliminate a number of other transporters as candidates for Pro uptake in the oocyte: B^0AT1, B^0AT2, and NTT4 can be eliminated because Leu did not prevent Na^+-dependent Pro uptake [14,58,60,61]. We can also eliminate the ASC/asc transporter family and SNAT2 since Ser did not reduce Pro uptake [14,31,66]. GLYT1 is located in the oocyte plasma membrane and actively transports Gly at fertilisation and throughout preimplantation development [1,62,67]. As GLYT1 substrates Gly and Sar both reduced Pro uptake, we cannot rule out the possibility that GLYT1 takes up Pro and PA in oocytes. However, its weak affinity for Pro makes it unlikely to be the main contributor [63]. It should also be noted that some transporters in oocytes and early embryos have characteristics that differ from the corresponding cloned transporters [64,65]. However, based on the available evidence, this leaves PROT, which is Na^+-dependent, and PAT1 and PAT2, which are Na^+-independent, as candidate transporters for Pro in the oocyte [68].

PROT is a good candidate for Pro transport in the oocyte as it is a high-affinity Na^+-dependent Pro transporter which also transports PA, Sar, and His [69]. Immunostaining for PROT in oocytes showed localisation in cytoplasmic vesicles and limited plasma membrane staining concentrated near the MII chromosomes. In the central nervous system PROT takes up Pro for normal brain function, memory, and locomotor activity [67]. PROT-null mice exhibit brain dysfunction but have no reported fertility or embryological defects suggesting that there is Pro transporter redundancy [70]. It is also likely that compensatory mechanisms are activated when important genes are knocked out that are needed for normal development, as may be the case for PROT. For the Na^+-independent transporters PAT1 and PAT2, the substrates D-Pro, Gly, and Trp [50,71] all prevented Pro uptake. In addition, immunostaining for both PAT1 and PAT2 showed both are located in the plasma membrane of oocytes. Gly and Ala are found in high concentrations in oviductal fluid which would compete with the uptake of Pro, so it is unlikely that PAT1/2 are the only functional transporters of Pro in vivo [47,72]. Future studies could investigate whether the observed inhibition of Pro transport by the relatively high concentrations of Gly and Ala used in this study was competitive or non-competitive. Our data, therefore, suggest there are multiple transporters of Pro in the oocyte with overlapping substrate specificities. Under in vitro conditions, the relative activity of each of these transporters in oocytes will depend on the amino acid composition of the medium. In particular, the beneficial effects on later embryo development through the exogenous addition of Pro during fertilisation might be mitigated or eliminated through inappropriate addition of other amino acids which reduce its uptake.

Pro and PA may have an immediate impact during IVF by reducing mitochondrial activity and ROS levels in the oocyte. Pro acts as an antioxidant and cryoprotectant by reducing ROS and mitochondrial activity in vitrified mouse oocytes [73]. Amino acids such as Pro which contain a secondary amine group have a lower ionizing potential than those with only a primary amine group, enabling them to easily donate electrons and quench ROS by stabilisation of the free radical [20,74]. The structurally similar PA appears to perform the same functions, acting as a scavenger of ROS and preventing their build-up in the oocyte. Gly, Bet, and His do not have a secondary amine group and were not able to reduce ROS. Instead, all three prevented the Pro-mediated reduction in ROS due to their competitive inhibition of Pro uptake via PAT1/2 and/or GLYT1. This scavenging mechanism is one potential pathway by which Pro and PA reduce ROS in the oocyte.

Both Pro and PA reduced mitochondrial activity in the oocyte. Our data therefore support the theory that metabolically quiet oocytes and embryos have improved viability [21,22]. Metabolism of Pro by proline oxidase produces P5C, which is converted to Glu. Glu can then be either converted to α-ketoglutarate, which enters the TCA cycle, or combine with cysteine and glycine to form the antioxidant glutathione [18,75]. Since the presence of Pro caused a decrease in mitochondrial activity, our data suggest that Pro metabolism does not feed significantly into the TCA cycle. It did not feed high-energy electrons into the ETC, but instead increased glutathione (GSH) production. Pro-mediated increases in GSH would also contribute to reduced ROS [20,76], as has been shown in boar sperm [76] and in-vitro cultured mouse embryos (unpublished data). Alternately, the Pro-induced reduction in mitochondrial activity indicates reduced ETC activity, which may be the mechanism by which Pro reduces ROS. Although the exact mechanism by which Pro reduces mitochondrial activity has not yet been identified, a Pro-induced simultaneous reduction of mitochondrial activity and ROS levels has been observed in other cells [77].

PA is metabolised similarly to Pro, whereby it is broken down by pipecolate oxidase to piperideine-6-carboxylate (P6C) and then to α-aminoadipic semialdehyde [37], which can result in increased production of glutamate and thus potentially GSH [78]. As Pro and PA both improve embryo development after IVF and decrease ROS and mitochondrial activity, we suggest that increased GSH due to Pro and PA metabolism may be one of the mechanisms responsible for their beneficial effect on later embryo development.

There are, however, important differences in development when Pro is added at the time of fertilisation compared to when it is added to already fertilised oocytes [7]. In the latter case, Pro (i) first improves embryo development during cleavage stages (late two-cell to eight-cell) instead of from compaction, and (ii) has no effect on blastocyst cell numbers [7]. This implies that the Pro-mediated mechanisms required to 'set up' improvement in later development are importantly different between these two scenarios and are dependent on the timing of the exogenous addition of this amino acid. Nevertheless, the molecular mechanisms at play when Pro is added during fertilisation may well also include one or more of those established for Pro in related work [7,11,15,16,79]. Finally, Pro added during fertilisation in vitro increased the number of inner cell mass cells in blastocysts to the level seen for blastocysts derived from cultured zygotes fertilised in vivo. However, trophectoderm cell numbers were not similarly increased. This is consistent with mouse studies showing that IVF reduces trophoblast cell numbers and affects subsequent placental development due to alterations in gene expression including the downregulation of SLC genes [80,81]. This indicates that one or more factors are still missing in the in vitro fertilisation environment.

In conclusion, Pro and PA are beneficial to preimplantation embryo development when added to the fertilisation medium. Pro and PA reduce mitochondrial activity and ROS in oocytes. The reduction in ROS by Pro and PA may be due to one or more of a variety of mechanisms: (i) The reduction in mitochondrial activity per se, (ii) the direct scavenging of ROS by these molecules, and (iii) their metabolism leading to increased production of GSH. We suggest that these may be some of the mechanisms responsible for the beneficial effect of Pro and PA. The beneficial actions of Pro are prevented when other amino acids are present in the IVF medium due to competitive uptake by amino acid transporters. Very little attention has been given to the amino acid composition of media used during fertilisation; rather, research has focused on culture media used during embryo development. This study shows that specific amino acids present during the process of fertilisation have impacts on later development. Indeed, the composition of currently used commercial IVF media may prevent the beneficial effect of Pro by oversupplying amino acids which prevent Pro uptake. We suggest a simplified fertilisation medium that contains key elements required to drive development in vivo that is a favourable strategy for promoting oocyte competency and subsequent embryo viability in vitro.

Author Contributions: T.T., M.L.M.H., M.G.-J., M.B.M. and M.L.D. designed the research experiments and prepared the manuscript. T.T., M.L.M.H. and M.G.-J. performed the experiments. T.T. and M.L.M.H. analysed data and prepared figures. All authors have read and agreed to the published version of the manuscript.

Funding: This research did not receive any specific grant from any funding agency in the public, commercial, or not-for-profit sector. This research was supported by the Discipline of Physiology, School of Medical Sciences, University of Sydney.

Institutional Review Board Statement: Experiments were conducted in accordance with the Australian Code of Practice for the Care and Use of Animals for Scientific Purposes and approved by the University of Sydney Animal Ethics Committee as required by the NSW Animal Research Act (protocols 5583 and 824).

Informed Consent Statement: Not applicable.

Data Availability Statement: All data is included in the published article. Data and materials will be made available upon request.

Acknowledgments: We gratefully acknowledge the help of the Bosch Institute for the use of its facilities and the training provided by its technicians and managers. In particular we thank Sheng Hua and Donna Lai of the Bosch Institute Molecular Biology Facility, and Sadaf Kalam Facility Officer of the Bosch Advanced Microscopy Facility.

Conflicts of Interest: The authors declare that there is no conflict of interest that could be perceived as prejudicing the impartiality of the research reported.

Abbreviations

IVF	In vitro fertilisation
ESC	Embryonic stem cells
GSH	Glutathione
P5C	Pyrroline-5-carboxylic acid
P6C	Piperideine-6-carboxylate
PA	Pipecolic acid
POX	Proline oxidase
ROS	Reactive oxygen species
SLC	Solute carrier transporter

References

1. Van Winkle, L.J.; Haghighat, N.; Campione, A.L. Glycine protects preimplantation mouse conceptuses from a detrimental effect on development of the inorganic ions in oviductal fluid. *J. Exp. Zool.* **1990**, *253*, 215–219. [CrossRef]
2. Gardner, D.; Lane, M. Amino Acids and Ammonium Regulate Mouse Embryo Development in Culture1. *Biol. Reprod.* **1993**, *48*, 377–385. [CrossRef] [PubMed]
3. Biggers, J.D.; Lawitts, J.A.; Lechene, C.P. The protective action of betaine on the deleterious effects of NaCl on preimplantation mouse embryos in vitro. *Mol. Reprod. Dev.* **1993**, *34*, 380–390. [CrossRef] [PubMed]
4. Gardner, D.K.; Lane, M. Alleviation of the'2-cell block'and development to the blastocyst of CF1 mouse embryos: Role of amino acids, EDTA and physical parameters. *Hum. Reprod.* **1996**, *11*, 2703–2712. [CrossRef]
5. Dawson, K.M.; Baltz, J.M. Organic osmolytes and embryos: Substrates of the Gly and β transport systems protect mouse zygotes against the effects of raised osmolarity. *Biol. Reprod.* **1997**, *56*, 1550–1558. [CrossRef] [PubMed]
6. Gardner, D.; Lane, M. Culture of viable human blastocysts in defined sequential serum-free media. *Hum. Reprod.* **1998**, *13*, 148–159. [CrossRef]
7. Morris, M.B.; Ozsoy, S.; Zada, M.; Zada, M.; Zamfirescu, R.C.; Todorova, M.G.; Day, M.L. Selected Amino Acids Promote Mouse Pre-implantation Embryo Development in a Growth Factor-Like Manner. *Front. Physiol.* **2020**, *11*, 1–12. [CrossRef]
8. Ermisch, A.F.; Herrick, J.R.; Pasquariello, R.; Dyer, M.C.; Lyons, S.M.; Broeckling, C.D.; Rajput, S.K.; Schoolcraft, W.B.; Krisher, R.L. A novel culture medium with reduced nutrient concentrations supports the development and viability of mouse embryos. *Sci. Rep.* **2020**, *10*, 1–15. [CrossRef]
9. Summers, M.C.; McGinnis, L.K.; Lawitts, J.A.; Biggers, J.D. Mouse embryo development following IVF in media containing either l-glutamine or glycyl-l-glutamine. *Hum. Reprod.* **2005**, *20*, 1364–1371. [CrossRef]
10. Summers, M.C.; McGinnis, L.K.; Lawitts, J.A.; Raffin, M.; Biggers, J.D. IVF of mouse ova in a simplex optimized medium supplemented with amino acids. *Hum. Reprod.* **2000**, *15*, 1791–1801. [CrossRef]
11. Washington, J.M.; Rathjen, J.; Felquer, F.; Lonic, A.; Bettess, M.D.; Hamra, N.; Semendric, L.; Tan, B.S.N.; Lake, J.-A.; Keough, R.A.; et al. L-Proline induces differentiation of ES cells: A novel role for an amino acid in the regulation of pluripotent cells in culture. *Am. J. Physiol. Cell Physiol.* **2010**, *298*, C982–C992. [CrossRef]
12. Shparberg, R.; Glover, H.J.; Morris, M.B. Modeling Mammalian Commitment to the Neural Lineage Using Embryos and Embryonic Stem Cells. *Front. Physiol.* **2019**, *10*, 705. [CrossRef]
13. Lonic, A. Molecular Mechanism of L-Proline Induced EPL-Cell Formation. Ph.D. Thesis, University of Adelaide, Adelaide, Australia, 2007.
14. Tan, B.S.N.; Lonic, A.; Morris, M.B.; Rathjen, P.D.; Rathjen, J. The amino acid transporter SNAT2 mediates L-proline-induced differentiation of ES cells. *Am. J. Physiol. Cell Physiol.* **2011**, *300*, C1270–C1279. [CrossRef] [PubMed]
15. D'aniello, C.; Fico, A.; Casalino, L.P.; Guardiola, O.; Di Napoli, G.; Cermola, F.; De Cesare, D.; Tate, R.J.; Cobellis, G.; Patriarca, E.J.; et al. A novel autoregulatory loop between the Gcn2-Atf4 pathway and L-Proline metabolism controls stem cell identity. *Cell Death Differ.* **2015**, *22*, 1094–1105. [CrossRef] [PubMed]
16. Comes, S.; Gagliardi, M.; Laprano, N.; Fico, A.; Cimmino, A.; Palamidessi, A.; De Cesare, D.; De Falco, S.; Angelini, C.; Scita, G.; et al. L-Proline Induces a Mesenchymal-like Invasive Program in Embryonic Stem Cells by Remodeling H3K9 and H3K36 Methylation. *Stem Cell Rep.* **2013**, *1*, 307–321. [CrossRef] [PubMed]
17. Phang, J.M.; Pandhare, J.; Liu, Y. The Metabolism of Proline as Microenvironmental Stress Substrate. *J. Nutr.* **2008**, *138*, 2008S–2015S. [CrossRef]
18. Phang, J.M. Proline Metabolism in Cell Regulation and Cancer Biology: Recent Advances and Hypotheses. *Antioxidants Redox Signal.* **2019**, *30*, 635–649. [CrossRef]
19. Phang, J.M.; Yeh, G.C.; Hagedorn, C.H. The intercellular proline cycle. *Life Sci.* **1981**, *28*, 53–58. [CrossRef]
20. Krishnan, N.; Dickman, M.B.; Becker, D.F. Proline modulates the intracellular redox environment and protects mammalian cells against oxidative stress. *Free Radic. Biol. Med.* **2008**, *44*, 671–681. [CrossRef]

21. Leese, H.J.; Baumann, C.G.; Brison, D.R.; McEvoy, T.G.; Sturmey, R. Metabolism of the viable mammalian embryo: Quietness revisited. *Mol. Hum. Reprod.* **2008**, *14*, 667–672. [CrossRef]
22. Houghton, F.D.; Hawkhead, J.A.; Humpherson, P.G.; Hogg, J.E.; Balen, A.H.; Rutherford, A.J.; Leese, H.J. Non-invasive amino acid turnover predicts human embryo developmental capacity. *Hum. Reprod.* **2002**, *17*, 999–1005. [CrossRef]
23. Lane, M.; Gardner, D. Fertilization and early embryology: Selection of viable mouse blastocysts prior to transfer using a metabolic criterion. *Hum. Reprod.* **1996**, *11*, 1975–1978. [CrossRef]
24. Oyawoye, O.; Gadir, A.A.; Garner, A.; Constantinovici, N.; Perrett, C.; Hardiman, P. Antioxidants and reactive oxygen species in follicular fluid of women undergoing IVF: Relationship to outcome. *Hum. Reprod.* **2003**, *18*, 2270–2274. [CrossRef] [PubMed]
25. Baumann, C.G.; Morris, D.G.; Sreenan, J.M.; Leese, H.J. The quiet embryo hypothesis: Molecular characteristics favoring viability. *Mol. Reprod. Dev.* **2007**, *74*, 1345–1353. [CrossRef] [PubMed]
26. Harper, M.-E.; Bevilacqua, L.; Hagopian, K.; Weindruch, R.; Ramsey, J.J. Ageing, oxidative stress, and mitochondrial uncoupling. *Acta Physiol. Scand.* **2004**, *182*, 321–331. [CrossRef] [PubMed]
27. Du, C.; Gao, Z.; Venkatesha, V.A.; Kalen, A.L.; Chaudhuri, L.; Spitz, D.R.; Cullen, J.J.; Oberley, L.W.; Goswami, P.C. Mitochondrial ROS and radiation induced transformation in mouse embryonic fibroblasts. *Cancer Biol. Ther.* **2009**, *8*, 1962–1971. [CrossRef]
28. Nasr-Esfahani, M.; Aitken, J.; Johnson, M. Hydrogen peroxide levels in mouse oocytes and early cleavage stage embryos developed in vitro or in vivo. *Development* **1990**, *109*, 501–507. [CrossRef] [PubMed]
29. Nasr-Esfahani, M.; Johnson, M. The origin of reactive oxygen species in mouse embryos cultured in vitro. *Development* **1991**, *113*, 551–560. [CrossRef] [PubMed]
30. Das, S.; Chattopadhyay, R.; Ghosh, S.; Goswami, S.; Chakravarty, B.; Chaudhury, K. Reactive oxygen species level in follicular fluid–embryo quality marker in IVF? *Hum. Reprod.* **2006**, *21*, 2403–2407. [CrossRef] [PubMed]
31. Pelland, A.M.; Corbett, H.E.; Baltz, J.M. Amino Acid Transport Mechanisms in Mouse Oocytes During Growth and Meiotic Maturation1. *Biol. Reprod.* **2009**, *81*, 1041–1054. [CrossRef]
32. Colonna, R.; Mangia, F. Mechanisms of Amino Acid Uptake in Cumulus-Enclosed Mouse Oocytes. *Biol. Reprod.* **1983**, *28*, 797–803. [CrossRef]
33. Corbett, H.E.; Dubé, C.D.; Slow, S.; Lever, M.; Trasler, J.M.; Baltz, J.M. Uptake of Betaine into Mouse Cumulus-Oocyte Complexes via the SLC7A6 Isoform of y+L Transporter1. *Biol. Reprod.* **2014**, *90*, 81. [CrossRef]
34. Anas, M.-K.I.; Lee, M.B.; Zhou, C.; Hammer, M.-A.; Slow, S.; Karmouch, J.; Liu, X.J.; Bröer, S.; Lever, M.; Baltz, J.M. SIT1 is a betaine/proline transporter that is activated in mouse eggs after fertilization and functions until the 2-cell stage. *Development* **2008**, *135*, 4123–4130. [CrossRef] [PubMed]
35. Anas, M.-K.I.; Hammer, M.-A.; Lever, M.; Stanton, J.-A.L.; Baltz, J.M. The organic osmolytes betaine and proline are transported by a shared system in early preimplantation mouse embryos. *J. Cell. Physiol.* **2006**, *210*, 266–277. [CrossRef] [PubMed]
36. Jamshidi, M.B.; Kaye, P.L. Glutamine transport by mouse inner cell masses. *Reproduction* **1995**, *104*, 91–97. [CrossRef] [PubMed]
37. Natarajan, S.K.; Muthukrishnan, E.; Khalimonchuk, O.; Mott, J.L.; Becker, D.F. Evidence for Pipecolate Oxidase in Mediating Protection Against Hydrogen Peroxide Stress. *J. Cell. Biochem.* **2017**, *118*, 1678–1688. [CrossRef]
38. Fremeau, R.T.; Caron, M.G.; Blakely, R.D. Molecular cloning and expression of a high affinity L-proline transporter expressed in putative glutamatergic pathways of rat brain. *Neuron* **1992**, *8*, 915–926. [CrossRef]
39. Galli, A.; Jayanthi, L.D.; Ramsey, I.S.; Miller, J.W.; Fremeau, R.T.; DeFelice, L.J. L-proline and L-pipecolate induce enkephalin-sensitive currents in human embryonic kidney 293 cells transfected with the high-affinity mammalian brain L-proline transporter. *J. Neurosci.* **1999**, *19*, 6290–6297. [CrossRef]
40. Nishio, H.; Ortiz, J.; Giacobini, E. Accumulation and metabolism of pipecolic acid in the brain and other organs of the mouse. *Neurochem. Res.* **1981**, *6*, 1241–1252. [CrossRef]
41. Kowalczuk, S.; Bröer, A.; Munzinger, M.; Tietze, N.; Klingel, K.; Bröer, S. Molecular cloning of the mouse IMINO system: An Na+- and Cl−-dependent proline transporter. *Biochem. J.* **2005**, *386*, 417–422. [CrossRef]
42. Takanaga, H.; Mackenzie, B.; Suzuki, Y.; Hediger, M. Identification of Mammalian Proline Transporter SIT1 (SLC6A20) with Characteristics of Classical System Imino. *J. Biol. Chem.* **2005**, *280*, 8974–8984. [CrossRef]
43. Pérez-García, F.; de Brito, L.F.; Wendisch, V.F. Function of L-Pipecolic Acid as Compatible Solute in Corynebacterium glutamicum as Basis for Its Production Under Hyperosmolar Conditions. *Front. Microbiol.* **2019**, *10*, 340. [CrossRef]
44. Green, C.J.; Day, M.L. Insulin-like growth factor 1 acts as an autocrine factor to improve early embryogenesis in vitro. *Int. J. Dev. Biol.* **2013**, *57*, 837–844. [CrossRef] [PubMed]
45. Fraser, L.R.; Drury, L.M. The Relationship Between Sperm Concentration and Fertilization in vitro of Mouse Eggs. *Biol. Reprod.* **1975**, *13*, 513–518. [CrossRef] [PubMed]
46. Whittingham, D.G. Culture of mouse ova. *J. Reprod. Fertil. Suppl.* **1971**, *14*, 7–21. [PubMed]
47. Guérin, P.; Ménézo, Y. Hypotaurine and taurine in gamete and embryo environments: De novo synthesis via the cysteine sulfinic acid pathway in oviduct cells. *Zygote* **1995**, *3*, 333–343. [CrossRef]
48. Vanslambrouck, J.M.; Bröer, A.; Thavyogarajah, T.; Holst, J.; Bailey, C.G.; Bröer, S.; Rasko, J.E.J. Renal imino acid and glycine transport system ontogeny and involvement in developmental iminoglycinuria. *Biochem. J.* **2010**, *428*, 397–407. [CrossRef] [PubMed]

49. Fernandez-Cardenas, L.; Villanueva-Chimal, E.; Salinas, L.S.; José-Nuñez, C.; Tuena de Gómez Puyou, M.; Navarro, R.E. Caenorhabditis elegans ATPase inhibitor factor 1 (IF1) MAI-2 preserves the mitochondrial membrane potential (Δ ψ m) and is important to induce germ cell apoptosis. *PLoS ONE* **2017**, *12*, 1–25. [CrossRef] [PubMed]
50. Metzner, L.; Neubert, K.; Brandsch, M. Substrate specificity of the amino acid transporter PAT1. *Amino Acids* **2006**, *31*, 111–117. [CrossRef] [PubMed]
51. Edwards, N.; Anderson, C.M.; Gatfield, K.M.; Jevons, M.P.; Ganapathy, V.; Thwaites, D.T. Amino acid derivatives are substrates or non-transported inhibitors of the amino acid transporter PAT2 (slc36a2). *Biochim. Biophys. Acta (BBA) Biomembr.* **2011**, *1808*, 260–270. [CrossRef]
52. Eruslanov, E.; Kusmartsev, S. Identification of ROS Using Oxidized DCFDA and Flow-Cytometry. *Methods Mol. Biol.* **2010**, *594*, 57–72. [CrossRef]
53. Bavister, B.D. Culture of preimplantation embryos: Facts and artifacts. *Hum. Reprod. Updat.* **1995**, *1*, 91–148. [CrossRef]
54. Tarahomi, M.; Vaz, F.; Van Straalen, J.P.; Schrauwen, F.A.P.; Van Wely, M.; Hamer, G.; Repping, S.; Mastenbroek, S. The composition of human preimplantation embryo culture media and their stability during storage and culture. *Hum. Reprod.* **2019**, *34*, 1450–1461. [CrossRef] [PubMed]
55. Gardner, D.K. Enhanced rates of cleavage and development for sheep zygotes cultured to the blastocyst stage in vitro in the absence of serum and somatic cells: Amino acids, vitamins, and culturing embryos in groups stimulate development. *Biol. Reprod.* **1994**, *50*, 390–400. [CrossRef] [PubMed]
56. Devreker, F.; Winston, R.M.; Hardy, K. Glutamine improves human preimplantation development in vitro. *Fertil. Steril.* **1998**, *69*, 293–299. [CrossRef]
57. Lim, J.M.; Lee, B.C.; Lee, E.S.; Chung, H.M.; Ko, J.J.; Park, S.E.; Cha, K.Y.; Hwang, W.S. In vitro maturation and in vitro fertilization of bovine oocytes cultured in a chemically defined, protein-free medium: Effects of carbohydrates and amino acids. *Reprod. Fertil. Dev.* **1999**, *11*, 127–132. [CrossRef] [PubMed]
58. Böhmer, C.; Bröer, A.; Munzinger, M.; Kowalczuk, S.; Rasko, J.E.J.; Lang, F.; Bröer, S. Characterization of mouse amino acid transporter B0AT1 (slc6a19). *Biochem. J.* **2005**, *389*, 745–751. [CrossRef] [PubMed]
59. Franchi-Gazzola, R.; Dall'Asta, V.; Sala, R.; Visigalli, R.; Bevilacqua, E.; Gaccioli, F.; Gazzola, G.C.; Bussolati, O. The role of the neutral amino acid transporter SNAT2 in cell volume regulation. *Acta Physiol.* **2006**, *187*, 273–283. [CrossRef]
60. Bröer, A.; Tietze, N.; Kowalczuk, S.; Chubb, S.; Munzinger, M.; Bak, L.K.; Bröer, S. The orphan transporter v7-3 (slc6a15) is a Na+-dependent neutral amino acid transporter (B0AT2). *Biochem. J.* **2005**, *393*, 421–430. [CrossRef]
61. Zaia, K.A.; Reimer, R.J. Synaptic Vesicle Protein NTT4/XT1 (SLC6A17) Catalyzes Na+-coupled Neutral Amino Acid Transport. *J. Biol. Chem.* **2009**, *284*, 8439–8448. [CrossRef]
62. Steeves, C.L.; Hammer, M.-A.; Walker, G.B.; Rae, D.; Stewart, N.; Baltz, J.M. The glycine neurotransmitter transporter GLYT1 is an organic osmolyte transporter regulating cell volume in cleavage-stage embryos. *Proc. Natl. Acad. Sci. USA* **2003**, *100*, 13982–13987. [CrossRef] [PubMed]
63. Van Winkle, L.J.; Campione, A.L.; Gorman, J.M. Na+-independent transport of basic and zwitterionic amino acids in mouse blastocysts by a shared system and by processes which distinguish between these substrates. *J. Biol. Chem.* **1988**, *263*, 3150–3163. [CrossRef]
64. Van Winkle, L.J.; Ryznar, R. Amino Acid Transporters: Roles for Nutrition, Signalling and Epigenetic Modifications in Embryonic Stem Cells and Their Progenitors. *eLS* **2019**, 1–13. [CrossRef]
65. Tan, B.S.N.; Rathjen, P.D.; Harvey, A.J.; Gardner, D.K.; Rathjen, J. Regulation of amino acid transporters in pluripotent cell populations in the embryo and in culture; novel roles for sodium-coupled neutral amino acid transporters. *Mech. Dev.* **2016**, *141*, 32–39. [CrossRef] [PubMed]
66. Pinilla-Tenas, J.; Barber, A.; Lostao, M.P. Transport of Proline and Hydroxyproline by the Neutral Amino-acid Exchanger ASCT1. *J. Membr. Biol.* **2003**, *195*, 27–32. [CrossRef] [PubMed]
67. Richard, S.; Tartia, A.P.; Boison, D.; Baltz, J.M. Mouse Oocytes Acquire Mechanisms That Permit Independent Cell Volume Regulation at the End of Oogenesis. *J. Cell. Physiol.* **2017**, *232*, 2436–2446. [CrossRef] [PubMed]
68. Guastella, J.; Brecha, N.; Weigmann, C.; Lester, H.A.; Davidson, N. Cloning, expression, and localization of a rat brain high-affinity glycine transporter. *Proc. Natl. Acad. Sci. USA* **1992**, *89*, 7189–7193. [CrossRef]
69. Bröer, S.; Bröer, A. Amino acid homeostasis and signalling in mammalian cells and organisms. *Biochem. J.* **2017**, *474*, 1935–1963. [CrossRef] [PubMed]
70. Gogos, J.; Santha, M.; Takacs, Z.; Beck, K.D.; Luine, V.; Lucas, L.R.; Nadler, J.V.; Karayiorgou, M. The gene encoding proline dehydrogenase modulates sensorimotor gating in mice. *Nat. Genet.* **1999**, *21*, 434–439. [CrossRef]
71. Thwaites, D.T.; Anderson, C. The SLC36 family of proton-coupled amino acid transporters and their potential role in drug transport. *Br. J. Pharmacol.* **2011**, *164*, 1802–1816. [CrossRef]
72. Harris, S.E.; Gopichandran, N.; Picton, H.M.; Leese, H.J.; Orsi, N.M. Nutrient concentrations in murine follicular fluid and the female reproductive tract. *Theriogenology* **2005**, *64*, 992–1006. [CrossRef]
73. Zhang, L.; Xue, X.; Yan, J.; Yan, L.-Y.; Jin, X.-H.; Zhu, X.-H.; He, Z.-Z.; Liu, J.; Li, R.; Qiao, J. L-proline: A highly effective cryoprotectant for mouse oocyte vitrification. *Sci. Rep.* **2016**, *6*, 26326. [CrossRef] [PubMed]
74. Matysik, J.; Alia; Bhalu, B.; Mohanty, P. Molecular mechanisms of quenching of reactive oxygen species by proline under stress in plants. *Curr. Sci.* **2002**, *82*, 525–532.

75. Arrieta-Cruz, I.; Su, Y.; Knight, C.M.; Lam, T.K.; Gutierrez-Juarez, R. Evidence for a Role of Proline and Hypothalamic Astrocytes in the Regulation of Glucose Metabolism in Rats. *Diabetes* **2012**, *62*, 1152–1158. [CrossRef] [PubMed]
76. Feng, C.; Zhu, Z.; Bai, W.; Li, R.; Zheng, Y.; Tian, X.; Wu, D.; Lu, H.; Wang, Y.; Zeng, W. Proline Protects Boar Sperm Against Oxidative Stress through Proline Dehydrogenase-Mediated Metabolism and the Amine Structure of Pyrrolidine. *Animals* **2020**, *10*, 1549. [CrossRef]
77. Delic, V.; Griffin, J.W.D.; Zivkovic, S.; Zhang, Y.; Phan, T.-A.; Gong, H.; Chaput, D.; Reynes, C.; Dinh, V.B.; Cruz, J.; et al. Individual Amino Acid Supplementation Can Improve Energy Metabolism and Decrease ROS Production in Neuronal Cells Overexpressing Alpha-Synuclein. *Neuro Mol. Med.* **2017**, *19*, 322–344. [CrossRef]
78. Broquist, H.P. Lysine-pipecolic acid metabolic relationships in microbes and mammals. *Annu. Rev. Nutr.* **1991**, *11*, 435–448. [CrossRef]
79. Casalino, L.; Comes, S.; Lambazzi, G.; De Stefano, B.; Filosa, S.; De Falco, S.; De Cesare, D.; Minchiotti, G.; Patriarca, E.J. Control of embryonic stem cell metastability by L-proline catabolism. *J. Mol. Cell Biol.* **2011**, *3*, 108–122. [CrossRef]
80. Fauque, P.; Mondon, F.; Letourneur, F.; Ripoche, M.-A.; Journot, L.; Barbaux, S.; Dandolo, L.; Patrat, C.; Wolf, J.-P.; Jouannet, P.; et al. In Vitro Fertilization and Embryo Culture Strongly Impact the Placental Transcriptome in the Mouse Model. *PLoS ONE* **2010**, *5*, e9218. [CrossRef]
81. Giritharan, G.; Piane, L.D.; Donjacour, A.; Esteban, F.J.; Horcajadas, J.A.; Maltepe, E.; Rinaudo, P. In Vitro Culture of Mouse Embryos Reduces Differential Gene Expression Between Inner Cell Mass and Trophectoderm. *Reprod. Sci.* **2012**, *19*, 243–252. [CrossRef] [PubMed]

MDPI

Review

The Role of Mitochondria in Human Fertility and Early Embryo Development: What Can We Learn for Clinical Application of Assessing and Improving Mitochondrial DNA?

Amira Podolak [1,2,*], Izabela Woclawek-Potocka [3,*] and Krzysztof Lukaszuk [1,2]

1 Invicta Research and Development Center, 81-740 Sopot, Poland; luka@gumed.edu.pl
2 Department of Obstetrics and Gynecological Nursing, Faculty of Health Sciences, Medical University of Gdansk, 80-210 Gdansk, Poland
3 Department of Gamete and Embryo Biology, Institute of Animal Reproduction and Food Research, Polish Academy of Sciences, 10-748 Olsztyn, Poland
* Correspondence: amira.podolak@invicta.pl (A.P.); i.woclawek-potocka@pan.olsztyn.pl (I.W.-P.)

Abstract: Mitochondria are well known as 'the powerhouses of the cell'. Indeed, their major role is cellular energy production driven by both mitochondrial and nuclear DNA. Such a feature makes these organelles essential for successful fertilisation and proper embryo implantation and development. Generally, mitochondrial DNA is exclusively maternally inherited; oocyte's mitochondrial DNA level is crucial to provide sufficient ATP content for the developing embryo until the blastocyst stage of development. Additionally, human fertility and early embryogenesis may be affected by either point mutations or deletions in mitochondrial DNA. It was suggested that their accumulation may be associated with ovarian ageing. If so, is mitochondrial dysfunction the cause or consequence of ovarian ageing? Moreover, such an obvious relationship of mitochondria and mitochondrial genome with human fertility and early embryo development gives the field of mitochondrial research a great potential to be of use in clinical application. However, even now, the area of assessing and improving DNA quantity and function in reproductive medicine drives many questions and uncertainties. This review summarises the role of mitochondria and mitochondrial DNA in human reproduction and gives an insight into the utility of their clinical use.

Keywords: embryo; embryogenesis; oocyte; oogenesis; fertility; mitochondria; mitochondrial DNA (mtDNA); mitochondrial score; mitochondrial replacement therapy (MRT); autologous mitochondrial transfer

Citation: Podolak, A.; Woclawek-Potocka, I.; Lukaszuk, K. The Role of Mitochondria in Human Fertility and Early Embryo Development: What Can We Learn for Clinical Application of Assessing and Improving Mitochondrial DNA? *Cells* **2022**, *11*, 797. https://doi.org/10.3390/cells11050797

Academic Editor: Lon J. van Winkle

Received: 25 January 2022
Accepted: 23 February 2022
Published: 24 February 2022

1. Introduction

Mitochondria are well known as 'the powerhouses of the cell'; however, in recent years, our understanding of its biology has vastly increased. Mitochondria have their own genome which is replicated independently of the nuclear genome. Human mitochondrial DNA (mtDNA) is 16.6 kbp in length and encodes 13 peptides which contribute to all complexes required for oxidative phosphorylation (OXPHOS) except complex 2 [1–3]. The remaining mitochondrial proteins are encoded by the nuclear genome and are imported into the mitochondria. The major role of mitochondria is to produce the ATP required by cells. This process relies on OXPHOS whose by-product is the generation of reactive oxygen species (ROS). ROS are generated in approximately 90% of cases by OXPHOS [3]. In addition to this, mitochondria can sequester and release Ca^{2+} regulating calcium responses [4,5]. Moreover, they mediate cell proliferation, differentiation, and apoptosis [3–6].

Human preimplantation development and embryo implantation is an energy-demanding process that involves a range of energetic cellular processes, requiring significant quantities of ATP [1,2]. Therefore, mitochondria must play a crucial role in proper fertilisation and embryogenesis. This review summarises the role of mitochondria and mtDNA in human oocytes and embryos, showing its influence on fertility and early embryo development. Moreover, we make an insight into the clinical usefulness of assessing and improving mtDNA quantity and function.

2. Mitochondria and the Cell Cycle

It is thought that mitochondria are derived from the symbiosis of the prokaryotic α-proteobacteria with the ancient archaea species. From their putative ancestors, they maintain some phenotypic features as a double-membrane, a similar proteome, and the ability to produce ATP via a proton gradient created across its inner membrane [7,8]. However, during the evolution process which led them to the current eukaryotic cells, they lost the capability to synthesize most of the proteins encoded by the primitive bacteria. *Bartonella henselae* is an α-proteobacteria with relatively small DNA homologous to mitochondrial DNA which encodes more than 1600 proteins [7,9]. Additionally, 16,6 kbp mtDNA controls the synthesis of 13 proteins of the OXPHOS, while the rest of the bacterial genes were transferred to the nuclear genome [3,8]. Approximately 1500 nuclear proteins contribute to the mitochondrial proteome, including, e.g., transcription factors, mtDNA polymerase, and ribosomal proteins, as well as enzymes required for the citric acid cycle [6].

The proper inheritance of the genetic material between daughter cells during mitosis requires major cell reorganization. To complete a successful division cycle, the cell follows a specific and coordinated series of events. Moreover, to ensure the correct mitosis, the cellular organelles are also needed to reorganize and segregate [7,10]. Therefore, how do mitochondrial factors influence mitosis? A dynamic equilibrium of fusion and fission is an important feature of the mitochondrial network. The fusion is operated by the dynamin-like GTPases mitofusins 1 and 2 (Mfn1 and Mfn2) at the outer membrane, and optic atrophy 1 (Opa1) localized on the inner membrane. If the cell lacks Mfn1 and Mfn2, no outer membrane fusion occurs [7,8,10,11]. Mice models demonstrated that targeted deletion of either Mfn1 or Mfn2 leads to phenotype consistent with female reproductive aging (e.g., apoptotic cell loss resulting in accelerated follicular depletion). The absence of Mfn1 additionally caused the interruption of oocyte growth and ovulation due to a block in folliculogenesis, whereas Mfn2-lacking oocytes revealed shortened telomeres [12–14]. On the other hand, cells lacking Opa1 do undergo outer membrane fusion, however cannot progress to inner membrane fusion. The fission process is mediated by the dynamin-related protein 1 (DRP1) recruited from the cytosol to the outer mitochondrial membrane. Its assembly on the mitochondrial surface causes constriction of the mitochondria and eventual division of the organelle into two separate entities [7,8,10,11]. There are four DRP1 receptors at the mitochondrial outer membrane: Fis1 (Anti-Mitochondrial fission 1 protein), Mff (Mitochondrial fission factor), Mid49 (Mitochondrial dynamics protein of 49 kDa), and Mid51 (Mitochondrial dynamics protein of 51 kDa), from which the latter three play the major role in fission. Repeated cycles of mentioned processes result in the intermixing of the mitochondrial population in the cell and determine mitochondrial morphology. While either increased fusion or decreased fission promotes the formation of elongated mitochondrial networks, increased fission and decreased fusion causes mitochondrial fragmentation [8]. During mitosis, Aurora-A phosphorylates the small Ras-like GTPase RALA (Ras-related protein Ral-A), which localizes to mitochondria and triggers the formation of a complex with RALBP1 (RalA-binding protein 1) and CDK1 (Cyclin-dependent kinase 1)/CyclinB, inducing the phosphorylation of DRP1 to stimulate mitochondrial fission. The knockdown of either RALA or RALBP1 leads to the inhibition of mitochondrial division which results in the inability of even distribution of mitochondria between the daughter cells and, in consequence, in cytokinesis defects [10]. It is noteworthy that experiments conducted with mammalian stem-like cells revealed their ability to asymmetrically sort young and old mitochondria. The daughter cells retaining stem-like features were found to receive most of the new mitochondria, thus, asymmetric portioning of aged mitochondria is required for stemness [15].

3. Mitochondrial Genetics

More than 40 years have passed since the first draft human mitochondrial DNA sequence was published [16]. During the last decades, a number of basic and clinical studies were conducted to investigate the mitochondrial genome.

It is generally accepted that in humans, similarly to most mammal species, mitochondria and mtDNA are exclusively maternally inherited. The paternal mitochondria and their DNA enter the oocyte cytoplasm upon fertilisation; however, the selective elimination of the paternal mtDNA from the oocyte may occur due to some tissue-specific mechanisms of their degradation [17–22]. It was demonstrated in mice models that liver mitochondria microinjected into pronucleus-stage embryos were not removed efficiently [19]. On the other hand, the ubiquitination of sperm mitochondria may be the reason for their degradation. It was shown that in some mammals (e.g., rhesus monkeys) the paternal mitochondria in fertilised oocytes are modified with ubiquitin and disappear between the 4-cell stage and 8-cell stage of development. In humans, the sperm mitochondria are tagged with ubiquitin ahead of fertilisation [17,23]. PINK1 (PTEN-induced kinase 1) phosphorylates ubiquitin, and ubiquitin ligase Parkin, leading to the recruitment of Parkin to the outer mitochondrial membrane where it polyubiquitinates multiple proteins. Polyubiquitination of proteins on the outer mitochondrial membrane initiates the recruitment of the machinery that causes the organelle to be engulfed into an autophagosome which, in turn, is directed to lysosomes for degradation, thus leading to the removal of the mitochondria [21,22]. According to another hypothesis, paternal mitochondria are just removed due to the dilution effect derived from far higher numbers of mitochondria in oocytes compared to those in spermatozoa or alternatively, are degraded even before reaching the oocytes [22]. Most likely, the degradation of the paternal mitochondria is regulated by multiple mechanisms.

A crucial role in the maintenance of the maternal inheritance of mtDNA plays the replication of the mitochondrial genome. The population of mtDNA that is inherited is present in the metaphase II oocyte just before fertilisation. A thousand-fold increase in mtDNA copy number is observed—from about 200 copies present in the primordial germ cell (PGC) to over 200,000 copies in mature fertilisable oocytes [24]. Therefore, copies present in the metaphase II oocyte may be either clonal expansion of those in PGCs if the genome is identical or, if the mtDNA variants are present, mutant and wild type molecules can be preferentially selected depending on their frequency and distribution across the oocyte's cytoplasm (Figure 1) [24]. The presence of more than one mtDNA variant is called heteroplasmy (Figure 1). Apart from the oocyte's mutations, it may also occur due to either some exceptional cases where paternal mtDNA could be replication-competent and passed to the offspring, or the defect of the above-mentioned mechanisms causing the oocyte unable to remove sperm mitochondria after fertilisation [16,17,25–28]. It is generally accepted that low levels of mutant mtDNA do not negatively influence the offspring as the coexistence of wild type mtDNA usually neutralises the effect of a mutant load [29]. Some studies reported that around 90% of the individuals in the general healthy population carry at least one mtDNA heteroplasmic mutation [30]. At some levels of mutation content, the offspring may present a mildly affected phenotype [31]. Reaching the threshold of disease-causing mutation frequency leads to mitochondrial dysfunction and, thus, severely affected offspring (Figure 1) [30,31]. Clearly, it is not regarded to every variant in mtDNA, as not all of them is disease-causing. For instance, some of them define the mtDNA haplogroups. Currently, global mtDNA phylogenetic tree contains over 4000 haplogroups [32].

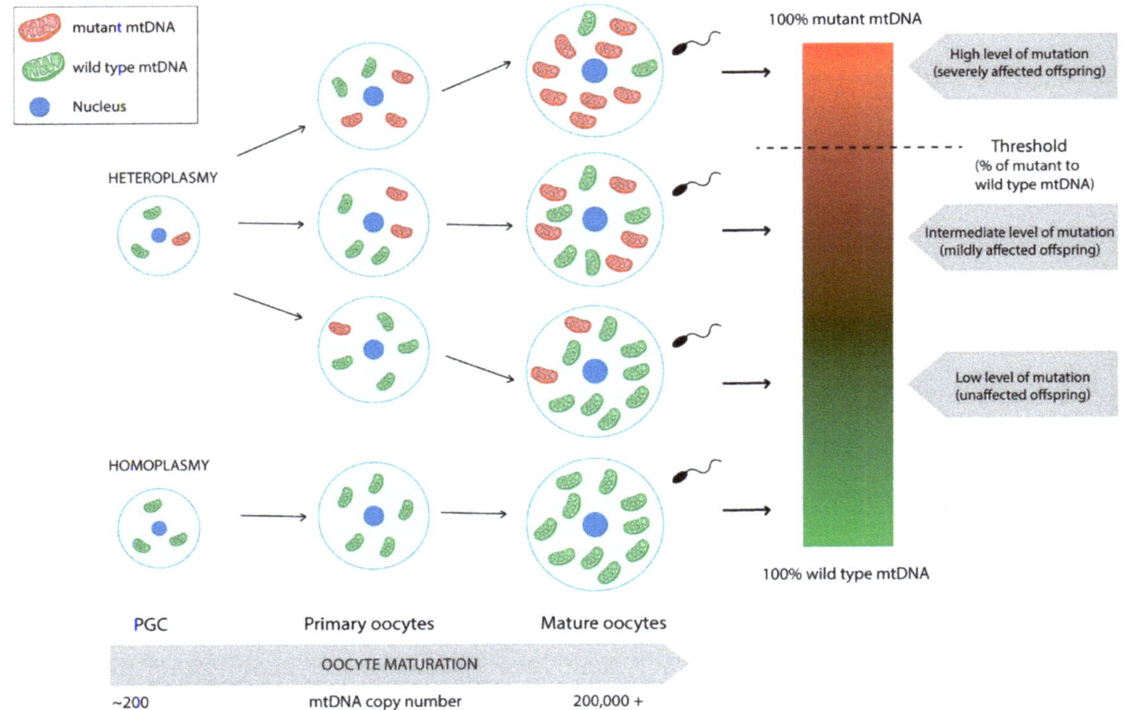

Figure 1. Differentiation of homoplasmic and heteroplasmic PGCs into mature oocytes with mtDNA copy number expansion. Mature oocytes derived from heteroplasmic PGC may present varying levels of mutation frequency leading to unaffected or mildly or severely affected offspring. Paternal leakage was not taken into consideration.

The phenotypic threshold is thought to usually be around 60% for deletions and around 90% for point mutations. However, its specific value depends on the type of mutation and also the tissue being evaluated. For instance, MERF (myoclonic epilepsy associated with ragged red fibres) syndrome occurs only when 90% of mutant mtDNA in muscle is reached. Moreover, experiments conducted directly on individual muscle fibres proved the existence of the phenotypic threshold effect also at the single-cell level [33]. On the other hand, the m.8993T>G mutation in *MT-ATP6* gene reaching the 60% load may result in mild symptoms such as headaches or mild pigmentary retinopathy or drive no effect. Gaining 70–90% of mutation content causes phenotype consistent with NARP (neurogenic muscle weakness, ataxia, and retinitis pigmentosa) syndrome, whereas levels above 90% lead to more severe Leigh syndrome. Contrarily, another point mutation in the same gene, m.8993T>C was demonstrated as less severe, as it affects only those with loads greater than 90% [31]. The phenotypic threshold is rather a phenomenon than a specific value to be obtained.

An important aspect of mtDNA inheritance is the bottleneck theory [16,26,28]. An absence of mtDNA replication between fertilisation and PGC formation implies a reduction from about 200,000 copies of mtDNA in mature oocytes to about 200 in PGCs. In consequence, cell division and differentiation drive the segregation of mtDNA variants among daughter cells, leading to varying levels within tissues and cells of a given tissue. First of all, the absence of mtDNA replication prevents mtDNA variants from increasing in their frequency. Secondly, it enables selection at the cellular level by enhancing the influence of mtDNA deletions on cell proliferation. Additionally, only a few dozen cells among thousands in the embryo give rise to PGCs, which supports the occurrence of a second bottleneck [28]. The reduction in mtDNA copy number remains until blastocyst formation [24]. Nevertheless, the reduction in mtDNA copy number in PGCs is not the one and only factor that can contribute to the observed germline mtDNA bottleneck. Similarly to population genetics in which the variance among descendants is inversely related to the effective population size, the shifts in frequency of heteroplasmy is observed between generations and also between offspring [34].

4. Mitochondria in Human Gametes and Embryos

4.1. Mitochondrial Distribution in Human Oocytes and Embryos

The heterogeneity in the number of mitochondria present in mature cells is well established. For instance, about 22 to 75 of these organelles are packed around the midpiece in mature human spermatozoa, whereas the high numbers of mitochondria are distributed across the cytoplasm of the oocyte [24]. After fertilisation, the oocyte's mitochondria are segregated asymmetrically. It seems that a newly fertilised oocyte adjusts mitochondrial density to different intracellular regions; however, its functional significance remains unknown [35–37]. It is possible that mitochondrial clustering may serve to supply energy directly and rapidly to the nucleus. This assumption was based on the large size of the oocyte and a hypothesis that the time required for ATP to diffuse across the cytoplasm could be too slow to provide energy [35]. Furthermore, mitochondria are reported to distribute disproportionately among formed blastomeres of developing embryos, though the significance of this phenomenon is also unknown [38].

4.2. Mitochondria and Energy Production in the Embryo

As mentioned above, mtDNA does not replicate until the blastocyst stage [24,39]. During that time, the embryo metabolism depends mainly on pyruvate and additionally lactate and amino acids [39–43]. The crucial role of pyruvate as the major energy substrate was confirmed in mice using imaging techniques in living oocytes and embryos. Pyruvate was found to be rapidly metabolised by mitochondria, whereas glucose was not [41,42]. The mitochondrial population in the oocyte must be sufficient to be distributed among formed blastomeres to allow ATP production for functioning until the next mitochondrial biogenesis [39]. However, as it was mentioned before, mitochondria content differs between blastomeres, therefore, the level of ATP production must vary also.

The important role of glucose starts at the blastocyst stage. Though glucose makes a moderate contribution to ATP production, it becomes a crucial substrate at this point of development [39–43]. Due to increases in glucose as a carbohydrate substrate as well as compacting of mitochondria cristae and initiation of replication, aerobic respiration appears to be upregulated at the blastocyst stage of development. It is estimated that 10% of glucose is metabolised through aerobic respiration in the early stages of development, whereas it increases to 85% in the blastocyst. While the elimination of pyruvate drastically decreases the rate of embryo development, such an effect is not observed for the elimination of either glucose or lactate [41]. Even though aerobic respiration may not be considered to be as essential as anaerobic respiration, the latter cannot be solely sufficient. Hence, both of them are required for proper embryo development.

4.3. Mitochondrial Activity versus Fertility and Early Embryogenesis

The importance of mitochondrial activity in proper fertilization and embryogenesis is demonstrated in many aspects. As mentioned above, the developing embryo is in high need of adequate ATP supply. Therefore, insufficient ATP content has been linked to fertilisation failure and abnormal embryo development [39,44]. Actual mitochondrial numbers vary in oocytes, though adequate amounts of mitochondria are required to provide the burst of activity essential up to the blastocyst stage. Thus, mitochondrial dysfunction is revealed when it drops below the necessary threshold. Nevertheless, mitochondrial activity should not be interpreted through the prism of copy number solely as it is strictly regulated by nuclear signals, intracellular ion concentrations, and the availability of substrates [44]. It was demonstrated in the in vitro study that embryonic metabolic response can be quite heterogeneous depending on early phenotypes and the composition of the culture media surrounding the embryos. It showed the ability to switch to other substrates if needed, and to modulate molecular machinery, ensuring their survival even in adverse external conditions [45]. On the other hand, dysfunction of mitochondria in oocytes leading to OXPHOS decrease results in abnormal embryo development [44].

Mitochondrial activity also plays a role in aspects of male fertility. The main role of mitochondria is ATP production from spermatogenesis to fertilization. In addition to this, they modulate several processes, such as spermatogonial stem cells differentiation, testicular somatic cell development, testosterone production in the testis, luminal acidification and sperm DNA condensation in the epididymis as well as ROS homeostasis for sperm capacitation and acrosome reaction in the female reproductive tract [22,46]. Therefore the dysfunction of mitochondria in spermatozoa may also be the reason for infertility.

4.4. *Effects of Ageing and Other Factors on Mitochondrial Insufficiency*

From a reproductive point of view, ageing of females is a progressive decline of ovarian function demonstrated in decreasing quantity and quality of oocytes [47]. It is thought that ovarian ageing may affect oocyte competence by targeting cytoplasmic components including mitochondria [48]. Hence, the quality of mitochondria in the oocyte determines the quality of the oocyte and, in consequence, the developing embryo [47]. However, does mitochondrial dysfunction induce ovarian ageing or ovarian ageing induce mitochondrial dysfunction?

Experiments both on humans and mice have proved the decrease in mtDNA content in oocytes with ovarian ageing [49–53]. Moreover, a 4977-bp deletion was indicated as another example of a quantitative dysfunction of mtDNA associated with ovarian ageing. Increased frequency of mentioned mutation in oocytes was observed in older women, suggesting the accumulation of this deletion with advancing ageing [48,54]. On the other hand, several studies were performed to evaluate the association of point mutations in mtDNA and ovarian ageing, though the topic remains highly debated [48,52,55–57]. Indeed, mtDNA may be more susceptible to damage than nuclear DNA (nDNA) due to its proximity with the free radical-respiratory chain, the lack of protective histones lower fidelity of mitochondrial polymerase γ (POLG), and limited repair mechanisms [48,54,58,59]. As the mitochondrial producing genome, similar to its bacterial ancestors, has no introns and only one major non-coding region (NCR), any point mutation or deletion could disrupt cellular respiration [60,61]. Oxidative stress is considered as one of the factors negatively influencing mtDNA. The hypothesised mechanism involves the damage induced by ROS, a by-product of OXPHOS, which can compromise the integrity of the respiratory chain leading to mitochondria-dependent ageing [47,48,58,59]. ROS are known to react with surroundings proteins, lipids, and DNA, leading to mutations and macromolecule damage [48]. Their mutagenicity was found in the modification of DNA bases. It has been shown that the somatic mutations of mtDNA of older subjects is characterised by a strong G>A mutation preference. Additionally, the ROS-induced formation of mtDNA double-strand breaks seems to be involved in the somatic mtDNA deletions generation [59]. Under physiological conditions, in response, the expression levels of antioxidants enzymes and intracellular proteins rapidly increase [48,58]. However, when ROS are overproduced, these compounds cause oxidative stress and cellular damage. Their high concentration in cells leads not only to mitochondrial and nuclear DNA damage but also apoptosis [58]. Furthermore, conducted experiments found the reducing defense against ROS with ovarian ageing due to the observed age-related down-regulation of genes *SOD1*, *SOD2*, and catalase mRNA encoding key antioxidant enzymes [62]. It is also noteworthy that mitochondria are reported to be targets of the main pituitary gonadotropins. Granulosa cells (GCs) are the main site of steroid hormones synthesis under the control of FSH, and mitochondria have a great contribution in this process. On the other hand, FSH regulates mitochondrial activity via stimulation of mitochondrial biogenesis in GCs under hypoxic conditions. In particular, FSH is considered as a mitophagy-reducing factor playing a crucial role in the maintenance of mitochondrial integrity under oxidative stress conditions. Such an effect can be obtained through FSH-dependent inhibition of the PINK1-Parkin pathway. LH is also regarded as a modulator of the mitochondrial steroidogenic activity and dynamics; however, more studies are needed to fully evaluate its mechanisms. Additionally, mitochondria are involved in oestrogens synthesis, and oestrogens regulate mitochondrial bioenergetics, calcium homeostasis, ROS-scavenger, and their dynamics [63]. Considering the above-mentioned relationship between pituitary–ovarian axis hormones and mitochondrial activity, it seems that age-related hormonal disorders may affect mitochondrial activity, and mitochondrial insufficiency may induce ovarian ageing.

Moreover, excessive Ca^{2+} influx can elevate mitochondrial oxidative stress and lead to apoptosis. Before fertilisation, oocytes at metaphase II stage require the sperm-triggered Ca^{2+} oscillations for several processes such as the resumption of meiosis, polyspermy block, male chromatin decondensation, recruitment of maternal mRNAs, and pronuclear formation. The Ca^{2+} homeostasis in postovulatory oocytes depends on the proper mitochondrial activity as mitochondria-associated membranes facilitate the transfer of Ca^{2+} ions from the endoplasmic reticulum (ER). An excess of Ca^{2+} transfer can either disrupt oxidative phosphorylation and redox homeostasis or trigger the mitochondrial permeability transition pore to open and in consequence affect mitochondrial function and induce apoptosis. That indicates the essentiality of mitochondrial calcium homeostasis for the maintenance of mitochondrial metabolic function; hence, its dysregulation can contribute to pathology [47].

Mitochondrial insufficiency was also reported to result from obesity [64,65]. Due to excessive intake of nutrients, mitochondria become overloaded with fatty acids and glucose, resulting in an increase in the production of Acetyl-CoA. This leads to the production of NADH in the Krebs cycle, which promotes the rise of electrons entering the mitochondrial intermembrane space and in consequence overproduction of ROS inducing oxidative stress. Subsequently, oxidative stress activates several transcription factors including the main mediator of the inflammatory response [65].

The production of ATP is also the primary aspect of mitochondrial function for supporting sperm motility. ROS were found to cause a loss of mitochondrial membrane potential (MMP), lipid peroxidation, impaired sperm motility, and sperm DNA integrity. There are several key pathways through which ROS may be generated by the sperm mitochondria, including disruption of the mitochondrial electron transport, formation of adducts with mitochondrial proteins, reduced mitochondrial expression of prohibitin, the opening of the mitochondrial permeability transition pore (PTP), and induction of apoptosis in spermatozoa. Furthermore, somatic alterations in mtDNA may impair OXPHOS and enhance ROS production, thereby hastening the rate of DNA mutation [60,66].

4.5. Impact of Mitochondrial Insufficiency on Fertility

The ageing-related oocyte's mtDNA copy number reduction is the reason for insufficient ATP levels leading to infertility and abnormal embryo development. Poor oocyte quality is not the only effect of diminished mtDNA content. With decreasing numbers of mitochondria, the proportion of mutant/wild type mtDNA in the oocyte may be increased to not allow the proper embryo development. Such an effect may be an explanation for the surprisingly high number of offspring affected with mitochondrial diseases. Moreover, a decline of ATP levels in the oocyte may be associated with the age-related high risk of aneuploidy. The process of completing meiotic division in the oocyte generally lasts from the preovulatory LH peak to its fertilisation. The proper positioning and segregation of chromosomes depend on the assembly and disassembly of microtubules which is one of the most energy-demanding processes within the oocyte. When oocytes age, their ability to produce adequate spindle microtubules decreases, which in consequence leads to increased incidence of aneuploidy [39,44,58].

Interestingly, mutations in the mitochondria-related genes (e.g., *MT-ATP6* and *MT-CO1*) were linked to primary ovarian insufficiency (POI); however, due to limited sample size, further studies are needed to determine whether mitochondrial dysfunction is a common contributing factor to the onset of POI. If the implication of mitochondrial dysfunction in POI was confirmed, it would indicate the possible therapeutic target for the treatment or prevention of such disorder [67].

Additionally, a recently performed study proved that the alteration of mitochondrial biogenesis of cumulus cells (CCs) could account for the impairment of oocyte quality observed in diminished ovarian reserve (DOR), which also suggests its major role in the determination of oocyte competence. If so, mitochondrial characteristics of CCs could serve as indicators of oocyte competence, and thus oocyte quality in DOR patients may be improved with mitochondrial biogenesis-enhancing therapies [68].

In the aspect of male fertility—the motility of human spermatozoa is entirely dependent on the functionality of the OXPHOS pathways. As sperm mtDNA partially encodes for OXPHOS-related proteins, any aberration in the mitochondrial genome may negatively influence sperm motility [60,66]. In addition to this, point mutations and deletions in mtDNA were linked to asthenozoospermia and oligoasthenozoospermia [69]. Furthermore, human spermatozoa with low MMP are less capable of undergoing the acrosome reaction and additionally show a direct and significant correlation with decreased sperm count, and negatively affected morphology, motility, and viability [60,66]. Probably, evaluation of the role of mitochondria in spermatogenesis may reveal new causes of male infertility. Moreover, when taking the effect of ageing, obesity, and metabolic health on the mitochondrial insufficiency into consideration, it is clear that infertility related to poor lifestyle may be settled into mitochondrial causes [66].

For both men and women, the pathogenesis of mitochondrial-related infertility may be direct through germ cells damage or indirect via diminished gonadotropin drive [70].

4.6. Do mtDNA Mutations Influence Early Embryo Development?

mtDNA mutations are a frequent cause of severe metabolic disorders. However, coexistence of wild type mtDNA usually neutralises the effect of a mutant load allowing normal phenotype to be maintained [29]. During early embryo development mtDNA content remains stable [1]. Therefore, the presence of mtDNA mutation in an oocyte or a preimplantation embryo may result in the reduction in the absolute number of wild type mtDNA copies. The protection against transgenerational transmission may be obtained either via inducing fertilisation defects and/or early embryogenesis failure or mechanisms eliminating damaged mitochondria during early embryo development.

The preimplantation mouse embryos were found to eliminate damaged mitochondria through a purifying selection during early embryogenesis; however, the precise mechanisms remain unknown. Such a phenomenon also suggested to take place to counter expansion of deleterious mtDNA mutations in the female germline [71]. On the other hand, a recently conducted study demonstrated that the presence of a pathogenic mutation was not the mtDNA metabolism modifier in human cleavage-stage embryos suggesting both no impact of mtDNA mutations and the absence of selection against them at this stage of development [29]. Contrarily, another study, focused on mtDNA mosaicism in early human development, identified a subgroup of low-level variants that may give rise to stable lineages of genetically diverse cells in the adult due to above-mentioned asymmetrical distribution of mitochondria within the oocyte [72].

Currently, the influence of mtDNA abnormalities on proper fertilisation and subsequent embryo development is not well known. Further studies (both animal and clinical) are needed to fully evaluate the relationship of mtDNA sequence and successful human embryogenesis.

5. Clinical Usefulness of Assessing and Improving Mitochondrial DNA Function and Quantity

5.1. Assessment of mtDNA Content in Human Embryos and Its Clinical Significance

The importance of mtDNA levels in the fertilisable oocyte was already demonstrated above. However, is there any relationship of mtDNA copy number with the embryo's ability to successfully implant and develop? In the past, dozens of experiments were performed to evaluate the significance of mtDNA content in developing embryos. The parameter, commonly called mitochondrial score, was defined as the ratio of mitochondrial/nuclear DNA copy number. However, almost every study group differently calculated its value [38,73–91]. Even now, there are a lot of controversies as conducted studies have reached conflicting results (Table 1).

Table 1. Comparison of results obtained by several study groups evaluating the relationship of mtDNA content in trophoectoderm biopsy (TE) or blastomeres (B) with embryos' features as aneuploidy, morphology, implantation, live birth and maternal age. Legend: N sample size, 🟢 positive correlation, 🔴 negative correlation, 🔵 no statistically significant correlation, - parameter not assessed.

	Material	N	Aneploidy	Morphology	Implantation	Live Birth	Maternal Age
Ritu et al., 2019	TE	287	🟢	🔵	🔵	🔵	🔵
Scott et al., 2020	TE	615	-	🔵	🔵	🔵	🔵
Wu et al., 2021	TE	1301	-	🔵	🟢	-	🔵
El-Damen et al., 2021	TE	355	-	🔴/🔵	🔵	🔵	🔵
Lee et al., 2019	B	39	🔵	-	-	-	-
	TE	998	🟢	-	🔵	-	🟢
De Munk et al., 2021	B	112	🔵	-	-	-	🔵
	TE	112	🔵	-	-	-	🔵
Diez-Juan et al., 2015	B	205	-	🔵	🔴	-	-
	TE	65	-	🔵	🔴	-	🔵
Arnanz et al., 2020	TE	504	🟢	🔴	-	-	-
Boynukalin et al., 2020	TE	707	-	-	-	🔴	-
Du et al., 2021	TE	246	🔵	🔵	🔴	-	-
Wang et al., 2021	TE	769	-	🔴	🔴	🔴	🔵
Klimczak et al., 2018	TE	1510	-	🔴	🔵	-	🔵
de Los Santos et al., 2018	TE	465	🟢	🔴	-	-	🔵
Victor et al., 2017	TE	1396	🔵	-	🔵	-	🔵
Fragouli et al., 2015	B	39	-	-	-	-	🔴
	TE	340	🟢	-	🔴	-	🟢
Fragouli et al., 2017	TE	199	-	-	-	-	🔵
Ravichandran et al., 2017	TE	1505	-	🔵	🔴	-	🔵
Treff et al., 2017	TE	374	-	🔴	🔵	-	🔴
Shang et al., 2018	B	149	🔵	🔵	🔵	-	🟢
	TE	250					
Podolak et al., 2022	B	314	🟢	🔵	🔵	🔵	🔵

Most study groups assessing the relationship of mtDNA content and embryo's ploidy status agree that increased levels of mtDNA correlate with aneuploidy [38,73,76,78,87,89]. Nevertheless, the mechanisms leading to such an effect are still not known. It may result from indirect indication of higher energy needs of aneuploid embryos, possibly stemming from the initiated repair mechanisms [38].

The primary aim of conducted experiments was to evaluate the link of mtDNA levels and the embryo's ability to implant [38,73–75,77–87,91]. Indeed, the proof of such a relationship would be a revolutionary discovery, giving the possibility to identify viable embryos and in consequence significantly improve the pregnancy rates of in vitro fertilisation treatment. Nevertheless, obtained results are highly conflicting [38,73–91]. Mitochondrial score was demonstrated to have either positive [85], negative [74,78,80,83,91] or no correlation [38,73,75,77,82,84,86,87] with implantation rate. A few research teams evaluated the link between mtDNA content with live birth ratio; however, most of them found no statistically significant difference between mitochondrial score values among embryos leading to live birth and those that did not [38,73,74,84,86,90]. Only twice were lower mtDNA copy number values demonstrated to correlate with live birth rate [74,90].

Divergent results were also obtained for assessment of the relationship of the mitochondrial score with embryos' morphology and especially maternal age [38,73–91].

5.2. Mitochondrial Score—The Debate under Its Usefulness as Embryo Selection Marker

Even though the mitochondrial score was evaluated numerous times by several study groups, its clinical usefulness remains unclear [38,73–91]. For the first time it was proposed as an embryo selection marker by Fragouli et al. [78]. Although they used both Next-Generation sequencing (NGS) and quantitative RT-PCR (qRT-PCR), and the mitochondrial score count was normalised based on the GC content and in-silico reference as well as taking into consideration the numbers of chromosomes, their study seems to be questionable. Interestingly, when assessing both TE and blastomeres they found positive correlation for mtDNA content with maternal age for TE and negative for blastomeres (Table 1). Such findings seem unlikely, and the reliability of these results should be reconsidered. Additionally, their subsequent study was claimed to echo their first findings, yet the observed relationship was statistically insignificant (Table 1). Even though the correlation for this parameter was not obtained again, they did not decide to reanalyse the rest of parameters, but only evaluated the pre-established thresholds [78,79]. The link of lower mtDNA levels with embryos' implantation potential was subsequently demonstrated by Diez-Juan et al. [83], Ravichandran et al. [80], Du et al. [91], and Wang et al. [74]. Diez-Juan et al. [83] assessed mtDNA content with qRT-PCR based on *ATP8* gene fragment representing mtDNA and β-actin gene fragment representing nuclear DNA. They did not take the ploidy status of chromosome 7 under consideration, which might influence results obtained for embryos with eventual trisomy or monosomy of chromosome 7. The correlation of mitochondrial score with implantation status was demonstrated for both TE and blastomeres (Table 1). Ravichandran et al. [80] conducted a study using NGS and qRT-PCR techniques. In fact, they did not assess the correlation between mtDNA content with implantation potential, but used the threshold established by Fragouli et al. [78], revalidating it due to technical changes in the laboratory. They concluded 100% negative predictive value of mtDNA assessment based on 33 embryos containing elevated levels of mtDNA that did not produce pregnancy. The sample size seems far too small to draw such conclusions. Both Wang et al. [74] and Du et al. [91] used NGS to evaluate mtDNA content at the blastocyst stage of development. Wang et al. [74] applied the formula proposed by Victor et al. [77], taking into account variables such us ploidy status of each chromosomes and embryo's genetic sex. They found a statistically significant difference of mitochondrial score values depending on implantation rate and live birth rate (Table 1). However, this relationship was only observed for day-5 biopsied embryos, and no correlation was reported for day-6 biopsied blastocysts. Moreover, the mitochondrial score was doubted as an independent embryo selection marker due to receiver operating characteristic (ROC) curve analysis outcomes, performed to evaluate the potential predictive value of mtDNA content in embryo implantation and live birth outcomes. Contrarily, Du et al. [91], who applied the formula proposed by Shang et al. [82], regarded ROC analysis outcomes as a highly predictive value. Nevertheless, a future, multicentre study was suggested to evaluate the mtDNA content in developing embryos and to verify its clinical application. On the other hand, Wu et al. [85] who used MitoCalc analyzer software dedicated to sequencing coverage analysis, found a positive correlation of mitochondrial score with implantation rate. In fact, obtained results may differ depending on biopsy day of assessed embryos. Day-5 biopsied blastocysts are reported to have higher mtDNA content compared to day-6 biopsied embryos [74,75,84,85]. Thus, the number of embryos biopsied at day-5 or 6 may influence obtained results. Apart from those mentioned above, the rest of listed study groups found no statistically significance of mitochondrial score in the context of assessing implantation potential [38,73,75,77,82,84,86,87]. Moreover, even though most study teams agree that mitochondrial score correlates with ploidy status, it seems unlikely to find the use in such assessment as preimplantation genetic testing for aneuploidies (PGT-A) gives much more definitive diagnosis. Furthermore, evaluation of mtDNA content may be useful at the blastocyst stage of development as the cleavage-stage embryos were demonstrated to present asymmetrical distribution of mitochondria among the formed blastomeres. Therefore, as the embryo may present the heterogeneity of their blastomeres, the obtained results may not be reliable [38]. Indeed, during our recent study we observed extreme differences of the mtDNA levels for embryos from the same patient, which suggests that the choice of a blastomere for biopsy strongly affects the outcomes of mtDNA quantification. In consequence, the use of an arbitrary threshold of mtDNA content as embryo selection criteria, at least for cleavage-stage embryos, must be avoided [38].

As so many study teams evaluated the clinical significance of mtDNA content in early embryos during past years, one question arises—how is it possible that we still have no sure answer? First of all, conflicting outcomes may result from different technical approaches on detection of mtDNA content and used formulas to calculate it. Some study teams take ploidy status for consideration, whereas some do not [77,78,82,83]. Technical approaches seem to play a crucial role as even study groups who used the same formula obtained conflicting results (Victor et al. [77] versus Wang et al. [74] or Shang et al. [82] versus Du et al. [91]) (Table 1). For instance, there are several whole genome amplification (WGA) protocols which differently influence the amplification of mtDNA in comparison to nDNA [38]. Secondly, as mentioned above, the day of biopsy may additionally influence obtained results. Moreover, storage protocols or reagents used for culturing may affect the assessment as the incidence of blastocysts with elevated mtDNA was reported to vary widely between clinics in which in vitro fertilisation was undertaken [80]. Last but not least, the sample sizes vary between conducted research, and mostly it is too small when assessing blastomeres biopsied from cleavage-stage embryos; thus, it cannot deliver reliable outcomes [78,87].

What should be done then? The best way to investigate the clinical significance of mtDNA content in developing embryos seems to be the assessment of it in a large, multicenter study using modern laboratory techniques and precise calculations.

Although the clinical value of mtDNA content assessment is still not established well, it is commercially available [92], driving another question—does commercialisation overtake science in reproductive medicine?

5.3. Opportunities to Improve Mitochondrial DNA Function

The fundamental role of proper mitochondrial activity and functionality for human health and reproductive success was already summarised. Hence, are there any possibilities to improve mitochondrial DNA function? In the gene-editing era, mitochondrial genome editing (MGE) seems to be one of the possible solutions. MGE refers to the modification of human oocytes or embryos with intracytoplasmic microinjection or mitochondria-targeted nucleases in order to prevent transmission of mitochondrial diseases [93]. Yahata et al. [94] engineered platinum TALENs, which were transported into mitochondria, recognized the mtDNA sequence including the m.13513 position, and preferentially cleaved G13513A mutant mtDNA, and conducted an experiment with induced pluripotent stem cells (iPCs). The heteroplasmy level of m.13513G>A mutation was decreased short after the transduction. Moreover, Yang et al. [95] demonstrated successful elimination of m.3243A>G mutation in iPCs via injection of mitoTALEN mRNA. However, protein engineering and assembly processes for every genomic target seem to be a time-consuming task [93]. On the other hand, CRISPR/Cas9 genome editing technology was also suggested to target mutant mtDNA [96]. The issue of mitochondrial genome editing is definitely of increasing interest to scientists, although there is still a long way to go to introduce it in clinical practice. Nevertheless, there is already a way to overcome the transmission of maternally-inherited mtDNA mutations—the use of mitochondrial replacement technologies in assisted reproductive treatment (ART). Currently, there are two transfer techniques in clinical application—maternal spindle transfer and pronuclear transfer (Figure 2).

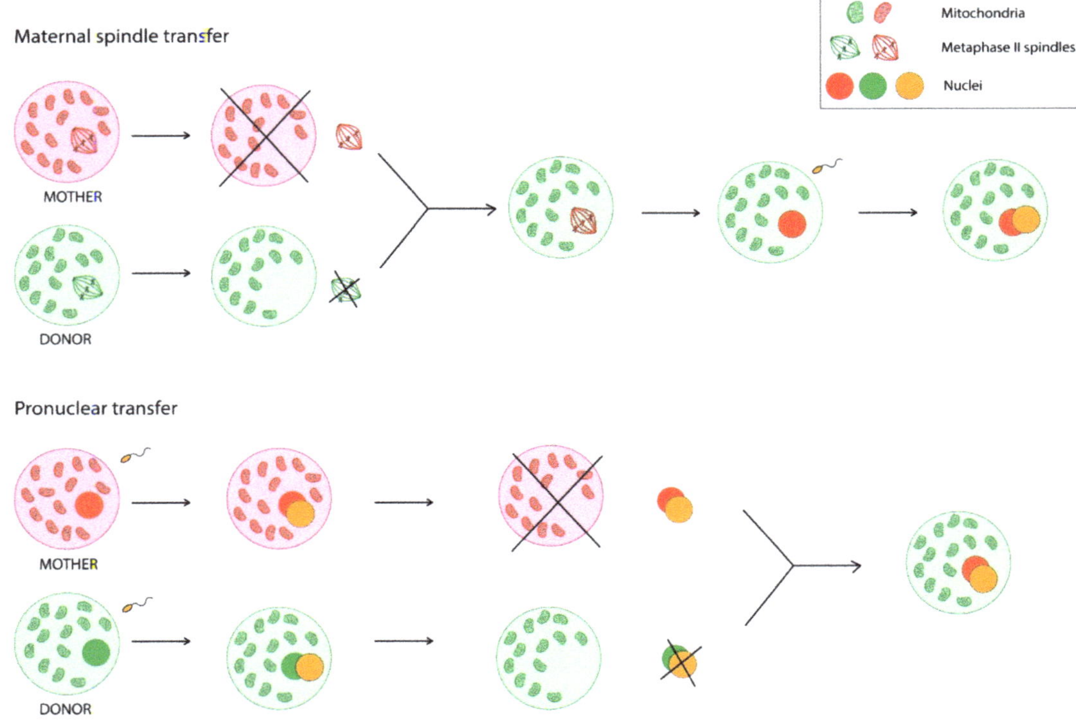

Figure 2. Comparison of two mitochondrial replacement technologies—maternal spindle transfer and pronuclear transfer.

Maternal spindle transfer refers to removing the nuclear DNA from the donor egg, leaving the part of the oocyte containing healthy mitochondria, and then inserting into this cell nuclear DNA from the mother's oocyte to finally fertilise it and implant into the mother's uterus. The pronuclear transfer is a similar process; however, it requires fertilisation of both—mother's and donor's—oocytes before transferring the nuclear DNA [97–100].

5.4. Mitochondrial DNA Transfer in Improving the Reproductive Potential of Oocytes

Mitochondrial replacement therapy is also an object of interest in the context of improving the quality, and hence the reproductive potential, of oocytes [100–102]. In 1997 it was performed for the first time using partial ooplasm transfer, a technique based on simultaneous injection of sperm with 1 to 5% of oocyte cytoplasm during intracytoplasmic sperm injection (ICSI) [103–105]. The ooplasm was derived from fresh or cryopreserved oocytes of young and healthy donors, and clinically abandoned trinuclear embryos. Although this technology achieved good clinical results, it was banned by the US Food and Drug Administration (FDA) for clinical use in the US due to the ethical and genetic safety issues caused by the involvement of a third-party genetic material [102]. A potential alternative approach could be autologous mitochondrial transfer, theoretically introducing much larger amounts of mitochondria into the oocyte. Mitochondria preferentially should be obtained from ovarian or oocyte origin [101,102]. Woods and Tilly [106] reported improving pregnancy outcomes in women with a previous history of assisted reproduction failure. They obtained autologous mitochondria from oogonial stem cells (OSCs) isolated from cortical biopsies. The injection was made via ICSI. The therapy was called Autologous Germline Mitochondrial Energy Transfer (AUGMENT) [106,107]. It aroused controversy from the very beginning [108–111]. Experts in the field were raising numerous questions about OSC (e.g., how many of them are recovered, how does their amplification affect them, are their mitochondria healthy) and AUGMENT protocol (e.g., how many mitochondria are being transferred; does transfer result in a significant boost to ATP). The therapy was considered to be based on the insecurities of the current knowledge about the character of the putative OSC and benefits claimed for mitochondrial transfer [111]. Moreover, there were also accusations of improper conduct of a clinical trial [108]. However, AUGMENT was commercially launched, raising a known question once again—does commercialisation overtake science in reproductive medicine? The introduction of the procedure turned out to give no benefit to the patients. That, in consequence, led the OvaScience—once worth over a billion dollars—to debt of millions and a reduction in workforce of around 50%. Finally, the company was sued by shareholders and closed its operations [112,113]. The AUGMENT history shows that, even though the field of reproductive medicine is highly privatised, it must have the same rules as the rest—first of all, be guided by evidence-based policy. Currently, there is one registered trial that aims to evaluate the effect of mitochondrial transfer from bone marrow mesenchymal stem cells on the quality of oocytes; however, the results are unknown [114].

6. Summary

Mitochondria play a crucial role in human fertility and early embryo development as it has a great contribution to several vital processes. Despite its great potential of clinical application, the field of mitochondrial research is still not well established. Further studies are needed to fully evaluate the clinical usefulness of assessing and improving mtDNA quantity and function.

Author Contributions: A.P. performed the literature review and designed and wrote the original manuscript, I.W.-P. revised and edited the manuscript, K.L. performed the literature review and designed, supervised, and edited the manuscript. All authors have read and agreed to the published version of the manuscript.

Funding: This research was supported by Grants-in-Aid for Scientific Research from the National Science Center (2018/31/B/NZ9/03412).

Institutional Review Board Statement: Not applicable.

Informed Consent Statement: Not applicable.

Data Availability Statement: Not applicable.

Acknowledgments: The authors would like to thank Magda Milewska for her help with graphical visualisation.

Conflicts of Interest: The authors declare no conflict of interest.

Abbreviations

ART—assisted reproductive treatment; ATP—adenosine triphosphate; ATP8—mitochondrially encoded ATP synthase membrane subunit 8; AUGMENT—Autologous Germline Mitochondrial Energy Transfer; B—blastomeres; CDK1—cyclin-dependent kinase 1; CCs—cumulus cells; DNA—deoxyribonucleic acid; DOR—diminished ovarian reserve; DRP1—dynamin-related protein 1; ER—endoplasmic reticulum; Fis1—anti-mitochondrial fission 1 protein; FSH—follicle stimulating hormone; GCs—granulosa cells; ICSI—intracytoplasmic sperm injection; iPCs—induced pluripotent stem cells; kbp—kilo base pairs; LH—luteinizing hormone; MERF—myoclonic epilepsy associated with ragged red fibres; Mff—mitochondrial fission factor; Mfn1—dynamin-like GTPases mitofusins 1; Mfn2—dynamin-like GTPases mitofusins 2; MGE—mitochondrial genome editing; Mid49—mitochondrial dynamics protein of 49 kDa; Mid51—mitochondrial dynamics protein of 51 kDa; MMP—mitochondrial membrane potential; MRT—mitochondrial replacement therapy; MT-ATP6—mitochondrially encoded ATP synthase membrane subunit 6; MT-CO1—mitochondrially encoded cytochrome C oxidase I; mtDNA—mitochondrial DNA; NARP—neurogenic muscle weakness, ataxia, and retinitis pigmentosa; NCR—non-coding region; nDNA—nuclear DNA; NGS—Next-Generation Sequencing; Opa1—optic atrophy 1; OSCs—oogonial stem cells; OXPHOS—oxidative phosphorylation; PGC—primordial germ cell; PINK1—PTEN-induced kinase 1; POI—primary ovarian insufficiency; POLG—DNA polymerase γ; PTP—permeability transition pore; qRT-PCR—quantitative reverse transcription polymerase chain reaction; RALA—Ras-related protein Ral-A; RALBP1—RalA-binding protein 1; ROC—receiver operating characteristic; ROS—reactive oxygen species; SOD1—superoxide dismutase 1; SOD2—superoxide dismutase 2; and TALEN—transcription activator-like effector nucleases.

References

1. St John, J.C.; Facucho-Oliveira, J.; Jiang, Y.; Kelly, R.; Salah, R. Mitochondrial DNA transmission, replication and inheritance: A journey from the gamete through the embryo and into offspring and embryonic stem cells. *Hum. Reprod. Update* **2010**, *16*, 488–509. [CrossRef] [PubMed]
2. Lenaz, G.; Genova, M.L. Supramolecular Organisation of the Mitochondrial Respiratory Chain: A New Challenge for the Mechanism and Control of Oxidative Phosphorylation. *Adv. Exp. Med. Biol.* **2012**, *748*, 107–144. [PubMed]
3. Annesley, S.J.; Fisher, P.R. Mitochondria in Health and Disease. *Cells* **2019**, *8*, 680. [CrossRef] [PubMed]
4. Duchen, M.R. Mitochondria and calcium: From cell signalling to cell death. *J. Physiol.* **2000**, *529 Pt 1*, 57–68. [CrossRef] [PubMed]
5. Kroemer, G. Mitochondrial control of apoptosis: An introduction. *Biochem. Biophys. Res. Commun.* **2003**, *304*, 433–435. [CrossRef]
6. Herst, P.M.; Rowe, M.R.; Carson, G.M.; Berridge, M.V. Functional Mitochondria in Health and Disease. *Front. Endocrinol. (Lausanne).* **2017**, *8*, 296. [CrossRef]
7. Arciuch, V.G.A.; Elguero, M.E.; Poderoso, J.J.; Carreras, M.C. Mitochondrial regulation of cell cycle and proliferation. *Antioxidants Redox Signal.* **2012**, *16*, 1150–1180. [CrossRef]
8. Mishra, P.; Chan, D.C. Mitochondrial dynamics and inheritance during cell division, development and disease. *Nat. Rev. Mol. Cell Biol.* **2014**, *15*, 634–646. [CrossRef] [PubMed]
9. Kurland, C.G.; Andersson, S.G. Origin and evolution of the mitochondrial proteome. *Microbiol. Mol. Biol. Rev.* **2000**, *64*, 786–820. [CrossRef]
10. Mascanzoni, F.; Ayala, I.; Colanzi, A. Organelle Inheritance Control of Mitotic Entry and Progression: Implications for Tissue Homeostasis and Disease. *Front. Cell Dev. Biol.* **2019**, *7*, 1–11. [CrossRef]
11. Pangou, E.; Sumara, I. The Multifaceted Regulation of Mitochondrial Dynamics During Mitosis. *Front. Cell Dev. Biol.* **2021**, *9*, 767221. [CrossRef] [PubMed]
12. Carvalho, K.F.; Machado, T.S.; Garcia, B.M.; Zangirolamo, A.F.; Macabelli, C.H.; Sugiyama, F.H.C.; Grejo, M.P.; Augusto Neto, J.D.; Tostes, K.; Ribeiro, F.K.S.; et al. Mitofusin 1 is required for oocyte growth and communication with follicular somatic cells. *FASEB J.* **2020**, *34*, 7644–7660. [CrossRef]
13. Zhang, M.; Bener, M.B.; Jiang, Z.; Wang, T.; Esencan, E.; Scott, R.; Horvath, T.; Seli, E. Mitofusin 2 plays a role in oocyte and follicle development, and is required to maintain ovarian follicular reserve during reproductive aging. *Aging (Albany. NY).* **2019**, *11*, 3919–3938. [CrossRef] [PubMed]
14. Zhang, M.; Bener, M.B.; Jiang, Z.; Wang, T.; Esencan, E.; Scott III, R.; Horvath, T.; Seli, E. Mitofusin 1 is required for female fertility and to maintain ovarian follicular reserve. *Cell Death Dis.* **2019**, *10*, 560. [CrossRef]
15. Katajisto, P.; Döhla, J.; Chaffer, C.L.; Pentinmikko, N.; Marjanovic, N.; Iqbal, S.; Zoncu, R.; Chen, W.; Weinberg, R.A.; Sabatini, D.M. Asymmetric apportioning of aged mitochondria between daughter cells is required for stemness. *Science* **2015**, *348*, 340–343. [CrossRef]

16. Wei, W.; Chinnery, P.F. Inheritance of mitochondrial DNA in humans: Implications for rare and common diseases. *J. Intern. Med.* **2020**, *287*, 634–644. [CrossRef]
17. Sato, M.; Sato, K. Maternal inheritance of mitochondrial DNA by diverse mechanisms to eliminate paternal mitochondrial DNA. *Biochim. Biophys. Acta - Mol. Cell Res.* **2013**, *1833*, 1979–1984. [CrossRef] [PubMed]
18. Kaneda, H.; Hayashi, J.I.; Takahama, S.; Taya, C.; Lindahl, K.F.; Yonekawa, H. Elimination of paternal mitochondrial DNA in intraspecific crosses during early mouse embryogenesis. *Proc. Natl. Acad. Sci. USA* **1995**, *92*, 4542–4546. [CrossRef]
19. Shitara, H.; Kaneda, H.; Sato, A.; Inoue, K.; Ogura, A.; Yonekawa, H.; Hayashi, J.I. Selective and continuous elimination of mitochondria microinjected into mouse eggs from spermatids, but not from liver cells, occurs throughout emborogenesis. *Genetics* **2000**, *156*, 1277–1284. [CrossRef]
20. Cummins, J.M.; Wakayama, T.; Yanagimachi, R. Fate of microinjected spermatid mitochondria in the mouse oocyte and embryo. *Zygote* **1998**, *6*, 213–222. [CrossRef]
21. Seli, E.; Wang, T.; Horvath, T.L. Mitochondrial unfolded protein response: A stress response with implications for fertility and reproductive aging. *Fertil. Steril.* **2019**, *111*, 197–204. [CrossRef] [PubMed]
22. Park, Y.J.; Pang, M.G. Mitochondrial functionality in male fertility: From spermatogenesis to fertilization. *Antioxidants* **2021**, *10*, 98. [CrossRef] [PubMed]
23. Sutovsky, P.; Moreno, R.D.; Ramalho-Santos, J.; Dominko, T.; Simerly, C.; Schatten, G. Ubiquitin tag for sperm mitochondria. *Nature* **1999**, *402*, 371–372. [CrossRef]
24. St. John, J. The control of mtDNA replication during differentiation and development. *Biochim. Biophys. Acta - Gen. Subj.* **2014**, *1840*, 1345–1354. [CrossRef] [PubMed]
25. Luo, S.; Valencia, C.A.; Zhang, J.; Lee, N.-C.; Slone, J.; Gui, B.; Wang, X.; Li, Z.; Dell, S.; Brown, J.; et al. Biparental Inheritance of Mitochondrial DNA in Humans. *Proc. Natl. Acad. Sci.* **2018**, *115*, 13039 LP–13044. [CrossRef]
26. Cree, L.M.; Samuels, D.C.; Chinnery, P.F. The inheritance of pathogenic mitochondrial DNA mutations. *Biochim. Biophys. Acta Mol. Basis Dis.* **2009**, *1792*, 1097–1102. [CrossRef]
27. Vaught, R.C.; Dowling, D.K. Maternal inheritance of mitochondria: Implications for male fertility? *Reproduction* **2018**, *155*, R159–R168. [CrossRef]
28. Chiaratti, M.R.; Garcia, B.M.; Carvalho, K.F.; Machado, S.; Karina, F.; Habermann, C. The role of mitochondria in the female germline: Implications to fertility and inheritance of mitochondrial diseases. *Cell Biol. Int.* **2018**, *42*, 1–39. [CrossRef]
29. Chatzovoulou, K.; Mayeur, A.; Gigarel, N.; Jabot-Hanin, F.; Hesters, L.; Munnich, A.; Frydman, N.; Bonnefont, J.P.; Steffann, J. Mitochondrial DNA mutations do not impact early human embryonic development. *Mitochondrion* **2021**, *58*, 59–63. [CrossRef]
30. Zhang, R.; Nakahira, K.; Choi, A.M.K.; Gu, Z. Heteroplasmy concordance between mitochondrial DNA and RNA. *Sci. Rep.* **2019**, *9*, 12942. [CrossRef]
31. McCormick, E.M.; Muraresku, C.C.; Falk, M.J. Mitochondrial Genomics: A Complex Field Now Coming of Age. *Curr. Genet. Med. Rep.* **2018**, *6*, 52–61. [CrossRef]
32. Chinnery, P.F.; Gomez-Duran, A. Oldies but Goldies mtDNA Population Variants and Neurodegenerative Diseases. *Front. Neurosci.* **2018**, 12. [CrossRef] [PubMed]
33. Rossignol, R.; Faustin, B.; Rocher, C.; Malgat, M.; Mazat, J.P.; Letellier, T. Mitochondrial threshold effects. *Biochem. J.* **2003**, *370*, 751–762. [CrossRef] [PubMed]
34. Stewart, J.B.; Chinnery, P.F. The dynamics of mitochondrial DNA heteroplasmy: Implications for human health and disease. *Nat. Rev. Genet.* **2015**, *16*, 530–542. [CrossRef]
35. Van Blerkom, J.; Davis, P.; Alexander, S. Differential mitochondrial distribution in human pronuclear embryos leads to disproportionate inheritance between blastomeres: Relationship to microtubular organization, ATP content and competence. *Hum. Reprod.* **2000**, *15*, 2621–2633. [CrossRef]
36. Squirrell, J.M.; Schramm, R.D.; Paprocki, A.M.; Wokosin, D.L.; Bavister, B.D. Imaging mitochondrial organization in living primate oocytes and embryos using multiphoton microscopy. *Microsc. Microanal.* **2003**, *9*, 190–201. [CrossRef]
37. Bavister, B.D.; Squirrell, J.M. Mitochondrial distribution and function in oocytes and early embryos. *Hum. Reprod.* **2000**, *15* (Suppl. 2), 189–198. [CrossRef] [PubMed]
38. Podolak, A.; Liss, J.; Kiewisz, J.; Pukszta, S.; Cybulska, C.; Rychlowski, M.; Lukaszuk, A.; Jakiel, G.; Lukaszuk, K. Mitochondrial DNA Copy Number in Cleavage Stage Human Embryos—Impact on Infertility Outcome. *Curr. Issues Mol. Biol.* **2022**, *44*, 273–287. [CrossRef]
39. May-Panloup, P.; Boguenet, M.; Hachem, H.E.; Bouet, P.E.; Reynier, P. Embryo and its mitochondria. *Antioxidants* **2021**, *10*, 139. [CrossRef]
40. Leese, H.J.; Houghton, F.D.; Macmillan, D.A.; Donnay, I. Metabolism of the Early Embryo: Energy Production and Utilization. *ART Hum. Blastocyst* **2001**, 61–68. [CrossRef]
41. Wilding, M.; Coppola, G.; Dale, B.; Di Matteo, L. Mitochondria and human preimplantation embryo development. *Reproduction* **2009**, *137*, 619–624. [CrossRef]
42. Remi, D.; Duchen, M.J.C. The Role of Mitochondrial Function in the Oocyte and Embryo. *Curr. Top. Dev. Biol.* **2007**, *77*. [CrossRef]
43. Rieger, D. Relationships between energy metabolism and development of early mammalian embryos. *Theriogenology* **1992**, *37*, 75–93. [CrossRef]

44. Chappel, S. The Role of Mitochondria from Mature Oocyte to Viable Blastocyst. *Obstet. Gynecol. Int.* **2013**, *2013*, 183024. [CrossRef] [PubMed]
45. de Lima, C.B.; dos Santos, É.C.; Ispada, J.; Fontes, P.K.; Nogueira, M.F.G.; dos Santos, C.M.D.; Milazzotto, M.P. The dynamics between in vitro culture and metabolism: Embryonic adaptation to environmental changes. *Sci. Rep.* **2020**, *10*, 15672. [CrossRef]
46. Moraes, C.R.; Meyers, S. The sperm mitochondrion: Organelle of many functions. *Anim. Reprod. Sci.* **2018**, *194*, 71–80. [CrossRef]
47. van der Reest, J.; Nardini Cecchino, G.; Haigis, M.C.; Kordowitzki, P. Mitochondria: Their relevance during oocyte ageing. *Ageing Res. Rev.* **2021**, *70*, 87–100. [CrossRef] [PubMed]
48. Chiang, J.L.; Shukla, P.; Pagidas, K.; Ahmed, N.S.; Karri, S.; Gunn, D.D.; Hurd, W.W.; Singh, K.K. Mitochondria in Ovarian Aging and Reproductive Longevity. *Ageing Res. Rev.* **2020**, *63*, 101168. [CrossRef]
49. Konstantinidis, M.; Alfarawati, S.; Hurd, D.; Paolucci, M.; Shovelton, J.; Fragouli, E.; Wells, D. Simultaneous assessment of aneuploidy, polymorphisms, and mitochondrial DNA content in human polar bodies and embryos with the use of a novel microarray platform. *Fertil. Steril.* **2014**, *102*, 1385–1392. [CrossRef]
50. Kushnir, V.A.; Ludaway, T.; Russ, R.B.; Fields, E.J.; Koczor, C.; Lewis, W. Reproductive aging is associated with decreased mitochondrial abundance and altered structure in murine oocytes. *J. Assist. Reprod. Genet.* **2012**, *29*, 637–642. [CrossRef]
51. May-Panloup, P.; Chrétien, M.F.; Jacques, C.; Vasseur, C.; Malthièry, Y.; Reynier, P. Low oocyte mitochondrial DNA content in ovarian insufficiency. *Hum. Reprod.* **2005**, *20*, 593–597. [CrossRef]
52. Boucret, L.; Bris, C.; Seegers, V.; Goudenège, D.; Desquiret-Dumas, V.; Domin-Bernhard, M.; Ferré-L'Hotellier, V.; Bouet, P.E.; Descamps, P.; Reynier, P.; et al. Deep sequencing shows that oocytes are not prone to accumulate mtDNA heteroplasmic mutations during ovarian ageing. *Hum. Reprod.* **2017**, *32*, 2101–2109. [CrossRef]
53. May-Panloup, P.; Brochard, V.; Hamel, J.F.; Desquiret-Dumas, V.; Chupin, S.; Reynier, P.; Duranthon, V. Maternal ageing impairs mitochondrial DNA kinetics during early embryogenesis in mice. *Hum. Reprod.* **2019**, *34*, 1313–1324. [CrossRef]
54. May-Panloup, P.; Boucret, L.; de la Barca, J.M.C.; Desquiret-Dumas, V.; Ferré-L'Hotellier1, V.; Morinière, C.; Descamps, P.; Procaccio, V.; Reynier, P. Ovarian ageing: The role of mitochondria in oocytes and follicles. *Hum. Reprod. Update* **2016**, *22*, 725–743. [CrossRef] [PubMed]
55. Barritt, J.A.; Kokot, M.; Cohen, J.; Steuerwald, N.; Brenner, C.A. Quantification of human ooplasmic mitochondria. *Reprod. Biomed. Online* **2002**, *4*, 243–247. [CrossRef]
56. Yang, L.; Lin, X.; Tang, H.; Fan, Y.; Zeng, S.; Jia, L.; Li, Y.; Shi, Y.; He, S.; Wang, H.; et al. Mitochondrial DNA mutation exacerbates female reproductive aging via impairment of the NADH/NAD+ redox. *Aging Cell* **2020**, *19*, 1–14. [CrossRef] [PubMed]
57. Trifunovic, A.; Wredenberg, A.; Falkenberg, M.; Spelbrink, J.N.; Rovio, A.T.; Bruder, C.E.; Bohlooly-Y, M.; Gidlöf, S.; Oldfors, A.; Wibom, R.; et al. Premature ageing in mice expressing defective mitochondrial DNA polymerase. *Nature* **2004**, *429*, 417–423. [CrossRef]
58. Park, S.U.; Walsh, L.; Berkowitz, K.M. Mechanisms of ovarian aging. *Reproduction* **2021**, *162*, R19–R33. [CrossRef] [PubMed]
59. Busnelli, A.; Navarra, A.; Levi-Setti, P.E. Qualitative and quantitative ovarian and peripheral blood mitochondrial dna (Mtdna) alterations: Mechanisms and implications for female fertility. *Antioxidants* **2021**, *10*, 55. [CrossRef]
60. Durairajanayagam, D.; Singh, D.; Agarwal, A.; Henkel, R. Causes and consequences of sperm mitochondrial dysfunction. *Andrologia* **2021**, *53*, 1–15. [CrossRef]
61. Menger, K.E.; Rodríguez-Luis, A.; Chapman, J.; Nicholls, T.J. Controlling the topology of mammalian mitochondrial DNA. *Open Biol.* **2021**, *11*, 210168. [CrossRef]
62. Tatone, C.; Carbone, M.C.; Falone, S.; Aimola, P.; Giardinelli, A.; Caserta, D.; Marci, R.; Pandolfi, A.; Ragnelli, A.M.; Amicarelli, F. Age-dependent changes in the expression of superoxide dismutases and catalase are associated with ultrastructural modifications in human granulosa cells. *Mol. Hum. Reprod.* **2006**, *12*, 655–660. [CrossRef]
63. Colella, M.; Cuomo, D.; Peluso, T.; Falanga, I.; Mallardo, M.; De Felice, M.; Ambrosino, C. Ovarian Aging: Role of Pituitary-Ovarian Axis Hormones and ncRNAs in Regulating Ovarian Mitochondrial Activity. *Front. Endocrinol. (Lausanne)* **2021**, *12*, 1–12. [CrossRef]
64. Heinonen, S.; Buzkova, J.; Muniandy, M.; Kaksonen, R.; Ollikainen, M.; Ismail, K.; Hakkarainen, A.; Lundbom, J.; Lundbom, N.; Vuolteenaho, K.; et al. Impaired Mitochondrial Biogenesis in Adipose Tissue in Acquired Obesity. *Diabetes* **2015**, *64*, 3135–3145. [CrossRef] [PubMed]
65. Cunarro, J.; Casado, S.; Lugilde, J.; Tovar, S. Hypothalamic Mitochondrial Dysfunction as a Target in Obesity and Metabolic Disease. *Front. Endocrinol. (Lausanne)*. **2018**, *9*, 283. [CrossRef] [PubMed]
66. Vertika, S.; Singh, K.K.; Rajender, S. Mitochondria, spermatogenesis, and male infertility – An update. *Mitochondrion* **2020**, *54*, 26–40. [CrossRef]
67. Tiosano, D.; Mears, J.A.; Buchner, D.A. Mitochondrial Dysfunction in Primary Ovarian Insufficiency. *Endocrinology* **2019**, *160*, 2353–2366. [CrossRef] [PubMed]
68. Boucret, L.; Chao De La Barca, J.M.; Morinière, C.; Desquiret, V.; Ferré-L'Hôtellier, V.; Descamps, P.; Marcaillou, C.; Reynier, P.; Procaccio, V.; May-Panloup, P. Relationship between diminished ovarian reserve and mitochondrial biogenesis in cumulus cells. *Hum. Reprod.* **2015**, *30*, 1653–1664. [CrossRef]
69. Wei, Y.H.; Kao, S.H. Mitochondrial DNA mutation and depletion are associated with decline of fertility and motility of human sperm. *Zool. Stud.* **2000**, *39*, 1–12.

70. Demain, L.A.M.; Conway, G.S.; Newman, W.G. Genetics of mitochondrial dysfunction and infertility. *Clin. Genet.* **2017**, *91*, 199–207. [CrossRef]
71. Machado, T.S.; Macabelli, C.H.; Collado, M.D.; Meirelles, F.V.; Guimarães, F.E.G.; Chiaratti, M.R. Evidence of Selection Against Damaged Mitochondria During Early Embryogenesis in the Mouse. *Front. Genet.* **2020**, *11*, 1–9. [CrossRef] [PubMed]
72. Mertens, J.; Regin, M.; De Munck, N.; De Deckersberg, E.C.; Belva, F.; Sermon, K.; Tournaye, H.; Blockeel, C.; Velde, H. Van De Mitochondrial DNA variants segregate during human preimplantation development into genetically different cell lineages that are maintained postnatally. *bioRxiv* **2021**, *3*.
73. Ritu, G.; Veerasigamani, G.; Ashraf, M.C.; Singh, S.; Laheri, S.; Modi, D. No association of mitochondrial DNA levels in trophectodermal cells with the developmental competence of the blastocyst and pregnancy outcomes. *bioRxiv* **2019**, 1–22. [CrossRef]
74. Wang, J.; Diao, Z.; Zhu, L.; Zhu, J.; Lin, F.; Jiang, W.; Fang, J.; Xu, Z.; Xing, J.; Zhou, J.; et al. Trophectoderm Mitochondrial DNA Content Associated with Embryo Quality and Day-5 Euploid Blastocyst Transfer Outcomes. *DNA Cell Biol.* **2021**, *40*, 643–651. [CrossRef]
75. Klimczak, A.M.; Pacheco, L.E.; Lewis, K.E.; Massahi, N.; Richards, J.P.; Kearns, W.G.; Saad, A.F.; Crochet, J.R. Embryonal mitochondrial DNA: Relationship to embryo quality and transfer outcomes. *J. Assist. Reprod. Genet.* **2018**, *35*, 871–877. [CrossRef] [PubMed]
76. de los Santos, M.J.; Diez Juan, A.; Mifsud, A.; Mercader, A.; Meseguer, M.; Rubio, C.; Pellicer, A. Variables associated with mitochondrial copy number in human blastocysts: What can we learn from trophectoderm biopsies? *Fertil. Steril.* **2018**, *109*, 110–117. [CrossRef]
77. Victor, A.R.; Brake, A.J.; Tyndall, J.C.; Griffin, D.K.; Zouves, C.G.; Barnes, F.L.; Viotti, M. Accurate quantitation of mitochondrial DNA reveals uniform levels in human blastocysts irrespective of ploidy, age, or implantation potential. *Fertil. Steril.* **2017**, *107*, 34–42.e3. [CrossRef] [PubMed]
78. Fragouli, E.; Spath, K.; Alfarawati, S.; Kaper, F.; Craig, A.; Michel, C.-E.; Kokocinski, F.; Cohen, J.; Munne, S.; Wells, D. Altered levels of mitochondrial DNA are associated with female age, aneuploidy, and provide an independent measure of embryonic implantation potential. *PLoS Genet.* **2015**, *11*, e1005241. [CrossRef]
79. Fragouli, E.; McCaffrey, C.; Ravichandran, K.; Spath, K.; Grifo, J.A.; Munné, S.; Wells, D. Clinical implications of mitochondrial DNA quantification on pregnancy outcomes: A blinded prospective non-selection study. *Hum. Reprod.* **2017**, *32*, 2340–2347. [CrossRef]
80. Ravichandran, K.; McCaffrey, C.; Grifo, J.; Morales, A.; Perloe, M.; Munne, S.; Wells, D.; Fragouli, E. Mitochondrial DNA quantification as a tool for embryo viability assessment: Retrospective analysis of data from single euploid blastocyst transfers. *Hum. Reprod.* **2017**, *32*, 1282–1292. [CrossRef]
81. Treff, N.R.; Zhan, Y.; Tao, X.; Olcha, M.; Han, M.; Rajchel, J.; Morrison, L.; Morin, S.J.; Scott, R.T.J. Levels of trophectoderm mitochondrial DNA do not predict the reproductive potential of sibling embryos. *Hum. Reprod.* **2017**, *32*, 954–962. [CrossRef] [PubMed]
82. Shang, W.; Zhang, Y.; Shu, M.; Wang, W.; Ren, L.; Chen, F.; Shao, L.; Lu, S.; Bo, S.; Ma, S.; et al. Comprehensive chromosomal and mitochondrial copy number profiling in human IVF embryos. *Reprod. Biomed. Online* **2018**, *36*, 67–74. [CrossRef] [PubMed]
83. Diez-Juan, A.; Rubio, C.; Marin, C.; Martinez, S.; Al-Asmar, N.; Riboldi, M.; Díaz-Gimeno, P.; Valbuena, D.; Simón, C. Mitochondrial DNA content as a viability score in human euploid embryos: Less is better. *Fertil. Steril.* **2015**, *104*, 534–541.e1. [CrossRef] [PubMed]
84. Scott, R.T.; Sun, L.; Zhan, Y.; Marin, D.; Tao, X.; Seli, E. Mitochondrial DNA content is not predictive of reproductive competence in euploid blastocysts. *Reprod. Biomed. Online* **2020**, *41*, 183–190. [CrossRef]
85. Wu, F.S.-Y.; Weng, S.-P.; Shen, M.-S.; Ma, P.-C.; Wu, P.-K.; Lee, N.-C. Suboptimal trophectoderm mitochondrial DNA level is associated with delayed blastocyst development. *J. Assist. Reprod. Genet.* **2021**, *38*, 587–594. [CrossRef] [PubMed]
86. El-Damen, A.; Elkhatib, I.; Bayram, A.; Arnanz, A.; Abdala, A.; Samir, S.; Lawrenz, B.; De Munck, N.; Fatemi, H.M. Does blastocyst mitochondrial DNA content affect miscarriage rate in patients undergoing single euploid frozen embryo transfer? *J. Assist. Reprod. Genet.* **2021**, *38*, 595–604. [CrossRef]
87. Lee, Y.X.; Chen, C.H.; Lin, S.Y.; Lin, Y.H.; Tzeng, C.R. Adjusted mitochondrial DNA quantification in human embryos may not be applicable as a biomarker of implantation potential. *J. Assist. Reprod. Genet.* **2019**, *36*, 1855–1865. [CrossRef]
88. De Munck, N.; Liñán, A.; Elkhatib, I.; Bayram, A.; Arnanz, A.; Rubio, C.; Garrido, N.; Lawrenz, B.; Fatemi, H.M. mtDNA dynamics between cleavage-stage embryos and blastocysts. *J. Assist. Reprod. Genet.* **2019**, *36*, 1867–1875. [CrossRef]
89. Arnanz, A.; De Munck, N.; Bayram, A.; El-Damen, A.; Abdalla, A.; ElKhatib, I.; Melado, L.; Lawrenz, B.; Fatemi, H.M. Blastocyst mitochondrial DNA (mtDNA) is not affected by oocyte vitrification: A sibling oocyte study. *J. Assist. Reprod. Genet.* **2020**, *37*, 1387–1397. [CrossRef]
90. Boynukalin, F.K.; Gultomruk, M.; Cavkaytar, S.; Turgut, E.; Findikli, N.; Serdarogullari, M.; Coban, O.; Yarkiner, Z.; Rubio, C.; Bahceci, M. Parameters impacting the live birth rate per transfer after frozen single euploid blastocyst transfer. *PLoS ONE* **2020**, *15*, e0227619. [CrossRef]
91. Du, S.; Huang, Z.; Lin, Y.; Sun, Y.; Chen, Q.; Pan, M.; Zheng, B. Mitochondrial DNA Copy Number in Human Blastocyst: A Novel Biomarker for the Prediction of Implantation Potential. *J. Mol. Diagn.* **2021**, *23*, 637–642. [CrossRef]

92. MitoScore: Mitochondrial Biomarker Developed by Igenomix Will Increase Implantation Rate of In Vitro Fertilization. Available online: https://carlos-simon.com/mitoscore-mitochondrial-biomarker-developed-by-igenomix-will-increase-implantation-rate-of-in-vitro-fertilization/ (accessed on 10 January 2022).
93. Fu, L.; Luo, Y.X.; Liu, Y.; Liu, H.; Li, H.Z.; Yu, Y. Potential of Mitochondrial Genome Editing for Human Fertility Health. *Front. Genet.* **2021**, *12*, 673951. [CrossRef] [PubMed]
94. Yahata, N.; Matsumoto, Y.; Omi, M.; Yamamoto, N.; Hata, R. TALEN-mediated shift of mitochondrial DNA heteroplasmy in MELAS-iPSCs with m.13513G>A mutation. *Sci. Rep.* **2017**, *7*, 15557. [CrossRef]
95. Yang, Y.; Wu, H.; Kang, X.; Liang, Y.; Lan, T.; Li, T.; Tan, T.; Peng, J.; Zhang, Q.; An, G.; et al. Targeted elimination of mutant mitochondrial DNA in MELAS-iPSCs by mitoTALENs. *Protein Cell* **2018**, *9*, 283–297. [CrossRef]
96. Wang, B.; Lv, X.; Wang, Y.; Wang, Z.; Liu, Q.; Lu, B.; Liu, Y.; Gu, F. CRISPR/Cas9-mediated mutagenesis at microhomologous regions of human mitochondrial genome. *Sci. China. Life Sci.* **2021**, *64*, 1463–1472. [CrossRef] [PubMed]
97. Craven, L.; Tang, M.X.; Gorman, G.S.; De Sutter, P.; Heindryckx, B. Novel reproductive technologies to prevent mitochondrial disease. *Hum. Reprod. Update* **2017**, *23*, 501–519. [CrossRef]
98. Reznichenko, A.S.; Huyser, C.; Pepper, M.S. Mitochondrial transfer: Implications for assisted reproductive technologies. *Appl. Transl. Genomics* **2016**, *11*, 40–47. [CrossRef]
99. Kristensen, S.G.; Humaidan, P.; Coetzee, K. Mitochondria and reproduction: Possibilities for testing and treatment. *Panminerva Med.* **2019**, 82–96. [CrossRef]
100. Mobarak, H.; Heidarpour, M.; Tsai, P.S.J.; Rezabakhsh, A.; Rahbarghazi, R.; Nouri, M.; Mahdipour, M. Autologous mitochondrial microinjection; A strategy to improve the oocyte quality and subsequent reproductive outcome during aging. *Cell Biosci.* **2019**, *9*, 1–15. [CrossRef] [PubMed]
101. Kristensen, S.G.; Pors, S.E.; Andersen, C.Y. Improving oocyte quality by transfer of autologous mitochondria from fully grown oocytes. *Hum. Reprod.* **2017**, *32*, 725–732. [CrossRef]
102. Jiang, Z.; Shen, H. Mitochondria: Emerging therapeutic strategies for oocyte rescue. *Reprod. Sci.* **2021**, *29*, 711–722. [CrossRef] [PubMed]
103. Cohen, J.; Scott, R.; Schimmel, T.; Levron, J.; Willadsen, S. Birth of infant after transfer of anucleate donor oocyte cytoplasm into recipient eggs. *Lancet (London, England)* **1997**, *350*, 186–187. [CrossRef]
104. Cohen, J.; Alikani, M.; Garrisi, J.G.; Willadsen, S. Micromanipulation of human gametes and embryos: Ooplasmic donation at fertilization VIDEO. *Hum. Reprod. Update* **1998**, *4*, 195–196. [CrossRef]
105. Cohen, J.; Scott, R.; Alikani, M.; Schimmel, T.; Munné, S.; Levron, J.; Wu, L.; Brenner, C.; Warner, C.; Willadsen, S. Ooplasmic transfer in mature human oocytes. *Mol. Hum. Reprod.* **1998**, *4*, 269–280. [CrossRef]
106. Woods, D.C.; Tilly, J.L. Autologous Germline Mitochondrial Energy Transfer (AUGMENT) in Human Assisted Reproduction. *Semin. Reprod. Med.* **2015**, *33*, 410–421. [CrossRef]
107. Woods, D.C.; Tilly, J.L. Isolation, characterization and propagation of mitotically active germ cells from adult mouse and human ovaries. *Nat. Protoc.* **2013**, *8*, 966–988. [CrossRef] [PubMed]
108. Lee, K. An ethical and legal analysis of ovascience – A publicly traded fertility company and its lead product AUGMENT. *Am. J. Law Med.* **2018**, *44*, 508–528. [CrossRef]
109. Powell, K. Born or made? Debate on mouse eggs reignites. *Nature* **2006**, *441*, 795. [CrossRef] [PubMed]
110. Grieve, K.M.; McLaughlin, M.; Dunlop, C.E.; Telfer, E.E.; Anderson, R.A. The controversial existence and functional potential of oogonial stem cells. *Maturitas* **2015**, *82*, 278–281. [CrossRef] [PubMed]
111. Gosden, R.G.; Johnson, M.H. Can oocyte quality be augmented? *Reprod. Biomed. Online* **2016**, *32*, 551–555. [CrossRef]
112. Troubled OvaScience cuts half of workforce. Available online: https://www.biopharmadive.com/news/troubled-ovascience-cuts-half-of-workforce/514160/ (accessed on 10 January 2022).
113. OvaScience Sued by Shareholders for Over-Hyped Fertility Treatment. Available online: https://www.courthousenews.com/ovascience-sued-by-shareholders-for-over-hyped-fertility-treatment/ (accessed on 10 January 2022).
114. Clinical Application of Autologous Mitochondria Transplantation for Improving Oocyte Quality. Available online: https://clinicaltrials.gov/ct2/show/NCT03639506?term=Mitochondrial+transfer&draw=2&rank=7 (accessed on 10 January 2022).

 *cells*

Review

Ultrastructural Evaluation of the Human Oocyte at the Germinal Vesicle Stage during the Application of Assisted Reproductive Technologies

Maria Grazia Palmerini [1,*,†], Sevastiani Antonouli [1,†], Guido Macchiarelli [1], Sandra Cecconi [1], Serena Bianchi [1], Mohammad Ali Khalili [2] and Stefania Annarita Nottola [3]

[1] Department of Life, Health and Environmental Sciences, University of L'Aquila, 67100 L'Aquila, Italy; arella_935@hotmail.com (S.A.); gmacchiarelli@univaq.it (G.M.); sandra.cecconi@univaq.it (S.C.); serena.bianchi@univaq.it (S.B.)
[2] Department of Reproductive Biology, Yazd Institute for Reproductive Sciences, Shahid Sadoughi University of Medical Sciences, Yazd 97514, Iran; khalili59@hotmail.com
[3] Department of Anatomical, Histological, Forensic Medicine and Orthopedic Science, La Sapienza University, 00161 Rome, Italy; stefania.nottola@uniroma1.it
* Correspondence: mariagrazia.palmerini@univaq.it
† These authors contributed equally to this work.

Abstract: After its discovery in 1825 by the physiologist J.E. Purkinje, the human germinal vesicle (GV) attracted the interest of scientists. Discarded after laparotomy or laparoscopic ovum pick up from the pool of retrieved mature oocytes, the leftover GV was mainly used for research purposes. After the discovery of Assisted Reproductive Technologies (ARTs) such as *in vitro* maturation (IVM), *in vitro* fertilization and embryo transfer (IVF-ET) and intracytoplasmic sperm injection (ICSI), its developing potential was explored, and recognized as an important source of germ cells, especially in the case of scarce availability of mature oocytes for pathological/clinical conditions or in the case of previous recurrent implantation failure. We here review the ultrastructural data available on GV-stage human oocytes and their application to ARTs.

Keywords: oocyte; germinal vesicle; human; electron microscopy; assisted reproductive technologies (ARTs)

Citation: Palmerini, M.G.; Antonouli, S.; Macchiarelli, G.; Cecconi, S.; Bianchi, S.; Khalili, M.A.; Nottola, S.A. Ultrastructural Evaluation of the Human Oocyte at the Germinal Vesicle Stage during the Application of Assisted Reproductive Technologies. *Cells* **2022**, *11*, 1636. https://doi.org/10.3390/cells11101636

Academic Editor: Lon J. van Winkle

Received: 20 April 2022
Accepted: 12 May 2022
Published: 13 May 2022

Publisher's Note: MDPI stays neutral with regard to jurisdictional claims in published maps and institutional affiliations.

1. Introduction

1.1. From the Ovum to the Cicatricula

Questions about the beginning of life have been raised by philosophers and scientists since the fourth century BC with Aristotle, who described embryonic development inside a chicken egg [1]. However, the discovery of the oocytes started in the 19th century: in 1651, the physician William Harvey employed the Latin word *"ovum"* to refer to the beginning of animal life [2], and with his study he introduced the theory *"omnia ab ovo"*, i.e., all animals are produced from ova. However, when he dissected a red deer during the rut, he could not find any visible evidence in the female *"testes"*, so-called at that time for the production of semen [3]. For three centuries (17th–19th), other researchers observed the reproductive organs in mammals following the "egg theory", trying to identify the ovum and the place of its generation [4]. Among them, the Dutch physician and anatomist Regner de Graaf (1672) mistakenly believed that the egg consisted of the ovarian follicle itself [5], but correctly discovered that a "seminal vapour" was responsible for fertilization after having reached the eggs. The Italian priest, biologist and physiologist Lazzaro Spallanzani further contrasted the theory of spontaneous generation (1765) [6] by demonstrating the importance of the contact between the sperm and the egg. However, he believed in the theory of preformed organs. The egg description was provided in 1824 by the Swiss physiologist Jean-Luis Prevost and the French chemist Jean-Baptiste Dumas, who wrote that, as long as the egg is in the ovary, it contains a brown zone within which a circular yellow spot can be seen [7]. This again surrounds a smaller circular outline, to which Prévost and Dumas gave the name, which it still bears, of "cicatricula". By putting a frog fertilized egg into the water they could follow the first phases of segmentation, describing a "granulation" ending in the formation of a structure similar to a "raspberry", including the presence of "black spots", probably the cell nuclei [8].

1.2. The Identification of the Germinal Vesicle

The germinal vesicle was identified in 1825 by the Bohemian physiologist Jan Evangelista Purkinje, who, a few years later, published at the University of Breslaw a treaty where he described the presence and consistency of a vesicular structure found in the hen's egg [9]. Within each oocyte or "germ", Purkinje saw a transparent, liquid-filled sphere, which he named the *"vesicula germinativa"* (germinal vesicle), being considered the germ cell. Two years later, the Prussian-Estonian embryologist Ernst Karl von Baer in Leipzig identified the egg cells in mammals [10]. Moreover, he described the process of ovulation and the initial stages of embryo development in the dog. Baer closely examined follicles under a microscope and saw that each one had a small "yellowish-white point" inside. Microscopic examination of the structure observed enabled him to discover that the ovum was hidden within the *"discus proligerus"*, now known as the cumulus oophorus. He removed one of the yellowish structures with the point of a knife and looked at it again under the microscope, revealing the oocyte. The idea that tissues were made up of cells was not yet known, and he called it the *"ovulum"* or "little egg" [11]. The cellular theory by Matthias Jacob Schleiden and Theodor Schwann in 1839 complemented von Baer's discovery of the mammalian ovum and endorsed the idea that all reproductive processes pass through a cell or comparable, fundamental organic unit [12]. Theodor Schwann argued that the ovum described by von Bear was probably a cell, and the structure named *"vesicula germinativa"* by Purkinje was the cell nucleus, as also confirmed by Vladislav Kruta (1956) [13]. In 1834, Adolph Bernhardt, a student and pupil of Purkinje and later one of the founding fathers of modern histology, observed in the mammalian ovum an analogous structure to the germinal vesicle, finally identifying the cell nucleus [14,15].

In 1835, Rudolph Wagner, while studying the Graafian follicle of the sheep, discovered the presence of a "spot" within the germinal vesicle, which he called *"macula germinativa"* (germinal spot). Wagner assumed that this "spot" was the origin or first stage in the development of the germinal vesicle. In 1839, Gabriel Gustav Valentin, Purkinje's close collaborator, was the one who confirmed and introduced the term nucleolus for the *macula germinativa*, according to Purkinje. Later, Valentin, based on Purkinje's and Wagner's descriptions, conducted his observations and referred to the nucleolus as a "rounded, transparent secondary nucleus" [14].

The long and conflictual history, as described in the 19th century, of the fully grown GV-stage oocyte, discovered a unique cell owning to its size (nearly 0.1 mm or 100 μm diameter), enclosed by a clear "glass-like shell" or zona pellucida provided with a large nucleus—or germinal vesicle—and containing a prominent spherical nucleolus [16].

1.3. GV-Stage Oocytes as a Potential Resource in ARTs

ARTs are now routinely applied for female fertility, also connected to delayed childbearing in western countries, and oncofertility problems. The use of GV-stage oocytes gains an option when the availability of mature oocytes is reduced. However, contrasting data about the right approach for using controlled ovarian stimulation (COS) protocols, IVM conditions and cryopreservation strategies limit the potentiality of this approach.

Data from morphological studies may be helpful to untangle the knot. A morpho-functional approach by confocal microscopy detected the presence of double-strand DNA breaks and the diverse activation of a DNA repair response in human GV-stage oocytes from COS or IVM cycles [17]. However, when immature oocytes collected from stimulated ovaries were subjected to vitrification, morpho-functional data evidenced that they survived the vitrification process and retained their potential to mature to the MII stage [18].

It should be noted that alterations to mitochondria redistribution, CG migration, global DNA methylation expression and, especially, spindle and chromosome organization were reported [18]. By contrast, a recent study of the metaphase plate by confocal microscopy revealed that immature human oocytes subjected first to vitrification and then to *in vitro* maturation up to the MII-stage showed a higher percentage of normal metaphase spindle configuration if compared to those matured *in vitro* first and then vitrified [19]. These discrepancies may be solved with the aid of electron microscopy (EM), the gold standard in revealing the ultrastructural characteristics of immature oocytes.

In this review, we aimed to revise available literature on immature oocytes, putting a particular emphasis on (1) the application of electron microscopy (EM) as an essential tool in ultrastructural studies to optimize and validate ARTs in humans and (2) the use of GV-stage oocytes as a useful source of germ cells in ARTs.

2. Materials and Methods

Inclusion and Exclusion Criteria for the Research in MEDLINE–PubMed, Scopus and ISI Web of Science Databases

All original articles were searched from international databases, including PubMed, Scopus and Web of Science. We searched these engines without language or time limitations. The search was performed using eight keywords in English, including "Immature oocytes, Germinal vesicle (GV), morphology, ultrastructure, electron microscopy, Assisted Reproductive Technology (ART), human". For the ultrastructural description of subcellular structures, the search was enriched by the following terms: nucleus, nuclear membrane, nucleolus, heterochromatin and euchromatin, cortical granules, endoplasmic reticulum, Golgi apparatus, mitochondria, vacuoles, ooplasm, microvilli, perivitelline space, zona pellucida. In this study, the inclusion criteria were: (1) morphological study; (2) the status of oocyte immaturity; (3) the use of immature oocytes for ARTs.

3. Ultrastructure of GV-Stage Oocytes

EM observations on human GV-stage oocytes are important for characterizing their ultrastructure, making an essential contribution in highlighting effects connected to patients' pathologies, or the detrimental action connected to ARTs as COS or cryopreservation. EM is, therefore, useful in the optimization and validation of ART protocols. Focusing on the application of EM on immature human oocytes from different ARTs we here review data available in the literature:

3.1. Nucleus

In immature oocytes aspirated from stimulated ovaries, the germinal vesicle, which was usually spherical and with a diameter of about 20 μm, was located centrally or, in a few oocytes, close to the cell membrane [20]. Similarly, in oocytes collected from patients enrolled in ICSI treatment cycles after COS, the nucleus normally occupied 1.2% of cell volume with a central position in the inner cytoplasm [21]. The nuclear position was, differently, found to be eccentrical in several studies, probably due to an early onset of meiosis resumption after harvesting or COS protocol timing. An eccentrical position of the nucleus in GV-stage human oocytes was described by LM and TEM in both fresh and vitrified/warmed oocytes collected from our group, after a long protocol [22] or by others in IVM cycles from stimulated patients [23–26]. In all cases, the nuclear membrane was continuous, well visible, two-layered and characterized by the presence of several nuclear pores. The nuclear envelope shape was roundish or folded and enclosed a nucleoplasm with a uniform and pale aspect of finely dispersed euchromatin fibrils, except for the presence of one or more large and electron-dense nucleoli and clusters of heterochromatin. The latter was connected to the nucleoli or organized in patches, sometimes located under the nuclear membrane [20,22,27,28]. Round and highly electron-dense nucleoli were composed almost entirely of fibrillar material; smaller nucleoli were evident in addition to the most evident nucleolus [25]. Dense spherical bodies could be present within their matrices [25]. An unpublished micrograph of the nuclear ultrastructure from a representative fully stimulated human leftover GV-stage oocytes is shown in Figure 1A.

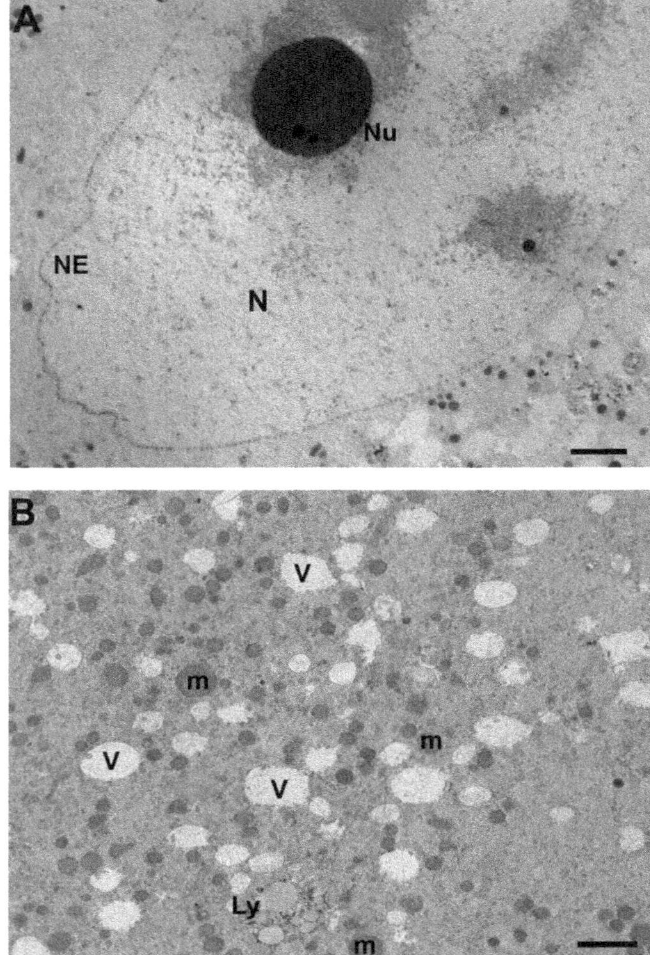

Figure 1. Representative unpublished archived TEM micrographs of GV-stage human oocytes from stimulated ovaries [22,29,30] showing in (**A**) the nucleus (N) delimited by a continuous nuclear envelope (NE) made by a double layer interrupted by electron pale nuclear pores. The nucleoplasm is characterized by a round and electron-dense nucleolus (Nu), dark fibrous clusters of heterochromatin and dispersed euchromatin. TEM, bar: 1 μm; (**B**) the ooplasm rich in electron-lucent vacuoles and small vesicles, delimited by a continuous membrane and sometimes presenting visible inclusions or debris. Numerous round-to-ovoid mitochondria (m) with slightly electron-dense cristae and big secondary lysosome (Ly) with numerous vesicles of different electron-density. TEM, bar: 1 μm.

3.2. *Ooplasm*

3.2.1. Mitochondria

GV-stage human oocytes recovered from ovaries from unstimulated patients for further IVM showed that most of the mitochondria were spherical and had sparse cristae [23]. Following ovarian stimulation, they were evenly distributed, although clusters were occasionally seen peripherally [20]. When GV-stage oocytes were collected from PCOS (polycystic ovarian syndrome) patients (no data available about the COS protocol, if any), mitochondria—fewer than at metaphase I—appeared small in size, mostly round but sometimes tubular [31]. In immature oocytes from COS cycles with GnRH (Gonadotropin-Releasing Hormones) agonists and antagonists, most mitochondria localized centrally in the cytoplasm of oocytes with a cluster-like structure; no significant amount of mitochondria was found to localize in their subcortical areas. In fact, the area ratio of "*mitochondria in cytoplasm at peripheral region*" was 0.98%, while that of "*mitochondria in cytoplasm at perinuclear region*" was 7.71% [32]. A similar perinuclear distribution of individual mitochondria and small mitochondrial clusters was found in human leftover GV-stage oocytes from young donors enrolled in an egg donation program [26]. A representative micrograph from our group of mitochondria from a fully stimulated GV-stage human oocyte is shown in Figure 1B.

3.2.2. Mitochondria-Smooth Endoplasmic Reticulum (M-SER) Aggregates

There was a close association between the mitochondria and the smooth endoplasmic reticulum (SER), which appeared as small irregular profiles or sometimes as flattened sacs with a size of about 0.5 to 1 μm [20]. We previously noted the presence of underdeveloped M-SER aggregates in human oocytes after full stimulation [22], while others noted the gradual affiliation of the mitochondria to a smooth membrane of variable ER elements only in MI-stage oocytes [26].

3.2.3. Golgi Apparatus and Smooth Endoplasmic Reticulum (ER)

The Golgi apparatus consisted of aggregates of tubuli and vesicles in human immature oocytes after retrieval from unstimulated patients [23]. This was confirmed by further studies from the group of Sathananthan on immature GV oocytes collected from large antral follicles at laparoscopy from stimulated patients, who described Golgi complexes as the most prominent organelles in the oocytes during the early phases of maturation. They were numerous, hypertrophied and appeared predominantly in the cortical ooplasm, where they were associated with cortical granules [25]. More recently, distinct Golgi apparatus, composed of an interconnected system of cisternae and vesicles, was regularly found in the cytoplasm of leftover human GV oocytes [26]. Interestingly, the cis-trans polarity was not easily distinguishable [26]. Extensive endoplasmic reticulum was visible in human GV oocytes [28], with the prevalence of vesicular smooth ER, as seen in PCOS patients [31].

3.2.4. Cortical Granules (CGs)

The cytoplasm of immature human oocytes collected from ovarian-stimulated women undergoing ARTs for tubal factor infertility contained membrane-bound CGs, with a diameter of 0.3 to 0.5 μm, in varying numbers located close to the plasma membrane and deep within the cytoplasm [20]. By contrast, GV-stage human oocytes obtained from very early antral follicles in unstimulated ovaries presented CGs in the subplasmalemma and cortical cytoplasm of fully grown, immature oocytes derived from very early antral follicles [33]. The group of Gethler [34] identified dark and light subpopulations of CGs in leftover human GV-stage oocytes collected after a "long" protocol (GnRH analogue and human menopausal gonadotrophin, hMG) for ovulation induction. Many clusters of light granules were distributed throughout the cytoplasm, with a few granules (mainly dark) at the cortex. Cryopreservation by slow-freezing induced a decrease in the dark granule's abundance, with membrane-coated electron-transparent vesicles, in some cases aggregated with CGs [34]. We, similarly, found scattered in the ooplasm CGs in both fresh and vitrified-warmed immature leftover GV-stage oocytes, with a higher counterpart of dark granules (indicative of a better maturity), probably due to the different protocol for cryopreservation [22]. In immature oocytes from PCOS patients, CGs were dispersed throughout the whole cytoplasm, without being prominent [31].

3.2.5. Vacuoles

Some vacuoles with a diameter of 1–2 μm were observed in the center of the oocytes [20]. Numerous membrane-bounded vacuoles, small and empty (medium diameter ± SD: 0.975 ± 0.113 μm), were found mainly in the region surrounding the nucleus of human leftover GV-stage oocytes after full stimulation [22]. Similarly, vacuolar structures were very abundant and with a variable ultrastructure in leftover human GV oocytes after superovulation [35]. When GV oocytes were subjected to vitrification by Cryotop, vacuoles were found also in the oocyte periphery. Numerous small MV complexes were identified in the cortex and subcortex; their number decreased in vitrified-warmed oocytes [22]. A centralized vacuolated ooplasm was connected to atresia [24]. A representative micrograph from our group of vacuoles and smaller vesicles from a fully stimulated human oocyte is shown in Figure 1B.

3.2.6. Lysosomes

One author reports that primary and secondary lysosomes were not present in GV oocytes [31]; by contrast, secondary and tertiary lysosomes were commonly found in GV oocytes from stimulated cycles, and bizarre images were considered a sign of atresia [24]. The lysosomes were also found in the cytoplasm of all analyzed leftover human GV from young stimulated donors [26]. They varied in size (0.2–1.5 μm) and were described as membrane-bound bodies containing aggregated vesicles with variable opacity, membrane remnants, or dark, dense patches (probably insoluble lipid droplets), residues of lysosomal degradation [26]. Lysosomes were also noticed in preovulatory human oocytes maturing *in vitro* at the metaphase I stage from stimulated patients [25]. A representative archived micrograph of lysosomes from fully stimulated human oocytes is shown in Figure 1B.

3.3. *Oolemma*

Microvilli, Perivitelline Space (PVS) and Zona Pellucida (ZP) Texture

Numerous fine microvilli were observed closely on the oocyte surface after enzymatic removal of the zona pellucida in GV oocytes from unstimulated patients. Twenty-eight hours of culture determined an extension from the surface of the plasma membrane into the ZP [23]. Numerous slender microvilli, 4–5 μm long, extended from the surface of the cells [20]. A continuous layer of thin and long microvilli covered the oocyte surface in leftover human GV-stage oocytes after full stimulation; by contrast, after vitrification, an irregular coverage of small and short microvilli was noted in the 40% of observed oocytes [22]. Moreover, when GV-stage oocytes were collected from a PCOS patient after COS, microvilli were rare and small [28]. In GV-stage oocytes collected from a PCOS patient, the PVS was very narrow, the outlines of the ZP were sharp and the ZP was less dense in the exterior and denser in the interior [28].

Immature oocytes retrieved from an *in vitro* fertilization program showed, by SEM, a smooth and compact surface of the zona pellucida, different from the spongy appearance in mature oocytes [36]. Human GV oocytes show corona radiata processes extending through the ZP, some ending in bulbous or club-shaped terminals containing microfilaments, granular vesicles, and a few lysosomes [25]. A representative micrograph of the cortical ooplasm, oolemma, microvilli, PVS and ZP from fully stimulated human oocytes is shown in Figure 2.

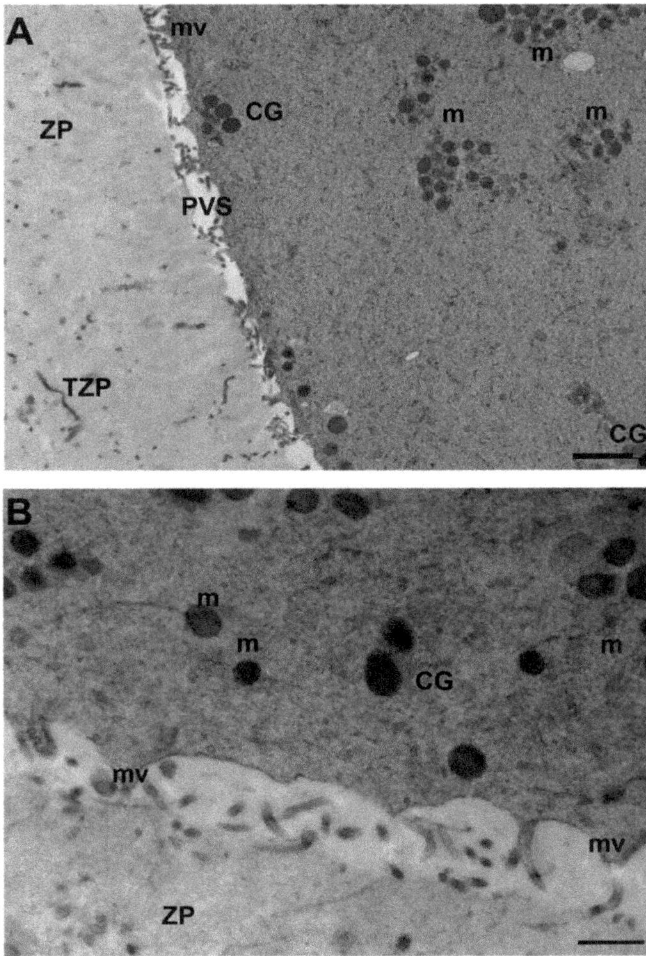

Figure 2. Representative unpublished archived TEM micrographs of GV-stage human oocytes from stimulated ovaries [22,29,30] showing in (**A**) the cortex with a pale and spongy zona pellucida (ZP) crossed by numerous residues of transzonal projections (TZP) and a thinner perivitelline space (PVS) in which numerous thin microvilli (mv) protruding from the oolemma are visible. The ooplasm is characterized by the rare presence of suboolemmal cortical granules (CG) and numerous mitochondria (m), which are often present in clusters in the inner ooplasm. TEM, bar: 1 μm; (**B**) higher magnification of the cortical ooplasm with round-to-ovoid mitochondria (m), slightly electron-dense cristae and rare cortical granules (CG). The oolemma is folded in numerous microvilli (mv) protruding in the perivitelline space. ZP: zona pellucida. TEM, bar: 800 nm.

4. Oocyte Quality and Early Embryo Development from ART Cycles

After ovulation triggering through hCG administration, the oocyte retrieval results in the collection of both mature MII-stage oocytes (~70–85%) and immature GV-/MI-stage oocytes (up to 30%) [37–39]. The number of immature oocytes, when high after transvaginal retrieval, is generally considered a marker of poor oocyte quality [40,41], and can increase exponentially in PCOS patients [42,43]. During the next steps of IVF, the oocyte quality covers a pivotal role in determining embryonic development and in increasing the chances of a successful clinical pregnancy. The definition of oocyte quality is dependent on its potential and ability to undergo meiotic maturation and fertilization process, to achieve the proper embryonic development and to further clinical pregnancy [44]. The quality of human GV-stage oocytes can be evaluated by morphological markers such as the surrounded nucleolus (SN) and non-SN, which identify high and low developmental potential cells and consequent IVM outcomes [45]. Literature data, although limited, provide the scientific and medical community with information about the importance of GV-oocyte quality utilized in IVF programs. Escrich and collaborators [46] identified two oocyte populations by real time-lapse recording, defined on the early or late timings of maturation (E-IVM oocytes: less than 20 h; and L-IVM oocytes: up to 30 h, respectively), which were mainly affected by the hours needed for the GVBD. While E-IVM oocytes reached the activation rate and pronucleus (PN) formation in a percentage comparable to *in vivo* matured oocytes, the L-IVM group showed abnormal patterns in chromatin segregation that impair normal PN formation and, thus, fertilization rates and embryo formation/development [46]. The importance of maturation timings was confirmed by Yang and collaborators, who additionally highlighted the role of serum progesterone levels, as well as GV-MI maturation, for a better developmental potential of IVM oocytes [47]. Furthermore, the quality of immature oocytes before and during IVM is pivotal for increasing the chances of success in women with low functional ovarian reserve. The application of a correct maturation of these oocytes can add the +60% of the total embryos available for transfer, providing an improvement in pregnancy and delivery chances [48]. On the other hand, a non-randomized prospective trial [49] suggested double ovarian stimulation (DuoStim) as an alternative solution to the rescue IVM of oocytes retrieved from ovarian reserve poor-prognosis women. The application of DuoStim resulted in higher efficiency in terms of oocyte competence, embryo viability and quality [49]. Due to the reasons here discussed, the use of GV-stage oocytes is still limited in routine clinical practice; thus, more studies are needed on this topic to provide a deeper knowledge of its potential applications in ARTs.

5. Unstimulated Vs. Stimulated GV-Stage Oocytes

The oocyte retrieval in natural cycles has been the main source of female germ cells since the beginning of IVF history [50,51]. This procedure consists of the transvaginal aspiration of oocytes from mature follicles and, to date, is mainly applied in PCOS patients and poor responders, or highly recommended to couples in which the woman is under the age of 40 and there is no male factor infertility. Nowadays, women undergoing ARTs are subjected to COS, which allows the maturation and ovulation of multiple follicles through the injection of exogenous gonadotropins [40]. Both unstimulated and stimulated cycles display a plethora of advantages and disadvantages for the successful retrieval of mature oocytes. Regarding natural unstimulated cycles, the advantages reside in decreasing the risk of ovarian hyperstimulation syndrome (OHSS), costs, and in avoiding the use of hormones for ovarian stimulation; moreover, IVF cycles may be frequently repeated, thus reducing the perinatal adverse outcomes and side effects of endometrial functions [52–54]. However, the efficiency of unstimulated cycles is counterbalanced by a lower number of available oocytes and, therefore, of embryos to be transferred, together with decreased pregnancy rates per cycle and increased cancellation rate (more than 30%) due to premature ovulation [55]. On the other hand, the advantages of using COS in IVF cycles include an increased number of oocytes available during follicular aspiration, and thus a greater number of embryos to be transferred to the uterine cavity [56]. Nevertheless, the administration of exogenous hormones, especially at high doses or in the case of repeated IVF cycles, can decrease endometrial receptivity, increase the risk of OHSS breast cancer and is associated with higher costs per cycle [57]. In both unstimulated and stimulated cycles, a high number of immature oocytes from secondary follicles are frequently retrieved and may undergo IVM [58]. This may increase pregnancy rates, although the percentages of successful full-term pregnancy and live birth rates from immature oocytes after IVM are still very low [48]. Particularly, a more numerous retrieval of immature oocytes may be correlated with the COS protocol utilized, i.e., when Follicle Stimulating Hormone (FSH), GnRH antagonist and short GnRH agonist protocols are preferred to FSH/LH, GnRH agonist and long GnRH agonist protocols, the former characterized by the administration of GnRH agonist on cycle-day 21 followed by gonadotropin from cycle-day 2 to 14 [59–61]. Therefore, GV-stages oocytes from both unstimulated and stimulated follicles have special importance for those patients with a history of unsuccessful treatments, representing an alternative solution for a tailored ART optimization in a patient-friendly and -specific manner. In this regard, IVM and oocyte cryopreservation may be applied, even if the potentiality of using immature oocytes is not yet assessed.

6. *In Vitro* Maturation (IVM)

Among the relevant ARTs, oocyte-cumulus complexes can be retrieved after the aspiration of mid-sized follicles and subsequently cultured for 24–48 h for undergoing the process of IVM. Moreover, GV-stage oocytes that did not undergo the transition to MII-stage after stimulation can be found in the pool of retrieved oocytes and subjected to IVM [62]. The IVM of these "leftover" immature (GV-stage) oocytes with developmental competence represents a potential tool for ART patients, in particular for poor responders, PCOS women, and for avoiding the side effects of OHSS [29,63,64]. While discarded in the past, today GV oocytes can be cryopreserved before or after IVM, thus increasing the opportunity for achieving a pregnancy [19,29]. However, the successful application of IVM on leftover oocytes is still facing issues, mainly due to defects in the cytoplasmic maturation or asynchronous cytoplasmic and nuclear maturation [30,46,65]. Additionally, the delayed timings of fertilization after MII arrest and meiotic spindle assembly significantly impair embryo developmental potential [30,66]. Notably, the IVM rates of GV-stage oocytes from unstimulated and stimulated cycles highlight differences that seem to be strictly associated with culture media, age, infertility factor and the presence of cumulus cells [67].

6.1. IVM of Unstimulated GV-Stage Oocytes

The first live birth after IVM resulted from oocytes retrieved from an unstimulated woman affected by PCOS and was reported in 1994 by Trounson and collaborators [68]. The IVM process is similar in stimulated and unstimulated cycles, although in the latter it appears to be less efficient, due to a lower number of retrieved oocytes and a decrease in both cleavage and pregnancy rates [69,70]. However, pilot studies highlighted differences between the IVM of stimulated and unstimulated oocytes, as structural and molecular changes [71], and a protein synthesis profile that leads to different germinal vesicle break-down (GVBD) time courses [72]. IVM of immature oocytes from unstimulated cycles evidenced smaller diameters after denudation (96–125 μm) and decreased *in vitro* competence than in stimulated cycles [73,74]. As a consequence, hCG priming before oocyte retrieval in control and PCO/PCOS patients increased oocyte maturation and fertilization rates, developmental potential and clinical pregnancy rates, when compared to unstimulated cycles [75,76]. Liu and collaborators scored GV- and MI-stage oocytes from unstimulated PCOS patients according to nuclear maturation, cumulus cells-oocyte shape and morphology. Maturation, fertilization and cleavage rates were similar between the two groups, while it is worth noting that the number of retrieved GV-stage oocytes showed a threefold increase in comparison with MI-stage oocytes [77]. This reinforces the need to better optimize ART settings for GV-stage oocytes.

6.2. Rescue IVM from Stimulated Oocytes

Nowadays, new strategies are still needed for low functional ovarian reserve patients with abnormally elevated FSH, together with abnormally low age-specific anti-mullerian hormone (AMH) levels [78], and for poor responder patients [64]. One of the clinical issues of these women is poor oocyte quantity and quality after retrieval. Approximately, up to 30% of the retrieved oocytes are immature (GV- or MI-stage) and need to be matured *in vitro* to increase the possibilities of a successful IVF cycle [79]. Oocytes that matured on Day 1 after retrieval were significantly higher in quality than those that matured later on [80]. Hormonal supplementation with key hormones as Epidermal Growth Factor (EGF) or EGF-like growth factors such as amphiregulin (AREG) and epiregulin (EREG), improved meiotic resumption and nuclear maturation [81]. Notably, a significantly higher percentage of GV-stage oocytes that matured, as evidenced by PBI extrusion, were detected when GVBD occurred in the central nucleolus rather than in the peripheral nucleolus [82]. As long as this procedure is considered a valuable tool for increasing the chances of successful fertilization and embryo development, it is worth mentioning that rescue IVM of GV oocytes could negatively influence embryo morphokinetics, with particular regard to the timings of the overall embryo formation and development and cleavage patterns [83].

7. Cryopreservation

One of the major debates regarding fertility preservation concerns the cryopreservation of GV-stage oocytes before or after IVM when they will reach the MII-stage. In 1998, Tucker and collaborators published the first birth from GV-stage oocytes that were first cryopreserved and, after thawing, matured *in vitro* [83]. After ICSI and the achievement of the cleavage stage, the newly formed embryos underwent embryo transfer and resulted in a live birth of a female infant [84]. The role of cumulus cells during immature oocyte cryopreservation is still unclear [85–87]. The presence of somatic cells during oocyte cryopreservation may exert a protective role against anisotonic conditions and stress derived from osmotic changes, mainly induced by the rapid influx and efflux of cryoprotectants during pre-freeze and post-thawing periods [88]. Among the different procedures of cryopreservation, the two methods mostly studied and applied in the clinical daily processes are the slow freezing and vitrification methods, further discussed in this section.

7.1. Slow Freezing

The slow-freezing method on GV-stage oocytes is an alternative strategy of cryopreservation aimed at avoiding spindle depolymerization, decreasing the risk of polyploidy and/or aneuploidies. The reason relies on the diffuse chromatin status at this stage of oocyte maturation and the presence of the nuclear membrane surrounding and protecting the DNA [89]. This method improves oocyte survival rates, and does not, concomitantly, increase the rates of maturation and fertilization or embryonic development [90,91]. The advent of vitrification, quick, economic and safe, progressively reduced the use of slow freezing, more subject to the formation of ice crystals in the intra- and extracellular spaces [92]. The warming/thawing process countersigned the main significant difference between slow freezing and vitrification, with rapid warming being more efficient in terms of survival rates than rapid thawing [93]. However, a study of GV and MII oocytes subjected to the two different cryopreservation methodologies found no difference in terms of mitochondrial distribution and viability rates [94].

7.2. Vitrification

Among cryopreservation protocols, vitrification is the gold standard for the oocyte and embryo "freeze-all" strategy [95]. It can be performed by using either open or closed systems, which are in direct and indirect contact with liquid nitrogen, respectively [96,97]. Regarding slow freezing, the use of vitrification achieved better oocyte survival and higher pregnancy rates [98,99]. Oocyte maturation rates were reduced in GV-stage oocytes subjected to IVM after cryopreservation in comparison with fresh *in vitro* matured oocytes [100]. Nevertheless, considering the statistics regarding healthy live births, the literature data demonstrate that vitrification of oocytes subjected to IVM is still less effective compared to its *in vivo* ovulated counterpart [29]. Due to the use of high concentrations of cryoprotectants for oocyte vitrification and the high cytotoxicity of these chemicals, linked to osmotic changes [101], the application of vitrification and the following clinical efficiency needs to be further investigated and be strengthened with multiple controlled clinical trials [102].

8. Mitochondrial Replacement

One of the latest ARTs is mitochondrial replacement therapy (MRT). The main function of mitochondria in eukaryotic cells is the production of ATP through oxidative phosphorylation. Being the only semi-autonomous organelle of the cell, the mitochondrion carries up to 10 copies of mitochondrial DNA (mtDNA), which represent around 0.1% of the genome present in a cell [103]. The mtDNA mutations, which can be inherited only from the maternal counterpart, may cause severely debilitating syndromes in newborns, or lead to abortion during pregnancy [104,105]. Besides the clinical difficulties in the MRT, it faces theological and ethical issues because of the involvement of a mitochondrial donor in embryo formation [106]. Thus, the newborn will have three distinct genetic materials: the paternal through the spermatozoon, the maternal through the nuclear DNA of the oocyte, and the donor through the healthy cytoplasmic mtDNA, ultimately generating a "three-parent baby" [107]. To date, several techniques are eligible for MRT, which consists preferably of the removal of the nuclear genome from the oocyte or zygote carrying the mtDNA mutation and transferring to a healthy enucleated oocyte [108,109]. Among the MRT techniques, ooplasm transfer (OT), spindle transfer (ST), pronuclear transfer (PNT) and polar body transfer (PBT) are the most widely known and used in ART procedures [107,110]. The OT technique consists of the transfer of 5–15% of the ooplasm of a healthy donated oocyte to the oocyte-carrying mutated mtDNA and is one of the most controversial among MRTs [111]. However, OT treatment resulted in 50 clinical pregnancies and more than 30 live births [109,112]. The transfer of the pathological maternal MII-stage cocytes spindle to a healthy donor's ooplasm previously deprived of its spindle has been applied to human oocytes by Tachibana and collaborators in 2013 [113]. The first clinical pregnancy leading to a live birth using ST was reported in 2017 [114]. Interestingly, germinal vesicle transfer (GVT) is also included among the MRT techniques, even if to date only two studies on mouse models have been reported, with unpromising results [115,116]. Briefly, GVT consists of the transplantation of the GV from a patient's immature oocyte to a healthy donor enucleated one. The pivotal issue of this technique resides in the need of subjecting the new immature oocytes to IVM for obtaining a mature oocyte ready for fertilization.

9. Conclusions

ART laboratories might improve the use of immature GV-stage oocytes for further applications during routine procedures due to an increasing percentage of women unable to achieve oocyte maturation in unstimulated or stimulated cycles. The current application of GV-stage oocytes in ARTs is still limited and a matter of debate, underlining the necessity of morphofunctional studies for shedding light on the correct use of these "leftover" oocytes. TEM and SEM are the most powerful tools for assessing the fine structure of immature human oocytes, allowing the qualitative evaluation of pivotal organelles such as the nucleus, mitochondria, vacuoles and lysosomes before or after COS protocols, IVM and freezing strategies. However, due to the limited number of available leftover oocytes for these purposes, more studies are necessary for clarifying the most efficient use of GV-stage oocytes for ART optimization.

Author Contributions: Conceptualization, S.A.N., G.M., S.C. and M.G.P.; methodology, S.A. and M.G.P.; formal analysis, S.A., M.G.P. and S.A.N.; writing—original draft preparation, M.G.P. and S.A.; writing—review and editing, M.G.P., S.A. and S.A.N.; Project administration: M.G.P., G.M. and S.A.N.; visualization, S.C., S.B. and M.A.K.; supervision, S.A.N. and G.M.; funding acquisition, G.M., M.G.P. and S.A.N. All authors have read and agreed to the published version of the manuscript.

Funding: This research was funded by G.M./M.G.P. (University Grants: FFO 2019–2021) and S.A.N. (University Grants).

Institutional Review Board Statement: Ethical review and approval were waived for this study, due to the use of unpublished archived micrographs from previously approved studies.

Informed Consent Statement: Informed consent was obtained from all subjects involved in the study.

Data Availability Statement: The data that support the findings of this study are available from the corresponding author upon reasonable request.

Conflicts of Interest: The authors declare no conflict of interest.

References

1. Aristotle. *Generation of Animals, with an English Translation by A.L. Peck*; William Heinemann LTD: London, UK; Harvard University Press: Cambridge, MA, USA, 1943.
2. Harvey, W. Exercises of Animal Generation. In *Exercitationes de Generatione Animalium*; Pulleyn, I.O., Ed.; Typis Du-Gardianis: London, UK, 1651.
3. Cobb, M. An amazing 10 years: The discovery of egg and sperm in the 17th century. *Reprod. Domest. Anim.* **2012**, *47*, 2–6. [CrossRef] [PubMed]
4. Turriziani Colonna, F. *De Ovi Mammalium et Hominis Genesi (1827), by Karl Ernst Von Baer*; Embryo Project Encyclopedia; Center for Biology and Society, School of Life Sciences, Arizona State University: Tempe, AZ, USA, 2017. Available online: http://embryo.asu.edu/handle/10776/11405 (accessed on 25 March 2022).
5. De Graaf, R. The Female Organs in the Service of Generation. In *De Mulierum Organis Generationi Inservientibus*; Ex Officina Hackiana: Leiden, The Netherlands, 1672.
6. Spallanzani, L. *Saggio di Osservazioni Microscopiche Concernenti il Sistema Della Generazione Dei Signori Needham e Buffon*; Società Tipografica Editrice Barese: Bari, Italy, 1914.
7. Prévost, J.L.; Dumas, J.B.A. Nouvelle Théorie de la Generation. In *Physiologie et Anatomie Animale*; Annales des Sciences Naturelles: Paris, France, 1984.
8. Buess, H. The contribution of Geneva physicians to the physiology of development in the 19th century. *Bull. Hist. Med.* **1947**, *21*, 871–897. [PubMed]
9. Purkinje, J.E. De evolutione vesiculae germinativae. In *Symbolae ad Ovi Avium Historiam ante Incubationem*; Leopold Vossi: Leipzig, Germany, 1830; pp. 1–24.
10. Von Baer, K.E. De Ovi Mammalium et Hominis Genesi. In *On the Genesis of the Ovum of Mammals and of Man*; Leopold Voss: Leipzig, Germany, 1827.
11. Altmäe, S.; Acharya, G.; Salumets, A. Celebrating Baer—A Nordic scientist who discovered the mammalian oocyte. *Acta Obstet. Gynecol. Scand.* **2017**, *96*, 1281–1282. [CrossRef] [PubMed]
12. Müller-Wille, S. Cell theory, specificity, and reproduction, 1837–1870. *Stud. Hist. Philos. Biol. Biomed. Sci.* **2010**, *41*, 225–231. [CrossRef]
13. Kruta, V. The letters of Purkinje. *Lancet* **1956**, *268*, 360. [CrossRef]
14. Harris, H. The Discovery of the Cell Nucleus. In *The Birth of the Cell*; Harris, H., Ed.; Yale University Press: New Haven, CT, USA, 1999; pp. 76–93.
15. Alexandre, H. A history of mammalian embryological research. *Int. J. Dev. Biol.* **2001**, *45*, 457–467.
16. Makabe, S.; Naguro, T.; Nottola, S.A.; Van Blerkom, J. *Atlas of Human Female Reproductive Function: Ovarian Development to Early Embryogenesis after In Vitro Fertilization*; Taylor & Francis: London, UK; New York, NY, USA, 2006; p. 89.
17. Coticchio, G.; Canto, M.D.; Guglielmo, M.C.; Albertini, D.F.; Renzini, M.M.; Merola, M.; Lain, M.; Sottocornola, M.; De Ponti, E.; Fadini, R. Double-strand DNA breaks and repair response in human immature oocytes and their relevance to meiotic resumption. *J. Assist. Reprod. Genet.* **2015**, *32*, 1509–1516. [CrossRef]
18. Liu, M.-H.; Zhou, W.-H.; Chu, D.-P.; Fu, L.; Sha, W.; Li, Y. Ultrastructural Changes and Methylation of Human Oocytes Vitrified at the Germinal Vesicle Stage and Matured in vitro after Thawing. *Gynecol. Obstet. Investig.* **2017**, *82*, 252–261. [CrossRef]
19. Peinado, I.; Moya, I.; Sáez-Espinosa, P.; Barrera, M.; García-Valverde, L.; Francés, R.; Torres, P.; Gómez-Torres, M. Impact of Maturation and Vitrification Time of Human GV Oocytes on the Metaphase Plate Configuration. *Int. J. Mol. Sci.* **2021**, *22*, 1125. [CrossRef]
20. Nilsson, B.O.; Liedholm, P.; Larsson, E. Ultrastructural characteristics of human oocytes fixed at follicular puncture or after culture. *J. Assist. Reprod. Genet.* **1985**, *2*, 195–206. [CrossRef]
21. Pires-Luís, A.S.; Rocha, E.; Bartosch, C.; Oliveira, E.; Silva, J.; Barros, A.; Sá, R.; Sousa, M. A stereological study on organelle distribution in human oocytes at prophase I. *Zygote* **2015**, *24*, 346–354. [CrossRef] [PubMed]
22. Palmerini, M.G.; Antinori, M.; Maione, M.; Cerusico, F.; Versaci, C.; Nottola, S.A.; Macchiarelli, G.; Khalili, M.A.; Antinori, S. Ultrastructure of immature and mature human oocytes after cryotop vitrification. *J. Reprod. Dev.* **2014**, *60*, 411–420. [CrossRef] [PubMed]
23. Suzuki, S.; Kitai, H.; Tojo, R.; Seki, K.; Oba, M.; Fujiwara, T.; Iizuka, R. Ultrastructure and some biologic properties of human oocytes and granulosa cells cultured in vitro. *Fertil. Steril.* **1981**, *35*, 142–148. [CrossRef]
24. Sathananthan, A.H.; Selvaraj, K.; Girijashankar, M.L.; Ganesh, V.; Selvaraj, P.; Trounson, A.O. From oogonia to mature oocytes: Inactivation of the maternal centrosome in humans. *Microsc. Res. Tech.* **2006**, *69*, 396–407. [CrossRef]
25. Sathananthan, A.H.; Trounson, A.; Wood, C. *Atlas of Fine Structure of Human Sperm Penetration, Eggs, and Embryos Cultured In Vitro*; Praeger: New York, NY, USA, 1985.

26. Trebichalská, Z.; Kyjovská, D.; Kloudová, S.; Otevřel, P.; Hampl, A.; Holubcová, Z. Cytoplasmic maturation in human oocytes: An ultrastructural study. *Biol. Reprod.* **2020**, *104*, 106–116. [CrossRef]
27. Szollosi, D.; Szöllösi, M.S.; Czolowska, R.; Tarkowski, A.K. Sperm penetration into immature mouse oocytes and nuclear changes during maturation: An EM study. *Biol. Cell* **1990**, *69*, 53–64. [CrossRef]
28. Yang, Y.-J.; Zhang, Y.-J.; Li, Y. Ultrastructure of human oocytes of different maturity stages and the alteration during in vitro maturation. *Fertil. Steril.* **2009**, *92*, 396.e1–396.e6. [CrossRef]
29. Khalili, M.A.; Shahedi, A.; Ashourzadeh, S.; Nottola, S.A.; Macchiarelli, G.; Palmerini, M.G. Vitrification of human immature oocytes before and after in vitro maturation: A review. *J. Assist. Reprod. Genet.* **2017**, *34*, 1413–1426. [CrossRef]
30. Coticchio, G.; Canto, M.D.; Fadini, R.; Renzini, M.M.; Guglielmo, M.C.; Miglietta, S.; Palmerini, M.G.; Macchiarelli, G.; Nottola, S.A. Ultrastructure of human oocytes after in vitro maturation. *Mol. Hum. Reprod.* **2016**, *22*, 110–118. [CrossRef]
31. Morimoto, Y. Ultrastructure of the Human Oocytes during in Vitro Maturation. *J. Mamm. Ova Res.* **2009**, *26*, 10–17. [CrossRef]
32. Takahashi, Y.; Hashimoto, S.; Yamochi, T.; Goto, H.; Yamanaka, M.; Amo, A.; Matsumoto, H.; Inoue, M.; Ito, K.; Nakaoka, Y.; et al. Dynamic changes in mitochondrial distribution in human oocytes during meiotic maturation. *J. Assist. Reprod. Genet.* **2016**, *33*, 929–938. [CrossRef] [PubMed]
33. Van Blerkom, J. Developmental Failure in Human Reproduction Associated with Preovulatory Oogenesis and Preimplantation Embryogenesis. In *Ultrastructure of Human Gametogenesis and Early Embryogenesis*; Van Blerkom, J., Motta, P.M., Eds.; Springer: New York, NY, USA, 1989. [CrossRef]
34. Ghetler, Y.; Skutelsky, E.; Ben Nun, I.; Ben Dor, L.; Amihai, D.; Shalgi, R. Human oocyte cryopreservation and the fate of cortical granules. *Fertil. Steril.* **2006**, *86*, 210–216. [CrossRef] [PubMed]
35. Monti, M.; Calligaro, A.; Behr, B.; Pera, R.R.; Redi, C.A.; Wossidlo, M. Functional topography of the fully grown human oocyte. *Eur. J. Histochem.* **2017**, *61*, 2769. [CrossRef] [PubMed]
36. Familiari, G.; Nottola, S.A.; Micara, G.; Aragona, C.; Motta, P.M. Is the sperm-binding capability of the zona pellucida linked to its surface structure? A scanning electron microscopic study of human in vitro fertilization. *J. Assist. Reprod. Genet.* **1988**, *5*, 134–143. [CrossRef] [PubMed]
37. Kim, B.-K.; Lee, S.-C.; Kim, K.-J.; Han, C.-H.; Kim, J.-H. In vitro maturation, fertilization, and development of human germinal vesicle oocytes collected from stimulated cycles. *Fertil. Steril.* **2000**, *74*, 1153–1158. [CrossRef]
38. Halvaei, I.; Khalili, M.A.; Razi, M.H.; Nottola, S.A. The effect of immature oocytes quantity on the rates of oocytes maturity and morphology, fertilization, and embryo development in ICSI cycles. *J. Assist. Reprod. Genet.* **2012**, *29*, 803–810. [CrossRef]
39. Alcoba, D.D.; Pimentel, A.M.; Brum, I.S.; Corleta, H.V.E. Developmental potential of in vitro or in vivo matured oocytes. *Zygote* **2013**, *23*, 93–98. [CrossRef]
40. Lee, J.E.; Kim, S.D.; Jee, B.C.; Suh, C.S.; Kim, S.H. Oocyte maturity in repeated ovarian stimulation. *Clin. Exp. Reprod. Med.* **2011**, *38*, 234–237. [CrossRef]
41. Astbury, P.; Subramanian, G.N.; Greaney, J.; Roling, C.; Irving, J.; Homer, H.A. The Presence of Immature GV–Stage Oocytes during IVF/ICSI Is a Marker of Poor Oocyte Quality: A Pilot Study. *Med. Sci.* **2020**, *8*, 4. [CrossRef]
42. Siristatidis, C.S.; Maheshwari, A.; Bhattacharya, S. In vitro maturation in sub fertile women with polycystic ovarian syndrome undergoing assisted reproduction. *Cochrane Database Syst. Rev.* **2009**, *1*, CD006606. [CrossRef]
43. Nikbakht, R.; Mohammadjafari, R.; Rajabalipour, M.; Moghadam, M.T. Evaluation of oocyte quality in Polycystic ovary syndrome patients undergoing ART cycles. *Fertil. Res. Pract.* **2021**, *7*, 2. [CrossRef] [PubMed]
44. Rienzi, L.; Balaban, B.; Ebner, T.; Mandelbaum, J. The oocyte. *Hum. Reprod.* **2012**, *27*, i2–i21. [CrossRef] [PubMed]
45. Hatırnaz, Ş.; Ata, B.; Hatırnaz, E.S.; Dahan, M.; Tannus, S.; Tan, J.; Tan, S.L. Oocyte in vitro maturation: A systematic review. *J. Turk. Soc. Obstet. Gynecol.* **2018**, *15*, 112–125. [CrossRef] [PubMed]
46. Escrich, L.; Grau, N.; Santos, M.J.D.L.; Romero, J.-L.; Pellicer, A.; Escribá, M.-J. The dynamics of in vitro maturation of germinal vesicle oocytes. *Fertil. Steril.* **2012**, *98*, 1147–1151. [CrossRef]
47. Yang, Q.; Zhu, L.; Wang, M.; Huang, B.; Li, Z.; Hu, J.; Xi, Q.; Liu, J.; Jin, L. Analysis of maturation dynamics and developmental competence of in vitro matured oocytes under time-lapse monitoring. *Reprod. Biol. Endocrinol.* **2021**, *19*, 183. [CrossRef]
48. Lee, H.-J.; Barad, D.H.; Kushnir, V.A.; Shohat-Tal, A.; Lazzaroni-Tealdi, E.; Wu, Y.-G.; Gleicher, N. Rescue in vitro maturation (IVM) of immature oocytes in stimulated cycles in women with low functional ovarian reserve (LFOR). *Endocrine* **2015**, *52*, 165–171. [CrossRef]
49. Liu, Y.; Jiang, H.; Du, X.; Huang, J.; Wang, X.; Hu, Y.; Ni, F.; Liu, C. Contribution of rescue in-vitro maturation versus double ovarian stimulation in ovarian stimulation cycles of poor-prognosis women. *Reprod. Biomed. Online* **2020**, *40*, 511–517. [CrossRef]
50. Steptoe, P.; Edwards, R. Birth after the reimplantation of a human embryo. *Lancet* **1978**, *312*, 366. [CrossRef]
51. Edwards, R.G.; Steptoe, P.C.; Purdy, J.M. Establishing full-term human pregnancies using cleaving embryos grown in vitro. *BJOG Int. J. Obstet. Gynaecol.* **1980**, *87*, 737–756. [CrossRef]
52. Pelinck, M.; Vogel, N.; Arts, E.; Simons, A.; Heineman, M.; Hoek, A. Cumulative pregnancy rates after a maximum of nine cycles of modified natural cycle IVF and analysis of patient drop-out: A cohort study. *Hum. Reprod.* **2007**, *22*, 2463–2470. [CrossRef]
53. Mak, W.; Kondapalli, L.A.; Celia, G.; Gordon, J.; DiMattina, M.; Payson, M. Natural cycle IVF reduces the risk of low birthweight infants compared with conventional stimulated IVF. *Hum. Reprod.* **2016**, *31*, 789–794. [CrossRef] [PubMed]
54. Von Wolff, M. The role of Natural Cycle IVF in assisted reproduction. *Best Pract. Res. Clin. Endocrinol. Metab.* **2018**, *33*, 35–45. [CrossRef] [PubMed]

55. Ingerslev, H.J.; Højgaard, A.; Hindkjær, J.; Kesmodel, U. A randomized study comparing IVF in the unstimulated cycle with IVF following clomiphene citrate. *Hum. Reprod.* **2001**, *16*, 696–702. [CrossRef] [PubMed]
56. Macklon, N.S.; Stouffer, R.L.; Giudice, L.C.; Fauser, B.C.J.M. The Science behind 25 Years of Ovarian Stimulation for in Vitro Fertilization. *Endocr. Rev.* **2006**, *27*, 170–207. [CrossRef]
57. Ubaldi, F.; Rienzi, L.; Baroni, E.; Ferrero, S.; Iacobelli, M.; Minasi, M.; Sapienza, F.; Romano, S.; Colasante, A.; Litwicka, K.; et al. Hopes and facts about mild ovarian stimulation. *Reprod. Biomed. Online* **2007**, *14*, 675–681. [CrossRef]
58. Escrich, L.; Galiana, Y.; Grau, N.; Insua, F.; Soler, N.; Pellicer, A.; Escribá, M. Do immature and mature sibling oocytes recovered from stimulated cycles have the same reproductive potential? *Reprod. Biomed. Online* **2018**, *37*, 667–676. [CrossRef]
59. Greenblatt, E.M.; Meriano, J.S.; Casper, R.F. Type of stimulation protocol affects oocyte maturity, fertilization rate, and cleavage rate after intracytoplasmic sperm injection. *Fertil. Steril.* **1995**, *64*, 557–563. [CrossRef]
60. Nogueira, D.; Friedler, S.; Schachter, M.; Raziel, A.; Ron-El, R.; Smitz, J. Oocyte maturity and preimplantation development in relation to follicle diameter in gonadotropin-releasing hormone agonist or antagonist treatments. *Fertil. Steril.* **2006**, *85*, 578–583. [CrossRef]
61. Huddleston, H.G.; Jackson, K.V.; Doyle, J.O.; Racowsky, C. hMG increases the yield of mature oocytes and excellent-quality embryos in patients with a previous cycle having a high incidence of oocyte immaturity. *Fertil. Steril.* **2009**, *92*, 946–949. [CrossRef]
62. Ming, T.X.; Nielsen, H.I.; Chen, Z.Q. Maturation arrest of human oocytes at germinal vesicle stage. *J. Hum. Reprod. Sci.* **2010**, *3*, 153–157. [CrossRef]
63. Cha, K.Y.; Koo, J.J.; Ko, J.J.; Choi, D.H.; Han, S.Y.; Yoon, T.K. Pregnancy after in vitro fertilization of human follicular oocytes collected from nonstimulated cycles, their culture in vitro and their transfer in a donor oocyte program. *Fertil. Steril.* **1991**, *55*, 109–113. [CrossRef]
64. Fatum, M.; Bergeron, M.-E.; Ross, C.; Bhevan, A.; Turner, K.; Child, T. Rescue In Vitro Maturation in Polycystic Ovarian Syndrome Patients Undergoing In Vitro Fertilization Treatment who Overrespond or Underrespond to Ovarian Stimulation: Is It a Viable Option? A Case Series Study. *Int. J. Fertil. Steril.* **2020**, *14*, 137–142. [CrossRef] [PubMed]
65. Jones, G.M.; Cram, D.S.; Song, B.; Magli, M.C.; Gianaroli, L.; Lacham-Kaplan, O.; Findlay, J.K.; Jenkin, G.; Trounson, A.O. Gene expression profiling of human oocytes following in vivo or in vitro maturation. *Hum. Reprod.* **2008**, *23*, 1138–1144. [CrossRef] [PubMed]
66. Gilchrist, R.B.; Smitz, J.E.J.; Thompson, J.G. Current Status and Future Trends of the Clinical Practice of Human Oocyte In Vitro Maturation. In *Human Assisted Reproductive Technology Future Trends in Laboratory and Clinical Practice*; Gardner, D.K., Rizk, B., Falcone, T., Eds.; Cambridge University Press: Cambridge, UK, 2011; pp. 186–198.
67. Khalili, M.A.; A Nottola, S.; Shahedi, A.; Macchiarelli, G. Contribution of human oocyte architecture to success of in vitro maturation technology. *Iran. J. Reprod. Med.* **2013**, *11*, 1–10.
68. Trounson, A.; Wood, C.; Kausche, A. In vitro maturation and the fertilization and developmental competence of oocytes recovered from untreated polycystic ovarian patients. *Fertil. Steril.* **1994**, *62*, 353–362. [CrossRef]
69. Barnes, F.L.; Kausche, A.; Tiglias, J.; Wood, C.; Wilton, L.; Trounson, A. Production of embryos from in vitro-matured primary human oocytes. *Fertil. Steril.* **1996**, *65*, 1151–1156. [CrossRef]
70. Russell, J.B.; Knezevich, K.M.; Fabian, K.F.; Dickson, J.A. Unstimulated immature oocyte retrieval: Early versus midfollicular endometrial priming. *Fertil. Steril.* **1997**, *67*, 616–620. [CrossRef]
71. Chian, R.C.; Park, S.E.; Park, E.H.; Son, W.Y.; Chung, H.M.; Lim, J.G.; Ko, J.J.; Cha, K.Y. Molecular and structural characteristics between immature human oocytes retrieved from stimulated and unstimulated ovaries. In *In Vitro Fertilization and Assisted Reproduction*; Gromel, V., Leung, P.C.K., Eds.; Monduzzi Editore: Bologna, Italy, 1997; pp. 315–319.
72. Chian, R.-C.; Buckett, W.M.; Tan, S.-L. In-vitro maturation of human oocytes. *Reprod. Biomed. Online* **2004**, *8*, 148–166. [CrossRef]
73. Choi, J.; Kim, D.; Cha, J.; Lee, S.; Yoon, S.; Lim, J. P-705: The relationship between oocyte size and developmental competence in unstimulated cycles. *Fertil. Steril.* **2006**, *86*, S395. [CrossRef]
74. Cavilla, J.; Byskov, A.; Hartshorne, G.; Kennedy, C. Human immature oocytes grow during culture for IVM. *Hum. Reprod.* **2007**, *23*, 37–45. [CrossRef]
75. Child, T.J.; Abdul-Jalil, A.K.; Gulekli, B.; Tan, S.L. In vitro maturation and fertilization of oocytes from unstimulated normal ovaries, polycystic ovaries, and women with polycystic ovary syndrome. *Fertil. Steril.* **2001**, *76*, 936–942. [CrossRef]
76. Mikkelsen, A.L. Strategies in human in-vitro maturation and their clinical outcome. *Reprod. Biomed. Online* **2005**, *10*, 593–599. [CrossRef]
77. Liu, S.; Jiang, J.-J.; Feng, H.-L.; Ma, S.-Y.; Li, M.; Li, Y. Evaluation of the immature human oocytes from unstimulated cycles in polycystic ovary syndrome patients using a novel scoring system. *Fertil. Steril.* **2010**, *93*, 2202–2209. [CrossRef] [PubMed]
78. Gleicher, N.; Weghofer, A.; Barad, D.H. Defining ovarian reserve to better understand ovarian aging. *Reprod. Biol. Endocrinol.* **2011**, *9*, 23. [CrossRef]
79. Álvarez, C.; García-Garrido, C.; Taronger, R.; González de Merlo, G. In vitro maturation, fertilization, embryo development & clinical outcome of human metaphase-I oocytes retrieved from stimulated intracytoplasmic sperm injection cycles. *Indian J. Med. Res.* **2013**, *137*, 331–338.
80. Son, W.-Y.; Lee, S.-Y.; Lim, J.-H. Fertilization, cleavage and blastocyst development according to the maturation timing of oocytes in in vitro maturation cycles. *Hum. Reprod.* **2005**, *20*, 3204–3207. [CrossRef]

81. Ben-Ami, I.; Komsky, A.; Bern, O.; Kasterstein, E.; Komarovsky, D.; Ron-El, R. In vitro maturation of human germinal vesicle-stage oocytes: Role of epidermal growth factor-like growth factors in the culture medium. *Hum. Reprod.* **2010**, *26*, 76–81. [CrossRef]
82. Levi, M.; Ghetler, Y.; Shulman, A.; Shalgi, R. Morphological and molecular markers are correlated with maturation-competence of human oocytes. *Hum. Reprod.* **2013**, *28*, 2482–2489. [CrossRef]
83. Faramarzi, A.; Khalili, M.A.; Ashourzadeh, S.; Palmerini, M.G. Does rescue in vitro maturation of germinal vesicle stage oocytes impair embryo morphokinetics development? *Zygote* **2018**, *26*, 430–434. [CrossRef]
84. Tucker, M.J.; Morton, P.C.; Wright, G.; Sweitzer, C.L.; Massey, J.B. Clinical application of human egg cryopreservation. *Hum. Reprod.* **1998**, *13*, 3156–3159. [CrossRef]
85. Mandelbaum, J.; Junca, A.M.; Tibi, C.; Plachot, M.; Alnot, M.O.; Rim, H.; Salat-Baroux, J.; Cohen, J. Cryopreservation of Immature and Mature Hamster and Human Oocytes. *Ann. N. Y. Acad. Sci.* **1988**, *541*, 550–561. [CrossRef] [PubMed]
86. Fabbri, R. Cryopreservation of Human Oocytes and Ovarian Tissue. *Cell Tissue Bank.* **2006**, *7*, 113–122. [CrossRef] [PubMed]
87. Minasi, M.G.; Fabozzi, G.; Casciani, V.; Ferrero, S.; Litwicka, K.; Greco, E. Efficiency of slush nitrogen vitrification of human oocytes vitrified with or without cumulus cells in relation to survival rate and meiotic spindle competence. *Fertil. Steril.* **2012**, *97*, 1220–1225. [CrossRef] [PubMed]
88. Fabbri, R.; Porcu, E.; Marsella, T.; Rocchetta, G.; Venturoli, S.; Flamigni, C. Human oocyte cryopreservation: New perspectives regarding oocyte survival. *Hum. Reprod.* **2001**, *16*, 411–416. [CrossRef] [PubMed]
89. Albertini, D.F.; Olsen, R. Effects of Fertility Preservation on Oocyte Genomic Integrity. In *Oocyte Biology in Fertility Preservation*; Springer: New York, NY, USA, 2013; Volume 761, pp. 19–27. [CrossRef]
90. Chian, R.C.; Xu, Y.; Keilty, D. Chapter 3 Current Challenges in Immature Oocyte Cryopreservation. *Methods Mol. Biol.* **2017**, *1568*, 33–44. [CrossRef] [PubMed]
91. Youm, H.S.; Choi, J.R.; Oh, D.; Rho, Y.H. Vitrfication and Slow Freezing for Cryopreservation of Germinal Vesicle-Stage Human Oocytes: A Bayesian Meta-Analysis. *Cryoletters* **2017**, *38*, 455–462.
92. Jang, T.H.; Park, S.C.; Yang, J.H.; Kim, J.Y.; Seok, J.H.; Park, U.S.; Choi, C.W.; Lee, S.R.; Han, J. Cryopreservation and its clinical applications. *Integr. Med. Res.* **2017**, *6*, 12–18. [CrossRef]
93. Parmegiani, L.; Tatone, C.; Cognigni, G.E.; Bernardi, S.; Troilo, E.; Arnone, A.; Maccarini, A.M.; Di Emidio, G.; Vitti, M.; Filicori, M. Rapid warming increases survival of slow-frozen sibling oocytes: A step towards a single warming procedure irrespective of the freezing protocol? *Reprod. Biomed. Online* **2014**, *28*, 614–623. [CrossRef]
94. Stimpfel, M.; Vrtacnik-Bokal, E.; Virant-Klun, I. No difference in mitochondrial distribution is observed in human oocytes after cryopreservation. *Arch. Gynecol. Obstet.* **2017**, *296*, 373–381. [CrossRef]
95. Bosch, E.; De Vos, M.; Humaidan, P. The Future of Cryopreservation in Assisted Reproductive Technologies. *Front. Endocrinol.* **2020**, *11*, 67. [CrossRef]
96. Vajta, G.; Rienzi, L.; Ubaldi, F.M. Open versus closed systems for vitrification of human oocytes and embryos. *Reprod. Biomed. Online* **2015**, *30*, 325–333. [CrossRef] [PubMed]
97. Gullo, G.; Petousis, S.; Papatheodorou, A.; Panagiotidis, Y.; Margioula-Siarkou, C.; Prapas, N.; D'Anna, R.; Perino, A.; Cucinella, G.; Prapas, Y. Closed vs. Open Oocyte Vitrification Methods Are Equally Effective for Blastocyst Embryo Transfers: Prospective Study from a Sibling Oocyte Donation Program. *Gynecol. Obstet. Investig.* **2020**, *85*, 206–212. [CrossRef] [PubMed]
98. Gosden, R.G. General principles of cryopreservation. *Methods Mol. Biol.* **2014**, *1154*, 261–268. [CrossRef]
99. Levi-Setti, P.E.; Patrizio, P.; Scaravelli, G. Evolution of human oocyte cryopreservation. *Curr. Opin. Endocrinol. Diabetes Obes.* **2016**, *23*, 445–450. [CrossRef] [PubMed]
100. Cao, Y.; Xing, Q.; Zhang, Z.-G.; Wei, Z.-L.; Zhou, P.; Cong, L. Cryopreservation of immature and in-vitro matured human oocytes by vitrification. *Reprod. Biomed. Online* **2009**, *19*, 369–373. [CrossRef]
101. Fahy, G.M. Theoretical considerations for oocyte cryopreservation by freezing. *Reprod. Biomed. Online* **2007**, *14*, 709–714. [CrossRef]
102. Cobo, A.; Diaz, C. Clinical application of oocyte vitrification: A systematic review and meta-analysis of randomized controlled trials. *Fertil. Steril.* **2011**, *96*, 277–285. [CrossRef]
103. Taylor, R.W.; Taylor, G.A.; Durham, S.E.; Turnbull, D.M. The determination of complete human mitochondrial DNA sequences in single cells: Implications for the study of somatic mitochondrial DNA point mutations. *Nucleic Acids Res.* **2001**, *29*, e74. [CrossRef]
104. Schon, E.A.; DiMauro, S.; Hirano, M. Human mitochondrial DNA: Roles of inherited and somatic mutations. *Nat. Rev. Genet.* **2012**, *13*, 878–890. [CrossRef]
105. Ye, M.; Shi, W.; Hao, Y.; Zhang, L.; Chen, S.; Wang, L.; He, X.; Li, S.; Xu, C. Associations of mitochondrial DNA copy number and deletion rate with early pregnancy loss. *Mitochondrion* **2020**, *55*, 48–53. [CrossRef]
106. Bianco, B.; Montagna, E. The advances and new technologies for the study of mitochondrial diseases. *Einstein* **2016**, *14*, 291–293. [CrossRef] [PubMed]
107. Farnezi, H.C.M.; Goulart, A.C.X.; Dos Santos, A.; Ramos, M.G.; Penna, M.L.F. Three-parent babies: Mitochondrial replacement therapies. *JBRA Assist. Reprod.* **2020**, *24*, 189–196. [CrossRef] [PubMed]
108. Richardson, J.; Irving, L.; Hyslop, L.A.; Choudhary, M.; Murdoch, A.; Turnbull, D.M.; Herbert, M. Concise Reviews: Assisted Reproductive Technologies to Prevent Transmission of Mitochondrial DNA Disease. *Stem. Cells* **2015**, *33*, 639–645. [CrossRef] [PubMed]

109. Craven, L.; Tang, M.-X.; Gorman, G.; De Sutter, P.; Heindryckx, B. Novel reproductive technologies to prevent mitochondrial disease. *Hum. Reprod. Updat.* **2017**, *23*, 501–519. [CrossRef]
110. Sendra, L.; García-Mares, A.; Herrero, M.J.; Aliño, S.F. Mitochondrial DNA Replacement Techniques to Prevent Human Mitochondrial Diseases. *Int. J. Mol. Sci.* **2021**, *22*, 551. [CrossRef]
111. Darbandi, S.; Darbandi, M.; Khorshid, H.R.K.; Sadeghi, M.R.; Agarwal, A.; Sengupta, P.; Al-Hasani, S.; Akhondi, M.M. Ooplasmic transfer in human oocytes: Efficacy and concerns in assisted reproduction. *Reprod. Biol. Endocrinol.* **2017**, *15*, 77. [CrossRef]
112. Cree, L.; Loi, P. Mitochondrial replacement: From basic research to assisted reproductive technology portfolio tool—technicalities and possible risks. *Mol. Hum. Reprod.* **2014**, *21*, 3–10. [CrossRef]
113. Tachibana, M.; Amato, P.; Sparman, M.; Woodward, J.; Sanchis, D.M.; Ma, H.; Gutierrez, N.M.; Tippner-Hedges, R.; Kang, E.; Lee, H.-S.; et al. Towards germline gene therapy of inherited mitochondrial diseases. *Nature* **2012**, *493*, 627–631. [CrossRef]
114. Zhang, J.; Liu, H.; Luo, S.; Lu, Z.; Chávez-Badiola, A.; Liu, Z.; Yang, M.; Merhi, Z.; Silber, S.J.; Munné, S.; et al. Live birth derived from oocyte spindle transfer to prevent mitochondrial disease. *Reprod. Biomed. Online* **2017**, *34*, 361–368. [CrossRef]
115. Cheng, Y.; Wang, K.; Kellam, L.D.; Lee, Y.S.; Liang, C.-G.; Han, Z.; Mtango, N.R.; Latham, K.E. Effects of Ooplasm Manipulation on DNA Methylation and Growth of Progeny in Mice1. *Biol. Reprod.* **2009**, *80*, 464–472. [CrossRef]
116. Neupane, J.; Vandewoestyne, M.; Ghimire, S.; Lu, Y.; Qian, C.; Van Coster, R.; Gerris, J.; Deroo, T.; Deforce, D.; De Sutter, P.; et al. Assessment of nuclear transfer techniques to prevent the transmission of heritable mitochondrial disorders without compromising embryonic development competence in mice. *Mitochondrion* **2014**, *18*, 27–33. [CrossRef] [PubMed]

 cells 

Article

Stage-Specific L-Proline Uptake by Amino Acid Transporter Slc6a19/B^{0}AT1 Is Required for Optimal Preimplantation Embryo Development in Mice

Tamara Treleaven [1], Matthew Zada [1], Rajini Nagarajah [2,3], Charles G. Bailey [2,3,4], John E. J. Rasko [2,3,5], Michael B. Morris [1,†] and Margot L. Day [1,*,†]

1 School of Medical Sciences, Faculty of Medicine and Health, The University of Sydney, Camperdown, NSW 2006, Australia
2 Gene & Stem Cell Therapy Program Centenary Institute, The University of Sydney, Camperdown, NSW 2050, Australia
3 Faculty of Medicine and Health, The University of Sydney, Camperdown, NSW 2006, Australia
4 Cancer & Gene Regulation Laboratory Centenary Institute, The University of Sydney, Camperdown, NSW 2050, Australia
5 Cell & Molecular Therapies, Royal Prince Alfred Hospital, Missenden Rd, Camperdown, NSW 2050, Australia
* Correspondence: margot.day@sydney.edu.au
† These authors contributed equally to this work.

Abstract: L-proline (Pro) has previously been shown to support normal development of mouse embryos. Recently we have shown that Pro improves subsequent embryo development when added to fertilisation medium during in vitro fertilisation of mouse oocytes. The mechanisms by which Pro improves embryo development are still being elucidated but likely involve signalling pathways that have been observed in Pro-mediated differentiation of mouse embryonic stem cells. In this study, we show that B^{0}AT1, a neutral amino acid transporter that accepts Pro, is expressed in mouse preimplantation embryos, along with the accessory protein ACE2. B^{0}AT1 knockout (*Slc6a19*$^{-/-}$) mice have decreased fertility, in terms of litter size and preimplantation embryo development in vitro. In embryos from wild-type (WT) mice, excess unlabelled Pro inhibited radiolabelled Pro uptake in oocytes and 4–8-cell stage embryos. Radiolabelled Pro uptake was reduced in 4–8-cell stage embryos, but not in oocytes, from *Slc6a19*$^{-/-}$ mice compared to those from WT mice. Other B^{0}AT1 substrates, such as alanine and leucine, reduced uptake of Pro in WT but not in B^{0}AT1 knockout embryos. Addition of Pro to culture medium improved embryo development. In WT embryos, Pro increased development to the cavitation stage (on day 4); whereas in B^{0}AT1 knockout embryos Pro improved development to the 5–8-cell (day 3) and blastocyst stages (day 6) but not at cavitation (day 4), suggesting B^{0}AT1 is the main contributor to Pro uptake on day 4 of development. Our results highlight transporter redundancy in the preimplantation embryo.

Keywords: proline; amino acid transporter; *Slc6a19*; B^{0}AT1; preimplantation embryo; mouse

Citation: Treleaven, T.; Zada, M.; Nagarajah, R.; Bailey, C.G.; Rasko, J.E.J.; Morris, M.B.; Day, M.L. Stage-Specific L-Proline Uptake by Amino Acid Transporter Slc6a19/B^{0}AT1 Is Required for Optimal Preimplantation Embryo Development in Mice. *Cells* **2023**, *12*, 18. https://doi.org/10.3390/cells12010018

Academic Editor: Lon J. van Winkle

Received: 19 October 2022
Revised: 13 December 2022
Accepted: 14 December 2022
Published: 21 December 2022

1. Introduction

Amino acids (AAs) play important roles in mammalian preimplantation development [1–4]. Reproductive fluid in the oviduct and uterus contains a mixture of all essential (EAA) and non-essential AAs (NEAA) which support embryo development from fertilisation to implantation [5,6]. During in vitro culture of mouse embryos, addition of NEAAs to the medium promotes cleavage to the 8-cell stage [1,7] whereas addition of EAAs has no effect until after the 8-cell stage, when they promote an increased number of cells in the inner cell mass (ICM) [1].

Some AAs, such as alanine, glycine, glutamine and taurine, are present at millimolar concentrations in reproductive tract fluid [5] and can be accumulated to the same or higher concentrations within the developing embryo to prevent cell volume changes caused by the high osmolality (300–310 mOsm/kg) of reproductive fluid [8]. In addition to their role as organic osmolytes, AAs are also important for protein synthesis [9], as an energy source for production of ATP [10], for buffering of intracellular pH, and regulation of reactive oxygen species (ROS) either by use in the production of glutathione or by direct ROS scavenging [11]. L-proline (Pro), for example, is able to directly scavenge ROS and when added to culture medium (in the absence of all other AAs), at a concentration similar to that found in reproductive fluid [6], improves mouse preimplantation embryo development [4]. Recently, we have shown that adding Pro and/or its analogue pipecolic acid to fertilisation medium during in vitro fertilisation (IVF) of oocytes increases subsequent embryo development, blastocyst formation and the number of cells in the ICM [12]. The improvement in development induced by Pro may be attributed to a reduction of mitochondrial activity and ROS levels in oocytes [12]. Collectively, these and other results show that Pro (and its metabolites) act in a growth factor-like manner by a variety of interconnected mechanisms including signal transduction activation, regulation of ROS levels and mitochondrial activity, and modulation of the epigenetic landscape as shown either in embryos themselves [4,12] or using embryonic stem (ES) cells as an in vitro model of embryo development [13–16].

For AAs to have a beneficial effect on preimplantation development they must be taken up into the embryo via AA transporters (AATs). During early embryo development, AATs are expressed in a stage-specific manner reflecting, in part, the changing nutritional needs of the developing embryo. For example, there are multiple transporters for Pro in the preimplantation embryo, which are expressed at various stages. In the cumulus oocyte complexes (COCs), Pro can be taken up by cumulus cells surrounding the oocyte [17] via y+LAT2 (*Slc7a6*) [18] and GlyT (*Slc6a5* and/or *Slc6a9*) [19] and Pro is then transferred to the oocyte via transzonal gap junctions. In oocytes themselves, Pro is taken up by one or more transporters, including GlyT1 (*Slc6a9*) [19,20], PROT (*Slc6a7*), and PAT1 and/or 2 (*Slc36a1*, *Slc36a2*) [12]. After fertilisation, Pro is transported via SIT1 (*Slc6a20*), which is active in the zygote and 2-cell stage [21], as well as by GlyT1, which is active in 8-cell and blastocyst stages [20]. System A transporters are the major transport system in the ICM [22]. In mouse ES cells, Pro uptake by SNAT2 (*Slc38a2*) promotes differentiation [13,23].

Improved development of zygotes to the blastocyst stage induced by Pro in culture medium is prevented by the addition of a molar excess of glycine (Gly), betaine (Bet) and leucine (Leu) [4]. This profile for AA transport is consistent with that for the transporter B^0AT1 (*Slc6a19*). In somatic cells, B^0AT1 is expressed in the kidney, intestine, colon and prostate, and is a major transporter of neutral AAs, including Pro, across the plasma membrane [24]. Autosomal recessive inheritance of *SLC6A19* mutations in humans leads to poor absorption of AAs from the gut, or poor retention of AAs by the kidneys, causing Hartnup disorder [24,25]. Mice in which *Slc6a19* has been knocked out (*Slc6a19*$^{-/-}$) [26] exhibit symptoms of neutral aminoaciduria characteristic of Hartnup disorder. However, the effect of the loss of B^0AT1 expression on reproductive capacity and in vitro development of embryos lacking B^0AT1 has not been reported.

In this study, we therefore investigated whether B^0AT1 is expressed during mouse preimplantation embryo development, along with known accessory proteins, and examined the role of B^0AT1 in fertility and embryo development by utilising B^0AT1 knockout (*Slc6a19*$^{-/-}$) mice. Our findings show that B^0AT1 is expressed throughout preimplantation development. The accessory protein ACE2 was also expressed, while its paralogue collectrin (TMEM27) was not present at any pre-implantation stage. *Slc6a19*$^{-/-}$ mice had decreased litter size compared to wild-type (WT). In addition, the development of embryos obtained from *Slc6a19*$^{-/-}$ mice to the blastocyst stage was decreased compared to embryos from WT mice, suggesting an important role for B^0AT1 in normal embryo development.

2. Methods

2.1. Animals (Mus musculus)

Generation N8 *Slc6a19*$^{-/-}$ mice [26] were fully backcrossed to N10 on C57Bl/6 background (Centenary Institute). Wild-type C57Bl/6 mice were obtained from Australian BioResources. Outbred Quackenbush Swiss (QS) mice (Animal Resource Centre, Perth, WA, Australia and Lab Animal Services, The University of Sydney, NSW, Australia) were housed under a 12 h light: 12 h dark cycle and experiments were performed under the Australian Code of Practice for the Care and Use of Animals for Scientific Purposes and in accordance with the University of Sydney Animal Ethics Committee, as required by the NSW Animal Research Act (5583 and 824) and by Sydney Local Health District Animal Ethics Committee (2016-031).

2.2. Isolation of Oocytes and Embryos

Female mice (4–8 weeks of age) were superovulated by intraperitoneal injection of 10 IU pregnant mare serum gonadotropin (PMSG; Intervet, Vic., Australia) followed by 10 IU human chronic gonadotropin (hCG; Intervet) 48 $\pm$ 2 h later. Immediately after hCG injection, female mice were individually housed and paired with QS, C57BL6 (WT) or *Slc6a19*$^{-/-}$ male mice (2–8 months of age). *Slc6a19*$^{-/-}$ embryos were obtained by mating *Slc6a19*$^{-/-}$ female with *Slc6a19*$^{-/-}$ male mice. Female mice were sacrificed by cervical dislocation at either 13, 24, 46, 56, 68, 80 or 92 h post-hCG to obtain oocytes, zygotes, 2–, 4–, 5–8-cell or morula stage embryos or blastocysts, respectively. Unfertilised oocytes and embryos were dissected from the oviduct or flushed from the reproductive tract into HEPES-buffered modified human tubal fluid (HEPES-modHTF). Bovine serum albumin (BSA; 0.3 mg/mL) was added to all media at a reduced amount to minimise the possibility of BSA breakdown into free AAs in the media. The concentration of NaCl in media was adjusted to give an osmolality of 270 mOsm/kg, as described previously [4]. Cumulus cells were removed by gently pipetting oocytes with a fine glass pipette for 1 min in HEPES-modHTF + 1 mg/mL hyaluronidase, followed by washing oocytes through three drops of modHTF + 0.3 mg/mL BSA. Denuded oocytes and embryos were allocated for embryo culture, L-[^{3}H]-Pro uptake or immunostaining experiments.

2.3. Culture of Zygotes from WT and Slc6a19$^{-/-}$ *Mice In Vitro*

Freshly isolated zygotes (day 1) were washed through at least three drops of HEPES-modHTF in the absence of AAs and then cultured in 96-well plates (Corning, NY, USA) to the blastocyst stage (120 h, day 6) at low density (1 embryo/100 µL) in modHTF + 0.3 mg/mL BSA $\pm$ 0.4 mM Pro at 37 °C in humidified air with 5% CO_2 under mineral oil. Embryos were scored for developmental stage every 24 h. The concentration of Pro used was previously shown to improve in vitro embryo development [4] and is similar to its physiological concentration in fluid of the reproductive tract [6]. Culture experiments were repeated 4–6 times with 5–21 embryos per treatment group per experiment.

2.4. Measurement of L-[^{3}H]-Pro Uptake in Oocytes and Embryos

To measure the uptake of L-[^{3}H]-Pro from medium, groups of 1–4 freshly isolated oocytes or embryos were allocated to 20 µL drops containing 1 µM L-[^{3}H]-Pro (L-[2,3,4,5-^{3}H]-proline (1 mCi/mL; Perkin-Elmer, Vic., Australia, NET483001MC) in HEPES-modHTF $\pm$ molar excess unlabelled AAs (5 mM for all AAs except Pro, which was used at 0.4 mM) covered with mineral oil and incubated for 60 min at 37 °C in humidified air with 5% CO_2. The L-isomer was used for all unlabelled chiral AAs (Sigma Aldrich, St. Louis, MO, USA). A time-course experiment was previously performed to show L-[^{3}H]-Pro uptake was linear for at least 60 min (data not shown; [12]). After incubation in each treatment, oocytes and embryos were immediately washed through cold HEPES-modHTF media (4 °C) and aspirated onto a filter mat (Perkin-Elmer, Vic., Australia, #1450-421) with 4 mL scintillation fluid (Perkin-Elmer, Vic., Australia, #6013371). Each sample group was analysed for 30 min in a MicroBeta TriLux Plate Counter (PerkinElmer, Vic., Australia). A standard curve was created from a serial dilution of 1 μM L-[^{3}H]-Pro and fitted by linear regression to enable calculation of uptake in fmol min^{-1} oocyte^{-1} (or embryo^{-1}) for each separate experiment.

2.5. Immunoflurescent Staining and Confocal Microscopy of COCs and Embryos

After isolation or following culture, oocytes and embryos were fixed in 4% (*w*/*v*) paraformaldehyde (PFA) for 30 min at room temperature then washed three times in PBS + 1 mg/mL polyvinyl acid (PBS + PVA; Sigma-Aldrich, St. Louis, MO, USA). Oocytes/embryos were permeabilised with PBS + PVA + 0.3% (*v*/*v*) Triton X-100 for 30 min and blocked by incubation in PBS + PVA + 0.1% (*v*/*v*) Tween-20 + 0.7% (*w*/*v*) BSA (PPTB) for 30 min at room temperature. Primary antibodies were diluted 1 in 200 in PPTB and included polyclonal chicken anti-human B^{0}AT1 (custom antibody from Aves Laboratories; immunising peptide CZ-DPNYEEFPKSQK representing aa 564–574 of B^{0}AT1 protein [27]), rabbit anti-human ACE2 (Abcam, Cambridge, UK, Cat. No. ab15348; RRID:AB_301861) and mouse anti-collectrin (Alexis Biochemicals, San Diego, CA, USA). Appropriate control sera for each staining included pre-immune IgY serum (chicken) and purified IgG (rabbit and mouse). Oocytes and embryos were incubated in primary antibody for 2 h at room temperature and washed three times in PPTB. Secondary antibodies used were Alexa Fluor 594-coupled goat anti-chicken IgG (diluted 1:500), Alexa Fluor 488-coupled goat anti-rabbit IgG (diluted 1:200) and anti-mouse IgG (diluted 1:200). For staining of embryos for only B^{0}AT1, the secondary antibody was diluted in PPTB containing FITC-conjugated phalloidin (1:200, Invitrogen, Carlsbad, CA, USA, Cat. No. F432). Oocytes and embryos were incubated in the dark for 1 h at room temperature and washed three times in PPTB then mounted in 3 μL Vectashield containing 1.5 μg/mL DAPI (Vector Laboratories, Newark, CA, USA). Images of oocytes and embryos were taken using confocal microscopy (LSM 510 Meta, Carl Zeiss, Oberkochen, Germany) using 405, 488 and 516 nm lasers and a 40× objective. Images were prepared using Fiji by Image J and are representative of 6–9 embryos stained from three separate embryo isolations.

2.6. Statistical Analyses

Embryo culture experiments were performed 4–6 times with 5–21 embryos per treatment group per experiment. The number of embryos developing to a particular stage was summed together from each replicate and the percentage development calculated. The total number of embryos in each group is provided in each figure and differences between treatment groups were determined by chi-square analysis. AA uptake experiments were conducted at least three times with 1–4 oocytes or embryos per treatment group. The average uptake of L-[^{3}H]-Pro for each treatment was calculated and the difference between groups compared by Student's *t*-test or 1-way ANOVA with Dunnett's post hoc test, as indicated in the figure legends. Data analysis was performed using GraphPad Prism v7.

3. Results

3.1. B^{0}AT1 Is Expressed in Preimplantation Embryos

The stage-specific expression of the B^{0}AT1 transporter in the preimplantation embryo was investigated by immunostaining. Embryos were counterstained with phalloidin, which binds to F-actin including subcortical F-actin, thereby enabling the identification of the periphery of each blastomere. Expression of B^{0}AT1 was detected with F-actin at or near the plasma-membrane of embryos at all developmental stages examined, but not at the surface of contact between blastomeres (Figure 1). In 1-cell through to 8-cell stages, B^{0}AT1 staining was observed in the cytoplasm close to the plasma membrane. B^{0}AT1 was also detected in the apical and basolateral membranes of trophoblast cells and at the plasma membrane of inner cell mass cells in blastocysts (Figure 1).

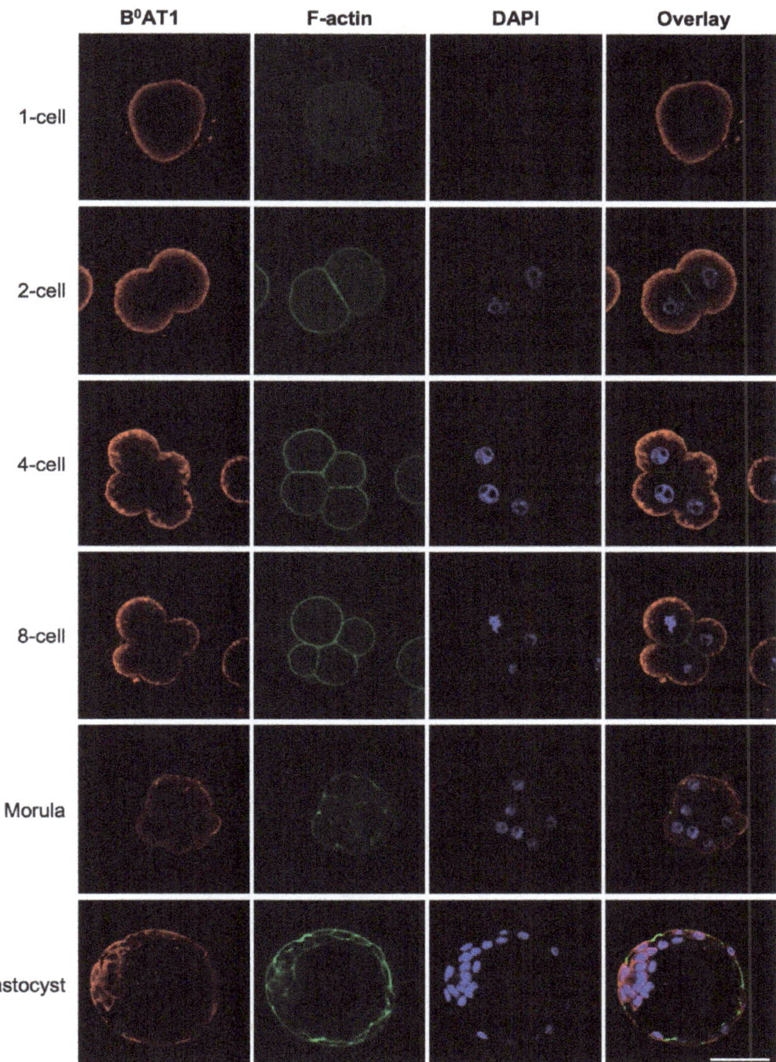

Figure 1. B^0AT1 is expressed throughout preimplantation mouse embryo development. Embryos from QS mice were freshly isolated at 1-, 2-, 4-, and 8-cell, morula and blastocyst stages, fixed and immunostained for B^0AT1 (red) and counterstained for F-actin with phalloidin (green) and nuclei with DAPI (blue). Fluorescent images were captured by confocal microscopy using a 40× objective. No staining was observed when a pre-immune IgY equivalent was used as a negative control for B^0AT1 staining (not shown). Scale bar = 50 μm for all images.

3.2. *The B^0AT1 Accessory Protein ACE2 Is Expressed in the Preimplantation Embryo*

Trafficking of B^0AT1 to the plasma membrane is dependent on the co-expression of an accessory protein, either angiotensin converting enzyme 2 (ACE2) or collectrin [28–30]. Embryos were therefore immunostained with the B^0AT1 antibody together with either ACE2 or collectrin antibodies to investigate whether one or both accessory proteins associated with B^0AT1 in preimplantation embryos. B^0AT1 colocalised with ACE2 in all preimplantation stages up to and including the blastocyst (Figure 2A). No collectrin staining was observed at any embryonic stage (Figure 2B).

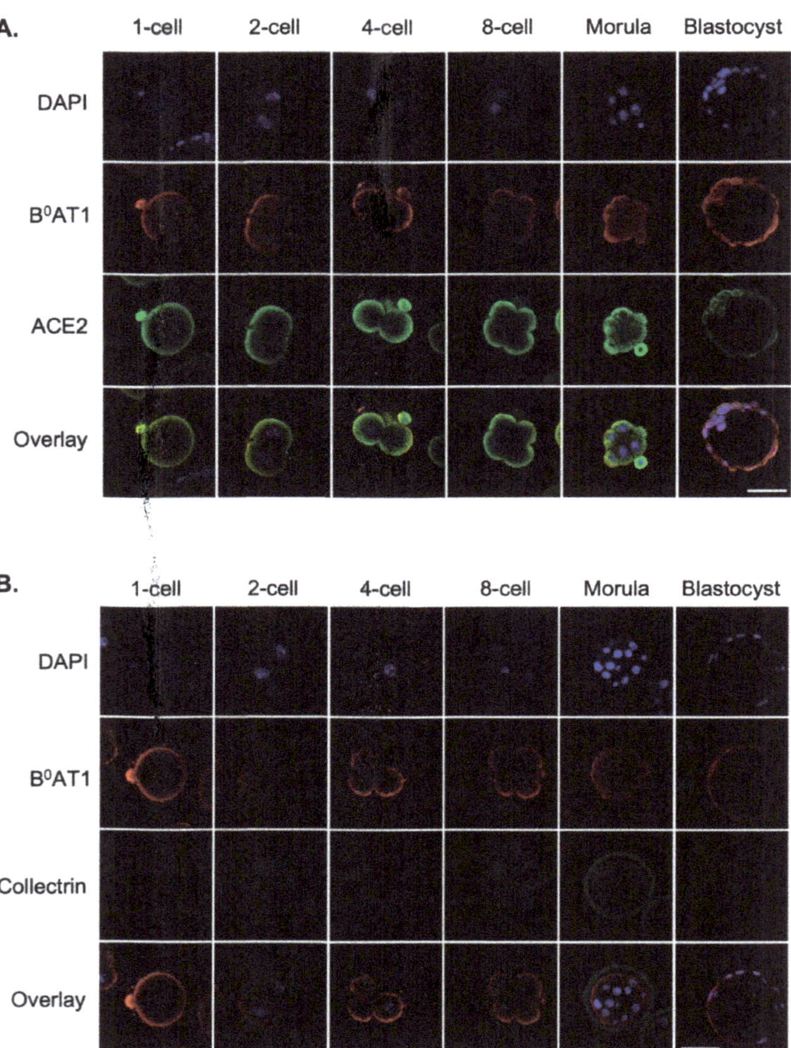

Figure 2. ACE2 colocalises with B^0AT1 in the mouse preimplantation embryo. Embryos were freshly isolated from QS mice at 1-, 2-, 4-, and 8-cell, morula and blastocyst stages, fixed and immunostained for B^0AT1 (red) and (**A**) ACE2 (green), or (**B**) collectrin (green). Nuclei were counterstained with DAPI (blue). Fluorescent images were captured by confocal microscopy using a 40× objective. No staining was observed when a pre-immune IgY or IgG equivalent was used as a negative control for the primary antibodies (not shown). Scale bar = 50 µm for all images.

3.3. Fertility of Slc6a19$^{-/-}$ Mice Is Reduced

To confirm the role of B^0AT1 in early embryonic development, we examined a *Slc6a19*-nullizygous mouse model previously reported by us [26]. Firstly, the absence of B^0AT1 protein in COCs and blastocysts from *Slc6a19*$^{-/-}$ mice was confirmed by immunostaining

(Figure 3A). In viable mice, litter size in *Slc6a19*$^{-/-}$ mice (KO × KO; 3.8 ± 1.5 (mean ± SD)) was significantly reduced compared to that in both WT (WT × WT; 6.4 ± 1.3 (mean ± SD), $p < 0.001$) and heterozygous mice (het × het; 6.2 ± 2.2 (mean ± SD), $p < 0.0001$) (Figure 3B), indicating B^{0}AT1 deficiency negatively impacts fertility and/or live birth rate.

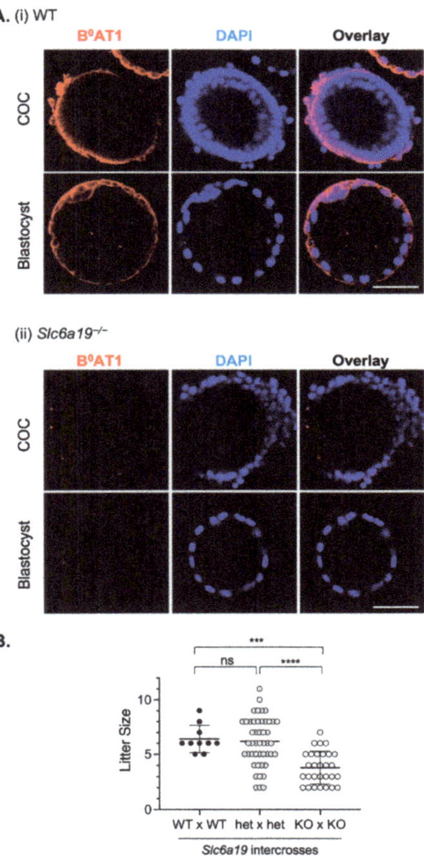

Figure 3. B^{0}AT1 deficiency impairs mouse fertility. (**A**) COCs and blastocysts were freshly isolated from (**i**) WT and (**ii**) *Slc6a19*$^{-/-}$ mice and immunostained for B^{0}AT1 (red) and DAPI (blue) to confirm loss of B^{0}AT1 protein in knockouts. Scale bar = 50 µm for all images. No staining was observed when a pre-immune IgY was used as a negative control for B^{0}AT1 staining (not shown). (**B**) The number of pups born from wild-type (WT × WT) [31], heterozygous (het × het) and homozygous *Slc6a19*$^{-/-}$ mice (KO × KO) intercrosses. Data show mean ± SD; *** $p < 0.001$, **** $p < 0.0001$, ns is not significant. Statistical analysis was performed using a one-way ANOVA with Tukey's post hoc test for multiple comparisons.

3.4. Pro Uptake by Embryos from WT and Slc6a19$^{-/-}$ Mice Is Dependent on Developmental Stage

The rate of uptake of L-[^{3}H]-Pro by oocytes and 4–8-cell stage embryos from WT and *Slc6a19*$^{-/-}$ mice was examined in the absence (No AA) and presence of excess unlabelled Pro. In both of these developmental stages from WT mice, excess unlabelled Pro significantly decreased the rate of L-[^{3}H]-Pro uptake (Figure 4), indicating the presence of one or more AAT(s) for Pro. Whereas in oocytes and 4–8 cell embryos from *Slc6a19*$^{-/-}$ mice, unlabelled Pro failed to prevent L-[^{3}H]-Pro uptake (Figure 4), suggesting B^{0}AT1 is normally an active AAT at these stages in WT mice. Any uptake now in the *Slc6a19*$^{-/-}$ mice is via a compensatory mechanism, which is not inhibited by the presence of excess unlabelled Pro (see Discussion).

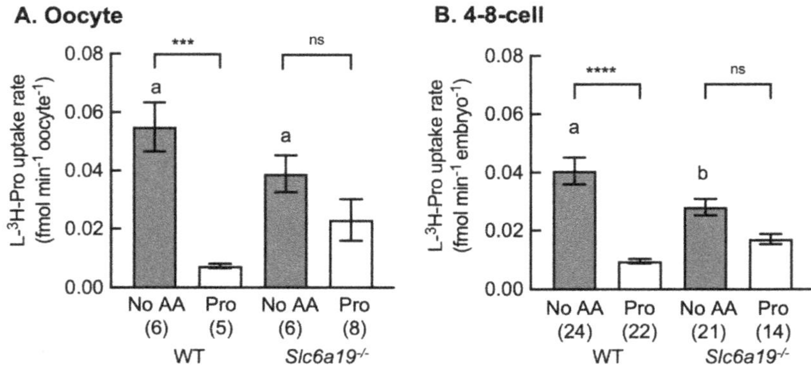

Figure 4. Excess unlabelled Pro reduces the rate of L-[^{3}H]-Pro uptake in WT but not *Slc6a19*$^{-/-}$ embryos. Rate of uptake of 1 μM L-[^{3}H]-Pro in (**A**) oocytes, and (**B**) 4–8-cell embryos from wild-type (WT) and *Slc6a19*$^{-/-}$ mice without (No AA) or with molar excess of unlabelled Pro (0.4 mM). Each bar represents the mean ± SEM with the number of replicates given in parentheses. Each replicate was performed on 1–4 oocytes or embryos over at least 3 independent experiments. Data were analysed by Student's *t*-test. 'No AA' bars with different letters (a,b) are significantly different ($p < 0.05$). *** and **** indicate $p < 0.001$ and $p < 0.0001$ respectively; ns, not significant ($p > 0.05$) for No AA vs Pro.

3.5. Pro Is Taken up by B^{0}AT1 and/or an Unknown Betaine-Sensitive Pro Transporter

B^{0}AT1 is a sodium-dependent neutral AAT which accepts Pro, Leu and Ala but not Bet [32]. Previously, we have shown that PROT and PAT1/2 transporters are responsible for the uptake of Pro in oocytes [12]. Here, we also investigated the competitive effect of B^{0}AT1 substrates on the rate of Pro uptake at the 4–8-cell stage. All 3 AAs significantly decreased the rate of L-[^{3}H]-Pro uptake in embryos from WT mice (Figure 5A) but not in embryos from *Slc6a19*$^{-/-}$ mice (Figure 5B).

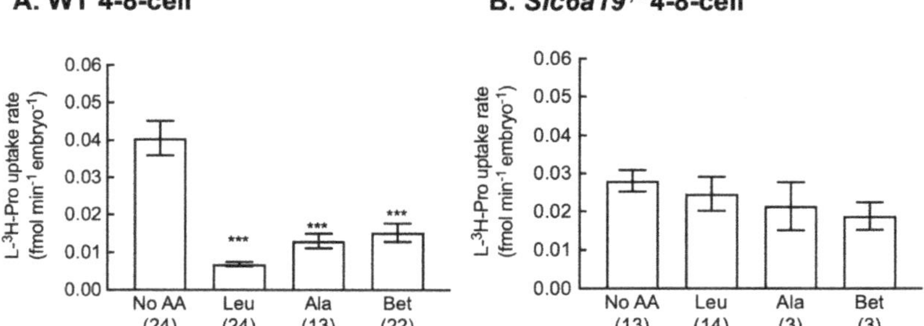

Figure 5. Characteristics of AA transport are different in embryos obtained from *Slc16a19*$^{-/-}$ mice compared to embryos from WT mice. Rate of uptake of L-[^{3}H]-Pro was measured in 4–8-cell stage embryos from (**A**) WT or (**B**) *Slc6a19*$^{-/-}$ mice without (No AA) or with molar excess (5 mM) of unlabelled AAs. Each bar represents the mean ± SEM with the number of replicates given in parentheses. Each replicate was performed on 1–4 embryos over at least 3 independent experiments. Data were analysed using one-way ANOVA with Dunnett's post hoc test. *** $p < 0.001$ compared to No AA. Bet = betaine, Leu = leucine, Ala = alanine. Note that bars for No AA and Pro in this figure are the same data as in Figure 4.

3.6. B^0AT1 Is Needed for Pro Uptake and Optimal Development at Day 4 of Development

The effect of loss of B^0AT1 on embryo development was examined by culturing embryos from WT and *Slc6a19*$^{-/-}$ mice from the zygote to blastocyst stage in medium ± 0.4 mM Pro (Figure 6). On day 2 (i.e., at the 2-cell stage), there was no significant difference in development between WT and *Slc6a19*$^{-/-}$ embryos with no AAs. Fewer *Slc6a19*$^{-/-}$ embryos developed in medium containing no AA to the ≥5-cell stage by the appropriate time (i.e., by day 3) and to the blastocyst stage (days 5 and 6) (Figure 6A). Addition of 0.4 mM Pro to the culture medium largely rescued the decreased development of embryos from *Slc6a19*$^{-/-}$ mice, except at day 4, when fewer embryos had cavitated, suggesting slower development at this point (Figure 6B). Note, that in embryos from WT mice, Pro significantly increased development to the appropriate stage on day 4 ($p < 0.001$). Whereas in *Slc6a19*$^{-/-}$ embryos, Pro increased development on days 3 and 6 ($p < 0.05$) compared to embryos cultured in medium containing no AA. These data suggest that in the absence of B^0AT1, Pro can be taken up by other transporter(s) on days 3 and 6 but not on day 4 of preimplantation development.

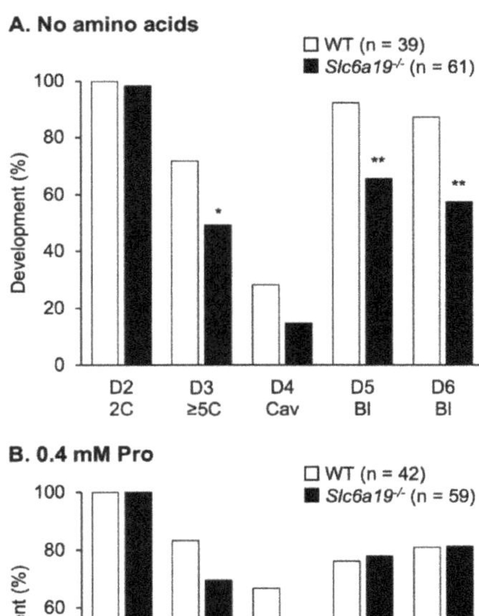

Figure 6. Pro improves development of *Slc6a19*$^{-/-}$ embryos at ≥5–8-cell and blastocyst stages. Zygotes were cultured in the (**A**) absence of AAs, or (**B**) in the presence of 0.4 mM Pro and scored for development to the appropriate stage every 24 h (i.e., to the 2-cell (2C) on day 2 (D2), 5–8-cell (≥5C) on D3, cavitation (Cav) on D4, and blastocyst stage (Bl) on D5 and D6. Data on each day of development were compared by chi-square analysis of pooled data from 4–6 replicate experiments, with the total number of embryos in each group given in parentheses. Bars with an asterix are significantly different to WT at each timepoint: * $p < 0.05$, ** $p < 0.01$ and *** $p < 0.001$.

4. Discussion

In this study, we investigated the effect of Pro on embryo development to determine: (1) if Pro is taken up by preimplantation embryos; and (2) if Pro transporters are expressed in embryos from WT and *Slc6a19*$^{-/-}$ mice. In vitro, the developmental potential of an oocyte and/or embryo is improved by the exogenous addition of Pro to the culture medium in a time-dependent manner. For example, when Pro is added to fertilisation medium, cleavage rates are not affected but improvement in development occurs at compaction through to the blastocyst stage, and blastocysts produced have more cells in their inner cell mass [12]. If Pro is added to culture medium from the zygote stage onwards, improved embryo development occurs from the 5-cell stage and the proportion proceeding to blastocyst development and hatching is increased [4]. Stage-specific uptake of Pro by AATs is needed for these Pro-mediated improvements in development. Since Gly and Leu both inhibit Pro-mediated improvement in development in culture [4], and are common substrates of the neutral AAT system B^0, we hypothesised that B^0AT1 (*Slc6a19*) is expressed in preimplantation embryos where it is responsible for Pro uptake.

B^0AT1 was detected in the plasma membrane of embryos at all stages from WT mice. B^0AT1 was also present in the cumulus cells of COCs and suggests a role for B^0AT1 in the increased Pro uptake into oocytes by coupled cumulus cells [19]. B^0AT1 staining was co-localised with ACE2 at all embryo stages, suggesting a role for ACE2, rather than collectrin, in the trafficking of B^0AT1 to the plasma membrane in embryos.

The functional role of B^0AT1 in uptake of Pro into embryos was determined by measuring the rate of radiolabelled Pro uptake in oocytes and 4–8-cell stage embryos and competition for Pro uptake by other substrates of B^0AT1, including Leu and Ala [32,33]. Due to the decreased fertility of *Slc6a19*$^{-/-}$ mice, we were unable to elucidate the complete inhibition profile of Pro transport in embryos from *Slc6a19*$^{-/-}$ mice. AAT-mediated uptake of Pro was present in oocytes from WT mice. However, despite the fact that the rate of Pro uptake into oocytes lacking B^0AT1 is the same as for WT, it was not inhibited by excess Pro, indicating Pro uptake in the knockout is not transporter-mediated. Other studies have shown that Pro uptake in WT oocytes can occur via a number of AATs including PROT (Slc6a7), GlyT1 (Slc6a9), PAT1 (Slc36a1) and/or PAT2 (Slc36a2) [12,19], but none of these seem to be active in this knockout since excess unlabeled Pro did not reduce uptake of labelled Pro. Thus, in this study, Pro was taken up in B^0AT1 knockout oocytes and embryos via a non-saturable route. This may be due to transport via channels that also conduct AAs, such as the volume-sensitive anion channel known to be present in oocytes [34–36].

At the 4–8-cell stage, Pro was taken up by embryos from both WT and *Slc6a19*$^{-/-}$ mice with the uptake reduced in *Slc6a19*$^{-/-}$ mice. The AAs Leu and Ala inhibited the uptake of Pro in WT but not in embryos from *Slc6a19*$^{-/-}$ mice. Bet, which is not a substrate for B^0AT1 [32], also inhibited Pro uptake in 4–8-cell WT embryos. PAT1 and PAT2 transport Pro, Bet and Ala [37] and are therefore good candidates for the Bet-sensitive portion of Pro uptake in 4–8 cell embryos. Recently, we have shown that PAT1 and 2 are present in the plasma membrane of mouse oocytes and may be responsible for Na^+-independent uptake of Pro in oocytes [12]. System A and BGT1 (*Slc6a12*), which also transport Bet, are not present in preimplantation embryos [20,38]. Further studies are needed to examine the expression profile of PAT1/2 in later embryo stages. A proportion of Pro uptake at the 4–8-cell stage could be via GlyT1, which is expressed in 4- and 8-cell embryos [39,40]. Again, none of these AATs seem to be active in the knockout. It is not clear why the Bet-sensitive uptake was also lost when B^0AT1 was knocked out. Further investigation is needed to determine whether the AAT profile in 4–8-cell embryos represents more than one transport system.

Development of embryos from *Slc6a19*$^{-/-}$ mice was decreased at the $\geq$5-cell and blastocyst stages when the medium contained no AA. These data suggest that B^0AT1 is required in vivo for uptake and storage of Pro in the oocyte and during fertilisation, which subsequently improves later embryo development [12]. Exogenous addition of Pro to culture improved development of embryos from *Slc6a19*$^{-/-}$ mice at the $\geq$5-cell and blastocyst stages but not on day 4 when cavitation occurs. In embryos from WT mice, Pro improved development at the cavitation stage. These findings indicate that B^0AT1 is normally expressed on day 4 of in vitro embryo development, or at a cleavage stage just before cavitation, to accumulate intracellular Pro for activation of Pro-mediated intracellular mechanisms, such as regulation of reactive oxygen species and mitochondrial activity [12,41]. Furthermore, the expression of AATs during pre-implantation development is very dynamic. We speculate that during or soon after day 4 in *Slc6a19*$^{-/-}$ mice, new Pro transporters are now expressed, and Pro is then taken up in sufficient quantity for cavitation and further development to proceed.

There was no decrease in litter size in Het x Het intercrosses and the appropriate Mendelian distribution of litter genotypes was observed [26], suggesting the maternal reproductive tract environment of *Slc6a19*$^{+/-}$ mice enables survival of *Slc6a19*$^{-/-}$ embryos, as seen with improved development in vitro of *Slc6a19*$^{-/-}$ embryos in the presence of Pro. Note though, a maternal diet low in vitamin B3 in pregnant *Slc6a19*$^{+/-}$ mice results in embryo loss and malformations [42]. We postulate that both intrinsic (i.e., *Slc6a19* KO embryo) and extrinsic factors (i.e., *Slc6a19* KO reproductive tract) are responsible for the decreased litter size in KO $\times$ KO intercrosses. This has implications for pregnancy in humans who carry loss-of-function mutant alleles in *SLC6A19*. These findings suggest that embryonic development and fertility could be improved in *Slc6a19*$^{-/-}$ mice by supplementation of the maternal diet with Pro.

This is the first study to demonstrate expression of the neutral AAT B^0AT1 in preimplantation embryos. Absence of this transporter adversely impacts fertility and embryo development, although the effect on embryo development can be overcome by the presence of Pro in embryo culture medium. This study demonstrates the importance of Pro for embryo development and that Pro can be taken up at each developmental stage by one or more transporters. The presence of multiple AATs for Pro and the ability to compensate for the loss of an individual AAT by upregulation of non-saturable transport mechanism indicates the critical importance of Pro for embryo development.

Author Contributions: Conceptualization and design of the experiments, M.L.D., M.B.M. and C.G.B.; performed the experiments, T.T., M.Z., M.L.D., C.G.B. and R.N.; data acquisition and analysis, T.T., M.Z., M.L.D. and C.G.B.; writing—original draft preparation, T.T. and M.L.D.; writing—review and editing, T.T., M.Z., M.L.D., M.B.M., C.G.B. and J.E.J.R. All authors have read and agreed to the published version of the manuscript.

Funding: C.G.B. and J.E.J.R. acknowledge funding from Cure the Future and an anonymous foundation. M.L.D. and M.B.M acknowledge funding from the School of Medical Sciences, University of Sydney.

Institutional Review Board Statement: The animal study protocol was approved by the Animal Ethics Committee of The University of Sydney as required by the NSW Animal Research Act (protocols 5583 and 824) and by Sydney Local Health District Animal Ethics Committee (2016-031).

Informed Consent Statement: Not applicable.

Data Availability Statement: All data are included in the published article. Data and materials will be made available upon request.

Conflicts of Interest: The authors declare no conflict of interest. The funders had no role in the design of the study; in the collection, analyses, or interpretation of data; in the writing of the manuscript; or in the decision to publish the results.

References

1. Lane, M.; Gardner, D.K. Differential regulation of mouse embryo development and viability by amino acids. *Reproduction* **1997**, *109*, 153–164. [CrossRef] [PubMed]
2. Leese, H.J.; McKeegan, P.J.; Sturmey, R.G. Amino Acids and the Early Mammalian Embryo: Origin, Fate, Function and Life-Long Legacy. *Int. J. Environ. Res. Public Health* **2021**, *18*, 9874. [CrossRef]
3. Van Winkle, L.J. Amino acid transport regulation and early embryo development. *Biol. Reprod.* **2001**, *64*, 1–12. [CrossRef]

4. Morris, M.B.; Ozsoy, S.; Zada, M.; Zada, M.; Zamfirescu, R.C.; Todorova, M.G.; Day, M.L. Selected Amino Acids Promote Mouse Pre-implantation Embryo Development in a Growth Factor-Like Manner. *Front. Physiol.* **2020**, *11*, 140. [CrossRef] [PubMed]
5. Harris, S.E.; Gopichandran, N.; Picton, H.M.; Leese, H.J.; Orsi, N.M. Nutrient concentrations in murine follicular fluid and the female reproductive tract. *Theriogenology* **2005**, *64*, 992–1006. [CrossRef] [PubMed]
6. Guerin, P.; Menezo, Y. Hypotaurine and taurine in gamete and embryo environments: De novo synthesis via the cysteine sulfinic acid pathway in oviduct cells. *Zygote* **1995**, *3*, 333–343. [CrossRef]
7. Gardner, D.K.; Lane, M. Amino acids and ammonium regulate mouse embryo development in culture. *Biol. Reprod.* **1993**, *48*, 377–385. [CrossRef]
8. Richards, T.; Wang, F.; Liu, L.; Baltz, J.M. Rescue of postcompaction-stage mouse embryo development from hypertonicity by amino acid transporter substrates that may function as organic osmolytes. *Biol. Reprod.* **2010**, *82*, 769–777. [CrossRef]
9. Epstein, C.J.; Smith, S.A. Amino acid uptake and protein synthesis in preimplanatation mouse embryos. *Dev. Biol.* **1973**, *33*, 171–184. [CrossRef]
10. Phang, J.M.; Pandhare, J.; Liu, Y. The metabolism of proline as microenvironmental stress substrate. *J. Nutr.* **2008**, *138*, 2008S–2015S. [CrossRef]
11. Krishnan, N.; Dickman, M.B.; Becker, D.F. Proline modulates the intracellular redox environment and protects mammalian cells against oxidative stress. *Free Radic. Biol. Med.* **2008**, *44*, 671–681. [CrossRef] [PubMed]
12. Treleaven, T.; Hardy, M.L.M.; Guttman-Jones, M.; Morris, M.B.; Day, M.L. In Vitro Fertilisation of Mouse Oocytes in L-Proline and L-Pipecolic Acid Improves Subsequent Development. *Cells* **2021**, *10*, 1352. [CrossRef] [PubMed]
13. Washington, J.M.; Rathjen, J.; Felquer, F.; Lonic, A.; Bettess, M.D.; Hamra, N.; Semendric, L.; Tan, B.S.; Lake, J.A.; Keough, R.A.; et al. L-Proline induces differentiation of ES cells: A novel role for an amino acid in the regulation of pluripotent cells in culture. *Am. J. Physiol.-Cell Physiol.* **2010**, *298*, C982–C992. [CrossRef] [PubMed]
14. Casalino, L.; Comes, S.; Lambazzi, G.; De Stefano, B.; Filosa, S.; De Falco, S.; De Cesare, D.; Minchiotti, G.; Patriarca, E.J. Control of embryonic stem cell metastability by L-proline catabolism. *J. Mol. Cell Biol.* **2011**, *3*, 108–122. [CrossRef] [PubMed]
15. D'Aniello, C.; Fico, A.; Casalino, L.; Guardiola, O.; Di Napoli, G.; Cermola, F.; De Cesare, D.; Tate, R.; Cobellis, G.; Patriarca, E.J.; et al. A novel autoregulatory loop between the Gcn2-Atf4 pathway and L-Proline metabolism controls stem cell identity. *Cell Death Differ.* **2015**, *22*, 1234. [CrossRef]
16. Comes, S.; Gagliardi, M.; Laprano, N.; Fico, A.; Cimmino, A.; Palamidessi, A.; De Cesare, D.; De Falco, S.; Angelini, C.; Scita, G.; et al. L-Proline induces a mesenchymal-like invasive program in embryonic stem cells by remodeling H3K9 and H3K36 methylation. *Stem Cell Rep.* **2013**, *1*, 307–321. [CrossRef]
17. Colonna, R.; Mangia, F. Mechanisms of amino acid uptake in cumulus-enclosed mouse oocytes. *Biol. Reprod.* **1983**, *28*, 797–803. [CrossRef]
18. Corbett, H.E.; Dube, C.D.; Slow, S.; Lever, M.; Trasler, J.M.; Baltz, J.M. Uptake of betaine into mouse cumulus-oocyte complexes via the SLC7A6 isoform of y+L transporter. *Biol. Reprod.* **2014**, *90*, 81. [CrossRef]
19. Haghighat, N.; Van Winkle, L.J. Developmental change in follicular cell-enhanced amino acid uptake into mouse oocytes that depends on intact gap junctions and transport system Gly. *J. Exp. Zool.* **1990**, *253*, 71–82. [CrossRef]
20. Van Winkle, L.J.; Campione, A.L.; Gorman, J.M. Na+-independent transport of basic and zwitterionic amino acids in mouse blastocysts by a shared system and by processes which distinguish between these substrates. *J Biol Chem* **1988**, *263*, 3150–3163. [CrossRef]
21. Anas, M.K.; Hammer, M.A.; Lever, M.; Stanton, J.A.; Baltz, J.M. The organic osmolytes betaine and proline are transported by a shared system in early preimplantation mouse embryos. *J. Cell. Physiol.* **2007**, *210*, 266–277. [CrossRef] [PubMed]
22. Jamshidi, M.B.; Kaye, P.L. Glutamine transport by mouse inner cell masses. *Reproduction* **1995**, *104*, 91–97. [CrossRef] [PubMed]
23. Tan, B.S.; Lonic, A.; Morris, M.B.; Rathjen, P.D.; Rathjen, J. The amino acid transporter SNAT2 mediates L-proline-induced differentiation of ES cells. *Am. J. Physiol.-Cell Physiol.* **2011**, *300*, C1270–C1279. [CrossRef] [PubMed]
24. Seow, H.F.; Broer, S.; Broer, A.; Bailey, C.G.; Potter, S.J.; Cavanaugh, J.A.; Rasko, J.E. Hartnup disorder is caused by mutations in the gene encoding the neutral amino acid transporter SLC6A19. *Nat. Genet.* **2004**, *36*, 1003–1007. [CrossRef]
25. Kleta, R.; Romeo, E.; Ristic, Z.; Ohura, T.; Stuart, C.; Arcos-Burgos, M.; Dave, M.H.; Wagner, C.A.; Camargo, S.R.; Inoue, S.; et al. Mutations in SLC6A19, encoding B0AT1, cause Hartnup disorder. *Nat. Genet.* **2004**, *36*, 999–1002. [CrossRef] [PubMed]
26. Broer, A.; Juelich, T.; Vanslambrouck, J.M.; Tietze, N.; Solomon, P.S.; Holst, J.; Bailey, C.G.; Rasko, J.E.; Broer, S. Impaired nutrient signaling and body weight control in a Na+ neutral amino acid cotransporter (Slc6a19)-deficient mouse. *J. Biol. Chem.* **2011**, *286*, 26638–26651. [CrossRef]
27. Vanslambrouck, J.M.; Broer, A.; Thavyogarajah, T.; Holst, J.; Bailey, C.G.; Broer, S.; Rasko, J.E. Renal imino acid and glycine transport system ontogeny and involvement in developmental iminoglycinuria. *Biochem. J.* **2010**, *428*, 397–407. [CrossRef]
28. Danilczyk, U.; Penninger, J.M. Angiotensin-converting enzyme II in the heart and the kidney. *Circ. Res.* **2006**, *98*, 463–471. [CrossRef]
29. Kowalczuk, S.; Broer, A.; Tietze, N.; Vanslambrouck, J.M.; Rasko, J.E.; Broer, S. A protein complex in the brush-border membrane explains a Hartnup disorder allele. *FASEB J.* **2008**, *22*, 2880–2887. [CrossRef]
30. Danilczyk, U.; Sarao, R.; Remy, C.; Benabbas, C.; Stange, G.; Richter, A.; Arya, S.; Pospisilik, J.A.; Singer, D.; Camargo, S.M.; et al. Essential role for collectrin in renal amino acid transport. *Nature* **2006**, *444*, 1088–1091. [CrossRef]

31. Bailey, C.G.; Metierre, C.; Feng, Y.; Baidya, K.; Filippova, G.N.; Loukinov, D.I.; Lobanenkov, V.V.; Semaan, C.; Rasko, J.E. CTCF Expression is Essential for Somatic Cell Viability and Protection Against Cancer. *Int. J. Mol. Sci.* **2018**, *19*, 3832. [CrossRef] [PubMed]
32. Camargo, S.M.; Makrides, V.; Virkki, L.V.; Forster, I.C.; Verrey, F. Steady-state kinetic characterization of the mouse B(0)AT1 sodium-dependent neutral amino acid transporter. *Pflug. Arch.* **2005**, *451*, 338–348. [CrossRef] [PubMed]
33. Broer, A.; Klingel, K.; Kowalczuk, S.; Rasko, J.E.; Cavanaugh, J.; Broer, S. Molecular cloning of mouse amino acid transport system B0, a neutral amino acid transporter related to Hartnup disorder. *J. Biol. Chem.* **2004**, *279*, 24467–24476. [CrossRef] [PubMed]
34. Arnaiz, I.; Johnson, M.H.; Cook, D.I.; Day, M.L. Changing expression of chloride channels during preimplantation mouse development. *Reproduction* **2013**, *145*, 73–84. [CrossRef] [PubMed]
35. Sonoda, M.; Okamoto, F.; Kajiya, H.; Inoue, Y.; Honjo, K.; Sumii, Y.; Kawarabayashi, T.; Okabe, K. Amino acid-permeable anion channels in early mouse embryos and their possible effects on cleavage. *Biol. Reprod.* **2003**, *68*, 947–953. [CrossRef]
36. Baltz, J.M. Osmoregulation and cell volume regulation in the preimplantation embryo. *Curr Top Dev Biol* **2001**, *52*, 55–106. [CrossRef]
37. Guastella, J.; Brecha, N.; Weigmann, C.; Lester, H.A.; Davidson, N. Cloning, expression, and localization of a rat brain high-affinity glycine transporter. *Proc. Natl. Acad. Sci. USA* **1992**, *89*, 7189–7193. [CrossRef]
38. Hammer, M.A.; Baltz, J.M. Betaine is a highly effective organic osmolyte but does not appear to be transported by established organic osmolyte transporters in mouse embryos. *Mol. Reprod. Dev.* **2002**, *62*, 195–202. [CrossRef]
39. Van Winkle, L.J.; Haghighat, N.; Campione, A.L.; Gorman, J.M. Glycine transport in mouse eggs and preimplantation conceptuses. *Biochim. Biophys. Acta* **1988**, *941*, 241–256. [CrossRef]
40. Hobbs, J.G.; Kaye, P.L. Glycine uptake in pre-implantation mouse embryos: Kinetics and the effects of external [Na+]. *Reprod. Fertil. Dev.* **1990**, *2*, 651–660. [CrossRef]
41. Hardy, M.L.M.; Day, M.L.; Morris, M.B. Redox Regulation and Oxidative Stress in Mammalian Oocytes and Embryos Developed In Vivo and In Vitro. *Int. J. Environ. Res. Public Health* **2021**, *18*, 11374. [CrossRef] [PubMed]
42. Cuny, H.; Bozon, K.; Kirk, R.B.; Sheng, D.Z.; Broer, S.; Dunwoodie, S.L. Maternal heterozygosity of Slc6a19 causes metabolic perturbation and congenital NAD deficiency disorder in mice. *Dis. Model. Mech.* **2022**, *16*, dmm049647. [CrossRef] [PubMed]

 cells

MDPI

Review

Challenges and Considerations during In Vitro Production of Porcine Embryos

Paula R. Chen [1], Bethany K. Redel [2], Karl C. Kerns [3], Lee D. Spate [1,4] and Randall S. Prather [1,4,*]

[1] Division of Animal Sciences, University of Missouri, Columbia, MO 65211, USA
[2] USDA-ARS, Plant Genetics Research Unit, Columbia, MO 65211, USA
[3] Department of Animal Science, Iowa State University, Ames, IA 50011, USA
[4] National Swine Resource and Research Center, University of Missouri, Columbia, MO 65211, USA
* Correspondence: PratherR@missouri.edu

Abstract: Genetically modified pigs have become valuable tools for generating advances in animal agriculture and human medicine. Importantly, in vitro production and manipulation of embryos is an essential step in the process of creating porcine models. As the in vitro environment is still suboptimal, it is imperative to examine the porcine embryo culture system from several angles to identify methods for improvement. Understanding metabolic characteristics of porcine embryos and considering comparisons with other mammalian species is useful for optimizing culture media formulations. Furthermore, stressors arising from the environment and maternal or paternal factors must be taken into consideration to produce healthy embryos in vitro. In this review, we progress stepwise through in vitro oocyte maturation, fertilization, and embryo culture in pigs to assess the status of current culture systems and address points where improvements can be made.

Keywords: porcine embryo culture; in vitro maturation; in vitro fertilization

Citation: Chen, P.R.; Redel, B.K.; Kerns, K.C.; Spate, L.D.; Prather, R.S. Challenges and Considerations during In Vitro Production of Porcine Embryos. *Cells* **2021**, *10*, 2770. https://doi.org/10.3390/cells10102770

Academic Editor: Lon J. van Winkle

Received: 22 September 2021
Accepted: 13 October 2021
Published: 15 October 2021

1. Introduction

In vitro production of embryos has several advantages over in vivo-derived embryo production, including efficient selection of superior genetics for transfer or genetic modification to rapidly obtain animals with desirable traits. Genetic engineering provides a powerful tool to aid in understanding basic mechanisms regulating animal physiology. Several studies using genetic engineering approaches have been conducted to ascertain the function of signaling molecules and enzymes during early pregnancy in pigs [1–4]. Importantly, interleukin 1 beta 2 (IL1B2) was shown to be required for conceptus elongation to proceed [1]. For production agriculture, embryos can be modified to produce animals with better carcass traits, such as improved fatty acid profiles with increased levels of omega-3 fatty acids [5], and resistance to diseases, such as porcine reproductive and respiratory syndrome (PRRS), that result in millions of dollars of losses per year [6,7]. Moreover, genetically modified swine have become powerful tools for studying genetic diseases in humans, such as cystic fibrosis [8] and phenylketonuria [9], and for xenotransplantation [10]. Therefore, production and manipulation of porcine embryos in vitro is crucial for advancements in agriculture and human medicine.

Each step of the culture system, from oocyte maturation to embryo culture, has been continuously optimized and adjusted based on the requirements of porcine oocytes and embryos. Our current system, from collection of cumulus-oocyte complexes (COCs) to surgical transfer into surrogates, is outlined in Figure 1 with media formulations described in the subsequent sections. However, in vitro culture of porcine embryos is still suboptimal as only about 40% of presumptive fertilized oocytes develop to the blastocyst stage for embryo transfer [11], and the number of cells in the blastocyst-stage embryos is decreased compared to those that developed in vivo [12]. Clues from other species have provided insight for culture of porcine embryos, but embryos from every species exhibit unique metabolic characteristics that need to be considered for media formulations. Even with advances in porcine embryo culture media formulations, stress from the environment and maternal or paternal factors can impact the developmental trajectory of the embryos. Thus, a significant amount of information has yet to be learned regarding the interaction of embryos with the culture environment, the ways in which the culture system can be modified to improve developmental viability, and important metabolic features that can be used to distinguish healthy embryos.

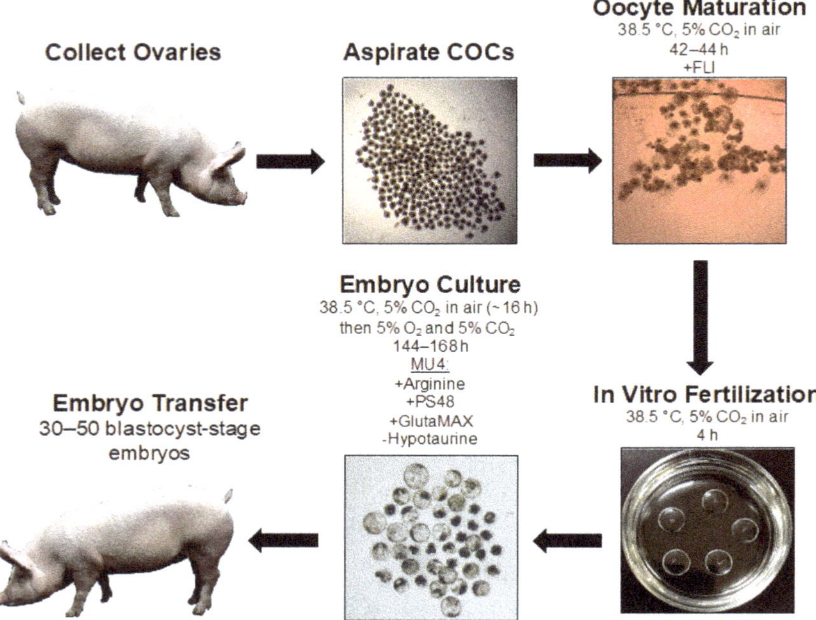

Figure 1. An overview of our current porcine embryo production pipeline. FLI: cocktail of fibroblast growth factor 2, leukemia inhibitory factor, and insulin-like growth factor 1. MU4 porcine embryo culture medium is a modified porcine zygote medium (PZM) that contains supplemental arginine, PS48, and GlutaMAX, and hypotaurine is removed from the formulation. Approximately 30–50 day 5 or 6 morula- and blastocyst-stage embryos are surgically transferred into a surrogate.

Herein, we provide an overview of the current porcine embryo culture system with a focus on the metabolic characteristics of porcine embryos. Moreover, measurements of embryo quality, stressors arising from the culture system, and parental factors that influence development are discussed.

2. Oocyte Quality and Maturation

2.1. Selection Criteria

In vitro maturation (IVM) of oocytes is the first critical step in the process of producing competent embryos that are able to produce viable offspring. While progress has been made to improve IVM of pig oocytes, the rate and the extent of maturation is considerably lower than in vivo-matured oocytes, suggesting a suboptimal maturation environment [13]. In vivo, the maternal ovarian follicle environment supplies the immature oocyte with the necessary components to develop and reach its developmental competence by maintaining an intimate relationship with its neighboring somatic cells. The bidirectional communication between the oocyte and cumulus cells allows for the oocyte to gradually acquire the necessary molecular and cytoplasmic machinery needed to support embryo development [14,15].

Evidence suggests that the key factor in determining the proportion of oocytes that develop to the blastocyst stage appears to be the intrinsic quality of the oocyte that begins the in vitro production process [16]. Given this, the ability to select the most competent oocytes that are capable of being fertilized in vitro and produce healthy young is imperative. General selection parameters to determine oocyte competence in the pig are the age of the oocyte donor [17,18], follicle size [19], and the layers of cumulus cells and subsequent cumulus cell expansion during culture [20]. Oocytes that originate from small (<3 mm) follicles are less likely to mature to metaphase II (MII) or develop to the blastocyst stage compared to oocytes derived from medium (3–5 mm) and large (>5 mm) follicles [21]. Moreover, larger follicles (>5 mm) have been shown to have higher β-estradiol concentrations, and high quality COCs from these follicles have increased abundance of transcripts involved in ovarian steroidogenesis [22]. Historically, cumulus cell expansion has been considered to be required for successful maturation in vitro and often the degree of cumulus cell expansion is correlated with improved oocyte maturation [23,24]. As another factor to consider, oocyte quality is also impacted by the energy status of the female as feed restriction (50% per day) during the last two weeks of lactation decreased follicle size, concentration of steroids in the follicular fluid, and maturation to MII [22].

2.2. Supplementation of Growth Factors

A cocktail of growth factors, fibroblast growth factor 2 (FGF2), leukemia inhibitory factor (LIF), and insulin-like growth factor 1 (IGF1) (termed FLI) supplemented only during oocyte maturation of prepubertal gilt COCs, was found to provide a four-fold increase in the number of piglets born per oocyte collected [11]. These FLI-cultured COCs had a distinct mitogen-activated protein kinase (MAPK) activation pattern in the cumulus cells, displayed a higher degree of cumulus cell expansion, and accelerated the disruption of gap junctions compared to control COCs. However, a complementary study found that in the absence of gonadotropins but with supplementation of FLI, cumulus cells showed little expansion but maturation to MII was not impeded in oocytes from prepubertal gilts, and those oocytes were competent to produce healthy piglets [25]. While cumulus cell expansion was not needed to produce competent oocytes, cumulus cells are required as their removal at the beginning of maturation and even up to 24 h post-start of maturation impedes oocyte development to MII and the blastocyst stage after fertilization. Our knowledge and understanding about the relationship of the oocyte, cumulus cells, growth factors, and hormones are constantly evolving as a highly complex interaction is revealed. More research is needed to understand the molecular mechanisms and signaling events by which FLI can promote oocyte competence.

3. In Vitro Fertilization

Historically, the lack of fertilization of porcine oocytes in vitro [26] was followed by tremendous polyspermy [27]. The inability to achieve consistent monospermy has been a major problem for in vitro fertilization (IVF) of the pig. In the mid-1990s, Billy Day from the University of Missouri was quoted as saying, "For years, we tried to get the sperm in, and now, we can't keep them out." Biologically, the block to polyspermy in mammals is via two distinct mechanisms. The first is a membrane block that acts relatively quickly [28], and the second is a block at the level of the zona pellucida that takes minutes. The relative importance of these two mechanisms is different between mammalian species. In pigs, humans, and mice, both mechanisms are used. The first relatively quick membrane block appears to act by changing the binding characteristic of the plasma membrane of the oocyte, whereas the second zona pellucida block is a result of the release of cortical granule contents into the space between the oolemma and the zona pellucida. These contents (e.g., ovastacin [29], among others [30]) modify the proteins of the zona pellucida such that sperm penetration is inhibited. Thus, these two mechanisms serve to complement each other in preventing multiple sperm from entering the oocyte cytoplasm.

In the pig, the block at the zona pellucida is relatively strong and is correlated with cortical granule release [31], while the membrane block is regulated by calreticulin [32]. Many efforts have been made to alter the fertilization conditions to improve monospermy rates. These efforts have included altering the concentration of sperm while varying the duration of exposure between the sperm and oocytes, adding follicular fluid [33] or substances found in oviductal fluid, such as deleted in malignant brain tumors 1 (DMBT1) [34], oviductal glycoprotein 1 (OVGP1) [35], secreted phosphoprotein 1 (SPP1) [36], or plasminogen [37,38], as well as trying to recreate the oviductal environment by adding oviductal fluid [39,40] or oviductal extracellular vesicles [41], among other things [42].

The problem of polyspermy encountered during IVF is generally confounded with in vitro-matured oocytes. Improvements in the quality of the oocytes will likely improve the success of monospermic fertilization. As in vitro-matured oocytes have a propensity to retain their transzonal projections, it was thought that these projections may interfere with sperm-oocyte binding and cortical granule exocytosis [43]. Retraction of those projections at the end of oocyte maturation may permit more timely and complete cortical granule exocytosis (the slower block to polyspermy), as well as altering the composition of the plasma membrane to provide a better fast block to polyspermy at the level of the plasma membrane. The aforementioned maturation system containing FLI has dramatically altered the timing of the retraction of the transzonal projections and improved the overall developmental competence of the oocytes [11]. As compared to other reports in the literature, polyspermy is less of a problem in our hands. A fertilization rate of 50–60% with monospermy of 70–80% is routinely achieved. Thus, on a per oocyte basis, the overall success is 35–50% monospermy. This range of monospermy corresponds well with our rates of development to the blastocyst stage (30–60%).

Another alternative to counteract polyspermy is to perform intracytoplasmic sperm injection (ICSI), which has led to the production of live piglets [44,45]. However, delays in cleavage divisions and decreased blastocyst formation of approximately 10–20% have been observed in ICSI-derived porcine embryos compared to those produced by IVF [46]. These decreases in development may be due to selection of damaged sperm and/or incomplete oocyte activation. Importantly, boar sperm selected by a Percoll gradient were shown to have higher abundance of phospholipase C-ζ (PLCζ), a well-studied sperm-borne oocyte activation factor, and the use of Percoll-selected sperm for ICSI increased monospermic fertilization and oocyte activation as well as development to the blastocyst stage [47]. Several protocols use polyvinylpyrrolidone (PVP) for sperm immobilization before ICSI; however, this compound is toxic when injected into oocytes. When boar sperm were selected by hyaluronic acid (HA), known as physiological intracytoplasmic sperm injection (PICSI), versus PVP (conventional ICSI) or IVF, increases in development to the blastocyst stage and total cell numbers were observed [48].

4. Comparing Metabolic Characteristics of Mammalian Embryos

4.1. Carbohydrates

The metabolism of preimplantation embryos differs drastically from that of adult somatic cells. In normal conditions, somatic cells metabolize glucose through glycolysis, yielding two ATP, two NADH, and two pyruvate molecules. Pyruvate is further metabolized through the TCA cycle to produce electron donors, NADH and $FADH_2$, for generation of 30–36 ATP molecules by oxidative phosphorylation. In the precompaction embryo, ATP is generated by oxidative phosphorylation at a reduced but optimized level, which is sustained through oxidation of pyruvate, amino acids, and fatty acids [49]. A low level of pyruvate oxidation is deemed ideal to avoid excess production of reactive oxygen species (ROS) by the electron transport chain. After compaction, the primary metabolic pathway is glycolysis that is characterized by the Warburg effect (WE), which is a phenomenon first observed in cancer cells by Otto Warburg [50,51]. Under the WE, glycolytic intermediates are shuttled towards the pentose phosphate pathway (PPP) and lactate production instead of the TCA cycle even in the presence of oxygen. The PPP supports rapid proliferation of cells by producing ribose-5-phophate, a precursor for nucleotides, and NADPH, which is involved in redox regulation and lipid synthesis [52,53]. Recent studies have shown that glucose does not drive ATP generation during most stages of preimplantation development, but that it is metabolized through the PPP and the hexosamine biosynthesis pathway to control localization of transcription factors, yes-associated protein 1 (YAP1) and transcription factor AP-2 gamma (TFAP2C), for cell fate specification [54]. Thus, generation of biomass, such as DNA and membrane lipids, and orchestration of key events during development are potentially the main roles of glucose to support rapid proliferation during these early developmental stages.

Classically, pyruvate and lactate are known to be important carbohydrates during early stages of preimplantation development, and glucose becomes a key energy source after compaction [55–57]. Pig embryos have decreased metabolism of pyruvate compared to mouse and sheep embryos, pointing to a reliance on other energy sources, such as glucose or lipids [58–61]. Studies in several species, including mouse, hamster, sheep, cow, and human, have reported that glucose inhibits the development of embryos in vitro [62–66]. Glucose does not appear to be inhibitory for porcine preimplantation embryo development in vitro, but glucose in the presence of inorganic phosphate has been shown to be deleterious [67,68]. In porcine embryos, glucose metabolism minimally increases from the 1-cell to 8-cell stage (0.4 ± 0.1 to 3.1 ± 0.4 pmol/embryo/4 h) but notably increases in the compacted morula (49.5 ± 2.7 pmol/embryo/4 h) and further by the blastocyst stage (147 ± 12.0 pmol/embryo/4 h) [69]. Subsequent studies have also shown that glucose metabolism increases after compaction with glycolysis becoming a key metabolic pathway, but metabolic activity is significantly decreased in in vitro-produced embryos compared to in vivo counterparts [56,61]. However, porcine embryos can be cultured to the blastocyst stage in medium with no glucose but containing only pyruvate and lactate [70].

4.2. Amino Acids

Amino acids serve numerous functions in promoting the development of preimplantation embryos. Oviductal and uterine fluids contain varying concentrations of amino acids, which has influenced media formulations for several species [71–73]. Aside from protein synthesis, amino acids are involved in other activities, such as ATP production, nucleotide synthesis, lipid synthesis, antioxidant production, cell signaling, osmolarity, and pH regulation. Uptake of amino acids from the microenvironment by preimplantation embryos relies on the activity of different transport systems, which are expressed during different stages [74]. For example, in porcine oocytes, Na^+-dependent transport of L-alanine and L-leucine was shown to be competitive and occurred by using the $B^{0,+}$ system [75]. At the blastocyst stage, L-leucine was transported in a Na^+-dependent manner through the L system. Transport of L-leucine into the embryo has been shown to be involved in activation of mechanistic target of rapamycin complex 1 (MTORC1) to promote trophoblast motility [76]. In support of this, treatment of mouse blastocyst-stage embryos with 200 nM rapamycin, an inhibitor of MTORC1, blocked outgrowth formation in vitro [74]. Sodium-dependent transport of L-aspartate and L-glutamate was not detected in porcine oocytes but was observed at the blastocyst stage, demonstrating that the X^-_{AG} system becomes active at later stages in development [77].

The addition of essential amino acids to embryo culture media has been shown to be inhibitory for development to the blastocyst stage in mouse and cattle embryos [78,79]. However, Steeves and Gardner [80] demonstrated that presence of essential amino acids in medium with nonessential amino acids and glutamine was not inhibitory during cattle embryo cleavage and increased development to the blastocyst stage. In porcine parthenogenetic embryos cultured in modified Whitten's medium with polyvinyl alcohol (PVA), essential amino acids were inhibitory if present at the 1-cell stage but supported development to the blastocyst stage when added after 48 h of culture. Specifically, presence of nonpolar essential amino acids, valine, leucine, isoleucine, and methionine, during the first 48 h of culture was shown to inhibit development past the 4-cell stage [81]. However, the porcine zygote medium (PZM) variants, containing both essential and nonessential amino acids, have been shown to support development of porcine embryos to the blastocyst stage [70].

Furthermore, transcriptional profiling of porcine blastocyst-stage embryos revealed that transcripts related to amino acid transport and metabolism were dysregulated in those cultured in vitro versus those derived in vivo [12]. In vitro-produced blastocyst-stage embryos had increased abundance of solute carrier 7A1 (*SLC7A1*), which encodes an arginine transporter, and supplementation of 1.69 mM arginine to PZM-3 decreased abundance of *SLC7A1* [82]. Arginine is metabolized for production of nitric oxide, and addition of nitric oxide synthase inhibitors to culture media has been shown to inhibit development of mouse and pig embryos [82,83]. Similarly, in vitro-produced porcine embryos had increased abundance of solute carrier 6A9 (*SLC6A9*), which encodes a glycine transporter, and supplementation of 10 mM glycine decreased *SLC6A9* abundance and improved development of blastocyst-stage embryos [84]. However, transfer of blastocyst-stage embryos cultured with 10 mM glycine did not result in pregnancies, unlike controls cultured with 0.1 mM glycine, showing that more cells at the blastocyst stage is not necessarily an indicator of developmental competence. Furthermore, early mouse and hamster embryo culture media formulations demonstrated the importance glutamine during cleavage stages to overcome the 2-cell block and for the development of blastocyst-stage embryos [62,85]. Glutamine has been shown to support development of porcine embryos from the 1- to 2-cell stage to the blastocyst stage in vitro without the presence of glucose [68]. Recently, glutamine supplementation to PZM-3 has been shown to correct abundance of transcripts related to glutamine transport and metabolism and increase activation of mTORC1 in porcine blastocyst-stage embryos [86,87]. Thus, RNA-sequencing is a powerful tool that can identify amino acid requirements of the embryo in culture, but viability in vivo must be assessed before supplementation of specific amino acids becomes common practice.

4.3. Lipids

In addition to carbohydrates, beta-oxidation of fatty acids is thought to contribute to sustaining oxidative phosphorylation in preimplantation embryos. Sturmey and Leese [60] demonstrated that oxygen consumption and ATP production increase at the early blastocyst stage, and ATP is mainly produced via aerobic metabolism. Treatment of mouse embryos with inhibitors of beta-oxidation, methyl palmoxirate and etomoxir, decreased the number of embryos reaching the blastocyst stage and total cell numbers within the embryos [88,89]. Interestingly, the inhibitory effects of etomoxir were decreased when porcine oocytes were matured in the presence of high glucose (4.0 mM) compared to low glucose (1.5 mM), thus other energy sources can compensate when beta-oxidation is blocked [90]. On the contrary, addition of L-carnitine, which is essential for uptake of fatty acids into the mitochondria, to embryo culture medium improved mitochondrial activity and development in embryos of several species [88,91–93].

Porcine oocytes and embryos have a dark appearance compared to other species due to presence of abundant lipid. Mouse oocytes were determined to have approximately 4 ng of fatty acids; sheep and cattle oocytes have a slightly darker appearance and about 89 and 63 ng of fatty acids, respectively; pig oocytes, almost black in appearance, have about 156 ng of fatty acids [49,94,95]. The most common fatty acids in pig oocytes are palmitic (16:0), stearic (18:0), and oleic (18:1n-9) acids [95]. Compared to other species, the large amount of lipid in pig oocytes is potentially the result of increased message for diacylglycerol acyltransferase 1 (*DGAT1*), which encodes an enzyme involved in the transfer of a fatty acyl CoA to a diacylglycerol to form triglycerides [96]. Moreover, inhibition of stearoyl-coenzyme A desaturase 1 (SCD1), which introduces a cis double bond in saturated fatty acids, decreased development to the blastocyst stage and lipid droplet formation in porcine parthenogenetic embryos; however, cotreatment with 100 μM oleic acid rescued development [97]. Sturmey and Leese [60] observed that triglyceride levels did not change from cleavage stages to the blastocyst stage, indicating that beta-oxidation might not be a key metabolic pathway in the presence of other energy sources. However, Romek et al. [98] noted that the volume of lipid droplets per unit of cytoplasm decreased at the blastocyst stage in both in vitro-produced and in vivo-derived porcine embryos, demonstrating controversy in this area.

The large amount of lipid in porcine oocytes and embryos has been identified as a barrier to vitrification and survival after thawing. In addition, porcine oocytes have a greater percentage of polyunsaturated fatty acids compared to ruminants, which further impairs survival after freezing [95]. L-carnitine supplementation, centrifugation in a high-osmolarity solution to separate lipids from the cytoplasm, and removal of lipids by micromanipulation have been shown to be effective methods of increasing the cryotolerance of porcine oocytes and embryos [99–102].

5. Strategies for Improving Current Culture Systems

5.1. Media Formulations and Supplements

Several media formulations exist that support development of porcine zygotes to the blastocyst stage, including modified Whitten's medium [103], North Carolina State University (NCSU)-23 medium [104], modified CZB medium [105], Beltsville embryo culture medium (BECM)-3 [106], and porcine zygote medium (PZM) variants [70]. Currently, NCSU-23 and PZM variants are most commonly used for porcine embryo culture. NSCU-23 contains glucose and glutamine as primary energy sources as well as taurine and hypotaurine to mediate osmolarity but lacks pyruvate, lactate, and all other amino acids [104]. PZM variants were formulated based on the composition of porcine oviductal fluid, containing pyruvate and lactate instead of glucose, and they lack taurine but include all other amino acids. Furthermore, PZM-3 contains bovine serum albumin (BSA), similar to NCSU-23, but culturing IVF or somatic cell nuclear transfer (SCNT)-derived embryos in PZM-3 was shown to improve development to the blastocyst stage and increase the number of cells in blastocyst-stage embryos compared to NCSU-23 [70,107]. PZM-4 substitutes PVA for BSA in an attempt to create a chemically defined medium [70]. Bovine serum albumin can contain undefined components, such as citrate and lipids, and different lots need to be tested before being routinely added to culture media. When embryos were cultured in PZM-4 compared to PZM-3, no difference in development was detected, and live piglets were born after embryo transfer [70]. However, subsequent studies demonstrated decreases in development to the blastocyst stage by using PZM-4 compared to PZM-3 [108].

Additives to PZM-3, such as amino acids and small molecules, have been shown to further improve porcine embryo development. For example, supplemental concentrations of arginine increased development to the blastocyst stage, improved embryo quality, and modulated transcript abundance related to arginine transport, resulting in a new medium named MU1 [82]. MU2 was developed after addition of 5-(4-Chloro-phenyl)-3-phenyl-pent-2-enoic acid (PS48) was shown to be able to replace BSA and improved development through stimulation of phosphoinositide 3 kinase (PI3K) to increase phosphorylation of v-akt murine thymoma viral oncogene homolog (AKT) [109]. Afterwards, MU3 was developed by replacing 1 mM glutamine with 3.75 mM GlutaMAX, an L-alanyl-L-glutamine dipeptide, which improved development to the blastocyst stage and increased mitochondrial activity [86]. Currently, MU4 has been implemented, which does not contain hypotaurine as removal of this component did not have a negative effect when embryo culture is conducted at 5% O_2 [110], and hypotaurine is a relatively expensive ingredient.

5.2. Morphological and Chromosomal Quality

Embryos developing in vivo are exposed to the oviductal and uterine epithelia, which are involved in provision of nutrients, removal of toxins, and production of antioxidant systems [111]. Embryos of several species fertilized and cultured in vitro demonstrate a delay of about one cell division to the blastocyst stage compared to in vivo-derived embryos [12,112–114]. Total cell numbers in porcine blastocyst-stage embryos produced in vitro are decreased and have higher ratios of trophectoderm to inner cell mass [12,112]. Conventionally, morphology of an embryo has been the main criterion for determining quality [115]. However, morphology should not necessarily be the primary selection standard as porcine embryos cultured with supplemental glycine exhibited improved development and cell numbers but failed to establish pregnancies after 11 embryo transfers [84]. When the synchronization protocol was altered to transfer embryos cultured with supplemental glycine on an earlier day of standing estrus (typically performed on day 3, 4, or 5 of standing estrus), one pregnancy was obtained (unpublished data), indicating that these embryos may have increased developmental competence. Similarly, addition of N-methyl-D-aspartic acid (NMDA) and homocysteine (HC) to PZM-4 increased the percentage of porcine embryos that developed to the blastocyst stage and sizes of the embryos over PZM-4 alone, but 16 transfers of embryos cultured with only NMDA or NMDA and HC did not result in any live piglets [108]. The use of time lapse monitoring of development in vitro has been shown to increase pregnancy rates over conventional morphological assessment [116]. This technology allows for the consideration of other factors, such as timing of first cleavage and number of blastomeres after the first cleavage, which have both been shown to be more reliable indicators of quality [117,118]. The ability of porcine embryos to reach the morula stage before 102 h was correlated with increases in reaching the blastocyst stage, and the percentage of fragmentation negatively correlated with developmental progression [119].

Furthermore, chromosomal abnormalities can occur during development that significantly increase the chances of pregnancy loss or defects after birth. By using comparative genomic hybridization, approximately 14% of in vivo-derived porcine embryos (72 h after insemination) were observed to be aneuploid [120]. However, about 39% of in vitro-produced porcine blastocyst-stage embryos demonstrated chromosomal abnormalities [121]. Assessment of chromosomes 6 and 7 in in vitro-produced bovine blastocyst-stage embryos revealed that 72% of the embryos contained cells that were polyploid; however, the proportion of abnormal cells within each embryo was generally low (<10% of the total cells) [122]. More recent studies have observed high occurrence chromosomal aberrations in at least one blastomere of IVM-IVF bovine embryos as compared to in vivo-derived embryos (85% vs. 19%) [123] and that entire parental genomes can segregate into different cell lineages during cleavage divisions [124].

5.3. Mitochondrial Function

Mitochondria are dynamic organelles with important roles during fertilization and preimplantation development. From the zygote to expanded blastocyst stage, mitochondrial morphology changes from spherical with minimal cristae to elongated with numerous transverse cristae, demonstrating a major shift in metabolic activity during preimplantation development [125]. Mitochondrial membrane potential ($\Delta\Psi m$) is formed by the pumping of protons across the inner membrane during oxidative phosphorylation and is commonly used as an indirect measure of mitochondrial function and developmental progression in preimplantation embryos. For example, JC-1 is a cationic, lipophilic dye that remains in its monomeric form at low (<100 mV) $\Delta\Psi m$ and fluoresces green but forms aggregates inside mitochondria at high (>140 mV) $\Delta\Psi m$ and fluoresces red [126]. Thus, the ratio of red to green fluorescence is an indicator of mitochondrial membrane potential and hence activity within an embryo. JC-10, which has better water solubility than JC-1, can effectively visualize mitochondrial activity in porcine blastocyst-stage embryos derived by IVF and SCNT [86,127], and glutamine supplementation into the culture medium increased the mitochondrial activity of the IVF-derived embryos [78]. Additionally, MitoTracker™ Green is non-fluorescent in aqueous solutions but accumulates in the mitochondrial matrix regardless of $\Delta\Psi m$ and fluoresces green to measure total mitochondrial mass. MitoTracker™ Red or Orange are oxidized by molecular oxygen in mitochondria to emit a red-orange fluorescence that can be used to measure oxidative activity [128]. MitoTracker™ Red has been used to determine mitochondrial activity of porcine IVF-derived blastocyst-stage embryos after culture with supplemental glycine; however, no difference in mitochondrial activity was observed [84].

5.4. Transcriptional Profiling

Transcriptional profiling has been used to elucidate "the needs" of the embryo. Studies using serial analysis of gene expression (SAGE) and microarrays revealed differences in abundance for transcripts related to cellular metabolism and transcriptional regulation in in vitro-produced porcine blastocyst-stage embryos compared to in vivo-derived counterparts [129,130]. By using next-generation sequencing (NGS), Bauer et al. [12] identified 1170 differentially abundant transcripts between in vivo-derived and in vitro-produced porcine blastocyst-stage embryos. Analysis of these data revealed that amino acid transport and metabolism were perturbed in in vitro-produced embryos. Specifically, message for an arginine transporter, *SLC7A1*, and a glycine transporter, *SLC6A9*, were upregulated in in vitro-produced embryos by 63- and 25-fold, respectively. As mentioned previously, supplementation of arginine to 1.69 mM increased development to the blastocyst stage and decreased abundance of *SLC7A1* to levels observed in in vivo-derived embryos [82]. Likewise, glycine supplementation to 10 mM improved developmental parameters of the embryos and decreased abundance of *SLC6A9*; however, pregnancies were not established after 11 transfers with embryos cultured in supplemental glycine [84]. Additionally, supplementation of glutamine improved development of porcine embryos and corrected abundance of transcripts related to glutamine transport and metabolism [78]. The use of RNA-sequencing datasets for defining and improving embryo culture is still in its infancy stages as there are numerous pathways yet to explore for improving media formulations.

5.5. Metabolomics

Consumption or production of metabolites from the culture medium has been used as a noninvasive marker of embryo viability. Metabolomics assays are largely applied in IVF clinics to select embryos for transfer; however, these techniques are also useful for selection of embryos from agricultural species for cryopreservation or transfer. Metabolism of glucose, pyruvate, and lactate has been a canonical measure of preimplantation embryo competence, which was studied by using ultramicrofluorescence assays that measure NADH oxidation to represent pyruvate conversion to lactate or NADPH formation to represent glucose uptake [131]. Hardy et al. [132] observed that increased pyruvate consumption by early cleavage-stage human embryos correlated with increased development to the blastocyst stage. Regarding in vitro-produced porcine embryos, metabolism of glucose for glycolysis and pyruvate for the TCA cycle increased after compaction, indicating greater metabolic activity [61]. Glutamine and arginine have been shown to be consumed from culture media by human and pig embryos [133,134]. However, increased arginine consumption was associated with increased development to the blastocyst stage in pig embryos [133] but was associated with decreased development to the blastocyst stage in human embryos [134].

Currently, different platforms are being used to study metabolomic profiles of preimplantation embryos. Gas or liquid chromatography (GC or LC) coupled with mass spectrometry (MS) are ideal for analyzing metabolites in small volumes of culture media (5–10 μL). After derivatization, the mass of compounds allows for the identification of metabolites based on database information [52]. This technique has been successfully applied to porcine embryos to determine the production and consumption of metabolites from the medium after glutamine supplementation [86]. Culturing embryos with supplemental glutamine increased leucine consumption from the medium as well as activation of mTORC1 [78,79]. Additionally, nuclear magnetic resonance (NMR) spectroscopy has been used to analyze metabolites in spent culture media; however, this technique requires larger sample volumes (25 μL) [135,136].

5.6. Microfluidics

Changing the type of culture system may provide methods to better customize the environment surrounding the developing embryos. Most current systems are static in a single atmosphere. The ability to continuously modulate the composition of the culture medium as the needs of the embryo change may provide dramatic improvements in developmental competence. Microfluidic devices are in development, and there are reports in the literature specific to pigs. Such devices have been reported to reduce osmotic stress [137], remove the zona pellucida [138], facilitate fertilization and reduce polyspermy [139,140], and customize the culture environment to improve development [141,142]. Moreover, microfluidic devices may improve tolerance of embryos to cryopreservation as addition and removal of cryoprotectants inflicts osmotic shock that may be detrimental to development. Use of a microfluidic device to gradually change the osmotic pressure during cryoprotectant addition or removal may increase the cryosurvival of pig oocytes or embryos [137].

5.7. Extended Culture

By day 12 of gestation, porcine embryos have undergone several changes in morphology, including spherical, ovoid, tubular, and filamentous forms, transitioning from tubular to approximately 100 mm in length in 1 to 2 h. Methods of culturing porcine embryos past the blastocyst stage as a measure of competence have been met with challenges that have hindered progress in this area. Spherical embryos encapsulated in double-layered alginate beads were able to attain a tubular form with increased estradiol-17β production in the culture medium [143]. However, the embryos did not elongate into filamentous forms; thus, maternally secreted factors and extracellular matrix components may be required to initiate this process. Chemical modifications to the alginate hydrogels, such as covalent attachment of RGD peptides or incorporation of 0.1 μg/mL SPP1, increased survival and promoted morphological changes [144]. Nonetheless, full elongation of porcine embryos in vitro has not been achieved and will require further investigation.

5.8. Cryopreservation

In many species, cryopreservation of sperm, embryos, and somatic cells is straightforward. Cryopreservation of pig somatic cells, such as fibroblast cells, can be achieved with conventional freezing systems used for other species. In contrast, pig sperm and embryos present some unique challenges. In all likelihood, the cell type with most variable survivability is boar sperm. Variation in viability is season-, breed-, boar- and ejaculate-specific. Commercial application of artificial insemination has historically been by using fresh semen. Since pigs are litter bearing, considerable genetic selection pressure can be placed on the offspring resulting from artificial insemination. As genetic progress can be made by using fresh semen, the lowered success rates from cryopreserved semen does not justify the lower farrowing rate (~20–30%) and decrease in litter size as compared to fresh semen [145].

While cryopreservation of semen presents difficulties, embryos are even more challenging. It is thought that the high lipid content of pig embryos presents the main obstacle to successful cryopreservation. To circumvent this problem, Nagashima [102,146] developed a method to centrifuge the early embryo, thus stratifying the lipids within the zona pellucida. Then, the lipids could be removed via micromanipulation, and the resulting embryos were cryopreserved by using conventional methods. One would reason that the lipids were necessary for development of the embryo; however, in at least one report, development was enhanced after removal [99]. Lipid removal by micromanipulation is labor-intensive. High speed centrifugation to completely separate the lipids within the zona pellucida has been used as a high throughput lipid removal technique for cryopreservation of in vitro-produced embryos [147]. Solid surface vitrification of oocytes or zygotes may be another alternative as 15 piglets were derived from 8 embryo transfers [148].

Although few reports on conventional cryopreservation of porcine embryos exist, those reports of successful cryopreservation of early pig embryos have not been widely repeatable, and the industry has not adopted the transfer of frozen embryos as a method of improving genetics or moving genetics around the world. As expected, in vitro-derived embryos are less viable than in vivo-derived embryos, and thus do not survive the rigors of cryopreservation to the same extent as in vivo-derived embryos. Continual efforts are being made to improve the quality of oocytes matured and embryos cultured in vitro. Improvements, such as adding FLI to the oocyte maturation system, may result in increases in cryosurvival in the pig as has been observed in cattle [149]. Two recent reports describe the production of piglets after cryopreservation and embryo transfer. After transferring 553 embryos to 35 surrogates, 59 piglets from 14 litters were produced (59/553 = 10%) [150]. In another study, 180 embryos were transferred to 12 surrogates, producing 37 piglets from 8 sows (37/180 = 21%) [151]. Unfortunately, the number of embryos cryopreserved is not well described in either case. Thus, the percentages listed above may not fully describe the success of cryopreservation as it may be concluded that more embryos were cryopreserved than were transferred. The question of viability is then raised, i.e., were only the embryos of good quality cryopreserved and transferred? If so, then the percentages above may represent an overestimation of the true success of these technologies.

6. Environmental Stressors Arising from the Culture System

6.1. *Oxygen Tension and Reactive Oxygen Species*

During oxidative phosphorylation, ROS, such as superoxides, form as a result of premature leaking of electrons from the electron transport chain to molecular oxygen [152]. Therefore, mitochondria are the main source of ROS within cells, which can act as signaling molecules for differentiation or cause membrane, protein, and DNA damage at higher concentrations, potentially leading to apoptosis. Interestingly, Dalvit et al. [153] observed a significant increase in ROS concentrations from the oocyte at MII to the two-cell stage in bovine. ROS levels continued to increase until the late morula stage but decreased significantly during blastocyst formation. In vitro-produced embryos are more susceptible to ROS formation and damage compared to in vivo-derived embryos because of external stressors, such as fluctuating oxygen tension, exposure to visible light, media contaminants, and absence of maternal antioxidant systems [111]. Supplementation of antioxidants to porcine embryo culture media has been shown to improve development and reduce ROS [154,155], but these studies were conducted by using atmospheric (21%) O_2 during the culture period which promotes ROS formation.

Oxygen tension during the culture period has an impact on subsequent embryonic development (Figure 2). One- and two-cell stage porcine embryos cultured in 5% O_2 and 5% CO_2 developed to the blastocyst stage at a rate of 50%, whereas culture in 2% or 21% O_2 and 5% CO_2 resulted in less than 10% development to the blastocyst stage [156]. Culture in 5% O_2 instead of atmospheric (21%) O_2 shifted the global patterns of gene expression in mouse embryos to be more similar to those derived in vivo [157]. Additionally, amino acid turnover, as an indicator of viability, by human embryos was lower when cultured at 5% O_2 compared to 21% O_2 [158]. Culture in low (5%) O_2 increased cell numbers in in vitro-produced porcine blastocyst-stage embryos and the abundance of transaldolase 1 (*TALDO1*) and pyruvate dehydrogenase kinase 1 (*PDK1*), which are signatures of a Warburg Effect-like metabolism [51]. As mentioned previously, lowering the oxygen tension to 5% O_2 eliminated the need to add hypotaurine, which functions as an antioxidant, to porcine embryo culture medium as removal of this component did not impair development nor increase apoptosis [110].

6.2. *Temperature*

Most porcine in vitro culture systems use a physiological temperature of 38.5 °C for the incubation (Figure 2). However, several studies have investigated the effects of increased temperature, or heat stress, on oocyte maturation and development in vitro. Heat stress during in vitro maturation may model the effects of summer months or fever in sows on oocyte or embryo quality. Porcine COCs exposed to heat stress (41.5 °C) for 24 h had decreased cumulus expansion, progression to MII, and development after fertilization [159]. Supplementation of antioxidants during maturation, including astaxanthin, melatonin, and resveratrol, has been shown to mitigate effects of heat stress and decrease apoptosis [160–162].

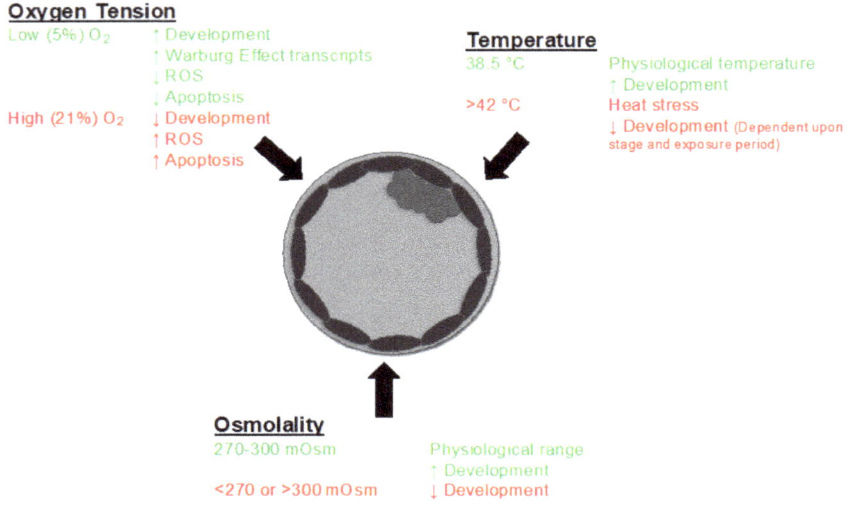

Figure 2. A summary of environmental stressors on porcine embryo development in vitro. Positive conditions are in green, and negative conditions are in red.

Increasing the incubator temperature above 43 °C during the embryo culture period has been shown to decrease development of porcine embryos in a temporal manner, but abundance of heat shock protein 70 was not increased after heat stress exposure [163]. When IVF-derived porcine embryos were exposed to 42 °C for 9 h after the late 1-cell stage, decreases in development to the blastocyst stage and increases in apoptotic nuclei were observed [164]. However, heat shock at the same conditions directly after fertilization increased the rate of cleavage and tended to increase development to the blastocyst stage compared to the non-heat shock controls [165]. Similarly, increases in development of parthenogenetic embryos was observed when heat shock occurred directly after oocyte activation [165], and subsequent analyses revealed that dephosphorylation of MAPK was increased in the heat-shocked embryos compared to controls [164].

6.3. Osmolality

The osmolalities of porcine oviductal and uterine fluids range from 290 to 320 mOsm during the first five days of gestation [73]. The osmolalities of the most commonly used culture media, PZM-3 and NCSU-23, are 288 and 291 mOsm, respectively [70]. Li et al. [73] demonstrated that culture of porcine embryos below 270 mOsm or above 300 mOsm, by altering the NaCl concentration, decreased development to the blastocyst stage (Figure 2) [73]. Contrarily, Hwang et al. [166] observed that increasing osmolality up to 320 mOsm with NaCl or sucrose did not impair development of IVF-derived embryos but surprisingly decreased transcript abundance of BCL2 associated X protein (*BAX*), a proapoptotic gene, compared to the PZM-3 control. Particular amino acids can also act as osmolytes, including glycine and proline. When 10 mM glycine was added to MU1 porcine embryo culture medium at 260, 275, or 300 mOsm, no difference in development to the blastocyst stage was detected [84]. However, culture with supplemental glycine increased the total cell numbers of blastocyst-stage embryos at 275 mOsm, demonstrating a beneficial effect on embryo quality at this osmolality [84].

7. Maternal and Paternal Factors Influencing Development

7.1. Gilt-Versus Sow-Derived Oocytes

The production of pig embryos in vitro starts with oocytes from either prepubertal gilts or sexually mature sows. Collection of ovaries at the abattoir is dependent upon the type of facility. Some plants slaughter only gilts while others focus on sows and sausage production. It is generally accepted that in vitro-matured oocytes derived from sows are more developmentally competent than oocytes derived from prepubertal gilts [10,167–169]. The diameter of the oocyte, thickness of the zona pellucida, and the perivitelline space were found to be larger in sow oocytes compared in gilt oocytes [170]. However, culture conditions can be modified to improve the quality of in vitro-matured oocytes derived from prepubertal gilts to a level similar to oocytes derived from sexually mature sows. For example, addition of fibroblast growth factor 2, leukemia inhibitory factor, and insulin like growth factor 1 (termed FLI) to the maturation medium quadrupled the production of offspring on a per oocyte basis [11]. Indeed, in the presence of FLI, gonadotropins are unnecessary during oocyte maturation [25].

7.2. Sperm Quality

Sperm quality is defined in three major categories: sperm motility, sperm morphology, and sperm biomarker-defined health. Sperm motility can be objectively measured using computer-assisted sperm analysis (CASA) systems [171]. These are typically phase-contrast microscopes fitted with cameras and attached to a computer to perform motility calculations. Software on these computers can likewise perform sperm morphology assessment. Common sperm morphology abnormalities include retained cytoplasmic droplets (proximal and distal in relationship to the sperm midpiece), distal midpiece refluxes, and coiled tails.

There has been increasing attention to evaluating sperm health and function from a biomarker perspective. The advantage to evaluating sperm with biomarkers is that it can reveal sperm quality characteristics that cannot be evaluated under optical microscopes (conventional light microscopy) alone. Biomarkers can be assessed by using microscopes equipped with epifluorescence or through flow cytometry. While there are several characteristics that can be analyzed, we will mention those that are relevant along with manufacturer in parentheses. A common sperm health attribute to assess is acrosomal integrity. Fluorescent conjugated lectins can be used for this. In pig, a common lectin to assess acrosome integrity is Arachis hypogaea/peanut agglutinin (Invitrogen Lectin PNA, Alexa Fluor™ 488) [172]. Plasma membrane integrity can give insight to sperm viability [173] and/or if there has been capacitation-associated plasma membrane remodeling [174]. This is commonly analyzed using propidium iodide (Invitrogen). Mitochondrial membrane potential can be assessed using JC-1 (Invitrogen) [175], while capacitation indicating calcium influx status can be evaluated using Fluo 4 NW (Invitrogen) [176]. An additional capacitation status indicator is the efflux of zinc, which can be analyzed using FluoZin™-3 AM (Invitrogen) [174]. Misfolded sperm protein content can be evaluated for with PROTEOSTAT aggresome kit (Enzo Life Sciences) [177] along with sperm proteins marked for recycling via the ubiquitin proteasome system pathway with antibodies against ubiquitin [178]. Reactive oxygen species can be indicated by 2′,7′-dichlorodihydrofluorescein diacetate (H2DCFDA, Invitrogen) [179] or membrane lipid peroxidation from ROS by 4, 4-difluoro-5-(4-phenyl-1,3-butadienyl)-4-bora-3a,4a-diaza-s-indacene-3-undecanoic acid, C11-BODIPY (BODIPY, Invitrogen) [180]. Lastly, DNA fragmentation can be evaluated by acridine orange (Acros Organics) [181] or TUNEL (Promega DeadEnd™ Colorimetric TUNEL System) [182].

While it is assumed that lower quality sperm can be used for successful fertilization in IVF compared to artificial insemination, sperm quality is still important. To emphasize this, recent studies have shown that flow cytometer (FC)-sorted sperm result in reduced IVF rates [183]. It is well understood that FC-sorted sperm have reduced sperm quality [184,185]. Even though sperm may undergo successful penetration of the zona pellucida, a series of events are necessary to activate the oocyte. Two sperm-oocyte activating factors implicated in activating the oocyte are phospholipase C zeta (PLCζ) [186] and postacrosomal WW domain-binding protein (PAWP) [187]. After sperm-oocyte activation, billions of zinc ions are released, termed the zinc spark (Xenopus [188], cattle [189], and mice [190]). While the zinc spark has not been reported in pigs, we previously reported that the zinc chelator, TPEN, activates pig oocytes [191]. Recent data have shown that zinc inhibits sperm zona pellucida proteinases, such as matrix metalloproteinase-2 (MMP2), and the 26S proteasome [192]. Altogether, sperm-oocyte zinc signaling might be a new polyspermy defense mechanism [174,193].

7.3. Epigenetics and In Vitro Production of Porcine Embryos

Creating or culturing pig embryos in vitro changes the developmental potential [112] as well as the transcriptional [12] and epigenetic profile [194–196] of the resulting embryos. Some would argue that the change in transcription is due to changes in DNA methylation. While there are reports of epigenetic control over transcription, it is not clear if the changes in DNA methylation, for example, are the result of or the cause of changes in transcription [197–200]. In pigs, epigenetic regulation of development is not limited to DNA methylation and includes histone modifications, such as acetylation [201], methylation [202,203], and RNA methylation [204]. Much work has been reported on epigenetic regulation in pig development and after SCNT [205,206]. Increasing histone acetylation by inhibiting deacetylase inhibitors improves the development of SCNT embryos and corrects the expression of some genes [201,207,208]. Presumably changes in the culture environment can bring the transcriptional profile and epigenetic marks, such as DNA methylation and histone acetylation, to a more 'in vivo' level.

8. Conclusions and Perspectives

Progress in the in vitro production of porcine embryos has been steady; however, with advances in technologies, such as deep sequencing and metabolomics, the conditions required for healthy oocytes and embryos will be revealed more rapidly. In turn, improvements in media formulations and implementation of dynamic culture systems have great potential for dramatically increasing developmental outcomes. Nevertheless, environmental stressors and parental factors that cannot be controlled will continuously pose challenges for the culture of porcine embryos. These challenges will be faced directly to improve embryo viability as porcine models are becoming more important and applicable in agriculture and human medicine.

Funding: Funding was provided by the United States Department of Agriculture, National Institute of Food and Agriculture (Award numbers 2019-67011-29543, 2019-67012-29714, and 2020-67015-31016), and a grant from the Rural Development Administration, Republic of Korea. Funding for the National Swine Resource and Research Center is from the National Institute of Allergy and Infectious Disease, the National Institute of Heart, Lung and Blood, and the Office of the Director (U42OD011140).

Conflicts of Interest: The authors declare no conflict of interest.

References

1. Whyte, J.J.; Meyer, A.E.; Spate, L.D.; Benne, J.A.; Cecil, R.; Samuel, M.S.; Murphy, C.N.; Prather, R.S.; Geisert, R.D. Inactivation of porcine interleukin-1beta results in failure of rapid conceptus elongation. *Proc. Natl. Acad. Sci. USA* **2018**, *115*, 307–312. [CrossRef] [PubMed]
2. Meyer, A.E.; Pfeiffer, C.A.; Brooks, K.E.; Spate, L.D.; Benne, J.A.; Cecil, R.; Samuel, M.S.; Murphy, C.N.; Behura, S.; McLean, M.K.; et al. New perspective on conceptus estrogens in maternal recognition and pregnancy establishment in the pigdagger. *Biol. Reprod.* **2019**, *101*, 148–161. [CrossRef] [PubMed]
3. Pfeiffer, C.A.; Meyer, A.E.; Brooks, K.E.; Chen, P.R.; Milano-Foster, J.; Spate, L.D.; Benne, J.A.; Cecil, R.F.; Samuel, M.S.; Ciernia, L.A.; et al. Ablation of conceptus PTGS2 expression does not alter early conceptus development and establishment of pregnancy in the pigdagger. *Biol. Reprod.* **2020**, *102*, 475–488. [CrossRef]

4. Chen, P.R.; Lucas, C.G.; Cecil, R.F.; Pfeiffer, C.A.; Fudge, M.A.; Samuel, M.S.; Zigo, M.; Seo, H.; Spate, L.D.; Whitworth, K.M.; et al. Disrupting porcine glutaminase does not block preimplantation development and elongation nor decrease mTORC1 activation in conceptuses. *Biol. Reprod.* **2021**, *165*. [CrossRef]
5. Lai, L.; Kang, J.X.; Li, R.; Wang, J.; Witt, W.T.; Yong, H.Y.; Hao, Y.; Wax, D.M.; Murphy, C.N.; Rieke, A.; et al. Generation of cloned transgenic pigs rich in omega-3 fatty acids. *Nat. Biotechnol.* **2006**, *24*, 435–436. [CrossRef] [PubMed]
6. Whitworth, K.M.; Rowland, R.R.; Ewen, C.L.; Trible, B.R.; Kerrigan, M.A.; Cino-Ozuna, A.G.; Samuel, M.S.; Lightner, J.E.; McLaren, D.G.; Mileham, A.J.; et al. Gene-edited pigs are protected from porcine reproductive and respiratory syndrome virus. *Nat. Biotechnol.* **2016**, *34*, 20–22. [CrossRef] [PubMed]
7. Whitworth, K.M.; Rowland, R.R.R.; Petrovan, V.; Sheahan, M.; Cino-Ozuna, A.G.; Fang, Y.; Hesse, R.; Mileham, A.; Samuel, M.S.; Wells, K.D.; et al. Resistance to coronavirus infection in amino peptidase N-deficient pigs. *Transgenic Res.* **2019**, *28*, 21–32. [CrossRef]
8. Rogers, C.S.; Hao, Y.; Rokhlina, T.; Samuel, M.; Stoltz, D.A.; Li, Y.; Petroff, E.; Vermeer, D.W.; Kabel, A.C.; Yan, Z.; et al. Production of CFTR-null and CFTR-DeltaF508 heterozygous pigs by adeno-associated virus-mediated gene targeting and somatic cell nuclear transfer. *J. Clin. Investig.* **2008**, *118*, 1571–1577. [CrossRef]
9. Koppes, E.A.; Redel, B.K.; Johnson, M.A.; Skvorak, K.J.; Ghaloul-Gonzalez, L.; Yates, M.E.; Lewis, D.W.; Gollin, S.M.; Wu, Y.L.; Christ, S.E.; et al. A porcine model of phenylketonuria generated by CRISPR/Cas9 genome editing. *JCI Insight* **2020**, *5*. [CrossRef]
10. Lai, L.; Kolber-Simonds, D.; Park, K.W.; Cheong, H.T.; Greenstein, J.L.; Im, G.S.; Samuel, M.; Bonk, A.; Rieke, A.; Day, B.N.; et al. Production of alpha-1,3-galactosyltransferase knockout pigs by nuclear transfer cloning. *Science* **2002**, *295*, 1089–1092. [CrossRef]
11. Yuan, Y.; Spate, L.D.; Redel, B.K.; Tian, Y.; Zhou, J.; Prather, R.S.; Roberts, R.M. Quadrupling efficiency in production of genetically modified pigs through improved oocyte maturation. *Proc. Natl. Acad. Sci. USA* **2017**, *114*, E5796–E5804. [CrossRef] [PubMed]
12. Bauer, B.K.; Isom, S.C.; Spate, L.D.; Whitworth, K.M.; Spollen, W.G.; Blake, S.M.; Springer, G.K.; Murphy, C.N.; Prather, R.S. Transcriptional profiling by deep sequencing identifies differences in mRNA transcript abundance in in vivo-derived versus in vitro-cultured porcine blastocyst stage embryos. *Biol. Reprod.* **2010**, *83*, 791–798. [CrossRef] [PubMed]
13. Spate, L.D.; Brown, A.N.; Redel, B.K.; Whitworth, K.M.; Murphy, C.N.; Prather, R.S. Dickkopf-related protein 1 inhibits the WNT signaling pathway and improves pig oocyte maturation. *PLoS ONE* **2014**, *9*, e95114. [CrossRef]
14. Russell, D.L.; Gilchrist, R.B.; Brown, H.M.; Thompson, J.G. Bidirectional communication between cumulus cells and the oocyte: Old hands and new players? *Theriogenology* **2016**, *86*, 62–68. [CrossRef]
15. Norris, R.P.; Ratzan, W.J.; Freudzon, M.; Mehlmann, L.M.; Krall, J.; Movsesian, M.A.; Wang, H.; Ke, H.; Nikolaev, V.O.; Jaffe, L.A. Cyclic GMP from the surrounding somatic cells regulates cyclic AMP and meiosis in the mouse oocyte. *Development* **2009**, *136*, 1869–1878. [CrossRef]
16. Lonergan, P.; Rizos, D.; Gutierrez-Adan, A.; Fair, T.; Boland, M.P. Oocyte and embryo quality: Effect of origin, culture conditions and gene expression patterns. *Reprod. Domest. Anim.* **2003**, *38*, 259–267. [CrossRef]
17. Armstrong, D.T. Effects of maternal age on oocyte developmental competence. *Theriogenology* **2001**, *55*, 1303–1322. [CrossRef]
18. Lechniak, D.; Warzych, E.; Pers-Kamczyc, E.; Sosnowski, J.; Antosik, P.; Rubes, J. Gilts and sows produce similar rate of diploid oocytes in vitro whereas the incidence of aneuploidy differs significantly. *Theriogenology* **2007**, *68*, 755–762. [CrossRef]
19. Lee, J.B.; Lee, M.G.; Lin, T.; Shin, H.Y.; Lee, J.E.; Kang, J.W.; Jin, D.-I. Effect of oocyte chromatin status in porcine follicles on the embryo development in vitro. *Asian-Australas J. Anim. Sci.* **2019**, *32*, 956–965. [CrossRef]
20. Chen, L.; Russell, P.T.; Larsen, W.J. Functional significance of cumulus expansion in the mouse: Roles for the preovulatory synthesis of hyaluronic acid within the cumulus mass. *Mol. Reprod. Dev.* **1993**, *34*, 87–93. [CrossRef]
21. Marchal, R.; Vigneron, C.; Perreau, C.; Bali-Papp, A.; Mermillod, P. Effect of follicular size on meiotic and developmental competence of porcine oocytes. *Theriogenology* **2002**, *57*, 1523–1532. [CrossRef]
22. Costermans, N.G.J.; Soede, N.M.; van Tricht, F.; Blokland, M.; Kemp, B.; Keijer, J.; Teerds, K.J. Follicular fluid steroid profile in sows: Relationship to follicle size and oocyte qualitydagger. *Biol. Reprod.* **2020**, *102*, 740–749. [CrossRef] [PubMed]
23. Yokoo, M.; Kimura, N.; Sato, E. Induction of oocyte maturation by hyaluronan-CD44 interaction in pigs. *J. Reprod. Dev.* **2010**, *56*, 15–19. [CrossRef] [PubMed]
24. Nevoral, J.; Orsák, M.; Klein, P.; Petr, J.; Dvořáková, M.; Weingartová, I.; Vyskočilová, A.; Zámostná, K.; Krejčová, T.; Jílek, F. Cumulus Cell Expansion, Its Role in Oocyte Biology and Perspectives of Measurement: A Review. *Sci. Agric. Bohem.* **2015**, *45*, 212–225. [CrossRef]
25. Redel, B.K.; Spate, L.D.; Yuan, Y.; Murphy, C.N.; Roberts, R.M.; Prather, R.S. Neither gonadotropin nor cumulus cell expansion is needed for the maturation of competent porcine oocytes in vitro. *Biol. Reprod.* **2021**. [CrossRef]
26. Nagai, T.; Niwa, K.; Iritani, A. Effect of sperm concentration during preincubation in a defined medium on fertilization in vitro of pig follicular oocytes. *J. Reprod. Fertil.* **1984**, *70*, 271–275. [CrossRef]
27. Abeydeera, L.R.; Day, B.N. In vitro penetration of pig oocytes in a modified Tris-buffered medium: Effect of BSA, caffeine and calcium. *Theriogenology* **1997**, *48*, 537–544. [CrossRef]
28. Evans, J.P. Preventing polyspermy in mammalian eggs-Contributions of the membrane block and other mechanisms. *Mol. Reprod. Dev.* **2020**, *87*, 341–349. [CrossRef]
29. Burkart, A.D.; Xiong, B.; Baibakov, B.; Jimenez-Movilla, M.; Dean, J. Ovastacin, a cortical granule protease, cleaves ZP2 in the zona pellucida to prevent polyspermy. *J. Cell Biol.* **2012**, *197*, 37–44. [CrossRef]

30. Fahrenkamp, E.; Algarra, B.; Jovine, L. Mammalian egg coat modifications and the block to polyspermy. *Mol. Reprod. Dev.* **2020**, *87*, 326–340. [CrossRef]
31. Wang, W.H.; Machaty, Z.; Abeydeera, L.R.; Prather, R.S.; Day, B.N. Parthenogenetic activation of pig oocytes with calcium ionophore and the block to sperm penetration after activation. *Biol. Reprod.* **1998**, *58*, 1357–1366. [CrossRef] [PubMed]
32. Saavedra, M.D.; Mondejar, I.; Coy, P.; Betancourt, M.; Gonzalez-Marquez, H.; Jimenez-Movilla, M.; Aviles, M.; Romar, R. Calreticulin from suboolemmal vesicles affects membrane regulation of polyspermy. *Reproduction* **2014**, *147*, 369–378. [CrossRef] [PubMed]
33. Funahashi, H.; Day, B.N. Effects of follicular fluid at fertilization in vitro on sperm penetration in pig oocytes. *J. Reprod. Fertil.* **1993**, *99*, 97–103. [CrossRef] [PubMed]
34. Teijeiro, J.M.; Marini, P.E. The effect of oviductal deleted in malignant brain tumor 1 over porcine sperm is mediated by a signal transduction pathway that involves pro-AKAP4 phosphorylation. *Reproduction* **2012**, *143*, 773–785. [CrossRef] [PubMed]
35. Coy, P.; Canovas, S.; Mondejar, I.; Saavedra, M.D.; Romar, R.; Grullon, L.; Matas, C.; Aviles, M. Oviduct-specific glycoprotein and heparin modulate sperm-zona pellucida interaction during fertilization and contribute to the control of polyspermy. *Proc. Natl. Acad. Sci. USA* **2008**, *105*, 15809–15814. [CrossRef]
36. Hao, Y.; Mathialagan, N.; Walters, E.; Mao, J.; Lai, L.; Becker, D.; Li, W.; Critser, J.; Prather, R.S. Osteopontin reduces polyspermy during in vitro fertilization of porcine oocytes. *Biol. Reprod.* **2006**, *75*, 726–733. [CrossRef]
37. Coy, P.; Jimenez-Movilla, M.; Garcia-Vazquez, F.A.; Mondejar, I.; Grullon, L.; Romar, R. Oocytes use the plasminogen-plasmin system to remove supernumerary spermatozoa. *Hum. Reprod.* **2012**, *27*, 1985–1993. [CrossRef]
38. Mondejar, I.; Grullon, L.A.; Garcia-Vazquez, F.A.; Romar, R.; Coy, P. Fertilization outcome could be regulated by binding of oviductal plasminogen to oocytes and by releasing of plasminogen activators during interplay between gametes. *Fertil. Steril.* **2012**, *97*, 453–461. [CrossRef]
39. Umehara, T.; Tsujita, N.; Goto, M.; Tonai, S.; Nakanishi, T.; Yamashita, Y.; Shimada, M. Methyl-beta cyclodextrin and creatine work synergistically under hypoxic conditions to improve the fertilization ability of boar ejaculated sperm. *Anim. Sci. J.* **2020**, *91*, e13493. [CrossRef]
40. Kim, N.H.; Funahashi, H.; Abeydeera, L.R.; Moon, S.J.; Prather, R.S.; Day, B.N. Effects of oviductal fluid on sperm penetration and cortical granule exocytosis during fertilization of pig oocytes in vitro. *J. Reprod. Fertil.* **1996**, *107*, 79–86. [CrossRef]
41. Harris, E.A.; Stephens, K.K.; Winuthayanon, W. Extracellular Vesicles and the Oviduct Function. *Int. J. Mol. Sci.* **2020**, *21*, 8280. [CrossRef]
42. Zigo, M.; Manaskova-Postlerova, P.; Zuidema, D.; Kerns, K.; Jonakova, V.; Tumova, L.; Bubenickova, F.; Sutovsky, P. Porcine model for the study of sperm capacitation, fertilization and male fertility. *Cell Tissue Res.* **2020**, *380*, 237–262. [CrossRef]
43. Galeati, G.; Modina, S.; Lauria, A.; Mattioli, M. Follicle somatic cells influence pig oocyte penetrability and cortical granule distribution. *Mol. Reprod. Dev.* **1991**, *29*, 40–46. [CrossRef] [PubMed]
44. Probst, S.; Rath, D. Production of piglets using intracytoplasmic sperm injection (ICSI) with flowcytometrically sorted boar semen and artificially activated oocytes. *Theriogenology* **2003**, *59*, 961–973. [CrossRef]
45. Yong, H.Y.; Hao, Y.; Lai, L.; Li, R.; Murphy, C.N.; Rieke, A.; Wax, D.; Samuel, M.; Prather, R.S. Production of a transgenic piglet by a sperm injection technique in which no chemical or physical treatments were used for oocytes or sperm. *Mol. Reprod. Dev.* **2006**, *73*, 595–599. [CrossRef] [PubMed]
46. Nakai, M.; Ozawa, M.; Maedomari, N.; Noguchi, J.; Kaneko, H.; Ito, J.; Onishi, A.; Kashiwazaki, N.; Kikuchi, K. Delay in cleavage of porcine embryos after intracytoplasmic sperm injection (ICSI) shows poorer embryonic development. *J. Reprod. Dev.* **2014**, *60*, 256–259. [CrossRef]
47. Nakai, M.; Suzuki, S.I.; Ito, J.; Fuchimoto, D.I.; Sembon, S.; Noguchi, J.; Onishi, A.; Kashiwazaki, N.; Kikuchi, K. Efficient pig ICSI using Percoll-selected spermatozoa; evidence for the essential role of phospholipase C-zeta in ICSI success. *J. Reprod. Dev.* **2016**, *62*, 639–643. [CrossRef]
48. Casillas, F.; Betancourt, M.; Cuello, C.; Ducolomb, Y.; Lopez, A.; Juarez-Rojas, L.; Retana-Marquez, S. An efficiency comparison of different in vitro fertilization methods: IVF, ICSI, and PICSI for embryo development to the blastocyst stage from vitrified porcine immature oocytes. *Porc. Heal. Manag.* **2018**, *4*, 16. [CrossRef] [PubMed]
49. Bradley, J.; Swann, K. Mitochondria and lipid metabolism in mammalian oocytes and early embryos. *Int. J. Dev. Biol.* **2019**, *63*, 93–103. [CrossRef]
50. Warburg, O. On the origin of cancer cells. *Science* **1956**, *123*, 309–314. [CrossRef]
51. Redel, B.K.; Brown, A.N.; Spate, L.D.; Whitworth, K.M.; Green, J.A.; Prather, R.S. Glycolysis in preimplantation development is partially controlled by the Warburg Effect. *Mol. Reprod. Dev.* **2012**, *79*, 262–271. [CrossRef] [PubMed]
52. Krisher, R.L.; Heuberger, A.L.; Paczkowski, M.; Stevens, J.; Pospisil, C.; Prather, R.S.; Sturmey, R.G.; Herrick, J.R.; Schoolcraft, W.B. Applying metabolomic analyses to the practice of embryology: Physiology, development and assisted reproductive technology. *Reprod. Fertil. Dev.* **2015**, *27*, 602–620. [CrossRef] [PubMed]
53. Rieger, D. Relationships between energy metabolism and development of early mammalian embryos. *Theriogenology* **1992**, *37*, 75–93. [CrossRef]
54. Chi, F.; Sharpley, M.S.; Nagaraj, R.; Roy, S.S.; Banerjee, U. Glycolysis-Independent Glucose Metabolism Distinguishes TE from ICM Fate during Mammalian Embryogenesis. *Dev. Cell* **2020**, *53*, 9–26.e24. [CrossRef]

55. Biggers, J.D.; Whittingham, D.G.; Donahue, R.P. The pattern of energy metabolism in the mouse oocyte and zygote. *Proc. Natl. Acad. Sci. USA* **1967**, *58*, 560–567. [CrossRef]
56. Gandhi, A.P.; Lane, M.; Gardner, D.K.; Krisher, R.L. Substrate utilization in porcine embryos cultured in NCSU23 and G1.2/G2.2 sequential culture media. *Mol. Reprod. Dev.* **2001**, *58*, 269–275. [CrossRef]
57. Martin, K.L.; Leese, H.J. Role of developmental factors in the switch from pyruvate to glucose as the major exogenous energy substrate in the preimplantation mouse embryo. *Reprod. Fertil. Dev.* **1999**, *11*, 425–433. [CrossRef]
58. Gardner, D.K.; Lane, M.; Batt, P. Uptake and metabolism of pyruvate and glucose by individual sheep preattachment embryos developed in vivo. *Mol. Reprod Dev.* **1993**, *36*, 313–319. [CrossRef]
59. Leese, H.J.; Barton, A.M. Pyruvate and glucose uptake by mouse ova and preimplantation embryos. *J. Reprod. Fertil.* **1984**, *72*, 9–13. [CrossRef]
60. Sturmey, R.G.; Leese, H.J. Energy metabolism in pig oocytes and early embryos. *Reproduction* **2003**, *126*, 197–204. [CrossRef]
61. Swain, J.E.; Bormann, C.L.; Clark, S.G.; Walters, E.M.; Wheeler, M.B.; Krisher, R.L. Use of energy substrates by various stage preimplantation pig embryos produced in vivo and in vitro. *Reproduction* **2002**, *123*, 253–260. [CrossRef]
62. Chatot, C.L.; Ziomek, C.A.; Bavister, B.D.; Lewis, J.L.; Torres, I. An improved culture medium supports development of random-bred 1-cell mouse embryos in vitro. *J. Reprod. Fertil.* **1989**, *86*, 679–688. [CrossRef]
63. Schini, S.A.; Bavister, B.D. Two-cell block to development of cultured hamster embryos is caused by phosphate and glucose. *Biol. Reprod.* **1988**, *39*, 1183–1192. [CrossRef]
64. Conaghan, J.; Hardy, K.; Handyside, A.H.; Winston, R.M.; Leese, H.J. Selection criteria for human embryo transfer: A comparison of pyruvate uptake and morphology. *J. Assist. Reprod. Genet.* **1993**, *10*, 21–30. [CrossRef] [PubMed]
65. Takahashi, Y.; First, N.L. In vitro development of bovine one-cell embryos: Influence of glucose, lactate, pyruvate, amino acids and vitamins. *Theriogenology* **1992**, *37*, 963–978. [CrossRef]
66. Thompson, J.G.; Simpson, A.C.; Pugh, P.A.; Tervit, H.R. Requirement for glucose during in vitro culture of sheep preimplantation embryos. *Mol. Reprod. Dev.* **1992**, *31*, 253–257. [CrossRef] [PubMed]
67. Hagen, D.R.; Prather, R.S.; Sims, M.M.; First, N.L. Development of one-cell porcine embryos to the blastocyst stage in simple media. *J. Anim. Sci.* **1991**, *69*, 1147–1150. [CrossRef]
68. Petters, R.M.; Johnson, B.H.; Reed, M.L.; Archibong, A.E. Glucose, glutamine and inorganic phosphate in early development of the pig embryo in vitro. *J. Reprod. Fertil.* **1990**, *89*, 269–275. [CrossRef]
69. Flood, M.R.; Wiebold, J.L. Glucose metabolism by preimplantation pig embryos. *J. Reprod. Fertil.* **1988**, *84*, 7–12. [CrossRef]
70. Yoshioka, K.; Suzuki, C.; Tanaka, A.; Anas, I.M.; Iwamura, S. Birth of piglets derived from porcine zygotes cultured in a chemically defined medium. *Biol. Reprod.* **2002**, *66*, 112–119. [CrossRef]
71. Gardner, D.K.; Leese, H.J. Concentrations of nutrients in mouse oviduct fluid and their effects on embryo development and metabolism in vitro. *J. Reprod. Fertil.* **1990**, *88*, 361–368. [CrossRef] [PubMed]
72. Hugentobler, S.A.; Diskin, M.G.; Leese, H.J.; Humpherson, P.G.; Watson, T.; Sreenan, J.M.; Morris, D.G. Amino acids in oviduct and uterine fluid and blood plasma during the estrous cycle in the bovine. *Mol. Reprod. Dev.* **2007**, *74*, 445–454. [CrossRef] [PubMed]
73. Li, R.; Whitworth, K.; Lai, L.; Wax, D.; Spate, L.; Murphy, C.N.; Rieke, A.; Isom, C.; Hao, Y.; Zhong, Z.; et al. Concentration and composition of free amino acids and osmolalities of porcine oviductal and uterine fluid and their effects on development of porcine IVF embryos. *Mol. Reprod. Dev.* **2007**, *74*, 1228–1235. [CrossRef]
74. Van Winkle, L.J. Amino acid transport regulation and early embryo development. *Biol. Reprod.* **2001**, *64*, 1–12. [CrossRef]
75. Prather, R.S.; Peters, M.S.; Van Winkle, L.J. Alanine and leucine transport in unfertilized pig oocytes and early blastocysts. *Mol. Reprod. Dev.* **1993**, *34*, 250–254. [CrossRef] [PubMed]
76. Van Winkle, L.J.; Tesch, J.K.; Shah, A.; Campione, A.L. System B0,+ amino acid transport regulates the penetration stage of blastocyst implantation with possible long-term developmental consequences through adulthood. *Hum. Reprod. Updat.* **2006**, *12*, 145–157. [CrossRef]
77. Prather, R.S.; Peters, M.S.; Van Winkle, L.J. Aspartate and glutamate transport in unfertilized pig oocytes and blastocysts. *Mol. Reprod. Dev.* **1993**, *36*, 49–52. [CrossRef]
78. Gardner, D.K.; Lane, M. Amino acids and ammonium regulate mouse embryo development in culture. *Biol. Reprod.* **1993**, *48*, 377–385. [CrossRef]
79. Pinyopummintr, T.; Bavister, B.D. Effects of amino acids on development in vitro of cleavage-stage bovine embryos into blastocysts. *Reprod. Fertil. Dev.* **1996**, *8*, 835–841. [CrossRef]
80. Steeves, T.E.; Gardner, D.K. Temporal and differential effects of amino acids on bovine embryo development in culture. *Biol. Reprod.* **1999**, *61*, 731–740. [CrossRef]
81. Van Thuan, N.; Harayama, H.; Miyake, M. Characteristics of preimplantational development of porcine parthenogenetic diploids relative to the existence of amino acids in vitro. *Biol. Reprod.* **2002**, *67*, 1688–1698. [CrossRef] [PubMed]
82. Redel, B.K.; Tessanne, K.J.; Spate, L.D.; Murphy, C.N.; Prather, R.S. Arginine increases development of in vitro-produced porcine embryos and affects the protein arginine methyltransferase-dimethylarginine dimethylaminohydrolase-nitric oxide axis. *Reprod. Fertil. Dev.* **2015**, *27*, 655–666. [CrossRef] [PubMed]
83. Tranguch, S.; Steuerwald, N.; Huet-Hudson, Y.M. Nitric oxide synthase production and nitric oxide regulation of preimplantation embryo development. *Biol. Reprod.* **2003**, *68*, 1538–1544. [CrossRef] [PubMed]

84. Redel, B.K.; Spate, L.D.; Lee, K.; Mao, J.; Whitworth, K.M.; Prather, R.S. Glycine supplementation in vitro enhances porcine preimplantation embryo cell number and decreases apoptosis but does not lead to live births. *Mol. Reprod. Dev.* **2016**, *83*, 246–258. [CrossRef]
85. Carney, E.W.; Bavister, B.D. Stimulatory and inhibitory effects of amino acids on the development of hamster eight-cell embryos in vitro. *J. Assist. Reprod. Genet.* **1987**, *4*, 162–167. [CrossRef]
86. Chen, P.R.; Redel, B.K.; Spate, L.D.; Ji, T.; Salazar, S.R.; Prather, R.S. Glutamine supplementation enhances development of in vitro-produced porcine embryos and increases leucine consumption from the medium. *Biol. Reprod.* **2018**, *99*, 938–948. [CrossRef]
87. Chen, P.R.; Lucas, C.G.; Spate, L.D.; Prather, R.S. Glutaminolysis is involved in the activation of mTORC1 in in vitro-produced porcine embryos. *Mol. Reprod. Dev.* **2021**, *88*, 490–499. [CrossRef]
88. Dunning, K.R.; Cashman, K.; Russell, D.L.; Thompson, J.G.; Norman, R.J.; Robker, R.L. Beta-oxidation is essential for mouse oocyte developmental competence and early embryo development. *Biol. Reprod.* **2010**, *83*, 909–918. [CrossRef]
89. Hewitson, L.C.; Martin, K.L.; Leese, H.J. Effects of metabolic inhibitors on mouse preimplantation embryo development and the energy metabolism of isolated inner cell masses. *Mol. Reprod. Dev.* **1996**, *43*, 323–330. [CrossRef]
90. Lowe, J.L.; Bathgate, R.; Grupen, C.G. Effect of carbohydrates on lipid metabolism during porcine oocyte IVM. *Reprod. Fertil. Dev.* **2019**, *31*, 557–569. [CrossRef]
91. Kim, M.K.; Park, J.K.; Paek, S.K.; Kim, J.W.; Kwak, I.P.; Lee, H.J.; Lyu, S.W.; Lee, W.S. Effects and pregnancy outcomes of L-carnitine supplementation in culture media for human embryo development from in vitro fertilization. *J. Obstet. Gynaecol. Res.* **2018**, *44*, 2059–2066. [CrossRef]
92. Somfai, T.; Kaneda, M.; Akagi, S.; Watanabe, S.; Haraguchi, S.; Mizutani, E.; Dang-Nguyen, T.Q.; Geshi, M.; Kikuchi, K.; Nagai, T. Enhancement of lipid metabolism with L-carnitine during in vitro maturation improves nuclear maturation and cleavage ability of follicular porcine oocytes. *Reprod. Fertil. Dev.* **2011**, *23*, 912–920. [CrossRef]
93. Knitlova, D.; Hulinska, P.; Jeseta, M.; Hanzalova, K.; Kempisty, B.; Machatkova, M. Supplementation of l-carnitine during in vitro maturation improves embryo development from less competent bovine oocytes. *Theriogenology* **2017**, *102*, 16–22. [CrossRef]
94. Loewenstein, J.E.; Cohen, A.I. Dry Mass, Lipid Content and Protein Content of the Intact and Zona-Free Mouse Ovum. *J. Embryol. Exp. Morphol.* **1964**, *12*, 113–121.
95. McEvoy, T.G.; Coull, G.D.; Broadbent, P.J.; Hutchinson, J.S.; Speake, B.K. Fatty acid composition of lipids in immature cattle, pig and sheep oocytes with intact zona pellucida. *J. Reprod. Fertil.* **2000**, *118*, 163–170. [CrossRef]
96. Jiang, Z.; Dong, H.; Zheng, X.; Marjani, S.L.; Donovan, D.M.; Chen, J.; Tian, X.C. mRNA Levels of Imprinted Genes in Bovine In Vivo Oocytes, Embryos and Cross Species Comparisons with Humans, Mice and Pigs. *Sci. Rep.* **2015**, *5*, 17898. [CrossRef] [PubMed]
97. Lee, D.K.; Choi, K.H.; Hwang, J.Y.; Oh, J.N.; Kim, S.H.; Lee, C.K. Stearoyl-coenzyme A desaturase 1 is required for lipid droplet formation in pig embryo. *Reproduction* **2019**, *157*, 235–243. [CrossRef]
98. Romek, M.; Gajda, B.; Krzysztofowicz, E.; Smorag, Z. Lipid content of non-cultured and cultured pig embryo. *Reprod. Domest. Anim.* **2009**, *44*, 24–32. [CrossRef] [PubMed]
99. Li, R.; Lai, L.; Wax, D.; Hao, Y.; Murphy, C.N.; Rieke, A.; Samuel, M.; Linville, M.L.; Korte, S.W.; Evans, R.W.; et al. Cloned transgenic swine via in vitro production and cryopreservation. *Biol. Reprod.* **2006**, *75*, 226–230. [CrossRef] [PubMed]
100. Li, R.; Murphy, C.N.; Spate, L.; Wax, D.; Isom, C.; Rieke, A.; Walters, E.M.; Samuel, M.; Prather, R.S. Production of piglets after cryopreservation of embryos using a centrifugation-based method for delipation without micromanipulation. *Biol. Reprod..* **2009**, *80*, 563–571. [CrossRef]
101. Lowe, J.L.; Bartolac, L.K.; Bathgate, R.; Grupen, C.G. Supplementation of culture medium with L-carnitine improves the development and cryotolerance of in vitro-produced porcine embryos. *Repro. Fertil. Dev.* **2017**, *29*, 2357–2366. [CrossRef] [PubMed]
102. Nagashima, H.; Kashiwazaki, N.; Ashman, R.J.; Grupen, C.G.; Seamark, R.F.; Nottle, M.B. Removal of cytoplasmic lipid enhances the tolerance of porcine embryos to chilling. *Biol. Reprod.* **1994**, *51*, 618–622. [CrossRef] [PubMed]
103. Beckman, L.S.; Day, B.N. Effects of media NaCl concentration and osmolarity on the culture of early-stage porcine embryos and the viability of embryos cultured in a selected superior medium. *Theriogenology* **1993**, *39*, 611–622. [CrossRef]
104. Petters, R.M.; Wells, K.D. Culture of pig embryos. *J. Reprod. Fertil. Suppl.* **1993**, *48*, 61–73. [PubMed]
105. Pollard, J.W.; Plante, C.; Leibo, S.P. Comparison of development of pig zygotes and embryos in simple and complex culture media. *J. Reprod. Fertil.* **1995**, *103*, 331–337. [CrossRef]
106. Dobrinsky, J.R.; Johnson, L.A.; Rath, D. Development of a culture medium (BECM-3) for porcine embryos: Effects of bovine serum albumin and fetal bovine serum on embryo development. *Biol. Reprod.* **1996**, *55*, 1069–1074. [CrossRef]
107. Im, G.S.; Lai, L.; Liu, Z.; Hao, Y.; Wax, D.; Bonk, A.; Prather, R.S. In vitro development of preimplantation porcine nuclear transfer embryos cultured in different media and gas atmospheres. *Theriogenology* **2004**, *61*, 1125–1135. [CrossRef]
108. Spate, L.D.; Redel, B.K.; Brown, A.N.; Murphy, C.N.; Prather, R.S. Replacement of bovine serum albumin with N-methyl-D-aspartic acid and homocysteine improves development, but not live birth. *Mol. Reprod. Dev.* **2012**, *79*, 310. [CrossRef]
109. Spate, L.D.; Brown, A.; Redel, B.K.; Whitworth, K.M.; Prather, R.S. PS48 can replace bovine serum albumin in pig embryo culture medium, and improve in vitro embryo development by phosphorylating AKT. *Mol. Reprod. Dev.* **2015**, *82*, 315–320. [CrossRef]

110. Chen, P.R.; Spate, L.D.; Leffeler, E.C.; Benne, J.A.; Cecil, R.F.; Hord, T.K.; Prather, R.S. Removal of hypotaurine from porcine embryo culture medium does not impair development of in vitro-fertilized or somatic cell nuclear transfer-derived embryos at low oxygen tension. *Mol. Reprod. Dev.* **2020**. [CrossRef]
111. Guérin, P.; Mouatassim, S.E.; Ménézo, Y. Oxidative stress and protection against reactive oxygen species in the pre-implantation embryo and its surroundings. *Hum. Reprod.* **2001**, *7*, 175–189. [CrossRef]
112. Machaty, Z.; Day, B.N.; Prather, R.S. Development of early porcine embryos in vitro and in vivo. *Biol. Reprod.* **1998**, *59*, 451–455. [CrossRef]
113. Sanfins, A.; Plancha, C.E.; Albertini, D.F. Pre-implantation developmental potential from in vivo and in vitro matured mouse oocytes: A cytoskeletal perspective on oocyte quality. *J. Assist. Reprod. Genet.* **2015**, *32*, 127–136. [CrossRef]
114. Ushijima, H.; Akiyama, K.; Tajima, T. Transition of cell numbers in bovine preimplantation embryos: In vivo collected and in vitro produced embryos. *J. Reprod. Dev.* **2008**, *54*, 239–243. [CrossRef]
115. Lindner, G.M.; Wright, R.W.J. Morphological evaluation of bovine embryos. *Theriogenology* **1983**, *20*, 407–416. [CrossRef]
116. Sugimura, S.; Akai, T.; Imai, K. Selection of viable in vitro-fertilized bovine embryos using time-lapse monitoring in microwell culture dishes. *J. Reprod. Dev.* **2017**, *63*, 353–357. [CrossRef]
117. Isom, S.C.; Li, R.F.; Whitworth, K.M.; Prather, R.S. Timing of first embryonic cleavage is a positive indicator of the in vitro developmental potential of porcine embryos derived from in vitro fertilization, somatic cell nuclear transfer and parthenogenesis. *Mol. Reprod. Dev.* **2012**, *79*, 197–207. [CrossRef] [PubMed]
118. Lundin, K.; Bergh, C.; Hardarson, T. Early embryo cleavage is a strong indicator of embryo quality in human IVF. *Hum. Reprod.* **2001**, *16*, 2652–2657. [CrossRef]
119. Mateusen, B.; Van Soom, A.; Maes, D.G.; Donnay, I.; Duchateau, L.; Lequarre, A.S. Porcine embryo development and fragmentation and their relation to apoptotic markers: A cinematographic and confocal laser scanning microscopic study. *Reproduction* **2005**, *129*, 443–452. [CrossRef] [PubMed]
120. Hornak, M.; Hulinska, P.; Musilova, P.; Kubickova, S.; Rubes, J. Investigation of chromosome aneuploidies in early porcine embryos using comparative genomic hybridization. *Cytogenet. Genome Res.* **2009**, *126*, 210–216. [CrossRef] [PubMed]
121. McCauley, T.C.; Mazza, M.R.; Didion, B.A.; Mao, J.; Wu, G.; Coppola, G.; Coppola, G.F.; Di Berardino, D.; Day, B.N. Chromosomal abnormalities in Day-6, in vitro-produced pig embryos. *Theriogenology* **2003**, *60*, 1569–1580. [CrossRef]
122. Viuff, D.; Rickords, L.; Offenberg, H.; Hyttel, P.; Avery, B.; Greve, T.; Olsaker, I.; Williams, J.L.; Callesen, H.; Thomsen, P.D. A high proportion of bovine blastocysts produced in vitro are mixoploid. *Biol. Reprod.* **1999**, *60*, 1273–1278. [CrossRef]
123. Tsuiko, O.; Catteeuw, M.; Zamani Esteki, M.; Destouni, A.; Bogado Pascottini, O.; Besenfelder, U.; Havlicek, V.; Smits, K.; Kurg, A.; Salumets, A.; et al. Genome stability of bovine in vivo-conceived cleavage-stage embryos is higher compared to in vitro-produced embryos. *Hum. Reprod.* **2017**, *32*, 2348–2357. [CrossRef]
124. Destouni, A.; Zamani Esteki, M.; Catteeuw, M.; Tsuiko, O.; Dimitriadou, E.; Smits, K.; Kurg, A.; Salumets, A.; Van Soom, A.; Voet, T.; et al. Zygotes segregate entire parental genomes in distinct blastomere lineages causing cleavage-stage chimerism and mixoploidy. *Genome Res.* **2016**, *26*, 567–578. [CrossRef]
125. Sathananthan, A.H.; Trounson, A.O. Mitochondrial morphology during preimplantational human embryogenesis. *Hum. Reprod.* **2000**, *15* (Suppl. 2), 148–159. [CrossRef]
126. Reers, M.; Smiley, S.T.; Mottola-Hartshorn, C.; Chen, A.; Lin, M.; Chen, L.B. Mitochondrial membrane potential monitored by JC-1 dye. *Methods Enzymol.* **1995**, *260*, 406–417. [CrossRef]
127. Mordhorst, B.R.; Murphy, S.L.; Ross, R.M.; Benne, J.A.; Samuel, M.S.; Cecil, R.F.; Redel, B.K.; Spate, L.D.; Murphy, C.N.; Wells, K.D.; et al. Pharmacologic treatment of donor cells induced to have a Warburg effect-like metabolism does not alter embryonic development in vitro or survival during early gestation when used in somatic cell nuclear transfer in pigs. *Mol. Reprod. Dev.* **2018**, *85*, 290–302. [CrossRef]
128. Agnello, M.; Morici, G.; Rinaldi, A.M. A method for measuring mitochondrial mass and activity. *Cytotechnology* **2008**, *56*, 145–149. [CrossRef]
129. Miles, J.R.; Blomberg, L.A.; Krisher, R.L.; Everts, R.E.; Sonstegard, T.S.; Van Tassell, C.P.; Zuelke, K.A. Comparative transcriptome analysis of in vivo- and in vitro-produced porcine blastocysts by small amplified RNA-serial analysis of gene expression (SAR-SAGE). *Mol. Reprod. Dev.* **2008**, *75*, 976–988. [CrossRef] [PubMed]
130. Whitworth, K.M.; Agca, C.; Kim, J.G.; Patel, R.V.; Springer, G.K.; Bivens, N.J.; Forrester, L.J.; Mathialagan, N.; Green, J.A.; Prather, R.S. Transcriptional profiling of pig embryogenesis by using a 15-K member unigene set specific for pig reproductive tissues and embryos. *Biol. Reprod.* **2005**, *72*, 1437–1451. [CrossRef] [PubMed]
131. Clarke, R.N.; Baltz, J.M.; Lechene, C.P.; Biggers, J.D. Use of ultramicrofluorometric methods for the study of single preimplantation embryos. *Poult. Sci.* **1989**, *68*, 972–978. [CrossRef] [PubMed]
132. Hardy, K.; Hooper, M.A.; Handyside, A.H.; Rutherford, A.J.; Winston, R.M.; Leese, H.J. Non-invasive measurement of glucose and pyruvate uptake by individual human oocytes and preimplantation embryos. *Hum. Reprod.* **1989**, *4*, 188–191. [CrossRef] [PubMed]
133. Booth, P.J.; Humpherson, P.G.; Watson, T.J.; Leese, H.J. Amino acid depletion and appearance during porcine preimplantation embryo development in vitro. *Reproduction* **2005**, *130*, 655–668. [CrossRef] [PubMed]

134. Stokes, P.J.; Hawkhead, J.A.; Fawthrop, R.K.; Picton, H.M.; Sharma, V.; Leese, H.J.; Houghton, F.D. Metabolism of human embryos following cryopreservation: Implications for the safety and selection of embryos for transfer in clinical IVF. *Hum. Reprod.* **2007**, *22*, 829–835. [CrossRef]
135. D'Souza, F.; Uppangala, S.; Asampille, G.; Salian, S.R.; Kalthur, G.; Talevi, R.; Atreya, H.S.; Adiga, S.K. Spent embryo culture medium metabolites are related to the in vitro attachment ability of blastocysts. *Sci. Rep.* **2018**, *8*, 17025. [CrossRef]
136. Seli, E.; Botros, L.; Sakkas, D.; Burns, D.H. Noninvasive metabolomic profiling of embryo culture media using proton nuclear magnetic resonance correlates with reproductive potential of embryos in women undergoing in vitro fertilization. *Fertil. Steril.* **2008**, *90*, 2183–2189. [CrossRef]
137. Guo, Y.; Yang, Y.; Yi, X.; Zhou, X. Microfluidic method reduces osmotic stress injury to oocytes during cryoprotectant addition and removal processes in porcine oocytes. *Cryobiology* **2019**, *90*, 63–70. [CrossRef]
138. Zeringue, H.C.; Wheeler, M.B.; Beebe, D.J. A microfluidic method for removal of the zona pellucida from mammalian embryos. *Lab. Chip.* **2005**, *5*, 108–110. [CrossRef]
139. Sano, H.; Matsuura, K.; Naruse, K.; Funahashi, H. Application of a microfluidic sperm sorter to the in-vitro fertilization of porcine oocytes reduced the incidence of polyspermic penetration. *Theriogenology* **2010**, *74*, 863–870. [CrossRef]
140. Clark, S.G.; Haubert, K.; Beebe, D.J.; Ferguson, C.E.; Wheeler, M.B. Reduction of polyspermic penetration using biomimetic microfluidic technology during in vitro fertilization. *Lab Chip* **2005**, *5*, 1229–1232. [CrossRef]
141. Krisher, R.L.; Wheeler, M.B. Towards the use of microfluidics for individual embryo culture. *Reprod. Fertil. Dev.* **2010**, *22*, 32–39. [CrossRef] [PubMed]
142. Wheeler, M.B.; Walters, E.M.; Beebe, D.J. Toward culture of single gametes: The development of microfluidic platforms for assisted reproduction. *Theriogenology* **2007**, *68* (Suppl. 1), S178–S189. [CrossRef] [PubMed]
143. Sargus-Patino, C.N.; Wright, E.C.; Plautz, S.A.; Miles, J.R.; Vallet, J.L.; Pannier, A.K. In vitro development of preimplantation porcine embryos using alginate hydrogels as a three-dimensional extracellular matrix. *Reprod. Fertil. Dev.* **2014**, *26*, 943–953. [CrossRef]
144. Laughlin, T.D.; Miles, J.R.; Wright-Johnson, E.C.; Rempel, L.A.; Lents, C.A.; Pannier, A.K. Development of pre-implantation porcine blastocysts cultured within alginate hydrogel systems either supplemented with secreted phosphoprotein 1 or conjugated with Arg-Gly-Asp Peptide. *Reprod. Fertil. Dev.* **2017**, *29*, 2345–2356. [CrossRef] [PubMed]
145. Silva, C.G.; Cunha, E.R.; Blume, G.R.; Malaquias, J.V.; Bao, S.N.; Martins, C.F. Cryopreservation of boar sperm comparing different cryoprotectants associated in media based on powdered coconut water, lactose and trehalose. *Cryobiology* **2015**, *70*, 90–94. [CrossRef] [PubMed]
146. Nagashima, H.; Kashiwazaki, N.; Ashman, R.J.; Grupen, C.G.; Nottle, M.B. Cryopreservation of porcine embryos. *Nature* **1995**, *374*, 416. [CrossRef] [PubMed]
147. Spate, L.D.; Murphy, C.N.; Prather, R.S. High-throughput cryopreservation of in vivo-derived swine embryos. *PLoS ONE* **2013**, *8*, e65545. [CrossRef]
148. Somfai, T.; Kikuchi, K. Vitrification of Porcine Oocytes and Zygotes in Microdrops on a Solid Metal Surface or Liquid Nitrogen. *Methods Mol. Biol.* **2021**, *2180*, 455–468. [CrossRef] [PubMed]
149. Stoecklein, K.S.; Ortega, M.S.; Spate, L.D.; Murphy, C.N.; Prather, R.S. Improved cryopreservation of in vitro produced bovine embryos using FGF2, LIF, and IGF1. *PLoS ONE* **2021**, *16*, e0243727. [CrossRef]
150. Tajima, S.; Motoyama, S.; Wakiya, Y.; Uchikura, K.; Misawa, H.; Takishita, R.; Hirayama, Y.; Kikuchi, K. Piglet production by non-surgical transfer of vitrified embryos, transported to commercial swine farms and warmed on site. *Anim. Sci. J.* **2020**, *91*, e13476. [CrossRef]
151. Hirayama, Y.; Takishita, R.; Misawa, H.; Kikuchi, K.; Misumi, K.; Egawa, S.; Motoyama, S.; Hasuta, Y.; Nakamura, Y.; Hashiyada, Y. Non-surgical transfer of vitrified porcine embryos using a catheter designed for a proximal site of the uterus. *Anim. Sci. J.* **2020**, *91*, e13457. [CrossRef]
152. Bigarella, C.L.; Liang, R.; Ghaffari, S. Stem cells and the impact of ROS signaling. *Development* **2014**, *141*, 4206–4218. [CrossRef] [PubMed]
153. Dalvit, G.C.; Cetica, P.D.; Pintos, L.N.; Beconi, M.T. Reactive oxygen species in bovine embryo in vitro production. *Biocell* **2005**, *29*, 209–212. [CrossRef] [PubMed]
154. Kang, H.G.; Lee, S.; Jeong, P.S.; Kim, M.J.; Park, S.H.; Joo, Y.E.; Park, S.H.; Song, B.S.; Kim, S.U.; Kim, M.K.; et al. Lycopene Improves In Vitro Development of Porcine Embryos by Reducing Oxidative Stress and Apoptosis. *Antioxidants* **2021**, *10*, 230. [CrossRef]
155. Yang, S.G.; Park, H.J.; Kim, J.W.; Jung, J.M.; Kim, M.J.; Jegal, H.G.; Kim, I.S.; Kang, M.J.; Wee, G.; Yang, H.Y.; et al. Mito-TEMPO improves development competence by reducing superoxide in preimplantation porcine embryos. *Sci. Rep.* **2018**, *8*, 10130. [CrossRef]
156. Berthelot, F.; Terqui, M. Effects of oxygen, CO_2/pH and medium on the in vitro development of individually cultured porcine one- and two-cell embryos. *Reprod. Nutr. Dev.* **1996**, *36*, 241–251. [CrossRef]
157. Rinaudo, P.F.; Giritharan, G.; Talbi, S.; Dobson, A.T.; Schultz, R.M. Effects of oxygen tension on gene expression in preimplantation mouse embryos. *Fertil. Steril.* **2006**, *86*, 1265.e1–1265.e36. [CrossRef] [PubMed]
158. Jeseta, M.; Cela, A.; Zakova, J.; Madr, A.; Crha, I.; Glatz, Z.; Kempisty, B.; Ventruba, P. Metabolic Activity of Human Embryos after Thawing Differs in Atmosphere with Different Oxygen Concentrations. *J. Clin. Med.* **2020**, *9*, 2609. [CrossRef]

159. Yin, C.; Liu, J.; He, B.; Jia, L.; Gong, Y.; Guo, H.; Zhao, R. Heat stress induces distinct responses in porcine cumulus cells and oocytes associated with disrupted gap junction and trans-zonal projection colocalization. *J. Cell. Physiol.* **2019**, *234*, 4787–4798. [CrossRef]
160. Do, L.T.; Luu, V.V.; Morita, Y.; Taniguchi, M.; Nii, M.; Peter, A.T.; Otoi, T. Astaxanthin present in the maturation medium reduces negative effects of heat shock on the developmental competence of porcine oocytes. *Reprod. Biol.* **2015**, *15*, 86–93. [CrossRef]
161. Li, Y.; Wang, J.; Zhang, Z.; Yi, J.; He, C.; Wang, F.; Tian, X.; Yang, M.; Song, Y.; He, P.; et al. Resveratrol compares with melatonin in improving in vitro porcine oocyte maturation under heat stress. *J. Anim. Sci. Biotechnol* **2016**, *7*, 33. [CrossRef]
162. Li, Y.; Zhang, Z.; He, C.; Zhu, K.; Xu, Z.; Ma, T.; Tao, J.; Liu, G. Melatonin protects porcine oocyte in vitro maturation from heat stress. *J. Pineal. Res.* **2015**, *59*, 365–375. [CrossRef]
163. Kojima, T.; Udagawa, K.; Onishi, A.; Iwahashi, H.; Komatsu, Y. Effect of heat stress on development in vitro and in vivo and on synthesis of heat shock proteins in porcine embryos. *Mol. Reprod. Dev.* **1996**, *43*, 452–457. [CrossRef]
164. Isom, S.C.; Lai, L.; Prather, R.S.; Rucker, E.B., 3rd. Heat shock of porcine zygotes immediately after oocyte activation increases viability. *Mol. Reprod. Dev.* **2009**, *76*, 548–554. [CrossRef] [PubMed]
165. Isom, S.C.; Prather, R.S.; Rucker Iii, E.B. Enhanced developmental potential of heat-shocked porcine parthenogenetic embryos is related to accelerated mitogen-activated protein kinase dephosphorylation. *Reprod. Fertil. Dev.* **2009**, *21*, 892–900. [CrossRef] [PubMed]
166. Hwang, I.S.; Park, M.R.; Moon, H.J.; Shim, J.H.; Kim, D.H.; Yang, B.C.; Ko, Y.G.; Yang, B.S.; Cheong, H.T.; Im, G.S. Osmolarity at early culture stage affects development and expression of apoptosis related genes (Bax-alpha and Bcl-xl) in pre-implantation porcine NT embryos. *Mol. Reprod. Dev.* **2008**, *75*, 464–471. [CrossRef] [PubMed]
167. Marchal, R.; Feugang, J.M.; Perreau, C.; Venturi, E.; Terqui, M.; Mermillod, P. Meiotic and developmental competence of prepubertal and adult swine oocytes. *Theriogenology* **2001**, *56*, 17–29. [CrossRef]
168. Bagg, M.A.; Nottle, M.B.; Grupen, C.G.; Armstrong, D.T. Effect of dibutyryl cAMP on the cAMP content, meiotic progression, and developmental potential of in vitro matured pre-pubertal and adult pig oocytes. *Mol. Reprod. Dev.* **2006**, *73*, 1326–1332. [CrossRef]
169. Bagg, M.A.; Nottle, M.B.; Armstrong, D.T.; Grupen, C.G. Relationship between follicle size and oocyte developmental competence in prepubertal and adult pigs. *Reprod. Fertil. Dev.* **2007**, *19*, 797–803. [CrossRef] [PubMed]
170. Hyun, S.H.; Lee, G.S.; Kim, D.Y.; Kim, H.S.; Lee, S.H.; Kim, S.; Lee, E.S.; Lim, J.M.; Kang, S.K.; Lee, B.C.; et al. Effect of maturation media and oocytes derived from sows or gilts on the development of cloned pig embryos. *Theriogenology* **2003**, *59*, 1641–1649. [CrossRef]
171. Amann, R.P.; Waberski, D. Computer-assisted sperm analysis (CASA): Capabilities and potential developments. *Theriogenology* **2014**, *81*, 5–17. [CrossRef]
172. Vazquez, J.M.; Martinez, E.; Pastor, L.M.; Roca, J.; Matas, C.; Calvo, A. Lectin histochemistry during in vitro capacitation and acrosome reaction in boar spermatozoa: New lectins for evaluating acrosomal status of boar spermatozoa. *Acta Histochem.* **1996**, *98*, 93–100. [CrossRef]
173. Garner, D.L.; Pinkel, D.; Johnson, L.A.; Pace, M.M. Assessment of spermatozoal function using dual fluorescent staining and flow cytometric analyses. *Biol. Reprod.* **1986**, *34*, 127–138. [CrossRef]
174. Kerns, K.; Zigo, M.; Drobnis, E.Z.; Sutovsky, M.; Sutovsky, P. Zinc ion flux during mammalian sperm capacitation. *Nature Communications* **2018**, *9*, 2061. [CrossRef]
175. Huo, L.-J.; Yue, K.-Z.; Yang, Z.-M. Characterization of viability, mitochondrial activity, acrosomal integrity and capacitation status in boar sperm during in vitro storage at different ambient temperatures. *Reprod. Fertil. Dev.* **2002**, *14*, 509. [CrossRef]
176. Hossain, M.S.; Johannisson, A.; Siqueira, A.P.; Wallgren, M.; Rodriguez-Martinez, H. Spermatozoa in the sperm-peak-fraction of the boar ejaculate show a lower flow of Ca2+ under capacitation conditions post-thaw which might account for their higher membrane stability after cryopreservation. *Anim. Reprod. Sci.* **2011**, *128*, 37–44. [CrossRef]
177. Kerns, K.; Jankovitz, J.; Robinson, J.; Minton, A.; Kuster, C.; Sutovsky, P. Relationship between the Length of Sperm Tail Mitochondrial Sheath and Fertility Traits in Boars Used for Artificial Insemination. *Antioxidants* **2020**, *9*, 1033. [CrossRef]
178. Lovercamp, K.W.; Safranski, T.J.; Fischer, K.A.; Manandhar, G.; Sutovsky, M.; Herring, W.; Sutovsky, P. Arachidonate 15-lipoxygenase and ubiquitin as fertility markers in boars. *Theriogenology* **2007**, *67*, 704–718. [CrossRef]
179. Guthrie, H.D.; Welch, G.R. Determination of intracellular reactive oxygen species and high mitochondrial membrane potential in Percoll-treated viable boar sperm using fluorescence-activated flow cytometry1. *J. Anim. Sci.* **2006**, *84*, 2089–2100. [CrossRef]
180. Brouwers, J.F.; Silva, P.F.N.; Gadella, B.M. New assays for detection and localization of endogenous lipid peroxidation products in living boar sperm after BTS dilution or after freeze–thawing. *Theriogenology* **2005**, *63*, 458–469. [CrossRef]
181. Evenson, D.P.; Thompson, L.; Jost, L. Flow cytometric evaluation of boar semen by the sperm chromatin structure assay as related to cryopreservation and fertility. *Theriogenology* **1994**, *41*, 637–651. [CrossRef]
182. Parrilla, I.; Vazquez, J.; Oliver-Bonet, M.; Navarro, J.; Yelamos, J.; Roca, J.; Martinez, E. Fluorescence in situ hybridization in diluted and flow cytometrically sorted boar spermatozoa using specific DNA direct probes labelled by nick translation. *Reproduction* **2003**, *126*, 317–325. [CrossRef] [PubMed]
183. Steele, H.; Makri, D.; Maalouf, W.E.; Reese, S.; Kölle, S. Bovine Sperm Sexing Alters Sperm Morphokinetics and Subsequent Early Embryonic Development. *Sci. Rep.* **2020**, *10*, 1–13. [CrossRef] [PubMed]
184. Suh, T.K.; Schenk, J.L.; Seidel, G.E. High pressure flow cytometric sorting damages sperm. *Theriogenology* **2005**, *64*, 1035–1048. [CrossRef] [PubMed]

185. Umezu, K.; Hiradate, Y.; Numabe, T.; Hara, K.; Tanemura, K. Effects on glycocalyx structures of frozen-thawed bovine sperm induced by flow cytometry and artificial capacitation. *J. Reprod. Dev.* **2017**, *63*, 473–480. [CrossRef] [PubMed]
186. Saunders, C.M.; Larman, M.G.; Parrington, J.; Cox, L.J.; Royse, J.; Blayney, L.M.; Swann, K.; Lai, F.A. PLCζ: A sperm-specific trigger of Ca2+ oscillations in eggs and embryo development. *Development* **2002**, *129*, 3533–3544. [CrossRef]
187. Wu, A.T.; Sutovsky, P.; Manandhar, G.; Xu, W.; Katayama, M.; Day, B.N.; Park, K.W.; Yi, Y.J.; Xi, Y.W.; Prather, R.S.; et al. PAWP, a Sperm-specific WW Domain-binding Protein, Promotes Meiotic Resumption and Pronuclear Development during Fertilization. *J. Biol. Chem.* **2007**, *282*. [CrossRef] [PubMed]
188. Seeler, J.F.; Sharma, A.; Zaluzec, N.J.; Bleher, R.; Lai, B.; Schultz, E.G.; Hoffman, B.M.; Labonne, C.; Woodruff, T.K.; O'Halloran, T.V. Metal ion fluxes controlling amphibian fertilization. *Nat. Chem.* **2021**, *13*, 683–691. [CrossRef]
189. Que, E.L.; Duncan, F.E.; Lee, H.C.; Hornick, J.E.; Vogt, S.; Fissore, R.A.; O'Halloran, T.V.; Woodruff, T.K. Bovine eggs release zinc in response to parthenogenetic and sperm-induced egg activation. *Theriogenology* **2019**, *127*, 41–48. [CrossRef]
190. Kim, A.M.; Bernhardt, M.L.; Kong, B.Y.; Ahn, R.W.; Vogt, S.; Woodruff, T.K.; O'Halloran, T.V. Zinc sparks are triggered by fertilization and facilitate cell cycle resumption in mammalian eggs. *ACS Chem. Bio* **2011**, *6*, 716–723. [CrossRef]
191. Lee, K.; Davis, A.; Zhang, L.; Ryu, J.; Spate, L.D.; Park, K.-W.; Samuel, M.S.; Walters, E.M.; Murphy, C.N.; Machaty, Z.; et al. Pig oocyte activation using a Zn2+ chelator, TPEN. *Theriogenology* **2015**, *84*, 1024–1032. [CrossRef] [PubMed]
192. Kerns, K.; Sharif, M.; Zigo, M.; Xu, W.; Hamilton, L.E.; Sutovsky, M.; Ellersieck, M.; Drobnis, E.Z.; Bovin, N.; Oko, R.; et al. Sperm Cohort-Specific Zinc Signature Acquisition and Capacitation-Induced Zinc Flux Regulate Sperm-Oviduct and Sperm-Zona Pellucida Interactions. *Int. J. Mol. Sci.* **2020**, *21*, 2121. [CrossRef] [PubMed]
193. Que, E.L.; Duncan, F.E.; Bayer, A.R.; Philips, S.J.; Roth, E.W.; Bleher, R.; Gleber, S.C.; Vogt, S.; Woodruff, T.K.; O'Halloran, T.V. Zinc sparks induce physiochemical changes in the egg zona pellucida that prevent polyspermy. *Integr. Biol.* **2017**, *9*, 135–144. [CrossRef] [PubMed]
194. Bonk, A.J.; Cheong, H.T.; Li, R.; Lai, L.; Hao, Y.; Liu, Z.; Samuel, M.; Fergason, E.A.; Whitworth, K.M.; Murphy, C.N.; et al. Correlation of developmental differences of nuclear transfer embryos cells to the methylation profiles of nuclear transfer donor cells in Swine. *Epigenetics* **2007**, *2*, 179–186. [CrossRef] [PubMed]
195. Bonk, A.J.; Li, R.; Lai, L.; Hao, Y.; Liu, Z.; Samuel, M.; Fergason, E.A.; Whitworth, K.M.; Murphy, C.N.; Antoniou, E.; et al. Aberrant DNA methylation in porcine in vitro-, parthenogenetic-, and somatic cell nuclear transfer-produced blastocysts. *Mol. Reprod. Dev.* **2008**, *75*, 250–264. [CrossRef] [PubMed]
196. Ivanova, E.; Canovas, S.; Garcia-Martinez, S.; Romar, R.; Lopes, J.S.; Rizos, D.; Sanchez-Calabuig, M.J.; Krueger, F.; Andrews, S.; Perez-Sanz, F.; et al. DNA methylation changes during preimplantation development reveal inter-species differences and reprogramming events at imprinted genes. *Clin. Epigenetics* **2020**, *12*, 64. [CrossRef] [PubMed]
197. Jost, J.P.; Saluz, H.P.; McEwan, I.; Feavers, I.M.; Hughes, M.; Reiber, S.; Liang, H.M.; Vaccaro, M. Tissue specific expression of avian vitellogenin gene is correlated with DNA hypomethylation and in vivo specific protein-DNA interactions. *Philos Trans. R Soc. Lond. B Biol. Sci.* **1990**, *326*, 231–240. [CrossRef]
198. Burbelo, P.D.; Horikoshi, S.; Yamada, Y. DNA methylation and collagen IV gene expression in F9 teratocarcinoma cells. *J. Biol. Chem.* **1990**, *265*, 4839–4843. [CrossRef]
199. Knust, B.; Bruggemann, U.; Doerfler, W. Reactivation of a methylation-silenced gene in adenovirus-transformed cells by 5-azacytidine or by E1A trans activation. *J. Virol.* **1989**, *63*, 3519–3524. [CrossRef]
200. Gerondakis, S.; Boyd, A.; Bernard, O.; Webb, E.; Adams, J.M. Activation of immunoglobulin mu gene expression involves stepwise demethylation. *EMBO J.* **1984**, *3*, 3013–3021. [CrossRef]
201. Zhao, J.; Ross, J.W.; Hao, Y.; Spate, L.D.; Walters, E.M.; Samuel, M.S.; Rieke, A.; Murphy, C.N.; Prather, R.S. Significant improvement in cloning efficiency of an inbred miniature pig by histone deacetylase inhibitor treatment after somatic cell nuclear transfer. *Biol. Reprod.* **2009**, *81*, 525–530. [CrossRef]
202. Jeong, P.S.; Sim, B.W.; Park, S.H.; Kim, M.J.; Kang, H.G.; Nanjidsuren, T.; Lee, S.; Song, B.S.; Koo, D.B.; Kim, S.U. Chaetocin Improves Pig Cloning Efficiency by Enhancing Epigenetic Reprogramming and Autophagic Activity. *Int. J. Mol. Sci* **2020**, *21*, 4836. [CrossRef]
203. Zhao, C.; Shi, J.; Zhou, R.; He, X.; Yang, H.; Wu, Z. DZNep and UNC0642 enhance in vitro developmental competence of cloned pig embryos. *Reproduction* **2018**, *157*, 359–369. [CrossRef] [PubMed]
204. Yu, T.; Qi, X.; Zhang, L.; Ning, W.; Gao, D.; Xu, T.; Ma, Y.; Knott, J.G.; Sathanawongs, A.; Cao, Z.; et al. Dynamic reprogramming and function of RNA N(6)-methyladenosine modification during porcine early embryonic development. *Zygote* **2021**, 1–10. [CrossRef] [PubMed]
205. Zhao, J.; Whyte, J.; Prather, R.S. Effect of epigenetic regulation during swine embryogenesis and on cloning by nuclear transfer. *Cell Tissue Res.* **2010**, *341*, 13–21. [CrossRef]
206. Li, G.; Jia, Q.; Zhao, J.; Li, X.; Yu, M.; Samuel, M.S.; Zhao, S.; Prather, R.S.; Li, C. Dysregulation of genome-wide gene expression and DNA methylation in abnormal cloned piglets. *BMC Genom.* **2014**, *15*, 811. [CrossRef] [PubMed]

207. Whitworth, K.M.; Mao, J.; Lee, K.; Spollen, W.G.; Samuel, M.S.; Walters, E.M.; Spate, L.D.; Prather, R.S. Transcriptome Analysis of Pig In Vivo, In Vitro-Fertilized, and Nuclear Transfer Blastocyst-Stage Embryos Treated with Histone Deacetylase Inhibitors Postfusion and Activation Reveals Changes in the Lysosomal Pathway. *Cell. Reprogram* **2015**, *17*, 243–258. [CrossRef]
208. Mao, J.; Zhao, M.T.; Whitworth, K.M.; Spate, L.D.; Walters, E.M.; O'Gorman, C.; Lee, K.; Samuel, M.S.; Murphy, C.N.; Wells, K.; et al. Oxamflatin treatment enhances cloned porcine embryo development and nuclear reprogramming. *Cell. Reprogram* **2015**, *17*, 28–40. [CrossRef]

 cells

Article

ROCK Inhibitor (Y-27632) Abolishes the Negative Impacts of miR-155 in the Endometrium-Derived Extracellular Vesicles and Supports Embryo Attachment

Islam M. Saadeldin [1,2], Bereket Molla Tanga [1], Seonggyu Bang [1], Chaerim Seo [1], Okjae Koo [3], Sung Ho Yun [4], Seung Il Kim [4], Sanghoon Lee [1] and Jongki Cho [1,*]

1 Laboratory of Theriogenology, College of Veterinary Medicine, Chungnam National University, Daejeon 34134, Korea
2 Research Institute of Veterinary Medicine, Chungnam National University, Daejeon 34134, Korea
3 Toolgen Inc., Seoul 08501, Korea
4 Korea Basic Science Institute (KBSI), Ochang 28119, Korea
* Correspondence: cjki@cnu.ac.kr; Tel.: +82-42-821-6788

Abstract: Extracellular vesicles (EVs) are nanosized vesicles that act as snapshots of cellular components and mediate cellular communications, but they may contain cargo contents with undesired effects. We developed a model to improve the effects of endometrium-derived EVs (Endo-EVs) on the porcine embryo attachment in feeder-free culture conditions. Endo-EVs cargo contents were analyzed using conventional and real-time PCR for micro-RNAs, messenger RNAs, and proteomics. Porcine embryos were generated by parthenogenetic electric activation in feeder-free culture conditions supplemented with or without Endo-EVs. The cellular uptake of Endo-EVs was confirmed using the lipophilic dye PKH26. Endo-EVs cargo contained miR-100, miR-132, and miR-155, together with the mRNAs of porcine endogenous retrovirus (PERV) and β-catenin. Targeting PERV with CRISPR/Cas9 resulted in reduced expression of PERV mRNA transcripts and increased miR-155 in the Endo-EVs, and supplementing these in embryos reduced embryo attachment. Supplementing the medium containing Endo-EVs with miR-155 inhibitor significantly improved the embryo attachment with a few outgrowths, while supplementing with Rho-kinase inhibitor (RI, Y-27632) dramatically improved both embryo attachment and outgrowths. Moreover, the expression of miR-100, miR-132, and the mRNA transcripts of BCL2, zinc finger E-box-binding homeobox 1, β-catenin, interferon-γ, protein tyrosine phosphatase non-receptor type 1, PERV, and cyclin-dependent kinase 2 were all increased in embryos supplemented with Endo-EVs + RI compared to those in the control group. Endo-EVs + RI reduced apoptosis and increased the expression of OCT4 and CDX2 and the cell number of embryonic outgrowths. We examined the individual and combined effects of RI compared to those of the miR-155 mimic and found that RI can alleviate the negative effects of the miR-155 mimic on embryo attachment and outgrowths. EVs can improve embryo attachment and the unwanted effects of the de trop cargo contents (miR-155) can be alleviated through anti-apoptotic molecules such as the ROCK inhibitor.

Keywords: miR-155; CRISPR/Cas9; extracellular vesicles; embryo; ROCK inhibitor

Citation: Saadeldin, I.M.; Tanga, B.M.; Bang, S.; Seo, C.; Koo, O.; Yun, S.H.; Kim, S.I.; Lee, S.; Cho, J. ROCK Inhibitor (Y-27632) Abolishes the Negative Impacts of miR-155 in the Endometrium-Derived Extracellular Vesicles and Supports Embryo Attachment. *Cells* **2022**, *11*, 3178. https://doi.org/10.3390/cells11193178

Academic Editor: Lon J. van Winkle

Received: 17 August 2022
Accepted: 7 October 2022
Published: 10 October 2022

Publisher's Note: MDPI stays neutral with regard to jurisdictional claims in published maps and institutional affiliations.

1. Introduction

The pig is considered a crucial model for transgenic animals and xenotransplantation; however, the process of embryo production in vitro is quite challenging due to a drastic decrease in the embryonic cell number and blastocyst formation as compared toother farm animal species [1,2]. This raises several questions about whether the effects are endogenous or lack exogenous supportive signals. Preimplantation embryos are more competent when co-cultured with other embryos or maternal cells due to the production of paracrine or juxtracrine factors that interact to support the inefficient culture conditions associated with individually cultured embryos [3–5]. The cell number of in vivo–derived embryos is twice that of those that are generated in vitro, and the oviduct significantly affects the cell number and results in a 1.5-fold increase in cell number and hatching rates [6]. It shows that exogenous maternal factors are important for acquiring embryo developmental competence, and there are some other endogenous factors from the embryos themselves that may hamper the development of the embryo. Recent transcriptomics studies have shown vast differences between the porcine blastocysts that are produced in vivo and in vitro and demonstrate that upregulated gene expression of metabolism and arginine transporter contribute to the low developmental competence in in vitro–derived embryos [7].

To reveal the possible factors that regulate the blastocyst development and the rate of attachment in porcine embryos, we designed experiments to investigate the molecular impact of exogenous and endogenous signals responsible for embryonic–maternal crosstalk. For instance, the endogenous factors are formed or released by the embryos themselves, while the exogenous factors are released by the maternal cells and hamper the developmental competence such as miRNAs. Several cargos of protein, mRNA, miRNA, and metabolites are carried through the extracellular vesicles (EVs) and affect the growth of the embryo [4,8,9]. Moreover, the interplay between the EVs derived from the endometrium during embryo implantation in humans and animals has been investigated [10–12]. The porcine endogenous retrovirus (PERV) is secreted by all porcine cells and is considered a natural inhabitant of cells and biological fluids including the uterine cells. The endogenous retroviruses establish interplay between maternal and embryonic cells and are present in the exosomes released by the endometrium [13–15]. Studies also revealed a supportive role of Rho-associated coiled-coil-containing kinases (ROCK) in the development of cleaved embryos, while ROCK inhibition is critical during embryonic and pluripotent stem cell development [16–18], particularly trophoblast adhesion and differentiation [19,20].

The mechanism behind this interplay between the endogenous and exogenous factors that affect porcine embryo developmental competence remains unclear. Therefore, our study is an attempt to understand the interplay between the exogenous factors represented in endometrial EVs and the endogenous factors represented in PERV and ROCK pathways in the developmental competence of porcine embryos to enhance the production of more competent embryos that can meet the needs of cloning and xenotransplantation.

2. Materials and Methods

2.1. Chemicals

Unless otherwise specified, chemicals and reagents were purchased from Sigma-Aldrich (St. Louis, MO, USA).

2.2. Generation of Porcine Parthenogenetic Embryos

Porcine embryos were obtained through chemical parthenogenetic activation of in vitro matured oocytes as per our previous reports [5,21,22]. Porcine ovaries were collected from a slaughterhouse and transferred to the laboratory within 4 h in saline (NaCl 0.9%) at 30 °C. Cumulus–oocyte complexes (COCs) were retrieved through aspiration by an 18-gauge needle connected with a 10 mL syringe. Oocytes surrounded by compact layers of cumulus cells were selected using a stereomicroscope (SMZ 745T, Nikon, Tokyo, Japan) and washed three times in HEPES buffered Tyrode's medium comprising 0.05% polyvinyl alcohol (TLH-PVA). COCs were cultured in 4-well dishes (Nunc, ThermoFisher Scientific, Roskilde, Denmark) containing 500 mL of a maturation medium comprising TCM-199 (Gibco, Waltham, MA, USA), 10% (*v/v*) porcine follicular fluid, cysteine (0.6 mM), sodium pyruvate (0.91 mM), epidermal growth factor (10 ng/mL), kanamycin (75 µg/mL), insulin (1 µg/mL), human chorionic gonadotrophin (10 IU/mL; Daesung Microbiological Labs; Uiwang, Korea), and equine chorionic gonadotrophin (10 IU/mL; Daesung Microbiological Labs) for 22 h. Then, the COCs were moved to the same culture conditions without the presence of the hormones for 22 h. Matured COCs were harvested, and cumulus cells were detached by gentle pipetting in hyaluronidase (0.6%) and then were washed in TLH-PVA and equilibrated in a pulsing medium consisting of mannitol (0.28 M), CaCl2 (0.1 mM), HEPES (0.5 mM), and MgSO4 (0.1 mM). Oocytes were then activated with a single direct current pulse of 1.5 kV/cm for 60 μs generated inside a glass chamber of two electrodes in an activation medium. The electric current was generated through a BTX Electro-Cell Manipulator 2001 (BTX Inc., San Diego, CA, USA). Activated oocytes were washed in TLH-PVA and cultured for 7 days in microdrops of porcine zygote medium-5 (PZM-5, Functional Peptides Research Institute Co. Ltd. (IFP), Yamagata, Japan) overlaid with mineral oil in a humidified atmosphere at 38.5 °C (5% O_2, 5% CO_2, and 90% N_2). Blastocysts were obtained and washed in PBS and zona pellucida was removed by 0.1% pronase (*w/v* in PBS) to obtain zona-free embryos ready for further experiments.

2.3. Endometrium Culture

Uterine tissues of diestrus multiparous sows were collected from the slaughterhouse and transported to the lab within 4 h. Endometrium was separated aseptically under a laminar flow hood [23]. Endometrium was chopped into 1 mm pieces and seeded on 100 mm tissue culture dishes with a minimal volume of culture medium that comprised DMEM, 10% fetal bovine serum, and penicillin/streptomycin (100 U/mL penicillin, 100 μg/mL streptomycin) at 38.5 °C in a humidified atmosphere of 5% CO_2. Tissue attachment and primary cell outgrowths were observed on day-5 of culture and the culture medium was then changed to a fresh one. Primary culture monolayer was maintained until day-8, and the tissue remnants were mechanically discarded.

2.4. Extracellular Vesicles Isolation and Characterization

On day-8, endometrial cell monolayers were cultured in a serum-free culture medium for 24 h and the conditioned medium was aspirated and centrifuged at $300\times g$ to discard cell debris pellets [24]. EVs were isolated through targeted protein binding and nanofiltration using PureExo Exosomes Isolation kits (101 Bio, Palo Alto, CA, USA) [25] to yield 100 μL of EVs in phosphate-buffered saline (PBS) solution. EVs were characterized through ZetaView PMX 110 (Particle Metrix, Meerbusch, Germany) nanoflow fluorescence cytometry and nanoparticle tracking analysis instrument associated with ZetaView 8.05.14 SP7 software and Microsoft Excel 365 (Microsoft Corp., Seattle, WA, USA) [26]. After calibration with 100 nm polystyrene particles (ThermoFisher Scientific), one mL of the sample (diluted 20X in 1× PBS) was loaded into the machine and eleven different positions and two reading cycles per position were automatically set to measure the mean, median, and mode sizes (indicated as diameter), concentrations, and outlier removal in each sample. EVs were examined through transmission electron microscopy (TEM) [4,27]. In brief, 4 μL of isolated EVs solution was stained with 2% uranyl acetate, mounted on the center of 200-mesh copper grids, dried, and visualized through an OMEGA-energy filtering TEM (ZEISS LEO 912, Carl Zeiss, Jena, Germany) at 120 kV. The EVs cargo contents of some selected mRNAs, miRNAs, and proteins were analyzed through reverse-transcription polymerase chain reaction and proteomics as discussed below.

2.5. Embryo Attachment Model

Embryo attachment in feeder-free culture condition was established according to our previous method [28] with some modifications. Fifty μL microdrops of Matrigel basement membrane matrix (BD Biosciences, San Jose, CA, USA) were placed on 4-well dishes (Nunc) and incubated for 30 min at 38.5 °C. Matrigel was removed and replaced with 50 μL of culture medium that was composed of DMEM/F-12 supplemented with 10% fetal bovine serum, β-mercaptoethanol (0.1 mM), 1% nonessential amino acids (Invitrogen, Waltham, MA, USA), and 1% penicillin/streptomycin (100 U/mL penicillin, 100 μg/mL streptomycin). The microdrops were covered with mineral oil and incubated for 30 min before embryo placement. On day-7, embryos were collected and zona pellucidae were removed by pronase (0.1% in PBS) for 1 min at 38.5 °C. Embryos were washed with the culture medium before placing them into the Matrigel-coated microdrops. Embryos were then incubated in a humidified atmosphere of 5% CO_2 at 38.5 °C and monitored for attachment and outgrowths on days 2–5 from culture.

2.6. Experimental Design

2.6.1. Effects of Endometrial-EVs and ROCK Inhibitor on Embryo Development and Attachment

First, embryos (*n* = 20 for 3 replicates) were divided into 4 groups: control, 10 µM ROCK (Rho-associated coiled-coil containing kinases) inhibitor (RI, Y-27632) [29], Endo-EVs (1.5×10^7 particles/mL) [25,30], or combined supplementation of RI and Endo-EVs for different durations (i.e., 36 h and extended to 5 days). The control group was cultured in a plain culture medium without supplementation. Embryos were monitored for attachment, cell number, apoptosis, outgrowths, immunofluorescence staining of pluripotency marker (Oct4) and trophoblast marker (Cdx2), and relative quantitation of some miRNA and mRNA transcripts' expression that are related to apoptosis, cell attachment, cell cycle, and embryo development.

2.6.2. Effect of miR-155 on Embryo Development and Attachment

Based on the findings of EVs analysis, we designed experiments to explore the roles of miR-155 in maternal-embryonic communications. The mir-155 inhibitor was supplied to examine its effect on Endo-EVs supplementation on embryonic development and attachment. Moreover, the effects of miR-155 mimic on embryonic development and attachment were studied in combination with or without RI.

2.6.3. Effect of Targeting PERV on Embryo Development and Attachment

Based on the findings of EVs analysis, we targeted PERV with CRISPR/Cas9 to explore the role and impact of EVs derived from PERV-depleted endometrium on embryo development and attachment.

2.7. EVs Labeling and Uptake

Before EVs isolation, a serum-free conditioned medium was mixed with the PKH26 lipophilic fluorescent stain (Invitrogen) according to the manufacturer's instructions, and EVs were isolated to remove the excess free PKH67 dye following the manufacturer's recommendations [31,32]. EVs were then supplemented (1.5×10^7 particles/mL) [25,30] with cultured embryos for 24 h to monitor their uptake through a fluorescent microscope (MshOt, Guangzhou Micro-shot Technology Co., Ltd., Guangzhou, China). For negative control staining, the plain conditioned medium was mixed with PKH26 and processed by the same EV labeling procedure.

2.8. Immunofluorescence

Immunofluorescence staining of OCT4 and CDX2 was performed according to our previous protocol [33] with some modifications as follows: attached embryos on day-5 were fixed in 4% paraformaldehyde (*w/v* in PBS), pH 7.4 for 15 min at room temperature. Fixed embryos were washed in PBS, permeabilized with 0.1% Triton-X100 (*v/v* in PBS) for 10 min, and then were blocked by 1% goat serum (*v/v*; Invitrogen) for 30 min at room temperature. Primary antibodies specified against Oct4 (mouse monoclonal IgG2b, sc-5279, Santa Cruz Biotech. Inc., Santa Cruz, CA, USA) and Cdx2 (rabbit monoclonal IgG, ab76541, Abcam, Seoul, Korea) were diluted (1:100) and prepared in PBS. The attached embryos were incubated with the primary antibodies (1 h at 38.5 °C), washed in PBS three times, then incubated with the secondary antibodies (Alexa Fluor 488 goat anti-mouse IgG, A11001 and Alexa Fluor 568 goat anti-rabbit IgG, A11011; Invitrogen, Life Technologies Corp., Eugene, OR, USA), and the resulting solution was diluted (1:200) and kept in PBS for 1 h at 38.5 °C before washing in PBS three times. Embryonic cell nuclei were counterstained with Vectashield antifade mounting medium containing 40,60 -diamidino-2-phenylindole (DAPI; Vector Laboratories, Vector Laboratories, Burlingame, CA, USA) for 5 min, and the fluorescence signals were examined with a fluorescent microscope at 488 nm, and 568 nm for Oct4 and Cdx2, respectively. Images were captured and the fluorescence intensity pixel analysis was analyzed with ImageJ 1.53k software (National Institute of Health, Bethesda, MD, USA).

2.9. TdT-Mediated dUTP-X Nick End Labeling (TUNEL) Assay

Labeling of DNA strand breaks and detection of apoptotic cells were examined through In Situ Cell Death Detection TUNEL assay Kit, Fluorescein (Roche Holding AG, Basel, Switzerland) according to the manufacturer's protocol. Embryos were fixed in 4% paraformaldehyde and permeabilized in 0.1% TritonX and then incubated with the working solution of an enzyme (TdT) and a label solution (fluorescein-dUTP) for 1 h at 38.5 °C. Nuclei were counterstained with Vectashield antifade mounting medium as mentioned above. Green fluorescence positive cells (apoptotic cells) were captured and counted with ImageJ software.

2.10. MiR-155 Mimic, miR-155 Inhibitor, and CRISPR/Cas9 Transfection

We used Campylobacter jejuni CRISPR/Cas9 (cjCas9) vector to cleave PERV mRNA. We cloned cjCas9-based sgRNA targeting env gene of PERV in our cjCas9 vector with slight modification (D8A for inactivating RuvC domain) (Supplementary Figure S1). CRISPR/Cas9 vector (1 mg), miR-155 mimic (100 nM), and miR-155 inhibitor (100 nM) oligonucleotide sequences (Table 1) [34] were transfected to the embryos [35] with some modifications. The nucleic acids were incubated with Lipidofect-P transfection reagent (Cat # LDL-P001, Lipidomia, Gachon University IT Center, Gyeonggi-do, Korea) for 30 min at room temperature, and then the mixture was supplemented to the embryo culture medium and incubated for the attachment and further development.

Table 1. Sequences for miR-155 mimic and miR-155 inhibitor [21].

Name	Sequence ($5' \rightarrow 3'$)
miRNA-155 inhibitor	UUAAUGCUAAUCGUGAUAGGGG
miRNA-155 mimic sense	UGGUGCAGGUUUAAUGCUAAUCGUGAUAGGGGUUUA
miRNA-155 mimic anti-sense	GUGCUGAUGAACACCUAUGCUGUUAGCAUUAAUCUUGCGCUA

2.11. Conventional and Real-Time Polymerase Chain Reaction

Total RNA was extracted from the embryos (n = 5, 4 replicates) using RNeasy Micro Kit (Qiagen GmbH, Hilden, Germany, Cat #74004) that included DNase I for removing any of DNA residuals. NanoDrop 2000 (Thermo Scientific) was used to determine the quality of the extracted RNA. Values of > 1.8 of OD 260/280 and 260/230 ratios were used for the reverse transcription. Complementary DNA (cDNA) was synthesized using 2X RT Pre-Mix of QuantiNova Reverse Transcription Kit (Qiagen) with a total volume of 20 μL (1 μg of total RNA, 4 μL of 5× RT buffer, 1 μL of reverse transcriptase enzyme mix, 1 μL of oligo dT primer for mRNA or universal stem-loop primer for miRNAs (Table 1), and RNase-free distilled water to reach the volume of 20 μL). Pulsed reverse transcription was conducted to generate complementary DNA (cDNA). Twenty nanograms of total RNA in a 20 μL reaction volume was used as a template for cDNA synthesis in 60 cycles of 2 min at 16 °C, 1 min at 37 °C, and 0.1 s at 50 °C, and a final inactivation at 85 °C for 5 min [36,37]. For conventional PCR, cDNA was amplified by using 2X Taq PCR Pre-Mix (BioFACT, Seoul, Korea) in the following conditions: initial denaturation (2 min at 95 °C), 40 amplification cycles of denaturation (95 °C for 20 s), annealing (60 °C for 30 s), and extension (72 °C for 30 s), followed by a final extension step of 5 min at 72 °C. The PCR products were analyzed by a Gel Doc XR+ UV transilluminator with Image Lab Software (Bio-Rad, Berkeley, CA, USA) on a 2.5% agarose gel (Amresco, Cleveland, OH, USA) stained with ethidium bromide (BioFACT). Gel electrophoresis was performed using Mupid®-One (TAKARA, Tokyo, Japan) at 135 V for 25 min. Relative quantitative

PCR was performed using the CFX Connect Real-Time PCR system (Bio-Rad) and SYBR 2X Real-Time PCR Pre-Mix (BioFACT). Details about the target genes, housekeeping mRNA and snRNA, primers, and product size are listed in Table 2. The reaction mixture (20 μL) comprised 10 μL of SYBR® Premix (2×), 2 μL of cDNA (100 ng), 2 μL of forward and reverse primers (10 μM), and 6 μL of distilled water. Cycling conditions were 95 °C (1 min) followed by 40 PCR cycles of 95 °C (5 s, DNA denaturation), 60 °C (30 s, primer annealing), and 72 °C (30 s, extension). Primer specificity was determined by melting curve protocol ranging from 65 to 95 °C and was confirmed with single peaks in the melt curves, gel electrophoresis, and cDNA-exempted samples. Transcripts of the target genes were compared to those of housekeeping genes (GAPDH-mRNA and U6-snRNA). The gene networks of these studied transcripts were analyzed through GeneMANIA webtool (https://genemania.org/, accessed on 30 September 2022) [21].

Table 2. Primers used for RT-PCR and RT-qPCR.

Name	Sequence 5′ → 3′		Product Size	Accession No.
	Forward	Reverse		
miR-100-p	AAACCCGTAGATCCGAACT	CAAGCTTGTGCGGACTAATA	43	NR_029515.1
miR-132-p	GTCTCCAGGGCAACCGTG	CGACCATGGCTGTAGACTGT	70	LM608489.1
miR-155	GCGGTTAATGCTAATCGTGATA	CGAGGAAGAAGACGGAAGAAT	65	LM608611.1
U6	GCTTCGGCAGCACATATACTAAAAT	CGCTTCACGAATTTGCGTGTCAT	89	NR_004394.1
Bax	GAGAGACACCTGAGCTGG	AGTTCATCTCCAATGCGC	165	XM_013998624.2
Bcl2	GTTGACTTTCTCTCCTACAAG	GGTACCTCAGTTCAAACTCAT	277	NM_214285.1
CDK2	GCTTCAGGGGCTAGCTTTTT	AGCCCAGAAGGATTTCAGGT	197	NM_001285465.1
CTNNB1	CCATTCCATTGTTTGTGCAG	GTTGCCACACCTTCATTCCT	175	NM_214367.1
GAPDH	ACACTCACTCTTCTACCTTTG	CAAATTCATTGTCGTACCAG	90	DQ845173.1
IFNG	CCATTCAAAGGAGCATGGAT	TTCAGTTTCCCAGAGCTACCA	76	NM_213948.1
PERV	TCCGTGCTTACGGGTTTTAC	TTTCTCCCAGAGCCTCCATA	388	XM_021074788.1
PTPN1	TCTCAAGAAACTCGAGAGAT	TCAGCCAGACAGAAGGTC	194	XM_021077277.1
Zeb1	ACGGATGCAGCAGATTGTGA	CCGGGTAACACTGTCTGGTC	71	XM_021064196.1
Zeb2	GACAATGTAGTGGACACGGGT	GGGGAGCACTCCTGGTT	131	XM_021076508.1
Universal stem-loop primer	GAAAGAAGGCGAGGAGCAGATCGAGGAAGAAGACGGAAGAATGTGCGTCTCGCCTTCTTTCNNNNNNNN			

U6: RNU6-1 RNA, U6 small nuclear 1 (house-keeping snRNA; https://www.ncbi.nlm.nih.gov/gene/26827, accessed on 16 August 2022); Bax: BCL2-associated X, apoptosis regulator (causes apoptosis; https://www.ncbi.nlm.nih.gov/gene/396633, accessed on 16 August 2022); Bcl2: BCL2 apoptosis regulator (antiapoptotic; https://www.ncbi.nlm.nih.gov/gene/100049703, accessed on 16 August 2022); CDK2: Cyclin-dependent kinase 2 (cell cycle regulation [38]); CTNNB1: β-catenin (cell attachment molecule and trophoblast invasion [39,40]); IFNG: interferon gamma (implantation signal produced by porcine embryo [41]); GAPDH: glyceraldehyde-3-phosphate dehydrogenase (house-keeping gene); PERV: porcine endogenous retrovirus; PTPN1: protein tyrosine phosphatase non-receptor type 1 (cell adhesion, migration, growth, differentiation, and mitotic cycle [42]); Zeb1: zinc finger E-box binding homeobox 1 (epithelial–mesenchymal transition and trophoblast cell differentiation [43]); Zeb2: zinc finger E-box binding homeobox 2 (epithelial–mesenchymal transition and trophoblast cell differentiation [44,45]).

2.12. Preparation of Endo-EVs Protein Fraction, In-Gel Digestion and Proteomic Analysis by LC–MS/MS

The protocol was performed according to our recent report [27]. In brief, EVs pellets were suspended and dialyzed against 10 volumes of 20 mM Tris-HCl (pH 8.0) (the molecular mass cutoff of was 10,000 Da). Protein concentration was estimated through the bicinchoninic acid method and then proteins were fractionated by sodium dodecyl sulfate-polyacrylamide gel electrophoresis (SDS-PAGE). For Coomassie Brilliant Blue staining, the gels were destained by 50% acetonitrile and 10 mM ammonium bicarbonate solution [46] and then gels were rinsed twice with distilled water, followed by 100% acetonitrile, respectively and then dried with a speed vacuum concentrator. The gels were treated with mixture of 10 mM dithiothreitol and 100 mM ammonium bicarbonate at 56 °C, before treatment with 100 nM iodoacetamide to minimize alkylate S–S bridges. The gels were vortexed in three volumes of distilled water for washing and then dried with a speed vacuum concentrator. The gels were incubated in 50 mM ammonium bicarbonate and

10 ng/mL trypsin at 37 °C for 12–16 h for tryptic digestion. Tryptic peptides were retrieved after treatment with 50 mM ammonium bicarbonate and 50% acetonitrile containing 5% trifluoroacetic acid. Peptide extract was lyophilized and stored at 4 °C until further analysis. Tryptic peptide extract was suspended in 0.5% trifluoroacetic acid and 10 μL from each sample was loaded onto MGU30-C18 trapping columns (LC Packings) to concentrate peptides and clear extra chemicals. Concentrated tryptic peptides were eluted from the column and loaded onto a 10 cm × 75 μm I.D. C18 reverse-phase column (PROXEON, Odense, Denmark) at adjusted flow rate (300 nL/min). Peptides were retrieved by a gradient of 0–65% acetonitrile for 80 min. MS and MS/MS spectrum was obtained by using LTQ-Velos ESI ion trap mass spectrometer (Thermo Scientific, Waltham, MA, USA). MASCOT 2.4 was used to analyze MS/MS data with a false discovery rate of 1% as a cutoff value. Protein quantities were estimated through the exponentially modified protein abundance index (emPAI) and were expressed as mol %. Three technical replicates were performed. Functional analysis and gene ontology were performed through the Functional Annotation Tool, DAVID Bioinformatics Resources (NIAID/NIH; https://david.ncifcrf.gov/home.jsp, accessed on 16 August 2022) [36,47].

2.13. Statistical Analysis

For each experiment, an average of 20 embryos and at least 3 replicates were used for the analysis. Lieven's test and Kolmogorov–Smirnov test were used to confirm the homogeneity of variance and the normality of distribution, respectively. Data were expressed as mean ± standard error of means (SEM) or standard deviation (SD) and examined through using an unpaired Student *t*-test or with univariate analysis of variance (ANOVA) followed by Tukey's multiple comparison test, respectively. Statistical significance was considered at $p < 0.05$ or $p < 0.01$. GraphPad Prism 5 (GraphPad Software Inc., San Diego, CA, USA) was used for the statistical analyses.

3. Results

3.1. Endo-EVs Isolation, Characteristics, and Cargo Contents

Endometrial cells were successfully cultured with the characteristic flattened epithelial properties and were maintained in culture until day-8 (Figure 1A,B). Cells were cultured in a serum-free culture medium for 24 h to retrieve the Endo-EVs. ZetaView analysis showed a presence of 1.1 × 108 particles/mL of average particle size 115.6 ± 28.4 nm in the isolated conditioned medium (Figure 1C). Furthermore, TEM images revealed the presence of lipid bilayer vesicles (Figure 1D) that characterize the appearance of EVs. The isolated Endo-EVs were analyzed with qPCR and showed the expression of certain miRNAs and mRNAs when compared to those of corresponding endometrial origin (Figure 2). Endo-EVs contained miR-100, miR-132, and miR-155, β-catenin, and PERV, but we could not observe GAPDH mRNA in the isolated EVs. Endo-EVs were further characterized through the profiling of protein contents by proteomics. We identified 82 proteins in the Endo-EVs (Supplementary Table S1), of which the top 20 proteins that constituted around 65% of the total proteins of the Endo-EVs included proteins associated with cytoskeleton structures (e.g., keratins), extracellular matrix (e.g., plasminogen), and calcium metabolism (e.g., vitamin-D-binding protein, and calcium-binding protein A2); however, they contain apoptosis-related proteins such as procathepsin (Table 3). Detailed information about the proteins and their functions are listed in Supplementary Tables S2 and S3.

3.2. Effect of Endo-EVs and ROCK-Inhibitor (RI) on Embryo Attachment and Outgrowths

The embryonic uptake of Endo-EVs was confirmed by the presence of intracytoplasmic fluorescence signals after labeling Endo-EVs with PKH26 stain and their incubation with the embryos for 30 h (Figure 3).

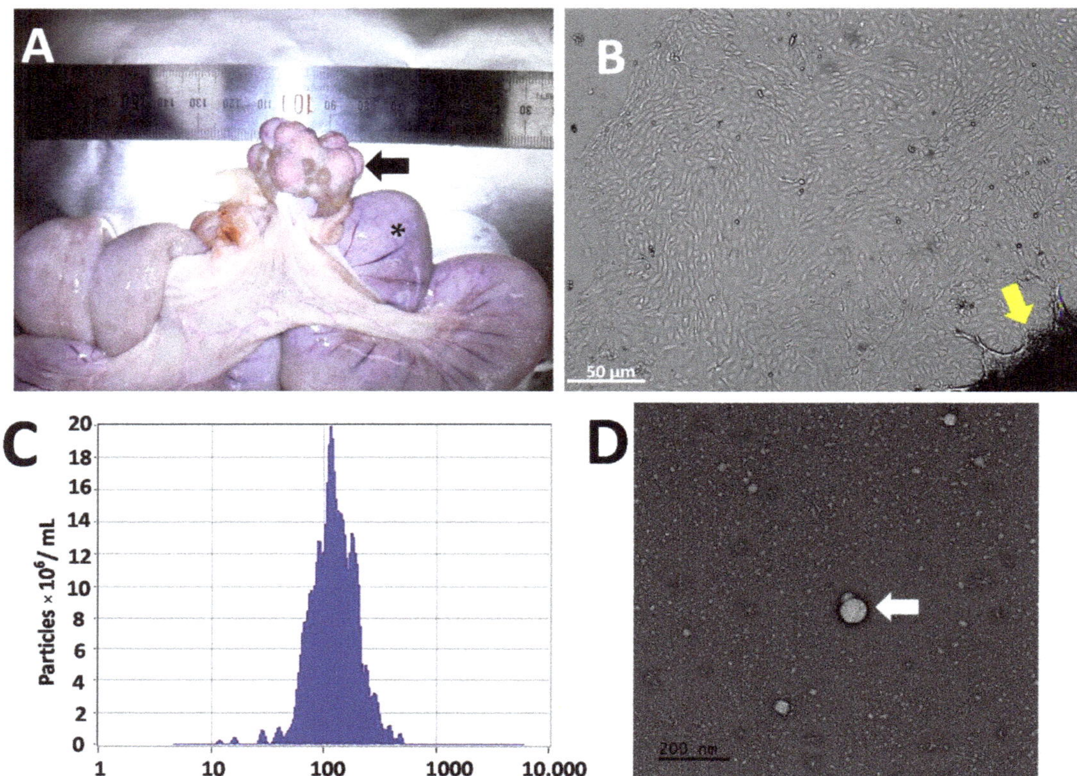

Figure 1. Obtaining the endometrial-derived extracellular vesicles (Endo-EVs). (**A**) Porcine endometrium (n = 6) was retrieved from the uterus (*) of diestrus sows (as indicated by corpora lutea, the black arrow); (**B**) primary endometrium cell culture was established (white arrow) from the endometrial tissue flakes (yellow arrow). On day-8 of primary outgrowths (Scale bar = 50 µm), the tissue chops were removed, and the cells were cultured in a serum-free culture medium to collect the conditioned medium for Endo-EVs isolation. (**C**) Endo-EVs were isolated by targeted nanofiltration method and were characterized by ZetaView nanoflow cytometry and nanoparticle tracking analysis and showed an average diameter of 115.6 ± 28.4 nm with a concentration of 1.1×10^8 particles/mL (dilution factor is 20X). (**D**) Endo-EVs were visualized by transmission electron microscope and showed bilipid vesicles (white arrow).

We then investigated the embryonic attachment and the development after supplementing the culture medium with EVs or with RI or with their combination together. When compared with the control group, Endo-EVs supplementation for 36 h showed significant improvement in embryo attachment (65.55% vs. 34.43%, $p < 0.01$), increased embryonic cell number (26 vs. 21.8, $p < 0.01$), and a significant reduction in apoptosis (2.37% vs. 26.1%, $p < 0.01$) (Figure 4A–D). Based on these preliminary experiments, we found that RI can reduce apoptosis in embryonic stem cells (Figure 4D); however, it could not support the embryonic development (Figure 4C) compared to control and Endo-EVs groups. Therefore, we examined the beneficial effects of both Endo-EVs and RI to support the embryonic development, which showed 100% embryonic attachment with a significant increase in cell numbers (mean = 33.6) and a significant reduction in the ratio of apoptotic cells (1.57%) when compared to those of the experimental groups (Figure 4A–D).

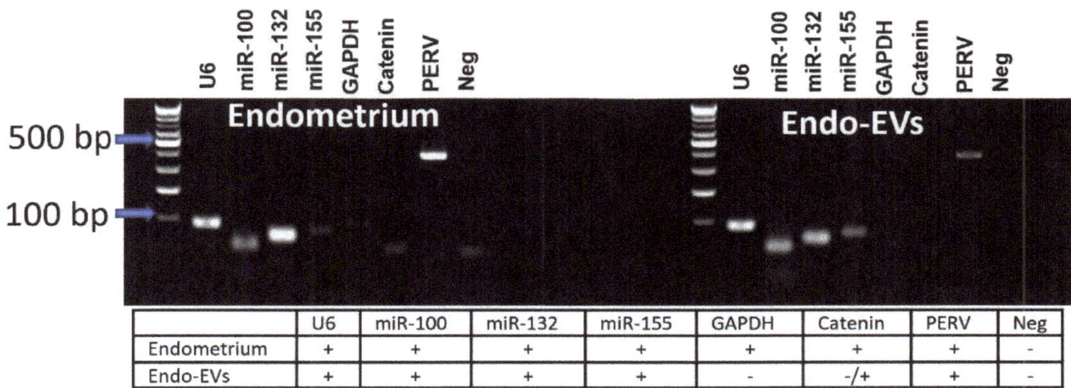

	U6	miR-100	miR-132	miR-155	GAPDH	Catenin	PERV	Neg
Endometrium	+	+	+	+	+	+	+	-
Endo-EVs	+	+	+	+	-	-/+	+	-

Figure 2. Images of gel electrophoresis of mRNA and miRNA of the endometrium and their derived extracellular vesicles (Endo-EVs). The PCR products were electrophorized in agarose gel (2%), and the bands were visualized using a 100 bp DNA ladder as reference. The band expression of the snRNA (U6) and GAPDH were used as housekeeping genes for miRNA and mRNA, respectively. We contrasted the expression in Endo-EVs and found that some mRNAs were not expressed in the Endo-EVs such as GAPDH and catenin. For more details about the PCR product size, please refer to Table 1 in Section 2.

Table 3. The top 20 proteins identified in endometrium-derived extracellular vesicles.

UniProt Accession	Description	Mol %
A0A287B5W2	Trypsinogen isoform X1	6.0325
A0A287AEL2	Keratin 14	5.0814
F1SGG6	Keratin 5	4.831
A0A287A0Q6	Tyrosine 3-monooxygenase/tryptophan 5-monooxygenase activation Protein zeta	4.6558
A0A287ANZ8	Thy-1 membrane glycoprotein	4.1051
A0A286ZT13	Albumin	4.03
P20112	SPARC	3.5544
Q28944	Procathepsin	3.0288
A0A5G2QTF5	Thioredoxin	2.8536
A0A287AHS0	Calmodulin 3	2.8285
A0A287BA49	Keratin 5	2.7785
A0A287A8S8	Phosphopyruvate hydratase	2.7284
F1SGG3	Keratin, type II cytoskeletal 1	2.7034
I3LDS3	Keratin 10	2.6783
A0A287AHK1	Vitamin D-binding protein	2.6283
K7GQ95	S100 calcium binding protein A2	2.3529
A0A5G2QSE8	Keratin 3	2.3029
A0A287BHY5	Keratin 2	1.7772
A0A5G2QUE0	Antithrombin-III	1.7522
A0A287ATD0	Keratin 75	1.577

We followed up the development of embryos supplanted with Endo-EVs and RI for 5 subsequent days and compared that with the control group. The results showed a significant increase in the embryonic outgrowths on day-5 in the embryos supplemented with Endo-EVs and RI (72.9% vs. 32%) (Figure 5A,B).

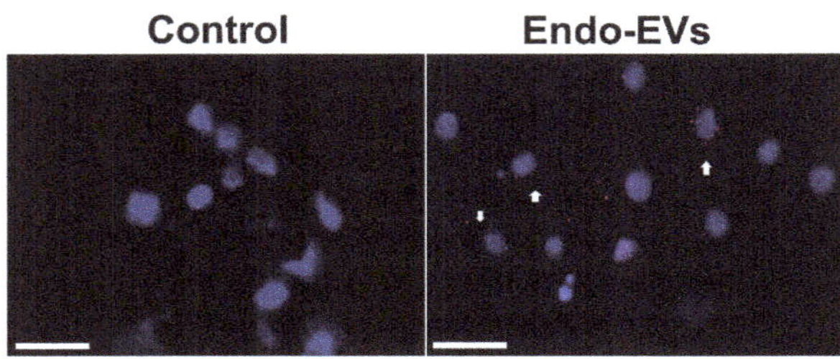

Figure 3. Cellular uptake of Endo-EVs. Endo-EVs were stained with a lipophilic live-imaging dye PKH26, and the free dye was removed by washing during isolation. Plain conditioned medium was processed in the same way as the isolated EVs and worked as negative control (Control). Attached embryonic cells were incubated with the stained control and Endo-EVs for 24 h and then were stained with DAPI and visualized by fluorescence microscope. White arrows indicate the presence of cytoplasmic stained EVs surrounding the nuclei. Scale bar = 20 μm.

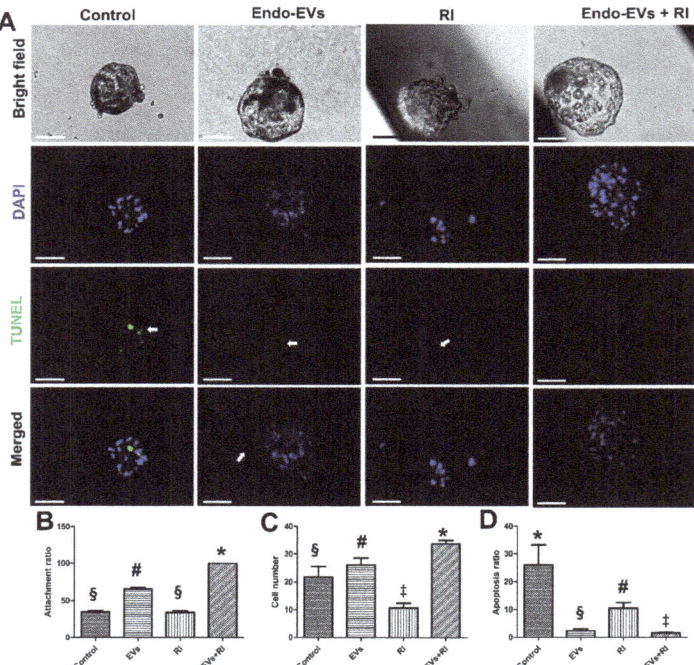

Figure 4. The effects of endometrial EVs (Endo-EVs) and ROCK inhibitor (RI) on porcine embryo development. (**A**) Day-7 zona-free embryos (n = 20, 3 replicates) were cultured on Matrigel-coated dishes in microdrops of culture medium in a humidified atmosphere of 5% CO_2 for 36 h. The control group was cultured in a plain culture medium while the RI group in a medium supplemented with Y-27632 (10 μg/mL) and EVs group in a medium supplemented with Endo-EVs of 2.6 × 106 particles/mL. In the combined group, embryos were cultured in a medium supplemented with both RI and EVs of the same working concentrations. Scale bar = 100 μm. All groups were imaged in a bright field before staining with TUNEL assay and contrasted with DAPI stain. White arrows indicate the apoptotic cells; (**B–D**) graphs show the embryo attachment, cell number, and apoptosis among the groups, respectively. Values (mean ± SD) were compared with ANOVA followed by Tukey's test to determine the difference among the groups. Values denoted by ‡, #, §, and * were considered statistically different ($p < 0.05$).

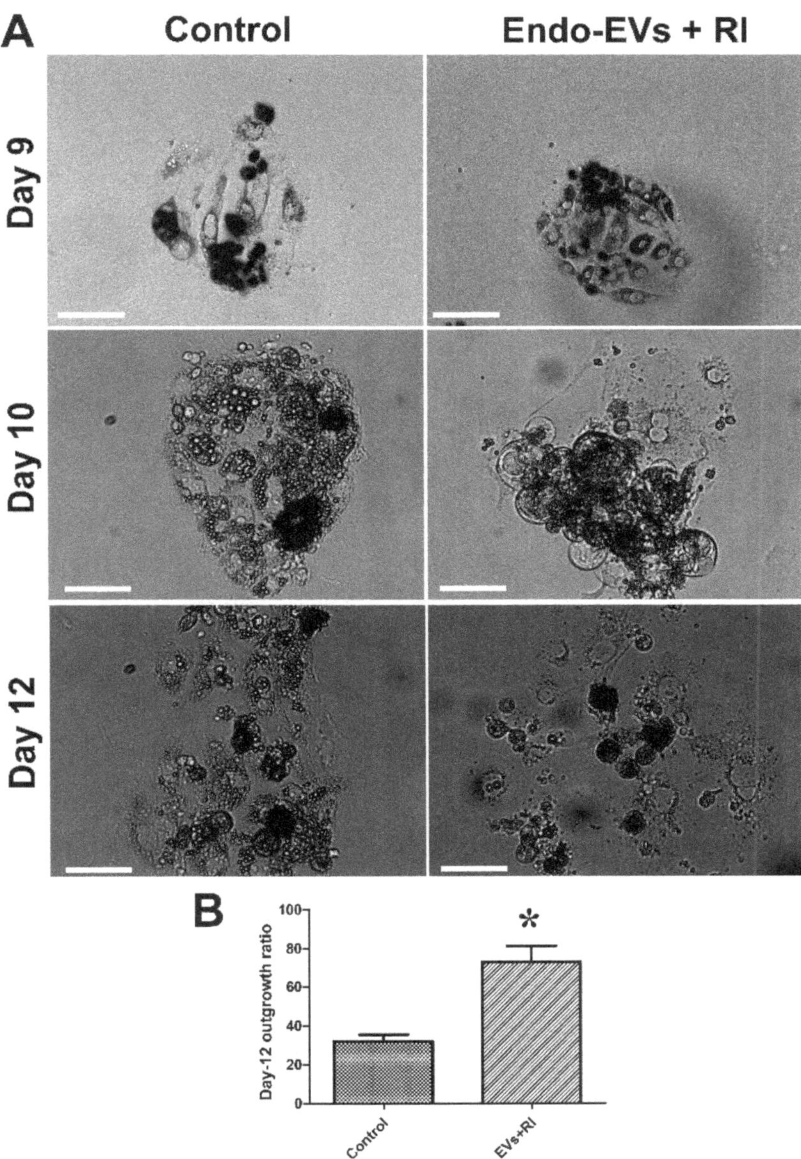

Figure 5. The effect of combined treatment of Endo-EVs and ROCK inhibitor (RI) on embryonic outgrowths. (**A**) Day-7 zona-free porcine embryos (n =18 for 3 replicates) were cultured on Matrigel-coated dishes in microdrops of culture medium in a humidified atmosphere of 5% CO_2 for 5 days. Scale bar = 50 µm; (**B**) the averages of percentages of embryonic cell outgrowths were calculated on day-2, day-3, and day-5 (mean ± S.E.M.) and compared with Student's t-test. Asterisk (*) indicates a significant difference ($p < 0.05$).

Furthermore, the expression of Oct4 and Cdx2 in the Endo-EVs and RI-treated embryonic cells was significantly increased by 2.45-fold and 3.48-fold, respectively, compared to those of the control group (Figure 6A–C).

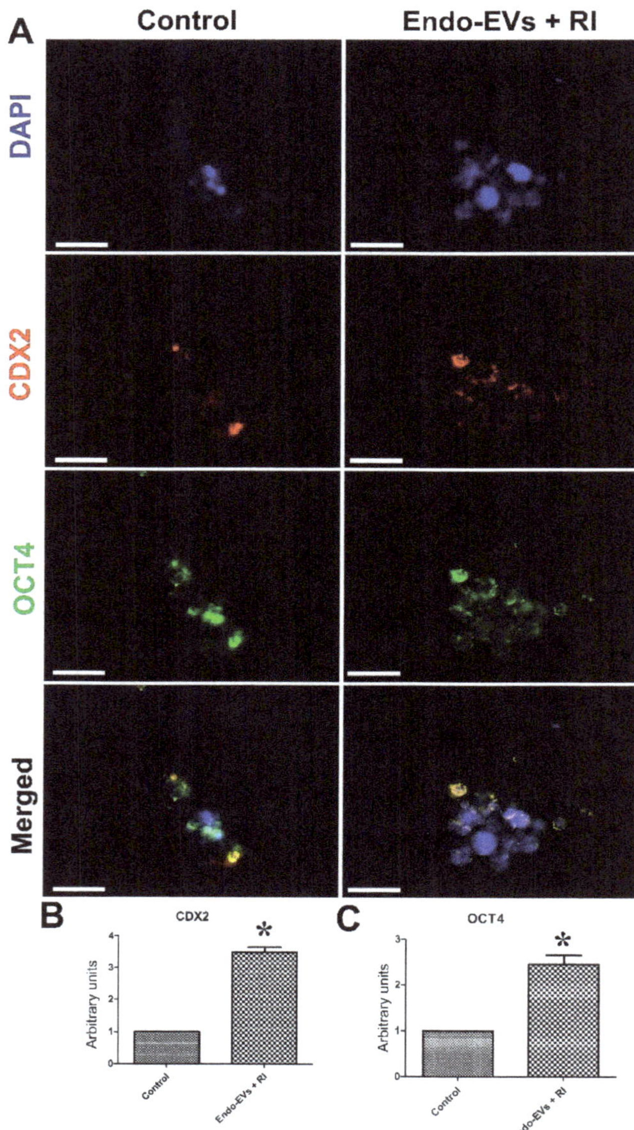

Figure 6. The effect of combined treatment of Endo-EVs with ROCK inhibitor (RI) on OCT4 and CDX2 expression in cultured embryonic cells. (**A**) Day-7 zona-free porcine embryos (n = 18 for 3 replicates) were cultured on Matrigel-coated dishes in microdrops of culture medium in a humidified atmosphere of 5% CO_2 for 5 days and then incubated with primary antibodies specific to OCT4 and CDX2 followed by corresponding specific secondary antibodies. Scale bar = 50 μm. (**B**,**C**) Images of CDX2 and OCT4, respectively, were analyzed with ImageJ software to compare the pixels of the fluorescence intensity in the same exposure time, contrast, and area of analysis. The values were normalized to the control group as an arbitrary unit to show the fold of change between the groups. Values (mean ± S.E.M.) were analyzed with Student's t-test. Asterisk (*) indicates a significant difference ($p < 0.05$).

Additionally, the qPCR analysis showed that combined supplementation of Endo-EVs with RI significantly reduced the expression of Bax (0.6-fold) and miR-155 (0.17-fold) but increased the expression of Bcl2 (4.73-fold), Cdk2 (4.33-fold), PERV (2.55-fold), β-catenin (7.13), interferon-gamma (1.7-fold), Zeb1 (1.9-fold), PTN mRNAs (9.83-fold), miR-100 (3.79-fold), and miR-132 (16.1-fold) compared to those of the control group (Figure 7).

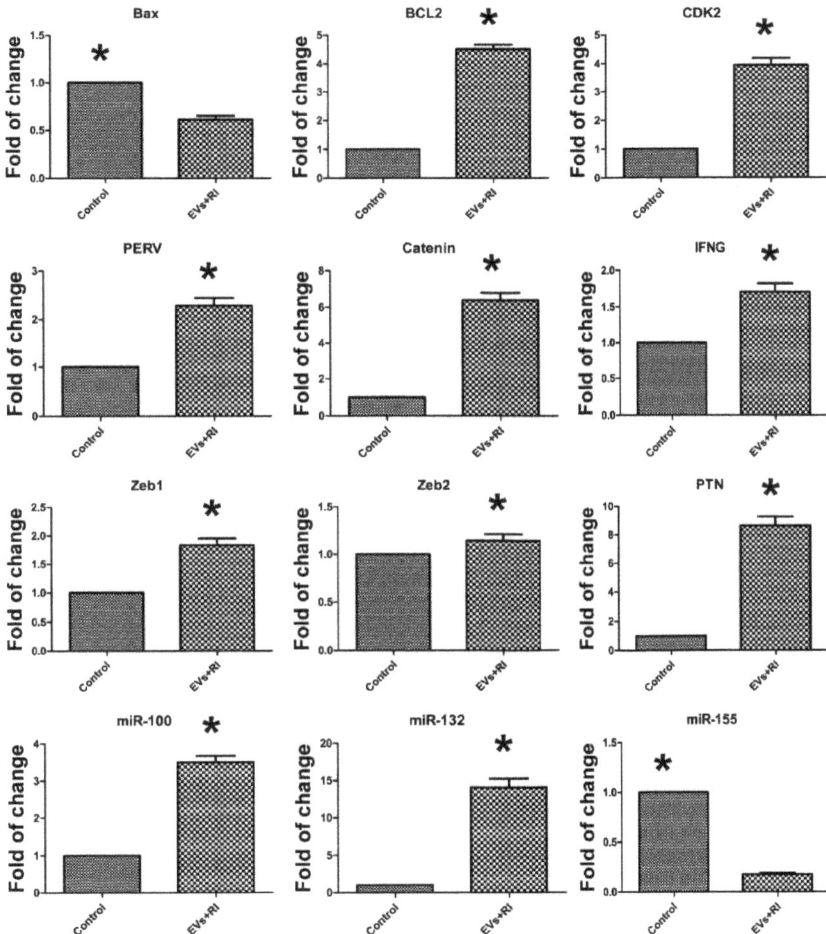

Figure 7. Relative quantitative analysis (RT-qPCR) of mRNA transcripts expressed in the embryos treated with Endo-EVs and RI. Five blastocysts from each group for 4 replicates were used for qPCR analysis. The means were normalized to the control group and expressed as arbitrary units. Data were expressed as mean ± S.E.M. and the difference between the two groups was compared with Student's *t*-test. Values denoted by an asterisk (*) were considered statistically significant ($p < 0.05$).

3.3. *Impact of miR-155 on Embryo Attachment and Development*

Based on the previous results, we speculated that the miR-155 contents of Endo-EVs could exert a negative impact on embryo attachment, and therefore we specifically targeted miR-155 with an inhibitor (miR-155 inhibitor). Treatment of embryos with Endo-EVs and miR-155 inhibitor significantly improved the attachment (90% vs. 50%, $p < 0.01$), increased the cell number (30 vs. 12, $p < 0.01$) but had no effects on the apoptosis ratio (2% vs. 3.16%, $p = 0.27$) as compared to those of Endo-EVs supplemented group (Figure 8A–D).

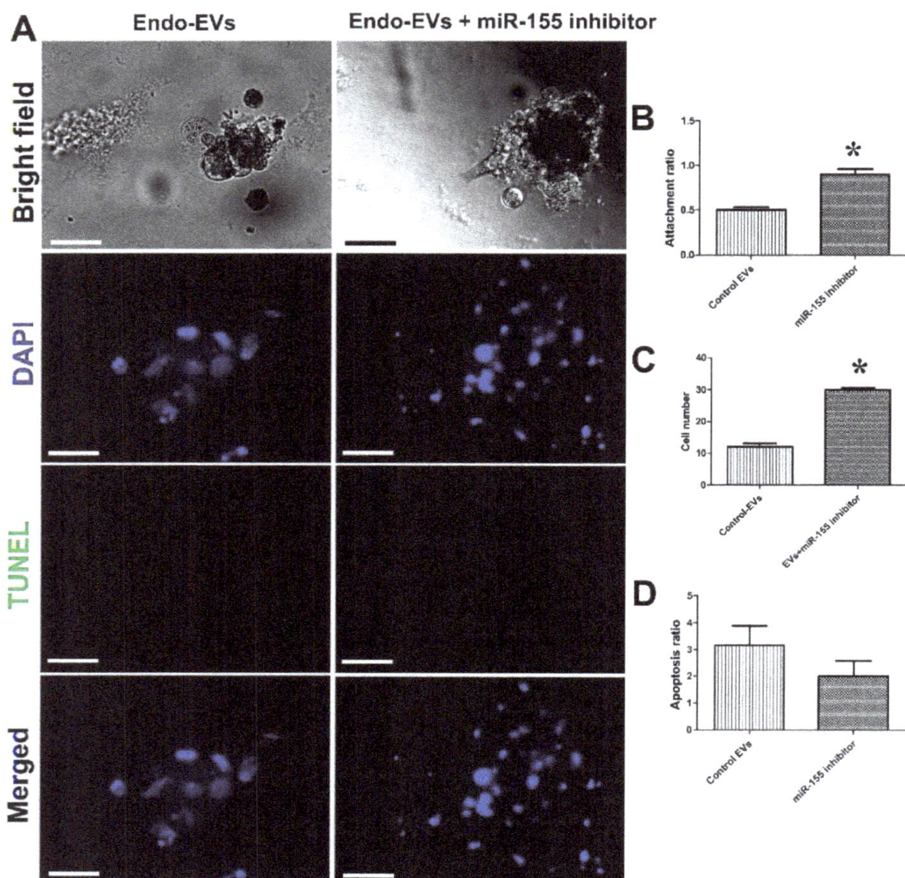

Figure 8. Investigating the effects of the miR-155 inhibitor on embryonic attachment and development. (**A**) MiR-155 inhibitor was designed (Table 1) and transfected to day-7 zona-free porcine embryos. Embryos were cultured on Matrigel-coated dishes in microdrops of culture medium containing Endo-EVs in a humidified atmosphere of 5% CO_2 for 5 days. Control Endo-EVs group was treated the same as the miR-155 group except for the absence of RNA sequence during transfection. Scale bar = 50 μm. All groups were imaged in a bright field before staining with TUNEL assay and contrasted with DAPI stain; (**B**–**D**) graphs show the embryo attachment, cell number, and apoptosis among the groups, and values (mean ± SEM) were compared with Student's *t*-test to determine the difference among the groups. Values denoted by an asterisk (*) were considered statistically significant ($p < 0.05$).

Additionally, individual treatment with miR-155 mimic showed a significant reduction ($p < 0.05$) in the attachment (20.8%) and cell number ($n = 6$) and a significantly increased apoptosis ratio (48.3%). Moreover, these effects were significantly alleviated with an individual treatment of RI (45.6%, 14, and 14.66% for attachment ratio, cell number, and apoptosis ratio, respectively; $p < 0.01$) (Figure 9A–D). Hence, we speculated that RI could antagonize the negative impact of miR-155 on embryonic attachment and development.

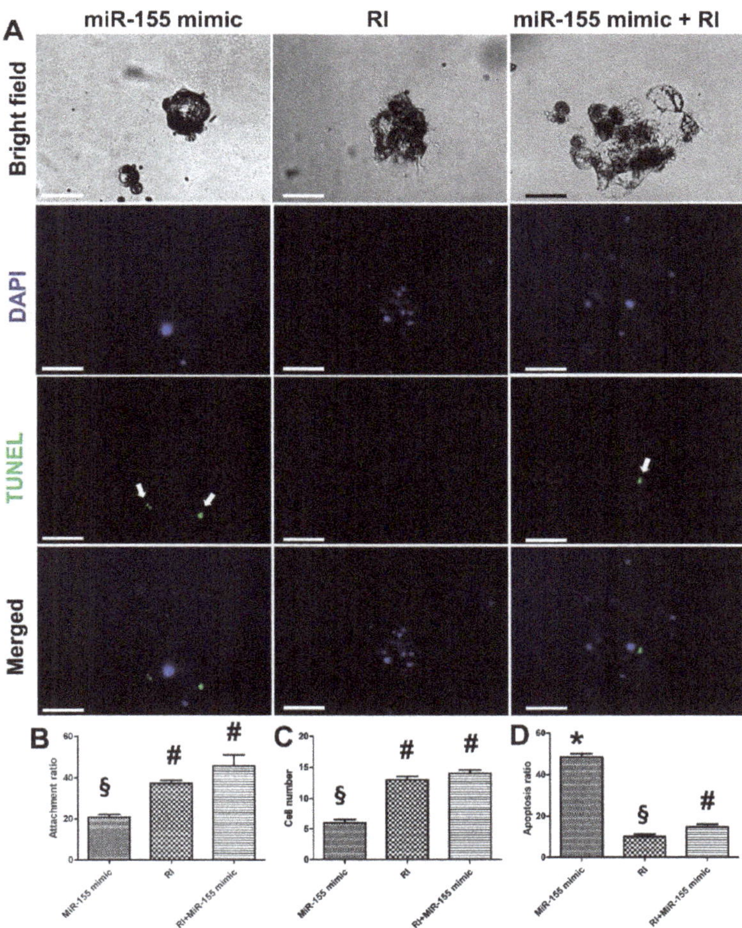

Figure 9. Investigating the effects of miR-155 mimic and ROCK inhibitor (RI) on embryonic attachment and development. (**A**) MiR-155 mimic duplex was designed (Table 1) and transfected to day-7 zona-free porcine embryos. Embryos were cultured on Matrigel-coated dishes in microdrops of culture medium in a humidified atmosphere of 5% CO_2 for 5 days. The three groups were treated the same as the miR-155 mimic group except for the absence of RNA sequence during transfection of the RI group. Scale bar = 50 μm. All groups were imaged in a bright field before staining with TUNEL assay and contrasted with DAPI stain. White arrows indicate the apoptotic cells; (**B–D**) graphs show the embryo attachment, cell number, and apoptosis among the groups. Data were expressed as means ± S.E.M. Values (mean ± SD) were compared with ANOVA followed by Tukey's test to determine the difference among the groups. Values denoted by symbols #, §, and * were considered statistically significant ($p < 0.05$).

3.4. Effects of Endo-EVs PERV Depletion by CRISPR/Cas9 on Embryo Attachment and Development

Furthermore, based on the Endo-EVs cargo contents of PERV, we targeted PERV with CRISPR/Cas9 to examine the effects of PERV reduction on embryonic attachment and development. The designed CRISP/Cas9 was successfully transfected and expressed in the cells as indicated by the green fluorescence protein expression in Figure 10A,A′,A″. PERV expression was significantly reduced (0.23-fold, $p < 0.05$) compared to that of control endometrium cells. Surprisingly, miR-155 expression showed a 6.16-fold increase ($p < 0.05$) in PERV-depleted endometrium compared with that of the control endometrium. Similarly, the derived EVs from PERV-depleted cells showed a significant reduction in PERV mRNA expression (0.27-fold, $p < 0.05$), and a significant increase in miR-155 (4.9-fold, $p < 0.01$). Moreover, supplementation of embryos with EVs from PERV-depleted cells significantly reduced the attachment and day-5 outgrowth ratios compared to those of control EVs (49% vs. 65.8% and 18% vs. 31%, respectively, $p < 0.05$).

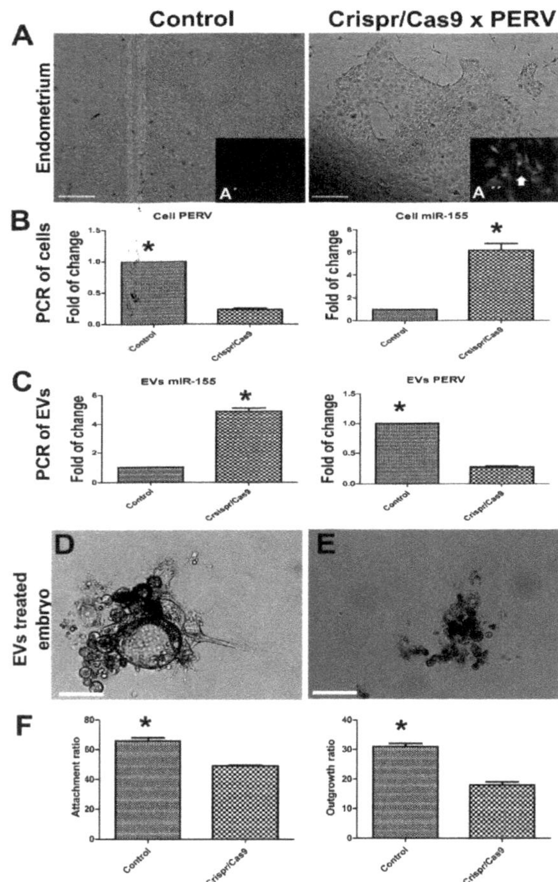

Figure 10. The impacts of targeting porcine endogenous retrovirus (PERV) expression in porcine endometrium through CRISPR/Cas9 on the derived EVs and the subsequent embryo development. (**A**) Endometrium was transfected with CRISP/Cas9 vector for 24 h;(**A′**,**A″**) to compare the green fluorescence protein expression (white arrow, see Supplementary Figure S1 for the vector details) in control and mutated cells, respectively. The resulting cells were used to isolate Endo-EVs as mentioned previously. Scale bar = 50 µm. PERV targeted endometrium showed 5-fold and 4.5-fold reduction in the mRNA expression in both endometrium (**B**) and their derived EVs (**C**), respectively, while they showed upregulated miR-155 about 6-fold and 5-fold in endometrium (**B**) and their derived EVs (**C**), respectively. (**D**); (**E**) On day-5, embryo treated with PERV-targeted and PERV-diminished EVs showed low percentages of embryo attachment and outgrowths (**F**). Values were expressed as means ± S.E.M. and the difference between the two groups was compared with Student's *t*-test. Values denoted by an asterisk (*) were considered statistically significant ($p < 0.05$).

4. Discussion

In this study, as a continuation of our work [28], we generated a model of culturing porcine embryos in feeder-free culture conditions using Matrigel basement membrane matrix but on a microdrop level. This model achieved 100% of blastocyst attachment and embryonic outgrowths with the help of Endo-EVs supplementation and ROCK pathway inhibition.

Recently, the interplay between EVs derived from the endometrium during the embryo implantation in humans has been investigated [10–12]. Our results showed that Endo-EVs enhanced embryonic attachment and adhesion to the Matrigel basement membrane matrix, which is in accordance with previous studies [11,48–51]. This improvement may be attributed to the transfer of proteins associated with cell attachment, cytoskeleton integrity, calcium metabolism as revealed by proteomics analysis. Additionally, embryonic development was improved because of β-catenin transfer through the Endo-EVs cargo that increased the expression of β-catenin in embryos. β-catenin plays a crucial role in endometrium functions and is considered an imperative signal in invasion and differentiation of trophoblasts and embryo implantation [39,52]. Moreover, Endo-EVs cargo contained miR-100, which is expressed in human endometrial cell-derived EVs and activates focal adhesion kinase (FAK) and c-Jun N-terminal kinase (JNK) and promotes the invasion and migration of human and goat trophoblasts [12,53,54]. Furthermore, Endo-EVs contained miR-132 that is expressed temporally in porcine endometrium at the time of embryo implantation [55] and is a potential factor for enhancing trophoblast invasion and embryo implantation by targeting death-associated protein kinase 1 (DAPK-1) [56].

In our study, several mRNA transcripts (antiapoptotic gene (BCL2), zinc finger E-box-binding homeobox 1 (Zeb1), β-catenin, interferon-γ (IFNG), protein tyrosine phosphatase non-receptor type 1 (PTPN1), and cyclin-dependent kinase 2 (CDK2)) were increased in the embryonic cells after EVs supplementation. Moreover, the pluripotency master Oct4 and the trophoblast associated gene CDX2 were also increased in the EVs supplementation which might be attributed to embryo developmental competence observed in this group [4,28,57–59]. These genes are of important roles in embryonic development, implantation, trophoblast attachment, and stem cell growth, as well as the cell cycle and survival as we revealed in our previous reports [5,21]. Furthermore, bioinformatics analysis indicated an existing physical interaction, shared protein domains, common pathways, and co-expression of the studies genes (Supplementary Figure S2). Detailed functions of the genes are listed in Table 2.

On the other hand, Endo-EVs contained miR-155, which inhibits trophoblast migration and proliferation [60,61], increases preeclampsia in patients, and induces trophoblast apoptosis by targeting BCL2 (apoptosis regulator) [62]. Furthermore, some of the cargo proteins in the Endo-EVs included proapoptotic signals such as cathepsin and procathepsin. Paradoxically, miR-155 showed a 1.6-fold increase in the mouse uterus during the receptive phase of embryo attachment, which suggests a modulatory role of miR-155 during this critical stage in mice [63].

In this study, the ROCK inhibitor (Y-27632) abolished all negative impacts of miR-155 on embryo attachment and development. RI remarkably reduces the tumor necrosis factor (TNFα)-induced upregulation of miR-155 [64]. RI interferes with the cargo transfer from microparticles and impairs their ability to mediate extracellular signaling [65]. Therefore, we speculate the likelihood of other mechanisms that are associated with miRNA export upstream to RhoA/ROCK signaling. Another mechanism is the antiapoptotic action of RI can antagonize the apoptotic action of miR-155 on embryonic cells [18,66–69]. Additionally, ROCK pathway inhibition enhances trophoblast adhesion and viability in humans. Paradoxically, RI can reduce the trophoblast migration of human extravillous trophoblasts [70]. This is in accordance with our findings regarding the ameliorative effects of RI on the negative impacts of both individual and EVs-transmitted miR-155; however, individual RI improved the embryonic attachment and development. Therefore, RI synergize the actions of Endo-EVs through antagonizing the effects of the non-useful cargo contents of Endo-EVs, such as miR-155.

Computational analysis of miR-155 targets (http://mirdb.org/, accessed on 16 August 2022) showed that they interfere with a cell-cycle-related gene (CDK2-associated cullin domain 1 (CACUL1)) (target score 82%) and an antiapoptotic gene (BCL2-associated athanogene 5 (BAG5)) (target score 70%), which had correlation with other proteins involved in cell apoptosis and growth, including BCL-2. Our qPCR data showed that RI ameliorates the negative effects on mRNA expression of CDK2 and BCL2, which could help the cell cycle and reduce apoptosis. Moreover, miR-155 targets catenin alpha 3 (CTNNA3) (target score 67%) which belongs to the catenin family and encodes a protein that plays a role in cell-to-cell adhesion. Similar findings in qPCR have been shown in β-catenin expression. Furthermore, miR-155 targets protein tyrosine phosphatase, non-receptor type 2 (PTPN2) (target score 84%), which regulates various cellular processes including cell growth, differentiation, and mitotic cycle. Moreover, we found positive effects of RI on the expression of PTPN1. Therefore, we inferred that miR-155 can interfere with several essential pathways related to cell growth and differentiation and cause apoptosis.

The EVs cargo contained PERV mRNA, which coincides with some recent reports showing exosomes that contain mRNAs for ovine endogenous jaagsiekte retroviruses (enJSRV-ENV) [13] and human endogenous retroviruses [71]. There is a consensus about the essential roles of endogenous retroviruses in the early stages of embryo attachment physiological functions of trophoblasts and placentation [15,72–75]; however, the role of PERV in porcine embryo attachment remains unclear. A recent report showed that targeting PERV with CRISPR/Cas9 at the zygote stage impaired the blastocyst development and indicated the essential roles of PERV for the preimplantation embryonic development [14]. Our results indicate that PERV targeting in EVs could have reduced the embryonic attachment and development. This reduction might be due to the increased levels of miR-155 in the transferred cargo contents of EVs. The reason behind these increased levels is unclear and might be attributed to the essential roles of PERV in cellular viability, normal physiological functions, and the indel mutations caused by CRISPR/Cas9 [14]. Furthermore, avian endogenous retrovirus shows negative regulation with miR-155, which is suggestive of the interplay between ERVs and miR-155 [76]. The human and zebrafish microRNA-155 target the corresponding HERV and ZFERV env sequence, which indicates that miR-155 targeting ERVs env is mostly conserved in animals and may regulate ERVs activity [76]. We speculate that EVs can carry both useful and harmful cargo contents and ameliorating the de trop cargo contents (such as miR-155) can maximize the useful effects of EVs, especially during embryo implantation and maternal recognition of pregnancy.

5. Conclusions

To our knowledge, this is the first attempt to understand the roles of EVs cargo in determining embryo developmental competence and mediating molecular signaling between the embryo and the endometrium in the pig. Endometrial EVs improved embryo attachment, increased cell numbers and reduced apoptosis, and the unwanted effects of their de trop cargo contents of miR-155 can be alleviated through anti-apoptotic molecules such as the ROCK inhibitor. This model would help in establishing an extended culture system to understand early embryonic stem cell differentiation. The current model would provide a paradigm for studying the embryonic–maternal crosstalk and to develop pharmaceutical criteria for improving pregnancy outcomes in porcine species.

Supplementary Materials: The following supporting information can be downloaded at: https://www.mdpi.com/article/10.3390/cells11193178/s1, Figure S1: The vector used for depletion of PERV through Crispr/Cas9; Figure S2: GeneMANIAbioinformatics analysis of gene-to-gene network. Table S1: The raw data of proteomics analysis. Table S2: The functional annotation of the Endo-EVs proteins, as indicated by DAVID analysis. Table S3: The involvement of the Endo-EVs in biological processes, as indicated by DAVID analysis.

Author Contributions: Conceptualization, I.M.S. and J.C.; methodology, I.M.S., B.M.T., S.B., C.S., O.K., S.H.Y., S.I.K. and S.L.; formal analysis, I.M.S. and J.C.; investigation, I.M.S. and J.C.; resources, O.K., S.H.Y. and S.I.K.; writing—original draft preparation, I.M.S., B.M.T., S.B., C.S., O.K., S.H.Y., S.I.K., S.L. and J.C.; writing—review and editing, I.M.S. and J.C.; supervision, J.C.; project administration, I.M.S. and J.C.; funding acquisition, I.M.S. and J.C. All authors have read and agreed to the published version of the manuscript.

Funding: This work was supported by the Ministry of Science and ICT through the National Research Foundation of Korea (NRF) (grants # 2021R1A2C2009294 and 2022R1I1A1A01065412) and the Brain Pool program (grant No.: 2021H1D3A2A02040098).

Institutional Review Board Statement: Not applicable.

Informed Consent Statement: Not applicable.

Data Availability Statement: The data supporting the findings of this study are included in the manuscript and the Supplementary Materials.

Acknowledgments: We would also like to thank Steve Park at WithInstrument Company (Seoul, Republic of Korea) for performing the ZetaView NTA analysis. We thank the staff members of Daejeon metropolitan slaughterhouse for the provision of porcine ovaries.

Conflicts of Interest: The authors declare no conflict of interest.

References

1. Hou, N.; Du, X.; Wu, S. Advances in pig models of human diseases. *Anim. Model. Exp. Med.* **2022**, *5*, 141–152. [CrossRef] [PubMed]
2. Eisenson, D.L.; Hisadome, Y.; Yamada, K. Progress in Xenotransplantation: Immunologic Barriers, Advances in Gene Editing, and Successful Tolerance Induction Strategies in Pig-To-Primate Transplantation. *Front. Immunol.* **2022**, *13*, 2308. [CrossRef] [PubMed]
3. Stokes, P.J.; Abeydeera, L.R.; Leese, H.J. Development of porcine embryos in vivo and in vitro; evidence for embryo 'cross talk' in vitro. *Dev. Biol.* **2005**, *284*, 62–71. [CrossRef] [PubMed]
4. Saadeldin, I.M.; Kim, S.J.; Bin Choi, Y.; Lee, B.C. Improvement of Cloned Embryos Development by Co-Culturing with Parthenotes: A Possible Role of Exosomes/Microvesicles for Embryos Paracrine Communication. *Cell Reprogram.* **2014**, *16*, 223–234. [CrossRef] [PubMed]
5. Fang, X.; Tanga, B.M.; Bang, S.; Seong, G.; Saadeldin, I.M.; Lee, S.; Cho, J. Oviduct epithelial cells-derived extracellular vesicles improve preimplantation developmental competence of in vitro produced porcine parthenogenetic and cloned embryos. *Mol. Reprod. Dev.* **2021**, *89*, 54–65. [CrossRef] [PubMed]
6. Kashiwazaki, N.; Kikuchi, K.; Suzuki, K.; Noguchi, J.; Nagai, T.; Kaneko, H.; Shino, M. Development In Vivo and In Vitro to Blastocysts of Porcine Oocytes Matured and Fertilized In Vitro. *J. Reprod. Dev.* **2001**, *47*, 303–310. [CrossRef]
7. van der Weijden, V.A.; Schmidhauser, M.; Kurome, M.; Knubben, J.; Flöter, V.L.; Wolf, E.; Ulbrich, S.E. Transcriptome dynamics in early in vivo developing and in vitro produced porcine embryos. *BMC. Genom.* **2021**, *22*, 1–13. [CrossRef]
8. Saadeldin, I.M.; Tanga, B.M.; Bang, S.; Fang, X.; Yoon, K.-Y.; Lee, S.; Cho, S.L.A.J. The theranostic roles of extracellular vesicles in pregnancy disorders. *J. Anim. Reprod. Biotechnol.* **2022**, *37*, 2–12. [CrossRef]
9. Saadeldin, I.; Oh, H.J.; Lee, B. Embryonic–maternal cross-talk via exosomes: Potential implications. *Stem. Cells Clon. Adv. Appl.* **2015**, *8*, 103–107. [CrossRef]
10. Godakumara, K.; Ord, J.; Lättekivi, F.; Dissanayake, K.; Viil, J.; Boggavarapu, N.R.; Faridani, O.R.; Jääger, K.; Velthut-Meikas, A.; Jaakma, Ü.; et al. Trophoblast derived extracellular vesicles specifically alter the transcriptome of endometrial cells and may constitute a critical component of embryo-maternal communication. *Reprod. Biol. Endocrinol.* **2021**, *19*, 1–14. [CrossRef]
11. Greening, D.W.; Nguyen, H.P.; Elgass, K.; Simpson, R.J.; Salamonsen, L.A. Human Endometrial Exosomes Contain Hormone-Specific Cargo Modulating Trophoblast Adhesive Capacity: Insights into Endometrial-Embryo Interactions. *Biol. Reprod.* **2016**, *94*, 38. [CrossRef] [PubMed]
12. Shi, S.; Tan, Q.; Liang, J.; Cao, D.; Wang, S.; Liang, J.; Chen, K.; Wang, Z. Placental trophoblast cell-derived exosomal microRNA-1290 promotes the interaction between endometrium and embryo by targeting LHX6. *Mol. Ther.-Nucleic. Acids* **2021**, *26*, 760–772. [CrossRef]
13. Ruiz-González, I.; Xu, J.; Wang, X.; Burghardt, R.; Dunlap, K.A.; Bazer, F.W. Exosomes, endogenous retroviruses and toll-like receptors: Pregnancy recognition in ewes. *Reproduction* **2015**, *149*, 281–291. [CrossRef] [PubMed]
14. Hirata, M.; Wittayarat, M.; Hirano, T.; Nguyen, N.T.; Le, Q.A.; Namula, Z.; Fahrudin, M.; Tanihara, F.; Otoi, T. The Relationship between Embryonic Development and the Efficiency of Target Mutations in Porcine Endogenous Retroviruses (PERVs) Pol Genes in Porcine Embryos. *Animals* **2019**, *9*, 593. [CrossRef] [PubMed]

15. Spencer, T.E.; Johnson, G.A.; Bazer, F.W.; Burghardt, R.C.; Palmarini, M. Pregnancy recognition and conceptus implantation in domestic ruminants: Roles of progesterone, interferons and endogenous retroviruses. *Reprod. Fertil. Dev.* **2007**, *19*, 65–78. [CrossRef]
16. Baek, S.-K.; Cho, Y.-S.; Kim, I.-S.; Jeon, S.-B.; Moon, D.-K.; Hwangbo, C.; Choi, J.-W.; Kim, T.-S.; Lee, J.-H. A Rho-Associated Coiled-Coil Containing Kinase Inhibitor, Y-27632, Improves Viability of Dissociated Single Cells, Efficiency of Colony Formation, and Cryopreservation in Porcine Pluripotent Stem Cells. *Cell Reprogram.* **2019**, *21*, 37–50. [CrossRef]
17. Kurosawa, H. Application of Rho-associated protein kinase (ROCK) inhibitor to human pluripotent stem cells. *J. Biosci. Bioeng.* **2012**, *114*, 577–581. [CrossRef]
18. Saadeldin, I.M.; Tukur, H.A.; Aljumaah, R.S.; Sindi, R.A. Rocking the Boat: The Decisive Roles of Rho Kinases During Oocyte, Blastocyst, and Stem Cell Development. *Front. Cell Dev. Biol.* **2021**, *8*, 616762. [CrossRef]
19. Motomura, K.; Okada, N.; Morita, H.; Hara, M.; Tamari, M.; Orimo, K.; Matsuda, G.; Imadome, K.-I.; Matsuda, A.; Nagamatsu, T.; et al. A Rho-associated coiled-coil containing kinases (ROCK) inhibitor, Y-27632, enhances adhesion, viability and differentiation of human term placenta-derived trophoblasts in vitro. *PLoS ONE* **2017**, *12*, e0177994. [CrossRef]
20. Hou, D.; Su, M.; Li, X.; Li, Z.; Yun, T.; Zhao, Y.; Zhang, M.; Zhao, L.; Li, R.; Yu, H.; et al. The Efficient Derivation of Trophoblast Cells from Porcine In Vitro Fertilized and Parthenogenetic Blastocysts and Culture with ROCK Inhibitor Y-27632. *PLoS ONE* **2015**, *10*, e0142442. [CrossRef]
21. Tanga, B.M.; Fang, X.; Bang, S.; Seong, G.; De Zoysa, M.; Saadeldin, I.M.; Lee, S.; Cho, J. MiRNA-155 inhibition enhances porcine embryo preimplantation developmental competence by upregulating ZEB2 and downregulating ATF4. *Theriogenology* **2022**, *183*, 90–97. [CrossRef] [PubMed]
22. Saadeldin, I.; Kim, S.; Choi, Y.; Lee, B. Post-maturation zona perforation improves porcine parthenogenetic trophoblast culture. *Placenta* **2014**, *35*, 286–288. [CrossRef] [PubMed]
23. Zhang, Z.; Paria, B.C.; Davis, D.L. Pig endometrial cells in primary culture: Morphology, secretion of prostaglandins and proteins, and effects of pregnancy. *J. Anim. Sci.* **1991**, *69*, 3005–3015. [CrossRef] [PubMed]
24. Ng, Y.H.; Rome, S.; Jalabert, A.; Forterre, A.; Singh, H.; Hincks, C.L.; Salamonsen, L.A. Endometrial Exosomes/Microvesicles in the Uterine Microenvironment: A New Paradigm for Embryo-Endometrial Cross Talk at Implantation. *PLoS ONE* **2013**, *8*, e58502. [CrossRef] [PubMed]
25. Abumaghaid, M.M.; Abdelazim, A.M.; Belali, T.M.; Alhujaily, M.; Saadeldin, I.M. Shuttle transfer of mRNA transcripts via extracellular vesicles from male reproductive tract cells to the cumulus–oocyte complex in rabbits (*Oryctolagus cuniculus*). *Front. Veter. Sci.* **2022**. [CrossRef]
26. Mehdiani, A.; Maier, A.; Pinto, A.; Barth, M.; Akhyari, P.; Lichtenberg, A. An Innovative Method for Exosome Quantification and Size Measurement. *J. Vis. Exp.* **2015**, *95*, e50974. [CrossRef]
27. Lee, H.; Yun, S.H.; Hyon, J.-Y.; Lee, S.-Y.; Yi, Y.-S.; Choi, C.-W.; Jun, S.; Park, E.C.; Kim, S.I. Streptococcus equi-derived Extracellular Vesicles as a Vaccine Candidate against Streptococcus equi Infection. *Veter. Microbiol.* **2021**, *259*, 109165. [CrossRef]
28. Saadeldin, I.M.; Kim, S.J.; Lee, B.C. Blastomeres aggregation as an efficient alternative for trophoblast culture from porcine parthenogenetic embryos. *Dev. Growth Differ.* **2015**, *57*, 362–368. [CrossRef]
29. Tukur, H.A.; Aljumaah, R.S.; Swelum, A.A.-A.; Alowaimer, A.N.; Abdelrahman, M.; Saadeldin, I.M. Effects of Short-Term Inhibition of Rho Kinase on Dromedary Camel Oocyte In Vitro Maturation. *Animals* **2020**, *10*, 750. [CrossRef]
30. Ferraz, M.; Fujihara, M.; Nagashima, J.B.; Noonan, M.J.; Inoue-Murayama, M.; Songsasen, N. Follicular extracellular vesicles enhance meiotic resumption of domestic cat vitrified oocytes. *Sci. Rep.* **2020**, *10*, 1–14. [CrossRef]
31. Simonsen, J.B. Pitfalls associated with lipophilic fluorophore staining of extracellular vesicles for uptake studies. *J. Extracell. Vesicles* **2019**, *8*, 1582237. [CrossRef] [PubMed]
32. Takov, K.; Yellon, D.; Davidson, S.M. Confounding factors in vesicle uptake studies using fluorescent lipophilic membrane dyes. *J. Extracell. Vesicles* **2017**, *6*, 1388731. [CrossRef] [PubMed]
33. Saadeldin, I.; Koo, O.J.; Kang, J.-T.; Kwon, D.K.; Park, S.J.; Kim, S.J.; Moon, J.H.; Oh, H.J.; Jang, G.; Lee, B.C. Paradoxical effects of kisspeptin: It enhances oocyte in vitro maturation but has an adverse impact on hatched blastocysts during in vitro culture. *Reprod. Fertil. Dev.* **2012**, *24*, 656–668. [CrossRef]
34. Zheng, Z.; Sun, R.; Zhao, H.-J.; Fu, D.; Zhong, H.-J.; Weng, X.-Q.; Qu, B.; Zhao, Y.; Wang, L.; Zhao, W.-L. MiR155 sensitized B-lymphoma cells to anti-PD-L1 antibody via PD-1/PD-L1-mediated lymphoma cell interaction with CD8+T cells. *Mol. Cancer* **2019**, *18*, 1–13. [CrossRef]
35. Qin, A.; Zhou, Y.; Sheng, M.; Fei, G.; Ren, T.; Xu, L. Effects of microRNA-155 on the growth of human lung cancer cell line 95D in vitro. *Zhongguo Fei Ai Za Zhi* **2011**, *14*, 575–580. [CrossRef]
36. Saadeldin, I.M.; Swelum, A.A.-A.; Elsafadi, M.; Mahmood, A.; Osama, A.; Shikshaky, H.; Alfayez, M.; Alowaimer, A.N.; Magdeldin, S. Thermotolerance and plasticity of camel somatic cells exposed to acute and chronic heat stress. *J. Adv. Res.* **2020**, *22*, 105–118. [CrossRef] [PubMed]
37. Varkonyi-Gasic, E.; Wu, R.; Wood, M.; Walton, E.F.; Hellens, R.P. Protocol: A highly sensitive RT-PCR method for detection and quantification of microRNAs. *Plant Methods* **2007**, *3*, 12. [CrossRef]
38. McKenzie, P.P.; Foster, J.S.; House, S.; Bukovsky, A.; Caudle, M.R.; Wimalasena, J. Expression of G1 Cyclins and Cyclin-Dependent Kinase-2 Activity during Terminal Differentiation of Cultured Human Trophoblast1. *Biol. Reprod.* **1998**, *58*, 1283–1289. [CrossRef] [PubMed]

39. Chen, L.; Wang, J.; Fan, X.; Zhang, Y.; Zhoua, M.; Li, X.; Wang, L. LASP2 inhibits trophoblast cell migration and invasion in preeclampsia through inactivation of the Wnt/β-catenin signaling pathway. *J. Recept. Signal Transduct.* **2020**, *41*, 67–73. [CrossRef]
40. Han, Q.; Zheng, L.; Liu, Z.; Luo, J.; Chen, R.; Yan, J. Expression of β-catenin in human trophoblast and its role in placenta accreta and placenta previa. *J. Int. Med. Res.* **2018**, *47*, 206–214. [CrossRef]
41. Johns, D.N.; Lucas, C.G.; Pfeiffer, A.C.; Chen, P.R.; Meyer, E.A.; Perry, S.D.; Spate, L.D.; Cecil, R.F.; Fudge, A.M.; Samuel, M.S.; et al. Conceptus interferon gamma is essential for establishment of pregnancy in the pig. *Biol. Reprod.* **2021**, *105*, 1577–1590. [CrossRef] [PubMed]
42. Arregui, C.O.; González, Á.; Burdisso, J.E.; González Wusener, A.E. Protein tyrosine phosphatase PTP1B in cell adhesion and migration. *Cell Adhes. Migr.* **2013**, *7*, 418–423. [CrossRef] [PubMed]
43. Lv, S.; Wang, N.; Lv, H.; Yang, J.; Liu, J.; Li, W.-P.; Zhang, C.; Chen, Z.-J. The Attenuation of Trophoblast Invasion Caused by the Downregulation of EZH2 Is Involved in the Pathogenesis of Human Recurrent Miscarriage. *Mol. Ther.-Nucleic. Acids* **2018**, *14*, 377–387. [CrossRef]
44. Hong, L.; Han, K.; Wu, K.; Liu, R.; Huang, J.; Lunney, J.K.; Zhao, S.; Yu, M. E-cadherin and ZEB2 modulate trophoblast cell differentiation during placental development in pigs. *Reproduction* **2017**, *154*, 765–775. [CrossRef] [PubMed]
45. DaSilva-Arnold, S.C.; Kuo, C.-Y.; Davra, V.; Remache, Y.; Kim, P.C.W.; Fisher, J.P.; Zamudio, S.; Al-Khan, A.; Birge, R.B.; Illsley, N.P. ZEB2, a master regulator of the epithelial–mesenchymal transition, mediates trophoblast differentiation. *Mol. Hum. Reprod.* **2018**, *25*, 61–75. [CrossRef]
46. Lee, S.-Y.; Yun, S.H.; Lee, H.; Yi, Y.-S.; Park, E.C.; Kim, W.; Kim, H.-Y.; Lee, J.C.; Kim, G.-H.; Kim, S.I. Analysis of the Extracellular Proteome of Colistin-Resistant Korean Acinetobacter baumannii Strains. *ACS Omega* **2020**, *5*, 5713–5720. [CrossRef]
47. Sherman, B.T.; Hao, M.; Qiu, J.; Jiao, X.; Baseler, M.W.; Lane, H.C.; Imamichi, T.; Chang, W. DAVID: A web server for functional enrichment analysis and functional annotation of gene lists (2021 update). *Nucleic Acids Res.* **2022**, *50*, W216–W221. [CrossRef]
48. Mishra, A.; Ashary, N.; Sharma, R.; Modi, D. Extracellular vesicles in embryo implantation and disorders of the endometrium. *Am. J. Reprod. Immunol.* **2020**, *85*, e13360. [CrossRef]
49. Evans, J.; Rai, A.; Nguyen, H.P.T.; Poh, Q.H.; Elglass, K.; Simpson, R.J.; Salamonsen, L.; Greening, D.W. Human Endometrial Extracellular Vesicles Functionally Prepare Human Trophectoderm Model for Implantation: Understanding Bidirectional Maternal-Embryo Communication. *Proteomics* **2019**, *19*, e1800423. [CrossRef]
50. Gurung, S.; Greening, D.; Catt, S.; Salamonsen, L.; Evans, J. Exosomes and soluble secretome from hormone-treated endometrial epithelial cells direct embryo implantation. *Mol. Hum. Reprod.* **2020**, *26*, 510–520. [CrossRef]
51. O'Neil, E.V.; Burns, G.W.; Ferreira, C.R.; Spencer, T.E. Characterization and regulation of extracellular vesicles in the lumen of the ovine uterusdagger. *Biol. Reprod.* **2020**, *102*, 1020–1032. [CrossRef] [PubMed]
52. Senol, S.; Sayar, I.; Ceyran, A.B.; Ibiloglu, I.; Akalin, I.; Firat, U.; Kosemetin, D.; Zerk, P.E.; Aydin, A. Stromal Clues in Endometrial Carcinoma: Loss of Expression of β-Catenin, Epithelial-Mesenchymal Transition Regulators, and Estrogen-Progesterone Receptor. *Int. J. Gynecol. Pathol.* **2016**, *35*, 238–248. [CrossRef] [PubMed]
53. Ma, L.; Zhang, M.; Cao, F.; Han, J.; Han, P.; Wu, Y.; Deng, R.; Zhang, G.; An, X.; Zhang, L.; et al. Effect of MiR-100-5p on proliferation and apoptosis of goat endometrial stromal cell in vitro and embryo implantation in vivo. *J. Cell Mol. Med.* **2022**, *26*, 2543–2556. [CrossRef] [PubMed]
54. Tan, Q.; Shi, S.; Liang, J.; Cao, D.; Wang, S.; Wang, Z. Endometrial cell-derived small extracellular vesicle miR-100-5p promotes functions of trophoblast during embryo implantation. *Mol. Ther.-Nucleic. Acids* **2020**, *23*, 217–231. [CrossRef]
55. Su, L.; Liu, R.; Cheng, W.; Zhu, M.; Li, X.; Zhao, S.; Yu, M. Expression Patterns of MicroRNAs in Porcine Endometrium and Their Potential Roles in Embryo Implantation and Placentation. *PLoS ONE* **2014**, *9*, e87867. [CrossRef]
56. Wang, Y.-P.; Zhao, P.; Liu, J.-Y.; Liu, S.-M.; Wang, Y.-X. MicroRNA-132 stimulates the growth and invasiveness of trophoblasts by targeting DAPK-1. *Eur. Rev. Med. Pharm. Sci.* **2020**, *24*, 9837–9843.
57. Saadeldin, I.M.; Kim, B.; Lee, B.; Jang, G. Effect of different culture media on the temporal gene expression in the bovine developing embryos. *Theriogenology* **2011**, *75*, 995–1004. [CrossRef]
58. du Puy, L.; Lopes, S.M.C.D.S.; Haagsman, H.P.; Roelen, B.A. Analysis of co-expression of OCT4, NANOG and SOX2 in pluripotent cells of the porcine embryo, in vivo and in vitro. *Theriogenology* **2011**, *75*, 513–526. [CrossRef]
59. Ezashi, T.; Matsuyama, H.; Telugu, B.P.V.; Roberts, R.M. Generation of Colonies of Induced Trophoblast Cells During Standard Reprogramming of Porcine Fibroblasts to Induced Pluripotent Stem Cells1. *Biol. Reprod.* **2011**, *85*, 779–787. [CrossRef]
60. Li, X.; Li, C.; Dong, X.; Gou, W. MicroRNA-155 inhibits migration of trophoblast cells and contributes to the pathogenesis of severe preeclampsia by regulating endothelial nitric oxide synthase. *Mol. Med. Rep.* **2014**, *10*, 550–554. [CrossRef]
61. Dai, Y.; Qiu, Z.; Diao, Z.; Shen, L.; Xue, P.; Sun, H.; Hu, Y. MicroRNA-155 inhibits proliferation and migration of human extravillous trophoblast derived HTR-8/SVneo cells via down-regulating cyclin D1. *Placenta* **2012**, *33*, 824–829. [CrossRef] [PubMed]
62. Wang, Z.; Shan, Y.; Yang, Y.; Wang, T.; Guo, Z. MicroRNA-155 is upregulated in the placentas of patients with preeclampsia and affects trophoblast apoptosis by targeting SHH/GLi1/BCL2. *Hum. Exp. Toxicol.* **2020**, *40*, 439–451. [CrossRef] [PubMed]
63. Chakrabarty, A.; Tranguch, S.; Daikoku, T.; Jensen, K.; Furneaux, H.; Dey, S.K. MicroRNA regulation of cyclooxygenase-2 during embryo implantation. *Proc. Natl. Acad. Sci. USA* **2007**, *104*, 15144–15149. [CrossRef] [PubMed]

64. Sun, H.-X.; Zeng, D.-Y.; Li, R.-T.; Pang, R.-P.; Yang, H.; Hu, Y.-L.; Zhang, Q.; Jiang, Y.; Huang, L.-Y.; Tang, Y.-B.; et al. Essential Role of MicroRNA-155 in Regulating Endothelium-Dependent Vasorelaxation by Targeting Endothelial Nitric Oxide Synthase. *Hypertension* **2012**, *60*, 1407–1414. [CrossRef]
65. Alexy, T.; Rooney, K.; Weber, M.; Gray, W.D.; Searles, C.D. TNF-α alters the release and transfer of microparticle-encapsulated miRNAs from endothelial cells. *Physiol. Genom.* **2014**, *46*, 833–840. [CrossRef]
66. Dakic, A.; DiVito, K.; Fang, S.; Suprynowicz, F.; Gaur, A.; Li, X.; Palechor-Ceron, N.; Simic, V.; Choudhury, S.; Yu, S.; et al. ROCK inhibitor reduces Myc-induced apoptosis and mediates immortalization of human keratinocytes. *Oncotarget* **2016**, *7*, 66740–66753. [CrossRef]
67. Kim, K.; Min, S.; Kim, D.; Kim, H.; Roh, S. A Rho Kinase (ROCK) Inhibitor, Y-27632, Inhibits the Dissociation-Induced Cell Death of Salivary Gland Stem Cells. *Molecules* **2021**, *26*, 2658. [CrossRef]
68. Street, C.A.; Bryan, B.A. Rho kinase proteins–pleiotropic modulators of cell survival and apoptosis. *Anticancer Res.* **2011**, *31*, 3645–3657.
69. Shi, J.; Wei, L. Rho kinase in the regulation of cell death and survival. *Arch. Immunol. Ther. Exp.* **2007**, *55*, 61–75. [CrossRef]
70. Shiokawa, S.; Iwashita, M.; Akimoto, Y.; Nagamatsu, S.; Sakai, K.; Hanashi, H.; Kabir-Salmani, M.; Nakamura, Y.; Uehata, M.; Yoshimura, Y. Small Guanosine TriphospataseRhoA and Rho-Associated Kinase as Regulators of Trophoblast Migration. *J. Clin. Endocrinol. Metab.* **2002**, *87*, 5808–5816. [CrossRef]
71. Lokossou, A.G.; Toudic, C.; Nguyen, P.T.; Elisseeff, X.; Vargas, A.; Rassart, É.; Lafond, J.; LeDuc, L.; Bourgault, S.; Gilbert, C.; et al. Endogenous retrovirus-encoded Syncytin-2 contributes to exosome-mediated immunosuppression of T cells†. *Biol. Reprod.* **2019**, *102*, 185–198. [CrossRef] [PubMed]
72. Dunlap, K.A.; Palmarini, M.; Varela, M.; Burghardt, R.C.; Hayashi, K.; Farmer, J.L.; Spencer, T.E. Endogenous retroviruses regulate periimplantation placental growth and differentiation. *Proc. Natl. Acad. Sci. USA* **2006**, *103*, 14390–14395. [CrossRef]
73. Fu, B.; Ma, H.; Liu, D. Endogenous Retroviruses Function as Gene Expression Regulatory Elements During Mammalian Pre-implantation Embryo Development. *Int. J. Mol. Sci.* **2019**, *20*, 790. [CrossRef]
74. Prudhomme, S.; Bonnaud, B.; Mallet, F. Endogenous retroviruses and animal reproduction. *Cytogenet. Genome. Res.* **2005**, *110*, 353–364. [CrossRef] [PubMed]
75. Meyer, T.J.; Rosenkrantz, J.; Carbone, L.; Chavez, S. Endogenous Retroviruses: With Us and against Us. *Front. Chem.* **2017**, *5*, 23. [CrossRef] [PubMed]
76. Hu, X.; Zhu, W.; Chen, S.; Liu, Y.; Sun, Z.; Geng, T.; Wang, X.; Gao, B.; Song, C.; Qin, A.; et al. Expression of the env gene from the avian endogenous retrovirus ALVE and regulation by miR-155. *Arch. Virol.* **2016**, *161*, 1623–1632. [CrossRef]

 cells

Article

Identification of the Inner Cell Mass and the Trophectoderm Responses after an In Vitro Exposure to Glucose and Insulin during the Preimplantation Period in the Rabbit Embryo

Romina Via y Rada [1,2], Nathalie Daniel [1,2], Catherine Archilla [1,2], Anne Frambourg [1,2], Luc Jouneau [1,2], Yan Jaszczyszyn [3], Gilles Charpigny [1,2], Véronique Duranthon [1,2] and Sophie Calderari [1,2,*]

1 BREED INRAE, UVSQ, Université Paris-Saclay, 78350 Jouy-en-Josas, France
2 Ecole Nationale Vétérinaire d'Alfort, BREED, 94700 Maisons-Alfort, France
3 Institute for Integrative Biology of the Cell (I2BC), UMR 9198 CNRS, CEA, Paris-Sud University F, 91190 Gif-sur-Yvette, France
* Correspondence: sophie.calderari@inrae.fr

Abstract: The prevalence of metabolic diseases is increasing, leading to more women entering pregnancy with alterations in the glucose-insulin axis. The aim of this work was to investigate the effect of a hyperglycemic and/or hyperinsulinemic environment on the development of the preimplantation embryo. In rabbit embryos developed in vitro in the presence of high insulin (HI), high glucose (HG), or both (HGI), we determined the transcriptomes of the inner cell mass (ICM) and the trophectoderm (TE). HI induced 10 differentially expressed genes (DEG) in ICM and 1 in TE. HG ICM exhibited 41 DEGs involved in oxidative phosphorylation (OXPHOS) and cell number regulation. In HG ICM, proliferation was decreased ($p < 0.01$) and apoptosis increased ($p < 0.001$). HG TE displayed 132 DEG linked to mTOR signaling and regulation of cell number. In HG TE, proliferation was increased ($p < 0.001$) and apoptosis decreased ($p < 0.001$). HGI ICM presented 39 DEG involved in OXPHOS and no differences in proliferation and apoptosis. HGI TE showed 16 DEG linked to OXPHOS and cell number regulation and exhibited increased proliferation ($p < 0.001$). Exposure to HG and HGI during preimplantation development results in common and specific ICM and TE responses that could compromise the development of the future individual and placenta.

Keywords: preimplantation embryo; diabetes; DOHaD; rabbit

Citation: Via y Rada, R.; Daniel, N.; Archilla, C.; Frambourg, A.; Jouneau, L.; Jaszczyszyn, Y.; Charpigny, G.; Duranthon, V.; Calderari, S. Identification of the Inner Cell Mass and the Trophectoderm Responses after an In Vitro Exposure to Glucose and Insulin during the Preimplantation Period in the Rabbit Embryo. *Cells* **2022**, *11*, 3766. https://doi.org/10.3390/cells11233766

Academic Editor: Lon J. van Winkle

Received: 5 October 2022
Accepted: 18 November 2022
Published: 25 November 2022

1. Introduction

The worldwide prevalence of metabolic diseases such as diabetes is increasing at an alarming rate [1]. In 2021, the International Diabetes Federation estimated that 1 in 10 adults live with diabetes [1]. Type 2 diabetes (T2D) is a chronic metabolic disease characterized by hyperglycemia, insulin resistance, and/or impaired insulin secretion and accounts for 90% of diabetes cases [1]. In prediabetes and the early stages of T2D, impaired glucose tolerance, or hyperglycemia, is accompanied by compensatory hyperinsulinemia due to decreasing insulin sensitivity [1,2]. Unfortunately, these first signs of metabolic dysregulation are often asymptomatic, resulting in nearly half of T2D patients going undiagnosed and untreated [1]. Known before as adult-onset diabetes, the prevalence of T2D is increasing in younger people, including women of childbearing age [1,3]. Type 1 diabetes (T1D), an immune-related disease characterized by the destruction of insulin-producing cells, affects a young population [4]. In T1D, the glucose-insulin axis is disrupted. Insulin is no longer produced, and insulin-stimulated glucose uptake is reduced, resulting in persistent hyperglycemia [4]. One in six pregnancies is estimated to be affected by hyperglycemia [1]. Exposure in utero to a perturbed glucose-insulin homeostasis increases the risk of birth defects and metabolic deregulations such as enhanced growth, higher fasting glucose, and lower insulin sensitivity in the offspring [5]. These metabolic dysregulations can be maintained throughout the life course of the individual, making it prone to developing cardiometabolic diseases such as obesity and T2D [3]. This is described by the Developmental Origins of Health and Disease (DOHaD) concept, which highlights that exposure to a suboptimal environment during critical periods of development predisposes the offspring to poor health later in life [6]. One key period of development sensitive to environmental insults is the preimplantation stage [7]. During the preimplantation stage, embryos undergo tightly regulated essential events such as the maternal-to-zygotic transition with the transcriptional activation of the embryonic genome (EGA) and the first lineage specification giving rise to the inner cell mass (ICM), the progenitor of the embryo proper, and the trophectoderm (TE), the progenitor of the embryonic portion of the placenta [8]. To sustain their development, embryos take advantage of the nutrients and growth factors present in the oviduct and uterine fluid [9,10]. The composition of these fluids varies according to maternal metabolic and hormonal status, as is the case for glucose and insulin, whose concentrations depend on maternal circulating plasma concentrations [9–11]. Preimplantation embryos are sensitive to perturbations in their surrounding microenvironment [7,8,11]. Variations in the environment of the early embryo, even restricted to the preimplantation period, result in irreversible defects in the adult offspring [12]. Studies in vivo and in vitro have demonstrated the susceptibility of preimplantation embryos to changes in glucose or insulin levels [13]. In diabetes-induced rabbit and mouse models, preimplantation embryos exposed to hyperglycemia resulted in perturbed insulin-mediated glucose metabolism, decreased glucose transport and utilization, reduced developmental competence and cell numbers, and increased apoptosis in the ICM [14–16]. In in vivo animal models, severe hyperglycemia was obtained by the chemical destruction of pancreatic β-cells, thus mimicking type 1 diabetes. Nevertheless, because insulin secretion was reduced or absent in these animals, frequent insulin injections were needed, which may have resulted in oscillating insulin levels in the intrauterine environment [15]. Unfortunately, insulin levels were not quantified in these studies; thus, it is impossible to identify whether the phenotypes described were the result of hyperglycemia or the combination of hyperglycemia and insulin. In vitro, exposure to high glucose alone led to impaired blastocyst development, reduced total cell numbers, decreased glycolytic activity, decreased insulin sensitivity, perturbed TE differentiation, and impaired capacity of trophoblast outgrowth in vitro—a marker of implantation potential [11,14,17]. Preimplantation embryos are exposed to insulin, which is present in the oviductal and uterine fluids at concentrations that depend on maternal insulin levels [14]. The extent of the cellular and molecular responses to insulin in early embryos has been less investigated [13]. Glucose and insulin, through the activation of signaling and metabolic pathways, are closely related [18]. In preimplantation embryos, glucose is used as an energy source, reaching the highest consumption rate at the blastocyst stage [19,20]. Furthermore, insulin receptors and insulin-responsive glucose transporters are expressed in mouse, rabbit, and human preimplantation embryos [21].

We hypothesized that the deregulation of glucose and insulin homeostasis present in an increasing number of women impacts the preimplantation embryo. Functionally different from the blastocyst stage, ICM and TE differ in their epigenetic, transcriptomic, and metabolic programs [20,22,23]. We hypothesized that exposure to this glucose-insulin altered environment affects ICM and TE differently and induces short- and long-term consequences not only in the future individual but also in the future placenta, a central element for fetal nutrition regulation, and whose structure and/or function adapt to suboptimal in utero environments [15,24,25]. Hence, to investigate the effects of high glucose and/or high insulin on preimplantation development, we used the rabbit model, a model with preimplantation development (i.e., EGA timing, gastrulation morphology), glucose metabolism at early stages, and a placental structure close to that of humans [26]. We established a model of one-cell stage rabbit embryos developed in vitro until the blastocyst stage with supplementation of glucose, insulin, or both to recreate a moderately hyperglycemic and/or hyperinsulinemic environment [11,14,17,27] and addressed the specific gene expression responses of the ICM and TE.

2. Materials and Methods

2.1. Embryo In-Vitro Development

New Zealand White female rabbits (INRA line 1077) were superovulated as previously described [28] and mated with New Zealand White male rabbits. At 19 h post-coïtum (hpc), does were euthanized, and one-cell embryos were recovered from oviducts by flushing with phosphate buffer saline (PBS, Gibco, Thermo Fisher Scientific, Waltham, MA, USA). One-cell embryos were sorted in M199 HEPES (Sigma-Aldrich, Saint-Louis, MO, USA) supplemented with 10% fetal bovine serum (FBS, Gibco) and rinsed in Global medium (LifeGlobal Group, Guilford, CT, USA) supplemented with 10% human serum albumin (HSA, LifeGlobal Group). Embryos were then placed in 10 µL microdrops of Global-10% HSA medium supplemented with either glucose (Sigma-Aldrich G6152) and/or insulin (Sigma-Aldrich I9278) and covered with mineral oil (Sigma-Aldrich M8410) for a 72h culture at 38 °C, 5% CO_2, and 5% O_2 until the blastocyst stage. Four experimental groups were designed: Control (CNTRL): 0.18 mM of glucose without insulin; high insulin (HI): 0.18 mM of glucose and 1.7 µM of insulin; high glucose (HG): 15 mM of glucose without insulin; and high glucose and high insulin (HGI): 15 mM of glucose and 1.7 µM of insulin. After 72 h of culture, to determine the embryo's developmental competence in each group, embryos were classified into three categories: (i) arrested embryos; (ii) compacted embryos; (iii) blastocysts or cavitated embryos. The rate of arrested embryos (developmental arrest), compacted embryos, and blastocysts/cavitated embryos reported in percentage was calculated in fifteen to twenty-nine independent experiments from the total of one-cell embryos placed in culture. Blastocysts were recovered to proceed to ICM and TE isolation by moderate immunosurgery. To remove the zona pellucida, blastocysts were incubated for 1–3 min in 5 mg/mL Pronase (P5147, Sigma-Aldrich). Embryos were next incubated in anti-rabbit goat serum (R5131 Sigma-Aldrich) for 90 min at 37 °C and then incubated in guinea pig complement (S1639 Sigma-Aldrich) for 20 sec. The ICM was mechanically isolated from the TE by pipetting with a small-bore glass pipette (60–70-µm diameter). To clean the ICM to limit any contamination, several back and forth injections into the glass pipette were perfomed. ICM and their corresponding TE were then immediately stored at −80 °C for RNA sequencing analysis or fixed for microscopic analyses.

2.2. RNA Sequencing

ICM and their corresponding TE originating from the same blastocysts were used. Only one biological replicate from the HI group did not include the corresponding TE due to low total RNA quality. Total RNA was extracted from three biological replicates per culture condition, corresponding to pooled samples (n = 11–16 ICM or TE per replicate) using the Arcturus PicoPure RNA Isolation Kit (Applied Biosystems Life Technologies, Waltham, MA, USA). RNA quality was assessed using RNA 6000 Pico chips with an Agilent 2100 Bioanalyzer (Agilent Technologies, Santa Clara, CA, USA). All extracted samples had an RNA Integrity Number (RIN) $\geq$ 8 value. Seven hundred and fifty pictograms of total RNA were used for amplification using the SMART-Seq V4 ultra-low input RNA kit (Clontech, Takara, Saint-Germain-en-Laye, France) according to the manufacturer's recommendations with nine PCR cycles for cDNA pre-amplification. The cDNA quality was assessed with the Agilent Bioanalyzer 2100. Libraries were prepared as previously described [29]. Reads were mapped to the rabbit transcriptome reference (Ensembl 98 Oryctolagus cuniculus 2.0) using the splice junction mapper TopHat (v2.1.1) associated with the short-read aligner Bowtie2 (v2.3.4.1). To generate the gene count table, featureCounts (v1.6.0) was used. Hierarchical clustering was computed as previously described [29]. Data normalization and single-gene level analysis of the differential expression were performed using the DESeq2 package (v1.28.1) [30]. Differences were considered significant for adjusted p=values (Benjamini-Hochberg) < 0.05 and when the normalized expression counts were more than 20 in two of the three biological replicates. Heatmaps were generated with the pheatmaps R package (v1.0.12), with the z-score calculation of the normalized expression counts obtained with DESeq2. Logarithm 2 Fold Change (Log2FC) of differentially expressed genes (DEG) was used to generate horizontal bar plots with R studio software (v1.2.5019). InteractiVenn [31] software was used for Venn diagram generation. Functional annotation of DEG with their associated Gene ontology (GO) Biological Process (BP) terms was performed using DAVID [32] (v6.8). Gene Set Enrichment Analysis (GSEA) [33] was performed using the GSEA Java Desktop application (v4.0.3) from the Broad Institute. Enrichment analysis was calculated using the normalized expression counts obtained with DESesq2 and the Molecular Signature Database (MSigDB, v7.0) gene set collections (Hallmarks [34], KEGG (Kyoto Encyclopedia of Genes and Genomes), Reactome [35], and GO BP [36,37]) by gene-set permutation. Gene sets were considered significant when the false discovery rate (FDR) was less than 0.05. Enrichment analysis results were analyzed with the R package SUMER [38] (v1.1.5) for the reduction of redundancy and condensation of gene sets. For cluster visualization, the clusterMaker2 [39] plugin from Cytoscape [40] (v3.8.2) was used.

2.3. Quantification of Total Cell Number in Whole Embryos

To quantify the total cell number in whole embryos, DAPI staining was assessed in in vitro-developed blastocysts. Blastocysts were recovered from in vitro culture, and the zona pellucida was removed as detailed above. Blastocysts were fixed in 4% paraformaldehyde (PFA, EMS) in PBS at room temperature (RT) for 20 min. Permeabilization was performed with 0.5% Triton X-100 (Sigma-Aldrich) in PBS with 0.5% polyvinylpyrrolidone (PVP) for 1 h at 37 °C in a humidified chamber. DNA was counterstained with 0.2 mg/mL DAPI (Invitrogen) in PBS for 15 min at RT. Blastocysts were analyzed by an inverted ZEISS AxioObserver Z1 microscope (Zeiss, Rueil Malmaison, France) equipped with an ApoTome slider (Axiovision software 4.8) using a 20X objective and a z-distance of 1.5 μm between optical sections at the MIMA2 platform (https://doi.org/10.15454/1.5572348210007727E12, accessed on 4 October 2022). The total number of DAPI-labeled nuclei was quantified manually using ImageJ software (1.53.j). Each condition was analyzed in five to nine independent experiments.

2.4. Quantification of Apoptotic and Proliferating Cells in ICM and TE

To quantify apoptotic and proliferating cells in ICM and TE, we first considered distinguishing the ICM and TE on whole embryos by immunostaining of known lineage-specific markers. CDX2 (CDX-2-88, Biogenex, Fremont, CA, USA), SOX2 (ab97959, Abcam, Cambridge, UK), and NANOG (14-5761-80, Invitrogen, Thermo Fisher Scientific, Waltham, MA, USA) antibodies were tested, but none showed sufficient specificity to consider differential counting (data not shown). Thus, determination of apoptotic and proliferating cell numbers was performed on isolated ICM and TE.

Detection of apoptotic cells was performed using the DeadEnd Fluorometric TUNEL System (Promega, Madison, WI, USA) in two to six independent experiments. Isolated ICM or TE were fixed in 4% PFA in PBS at RT for 20 min. Permeabilization was performed with 0.5% Triton X-100 in PBS with 0.5% PVP for 1 h at 37 °C in a humidified chamber. After rinsing in PBS with 0.5% PVP, a second fixation was performed in 4% PFA and 0.2% glutaraldehyde for 15 min at RT. As a positive control, ICM and TE were treated with 2 units of RQ1 RNase-free DNase (Promega) for 30 min. The TUNEL reaction was performed according to the manufacturer's directions. DNA was counterstained with 0.2 mg/mL DAPI in PBS for 15 min at RT.

Detection of proliferating cells was performed using the Click-iT® Edu Imaging Kit (Fisher Scientific, Waltham, MA, USA) in at least three independent experiments. Briefly, the zona pellucida was removed as detailed above, and then blastocysts were incubated with 10 μM EdU for 15 min at 38 °C, 5% CO_2, and 5% O_2. ICM and TE separation were performed by moderate immunosurgery. ICM and TE were fixed with 4% PFA at RT for 20 min. EdU detection was performed according to the instructions provided by the manufacturer. DNA was counterstained with 0.2 mg/mL DAPI in PBS for 15 min at RT.

ICM and TE were analyzed by an inverted ZEISS AxioObserver Z1 microscope equipped with an ApoTome slider using a 20X objective and a z-distance of 1.5 µm between optical sections at the MIMA2 platform. The number of DAPI-labeled nuclei, TUNEL-positive nuclei, and EdU-positive nuclei were quantified manually using ImageJ software.

2.5. Statistical Analysis

Statistical analysis was carried out using the generalized linear mixed-effects model (GLMM) with the glmer function and the lme4 R package (v1.1-28). The total cell number was analyzed using the linear mixed-effects model (LMM) using the lmer function. The glucose and insulin concentrations were considered as fixed effects. No significant interaction between glucose and insulin was detected. The models applied in the analysis of developmental competence and total cell number did not include the interaction of glucose and insulin. The in vitro culture experiments and rabbits were considered to have random effects. Estimated marginal means (emmeans, also known as least-squares means) and post-hoc tests between conditions were performed using the emmeans R package (v1.7.3) with the emmeans and pairs functions. Results are shown as emmeans with standard errors. Differences were considered significant when p-values were < 0.05.

3. Results

To determine the effect of high glucose and/or high insulin during preimplantation development, one-cell rabbit embryos were cultured in vitro under control (CNTRL), high insulin (HI), high glucose (HG), or high glucose and high insulin (HGI) until the blastocyst stage (Figure 1). To evaluate the effect of these conditions on developmental competence, the mean percentage of arrested embryos, compacted morula, or expanded blastocysts at the end of the 72 h culture period was determined (Table 1). On blastocysts, ICM and their corresponding TE were separated by moderate immunosurgery (Figure 1), and specific transcriptomic responses to high glucose and/or high insulin were explored by RNA-sequencing. RNA-seq of three biological replicates per culture condition generated 102–145 million raw reads per sample. Clustering of the transcriptome datasets by Euclidean distance revealed a clear separation between the ICM and TE regardless of the condition (Figure 2A). Without excluding minimal contamination, these results underline the successful separation of these two compartments by immunosurgery. Principal component analysis (PCA) was performed separately on ICM (Figure 2B) and TE (Figure 2C) transcriptomic data. Comparison to the CNTRL resulted in the identification of differentially expressed genes (DEG) between ICM or TE from embryos developed in HI, HG, or HGI (Figure 3 and Supplementary Table S1). Functions of the identified DEGs were explored using GO terms annotations (Supplementary Table S1). To determine coordinated gene expression changes, we analyzed the gene expression datasets using GSEA with the Hallmarks gene set collections, KEGG, Reactome, and GO BP databases (Supplementary Table S2). Enrichment analysis results were then analyzed with SUMER for gene set condensation. The following paragraphs will describe the identified effects of high insulin or high glucose alone and then in combination in the ICM and TE of exposed embryos.

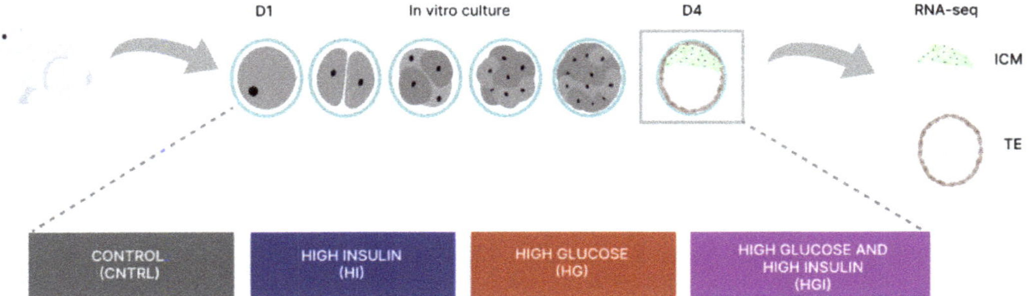

Figure 1. Schematic representation of the experimental workflow to analyze the in vitro exposure of preimplantation embryos from 1-cell to blastocyst stage for control, high insulin, high glucose, and high glucose and high insulin. The inner cell mass (ICM) and trophectoderm (TE) transcriptomes were determined by RNA-seq. D1, day 1. D4, day 4.

Table 1. Developmental competence of rabbit preimplantation embryos developed in vitro in CNTRL, HI, HG, or HGI conditions. Values are expressed as emmeans with standard errors in parenthesis. Different superscript letters (a, b) indicate significant differences within the same column ($p < 0.05$). CNTRL, control; HI, high insulin; HG, high glucose; HGI, high glucose and high insulin.

Condition	*N* Rabbits	*N* Embryos	Development Arrest Rate	Compacted Embryos Rate	Blastocyst Rate
CNTRL	60	1090	0.034 (0.009) [a]	0.303 (0.061) [a]	0.638 (0.057) [a]
HI	21	530	0.029 (0.009) [a]	0.309 (0.063) [a]	0.645 (0.059) [a]
HG	52	751	0.027 (0.008) [a]	0.228 (0.052) [b]	0.726 (0.051) [b]
HGI	35	519	0.023 (0.007) [a]	0.232 (0.053) [b]	0.732 (0.051) [b]

3.1. *Impact of High Insulin In Vitro Exposure*

The developmental competence of HI embryos showed no significant differences when compared to the CNTRL condition (Table 1). Quantification of total cell number by DAPI staining did not show significant changes in HI (262 ± 12, $n = 54$) versus CNTRL (240 ± 7, $n = 76$) blastocysts ($p < 0.05$, Supplementary Figure S1).

3.1.1. In ICM, High Insulin Induced Changes in Cellular Energy Metabolic Pathways

Transcriptome analysis by PCA and hierarchical clustering did not show a clear separation between HI ICM and CNTRL ICM (Figure 2). Differential analysis of HI ICM versus CNTRL ICM transcriptomes identified 10 DEG (3 overexpressed and 7 underexpressed) (Figure 3 and Supplementary Table S1). GSEA identified 37 significant positively enriched pathways (2 Hallmarks, 3 KEGG pathways, 14 GO BP, and 18 Reactome gene sets) and 8 negatively enriched (5 Hallmarks and 3 GO BP) pathways (Figure 4A and Supplementary Table S2).

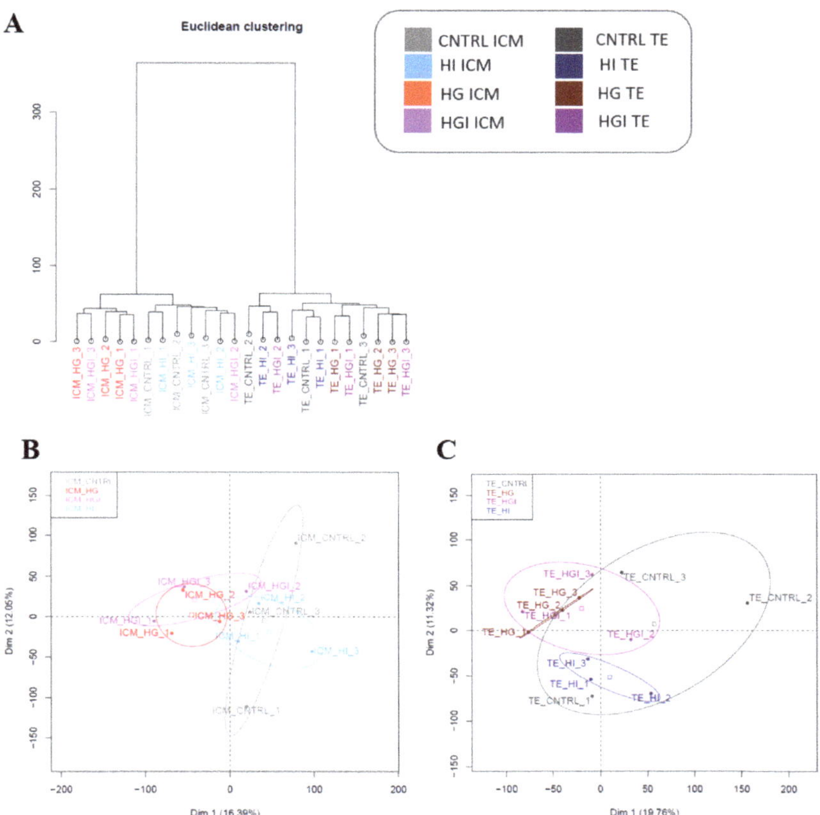

Figure 2. Transcriptome analysis of isolated ICM and TE from in vitro-developed blastocysts with high glucose and/or high insulin. (**A**). Clustering by Euclidean distance of the transcriptomic datasets of ICM and their corresponding TE developed in CNTRL, HI, HG, or HGI. Each group included three biological replicates which consisted of n = 11-16 ICM or TE. (**B**). Principal component analysis (PCA) of ICM groups. (**C**). PCA of TE groups. ICM, inner cell mass. TE, trophectoderm. CNTRL, control; HI, high insulin; HG, high glucose; HGI, high glucose and high insulin. Samples are color-coded according to the legend at the top (right).

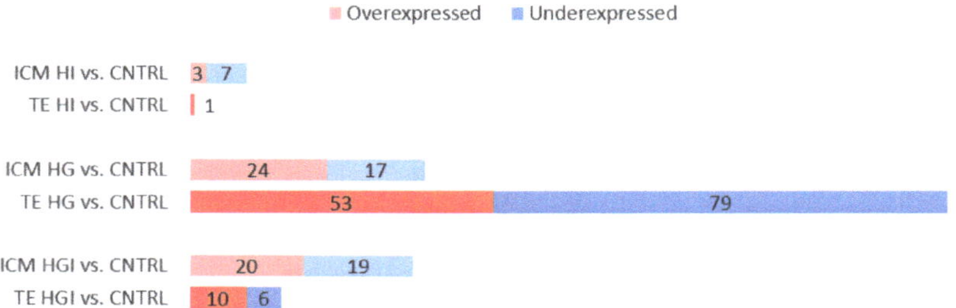

Figure 3. Differentially expressed genes (DEG) in ICM and TE of in vitro-developed blastocysts with HI, HG, or HGI compared to CNTRL. The number of overexpressed (red) and underexpressed (blue) DEGs with p-adjusted < 0.05 are shown. ICM, inner cell mass. TE, trophectoderm.

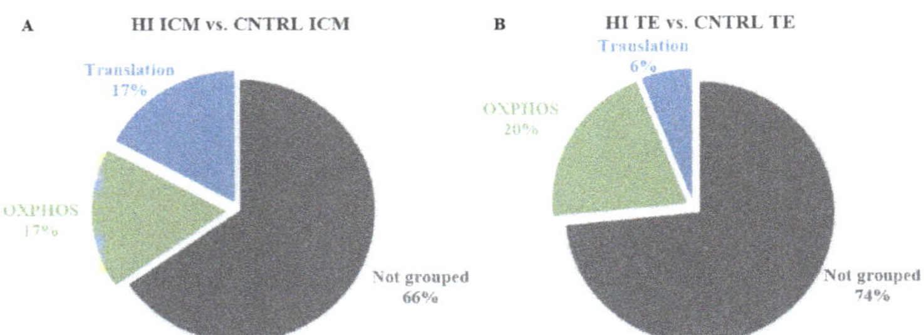

Figure 4. Significantly enriched gene sets (FDR < 0.05) in ICM and TE transcriptomes of in vitro-developed blastocysts with HI compared to CNTRL. Significantly enriched gene sets were identified by GSEA with the Molecular Signature Database (MSigDB) gene set collections: Hallmarks, KEGG, Reactome, and GO BP. GSEA was followed by SUMER analysis for gene set condensation. (**A**). Pie charts showing the enriched gene sets in HI ICM versus CNTRL ICM. (**B**). Pie charts showing the enriched gene sets in HI TE versus CNTRL TE. ICM, inner cell mass. TE, trophectoderm.

Enriched pathways included gene sets implicated in translation and oxidative phosphorylation (OXPHOS) (Figure 4A). Analysis of DEG and enrichment results in HI ICM transcriptomes compared to CNTRL ICM pointed out the perturbation of transcription and translation. Gene-by-gene statistical analysis identified DEG implicated in the regulation of transcription as *RC3H1* (ring finger and CCCH-type domains 1RC3H1, log2FC = −0.86) and *ICE1* (interactor of little elongation complex ELL subunit 1, log2FC = −0.71) (Supplementary Table S1). Concerning translation, enrichment analysis identified the overrepresentation of the "ribosome" KEGG pathway (normalized enrichment scores (NES) = 2.60), "translation" Reactome gene set (NES = 2.23), and the "translational elongation" and "translational termination" GO BP (NES = 1.98 and 2.06, respectively) (Supplementary Table S2). Enrichment results also highlighted the perturbation in OXPHOS. GSEA identified the significant positive enrichment of "oxidative phosphorylation" in Hallmark (NES = 1.98), KEGG (NES = 1.85), and GO terms (NES = 2.1), in addition to Reactome gene sets linked to OXPHOS as "NADH dehydrogenase complex assembly" (NES = 1.93) or "the citric acid cycle and respiratory electron transport" (NES = 2.22) (Supplementary Table S2). In line with these results, enrichment in "mitochondrial fatty acid (FA) β-oxidation" Reactome gene set (NES = 1.83) was also identified.

3.1.2. In TE, High Insulin Impacted Cellular Energy Metabolism and Oxidative Stress Pathways

Transcriptome analysis by PCA and hierarchical clustering showed no separation between HI TE and CNTRL TE (Figure 2). HI exposure resulted in the differential expression of only one gene, *PNLIP* (pancreatic lipase; log2FC = 5.1) (Figure 3 and Supplementary Table S1). However, GSEA analysis identified the significant positive enrichment of 83 gene sets (3 Hallmarks, 5 KEGG pathways, 20 GO BP, and 55 Reactome) and the significant negative enrichment of 7 gene sets (6 Hallmarks and 1 GO BP) (Supplementary Table S2).

Enriched pathways included gene sets implicated in translation and in OXPHOS (Figure 4B). Enrichment results highlighted a few gene sets implicated in translation, such as the "ribosome" KEGG pathway (NES = 1.82) or the "translation" Reactome gene set (NES = 1.82) (Figure 4B and Supplementary Table S2). Enrichment results related to OXPHOS included the overrepresentation of the "oxidative phosphorylation" gene sets in Hallmark (NES = 2.39), KEGG (NES = 2.27), and GO BP (NES = 2.11) or the Reactome gene set "respiratory electron transport" (NES = 2.30) (Figure 4B and Supplementary Table S2). In addition, reactive oxygen species (ROS) gene set "ROS and RNS production in phagocytes" (NES = 1.86) (Supplementary Table S2) and mitochondrial FA β-oxidation Reactome gene set (NES = 1.77) were identified. HI TE also showed the overrepresentation of activated NF-kB-related gene sets, including the Reactome FCERI-mediated NF-kB activation (NES = 1.94) (Supplementary Table S2).

3.1.3. High Insulin Induced Common Responses in ICM and TE

Between ICM and TE of high insulin-exposed embryos, whereas no common DEG was observed, 22 shared enriched GSEA gene sets were identified (Supplementary Tables S1 and S3). All shared gene sets exhibited the same level of enrichment. Among shared gene sets we highlighted translation, OXPHOS and FA β-oxidation. Enrichment of ROS and NF-kB signaling was only observed in HI TE.

3.2. *Impact of High Glucose In Vitro Exposure*

High glucose exposure led to a significant increase in blastocyst rate, mirrored by a significant reduction in the rate of compacted embryos compared to CNTRL embryos (Table 1). No significant differences were observed in the rate of arrested embryos after development with HG (Table 1). Quantification of total cell number showed a significantly increased cell number in HG (263 $\pm$ 9, n = 75) versus CNTRL (240 $\pm$ 7, n = 76) blastocysts ($p < 0.05$, Supplementary Figure S1).

3.2.1. In ICM, High Glucose Altered OXPHOS, Decreased Proliferation, Increased Apoptosis

Transcriptome analysis by PCA and hierarchical clustering showed the separation between HG ICM and CNTRL ICM (Figure 2). Differential analysis showed 41 DEG (24 upregulated and 17 downregulated) in the ICM of embryos exposed to HG compared to the CNTRL ICM (Figure 3 and Supplementary Table S1). GSEA analysis identified the significant positive enrichment of 73 functional gene sets (2 Hallmarks, 2 KEGG pathways, 11 GO BP, and 58 Reactome) (Supplementary Table S2).

Enrichment analyses identified 3 main clusters: translation, regulation of the cell number, and OXPHOS (Figure 5A). First, the protein translation cluster included KEGG "ribosome" (NES = 2.67), Reactome "metabolism of amino acids and derivatives" (NES = 1.79) and "translation" (NES = 2.51), and GO BP "translational initiation" (NES = 2.35) gene sets (Figure 5A, Supplementary Table S2). In addition, perturbation of transcription was also observed (Supplementary Table S1). Genes implicated in transcription were found to be differentially expressed, such as *KDM5A* (lysine demethylase 5A, log2FC = −0.35) and *GATA3* (GATA binding protein 3, log2FC = 1.05) (Supplementary Table S1). The second cluster highlighted alterations in the regulation of the cell number. GSEA revealed the enrichment of Hallmark "myc Target v1" (NES = 2.26), Reactome pathways such as "regulation of mitotic cell cycle" (NES = 1.89), and "regulation of apoptosis" (NES = 1.89). The differential analysis identified the overexpression of *LIN54* (lin-54 DREAM MuvB core complex component, log2FC = 0.65) and the underexpression of *CHP2* (calcineurin-like EF-hand protein 2, log2FC = −2.3) and *APC* (APC regulator of WNT signaling pathway, log2FC = −0.67) (Supplementary Table S1). To investigate cell proliferation and apoptosis at the cellular level, we performed EdU incorporation and the TUNEL assay (Figure 6 and Supplementary Figures S2–S5). Indeed, a reduced number of proliferating cells (Figure 6A and Supplementary Figure S2) and an increased proportion of apoptotic cells (Figure 6B and Supplementary Figure S4) were identified in HG ICM compared to CNTRL ICM. The third identified cluster concerning perturbations in energy metabolism included Hallmark "oxidative phosphorylation" (NES = 1.70), Reactome "respiratory electron transport" (NES = 2.23), GO BP "mitochondrial respiratory chain complex assembly" (NES = 2.02), and Reactome "cellular response to hypoxia" (NES = 1.81) (Figure 5A and Supplementary Table S2). In line with these results, HG ICM showed the overexpression of *FH* (fumarate hydratase, log2FC = 0.45) compared to CNTRL ICM (Supplementary Table S1).

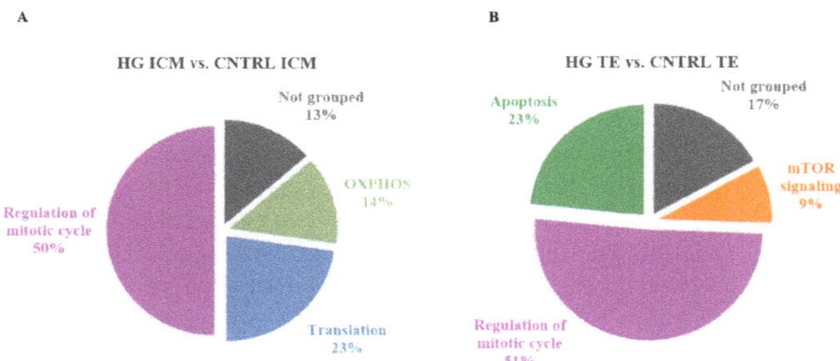

Figure 5. Significantly enriched gene sets (FDR < 0.05) in ICM and TE transcriptomes of in vitro-developed blastocysts with HG compared to CNTRL. Significantly enriched gene sets were identified by GSEA with the Molecular Signature Database (MSigDB) gene set collections: Hallmarks, KEGG, Reactome, and GO BP. GSEA was followed by SUMER analysis for gene set condensation. (**A**). Pie charts showing the enriched gene sets in HG ICM versus CNTRL ICM. (**B**). Pie charts showing the enriched gene sets in HG TE versus CNTRL TE. ICM, inner cell mass. TE, trophectoderm.

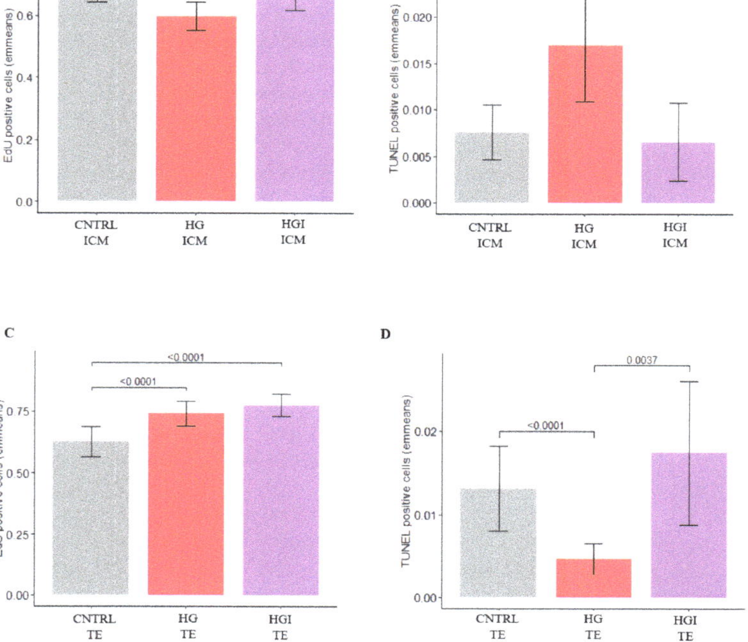

Figure 6. Quantification of proliferating and apoptotic cells in the ICM and TE of in-vitro-developed blastocysts with CNTRL, HG, and HGI by EdU incorporation and TUNEL assays. (**A**) Barplots showing the emmeans of proliferating cells in the ICM (n ICM = 16–38). (**B**) Barplots showing the emmeans percentage of apoptotic cells in the ICM (n ICM = 13–59). (**C**) Barplots showing the emmeans of proliferating cells in the TE (n TE = 16–24). (**D**) Barplots showing the emmeans of apoptotic cells in the TE (n TE = 18–52). Values are presented as emmeans ± S.E. Significant p values ($p < 0.05$) are shown. ICM, inner cell mass. TE, trophectoderm; CNTRL, control; HG, high glucose; HGI, high glucose and high insulin.

In addition to the main clusters, HG ICM transcriptomes showed perturbations in signaling pathways. HG ICM showed the overexpression of *REL* (REL proto-oncogene, NF-kB subunit, log2FC = 2.18) and the enrichment of gene sets related to NF-kB signaling, such as Reactome "FCERI mediated NF-kB activation" (NES = 1.84) (Supplementary Tables S1 and S2). Differential and enrichment analyses also showed dysregulations in the WNT signaling pathway. These results included the downregulation of *APC* (log2FC = −0.67) (Supplementary Table S1) and enrichment of Reactome "degradation of β-catenin by the destruction complex" (NES = 1.73) (Supplementary Table S2). Furthermore, gene-by-gene analysis of the HG ICM DEG revealed the overexpression of genes involved in the trophoblast lineage, such as GATA binding protein 3 (*GATA3*, log2FC = 1.05) and placenta expressed transcript 1 (*PLET1*, log2FC = 2.39) (Figure 7 and Supplementary Table S1).

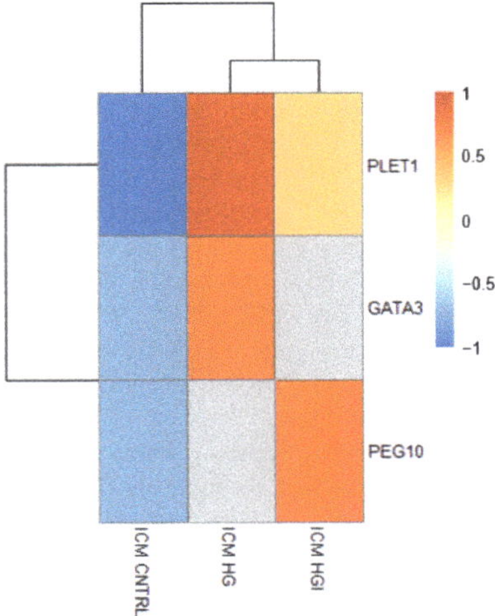

Figure 7. Heatmap showing the differential expression of genes (DEG) implicated in the TE lineage in HG and HGI ICM compared to CNTRL ICM. The mean normalized expression counts of n = 3 biological replicates, transformed to a Z-score, are represented by the color key. The gray color indicates the gene is not a DEG in that group. ICM, inner cell mass. TE, trophectoderm.

3.2.2. In TE, High Glucose Impacted Metabolic Pathways, Increased Proliferation, and Decreased Apoptosis

Transcriptome analysis by PCA and hierarchical clustering did not show a separation between HG TE and CNTRL TE (Figure 2). HG TE showed 132 DEG compared to CNTRL TE (53 overexpressed and 79 underexpressed) (Figure 3 and Supplementary Table S1). GSEA identified the enrichment of 78 functional gene sets, 76 of which were positively enriched (10 Hallmarks, 1 KEGG, and 65 Reactome), and 2 (GO BP) were negatively enriched (Supplementary Table S2).

Enriched pathways identified clusters related to metabolism and cell number regulation (Figure 5B). Concerning alterations in metabolism, enrichment results highlighted perturbations in mTOR signaling by the overrepresentation of Hallmarks "mTORC1 signaling" (NES = 1.68) and "PI3K-AKT-mTOR signaling" (NES = 1.61) (Figure 5B, Supplementary Table S2). Alteration of numerous genes involved in glycolysis, glycine, and lipid metabolism were identified such as *HK1* (hexokinase 1, log2FC = 0.53), *PHGDH* (phosphoglycerate dehydrogenase, log2FC = 0.85), *AACS* (acetoacetyl-CoA synthetase, log2FC = −1.48), *LDLR* (low density lipoprotein receptor, log2FC = −1.40, *GPAT3* (glycerol-3-phosphate acyltransferase 3, log2FC = 1.22), and *GPCPD1* (glycerophosphocholine phosphodiesterase 1, log2FC = 0.88) (Supplementary Table S1). The second main enrichment concerned the regulation of cell number (Figure 5B). Enriched pathways included Hallmark "myc Target v1" (NES = 2.01), Reactome "mitotic G1-G1/S phases" (NES = 2.02), and "regulation of mitotic cell cycle" (NES = 2.08) gene sets (Supplementary Table S2). Consistent with these results, the HG TE transcriptome showed the altered expression of cell cycle progression genes, such as the underexpression of *CDKL4* (cyclin dependent kinase like 4, log2FC = −1.82) or *SMARCD3* (SWI/SNF related matrix associated actin dependent regulator of chromatin subfamily d member 3, log2FC = −3.00) or the overexpression of *GADD45A* (growth arrest and DNA damage inducible alpha, log2FC = 0.85) or *WEE1* (WEE1 G2 Checkpoint Kinase, log2FC = 0.97) (Supplementary Table S1). Assessment of cell proliferation by the EdU incorporation assay in HG TE further confirmed an increased number of proliferating cells compared to CNTRL TE (Figure 6C and Supplementary Figure S3). In parallel, enrichment results showed the overrepresentation of apoptosis-related gene sets such as Hallmarks "apoptosis" (NES = 1.73), "P53 pathway" (NES = 1.75), and the Reactome "regulation of apoptosis" (NES = 1.77) gene set (Supplementary Table S2). HG TE exhibited the differential expression of genes implicated in apoptosis, such as *CASP7* (caspase 7; log2FC = 1.13), *PDCD6* (programmed cell death 6; log2FC = 0.59), and *TRADD* (TNFRSF1A associated via death domain; log2FC = −1.37) (Supplementary Table S1). Investigation of apoptosis in HG TE by TUNEL assay showed a decrease in the number of apoptotic cells compared to CNTRL TE (Figure 6D and Supplementary Figure S5).

Beyond these main clusters, HG TE transcriptomes exhibited several other perturbations, as in the immune response. Enrichment and differential analysis identified the Hallmark "TGF-β signalling" (NES = 1.64) and "TNF-α signalling via NF-kB" (NES = 1.52) gene sets, and the underexpression of *ERC1* (ELKS/RAB6-interacting/CAST family member 1, log2FC = −0.91) (Supplementary Tables S1 and S2). The WNT signaling was identified as deregulated as highlighted by the enrichment of the Reactome "degradation of β-catenin by the destruction complex" (NES = 1.81) and "β-catenin independent WNT signaling" (NES = 1.79), and the underexpression of *ANKRD10* (ankyrin repeat domain 10; log2FC = −1.13) (Supplementary Tables S1 and S2). In addition, HG TE showed the enrichment of gene sets implicated in transcription and translation, such as the Reactome "transcriptional activity of SMAD2 SMAD3:SMAD4 heterotrimer" (NES = 1.9) and "translation" (NES = 1.88) (Supplementary Table S2). Along with these gene sets, gene-by-gene analysis and functional annotation by DAVID showed several DEG associated with transcriptional regulation, chromatin remodeling, and epigenetic mechanisms (Figure 8 and Supplementary Tables S1 and S4). Among the HG TE DEG, we highlighted the overexpression of *GADD45A*, *WEE1*, and *NPM3* (nucleophosmin/nucleoplasmin 3), and the underexpression of *SMARCD3*, *PADI2* (peptidyl arginine deiminase 2), *MOV10L1*, *ATF7* (activating transcription factor 7), *RESF1* (retroelement silencing factor 1), *BPTF* (bromodomain PHD finger transcription factor), and *NSD3* (nuclear receptor binding SET domain protein 3) (Figure 8 and Supplementary Tables S1 and S4).

3.2.3. High Glucose Induced Common and Specific Responses in ICM and TE

From the DEG identified in HG ICM (n = 41) and HG TE (n = 132), only 3 were shared: *ARRDC4* (arrestin domain containing 4) (log2FC = 2.07 and 2.02, respectively), *FAM3D* (family with sequence similarity 3 member D) (log2FC = 0.96 and 0.54, respectively), and *MOV10L1* (Mov10 RISC complex RNA helicase like 1) (log2FC = −1.26 and −1.57, respectively) (Supplementary Table S1). Despite the small amount of shared DEG, several processes were common, as shown by the 37 shared gene sets identified by GSEA, which are overrepresented in both ICM and TE (Supplementary Table S3). These gene sets were mainly related to the regulation of cell number. However, we identified opposite responses: the HG ICM exhibited decreased proliferation and increased apoptosis, whereas the HG TE showed increased proliferation and reduced apoptosis. Specific responses included transcriptome changes related to OXPHOS and lineage commitment in HG ICM and metabolism and epigenetic regulation in HG TE.

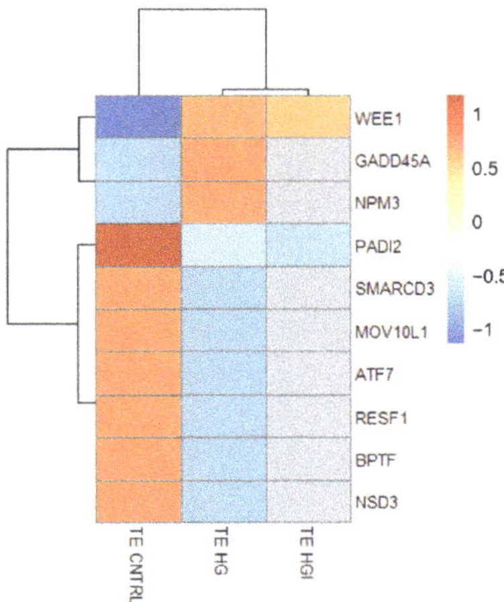

Figure 8. Heatmap showing the differential expression of genes (DEG) with a role in epigenetic regulation in HG and HGI TE compared to CNTRL TE. The mean normalized expression counts of $n = 3$ biological replicates, transformed to a z-score, are represented by the color key. The gray color indicates that the gene is not a DEG in that group. ICM, inner cell mass. TE, trophectoderm.

3.3. *Impact of High Glucose and High Insulin In Vitro Exposure*

High glucose and high insulin exposure led to a significant increase in blastocyst rate, mirrored by a significant reduction in the rate of compacted embryos compared to CNTRL embryos (Table 1). No significant differences were observed in the rate of arrested embryos after development with HGI (Table 1). Quantification of total cell number showed a significantly increased cell number in HGI (285 ± 14, $n = 50$) versus CNTRL (240 ± 7, $n = 76$) blastocysts ($p < 0.05$, Supplementary Figure S1).

As the development of HI embryos was not different from that of CNTRL embryos, a comparison of HGI vs. HI resulted in similar observations to HGI vs. CNTRL: no difference in the rate of arrested embryos, decrease in the rate of compacted embryos, and an increase of the rate of blastocysts and blastocyst total cell number (Table 1 and Supplementary Figure S1).

In comparison to high glucose, HGI embryos displayed similar development parameters: the rates of arrested, compacted, and blastocyst were similar in HGI compared to HG embryos. The cell number was also similar in HGI blastocysts in comparison to HG blastocysts (Table 1 and Supplementary Figure S1).

3.3.1. In ICM, Alteration of OXPHOS and ROS by High Glucose and High Insulin

Transcriptome analysis by PCA showed the separation between HGI ICM and CNTRL ICM (Figure 2). Differential analysis showed 39 DEG (20 overexpressed and 19 underexpressed) in comparison to CNTRL ICM (Figure 3 and Supplementary Table S1). GSEA analysis showed the significant positive enrichment of 107 gene sets (5 Hallmarks, 7 KEGG, 28 GO BP, and 66 Reactome) (Supplementary Table S2).

Enriched pathways highlighted two main clusters related to the regulation of gene expression and cellular energy metabolism (Figure 9A). Firstly, concerning the regulation of gene expression, several transcription factors were differentially expressed, such as *ELF2* (E74 like ETS transcription factor 2; log2FC = 0.71) and *SREBF2* (sterol regulatory element binding transcription factor 2; log2FC = −0.49). Overexpression of genes implicated in mRNA processing, such as *PTBP2* (polypyrimidine tract binding protein 2, log2FC = 0.96), was observed (Supplementary Table S1). Coherent with this, enrichment results highlighted transcriptome changes related to translation, including GO BP "translational initiation" (NES = 2.33), Reactome "mRNA splicing" (NES = 1.74), and KEGG "ribosome" (NES = 2.85) (Figure 9A and Supplementary Table S2). The second identified cluster was related to OXPHOS. HGI ICM showed transcriptomic changes including Hallmark "oxidative phosphorylation" (NES = 2.02), KEGG (NES = 2.26), GO BP (NES = 2.55), and Reactome "mitochondrial fatty acid β oxidation" (NES = 1.82) gene sets (Figure 9A and Supplementary Table S2). Additionally, HG ICM showed the enrichment of ROS pathways, including Reactome "cellular response to hypoxia" (NES = 1.78) and Hallmark "reactive oxygen species" (NES = 1.73) (Figure 9A and Supplementary Table S2).

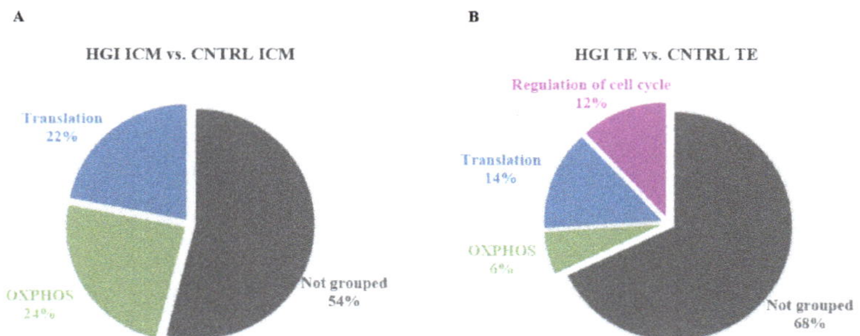

Figure 9. Significantly enriched gene sets (FDR < 0.05) in the ICM and TE transcriptomes of in vitro-developed blastocysts with HGI compared to CNTRL. Significantly enriched gene sets were identified by GSEA with the Molecular Signature Database (MSigDB) gene set collections: Hallmarks, KEGG, Reactome, and GO BP. GSEA was followed by SUMER analysis for gene set condensation. (**A**). Pie charts showing the enriched gene sets in HGI ICM versus CNTRL ICM. (**B**). Pie charts showing the enriched gene sets in HGI TE versus CNTRL TE. ICM, inner cell mass. TE, trophectoderm.

In addition to these main clusters, HGI ICM exhibited DEG and enriched gene sets implicated in cell number regulation. Among these, we can list the *APC* gene (log2FC = −0.74), the Hallmark "myc target v1" (NES = 2.02), and the Reactome "regulation of apoptosis" (NES = 1.73) (Supplementary Tables S1 and S3). We examined a possible imbalance in proliferation and apoptosis in HGI ICM and did not detect significant changes in the proportion of proliferating or apoptotic cells compared to CNTRL ICM (Figure 6A,B and Supplementary Figures S2 and S4). HGI ICM transcriptomes also showed enrichment in TNF-α signaling (Figure 9A and Supplementary Table S2). Among the DEG and gene sets implicated in this pathway, we highlighted the differential expression of *USP15* (ubiquitin specific peptidase 15, log2FC = 0.47), enriched Hallmark "TNFA signaling via NFKB" (NES = 1.70), and GO BP "cytokine metabolic process" (NES = 1.94) (Figure 9A and Supplementary Tables S1 and S2). Among the DEG identified, we also highlighted the differential expression of genes implicated in the trophoblast lineage and placenta development, such as *PLET1* (log2FC = 1.97) and *PEG10* (paternally expressed 10, log2FC = 1.79) (Figure 7 and Supplementary Table S1).

Then, the specific responses triggered by high glucose and high insulin in combination on the ICM were determined by the identification of common and specific transcriptomic changes between HGI versus CNTRL ICM and HI versus CNTRL ICM and between HGI versus CNTRL ICM and HG versus CNTRL ICM (Supplementary Figure S6A, Tables S1 and S3).

HGI ICM and HI ICM shared 2 DEG, including the protein-coding gene *RPS6KA3* (ribosomal protein S6 kinase A3) (log2FC = 0.77 and 0.88, respectively), and 29 gene sets, all with the same expression pattern (Supplementary Figure S6A, Tables S1 and S3). Shared changes in gene expression were related to the regulation of gene expression and OXPHOS (Supplementary Table S3).

HGI ICM and HG ICM shared 16 DEG and 62 gene sets, all with the same expression pattern (Supplementary Figure S6A, Tables S1 and S3). These shared transcriptome changes were related to the regulation of gene expression, OXPHOS, NF-kB signalling, and the aberrant expression of TE genes. Despite the few shared gene sets involved in cell number regulation, the increased apoptosis and decreased proliferation identified in HG ICM were not found in HGI ICM. The enrichment in the ROS pathway was exclusively identified in HGI ICM.

3.3.2. In TE, Alteration of OXPHOS, ROS, and Proliferation by High Glucose and High Insulin

Transcriptome analysis by PCA and hierarchical clustering did not show a separation between HGI TE and CNTRL TE (Figure 2C). HGI TE exhibited 16 DEG (10 overexpressed and 6 underexpressed) in comparison to CNTRL TE (Figure 3 and Supplementary Table S1). Enrichment analysis in HGI TE identified 108 gene sets positively enriched (8 Hallmarks, 2 KEGG pathways, 11 GO BP, and 87 Reactome) (Supplementary Table S2).

Enriched pathways included gene sets related to transcription, translation, cell number regulation, and OXPHOS, as highlighted by the three main clusters (Figure 9B and Supplementary Table S2). Transcription and translation enriched gene sets included the Reactome "transcriptional regulation by MECP2" (NES = 1.89), KEGG "ribosome" (NES = 2.53), Reactome "translation" (NES = 2.10), and KEGG "proteasome" (NES = 2.17) (Supplementary Table S2). Enriched pathways implicated in the regulation of cell number included Hallmark "myc target v1" (NES = 2.30) and Reactome "regulation of mitotic cell cycle" (NES = 2.22) (Figure 9B and Supplementary Table S2). In line with these enriched gene sets, the differential analysis identified the significant overexpression of *TUBB* (tubulin beta class I, log2FC = 0.50) and *WEE1* (log2FC = 0.92), 2 genes implicated in the G2/M transition of the mitotic cell cycle (Supplementary Table S1). The EdU incorporation assay in HGI TE confirmed a significant increase in the proportion of proliferating cells compared to CNTRL TE (Figure 6C and Supplementary Figure S3). Despite identifying a small enrichment of gene sets implicated in apoptosis (Supplementary Table S2), the TUNEL assay in HGI TE did not detect significant differences in the proportion of apoptotic cells compared to CNTRL TE (Figure 6D and Supplementary Figure S5). The third identified cluster was related to energy metabolism, as indicated by the Hallmark "oxidative phosphorylation" (NES = 1.54), "ROS pathway" (NES = 1.79), Reactome "cellular response to hypoxia" (NES = 2.00), and GO BP "regulation of transcription from RNA polymerase II promoter in response to hypoxia" (NES = 2.04) (Figure 9B and Supplementary Table S2). In line with these enrichments, DEG in HGI TE included the overexpression of *TXNIP* (thioredoxin-interacting protein; log2FC = 1.81) (Supplementary Table S1).

Besides these clusters, transcriptome analysis showed perturbations in the immune response in HGI TE. Enriched pathways included the Hallmark "TGF-α signaling via NF-kB" (NES = 1.73), GO BP "positive regulation of cytokine biosynthetic process" (NES = 2.02), and TGF-β signaling (NES = 1.67) (Supplementary Table S2). Additionally, enrichment results showed perturbations of the WNT signaling pathway by the Reactome "degradation of β-catenin by the destruction complex" (NES = 1.80) (Supplementary Table S2).

The combination of high glucose and high insulin triggered specific responses in the TE. HI TE and HGI TE shared 1 DEG (*PNLIP*, log2FC = 5.08 and 4.71, respectively) and 43 overrepresented gene sets (Supplementary Figure S6B, Tables S1 and S3). These gene sets were related to translation, OXPHOS, ROS, and NF-kB signaling (Supplementary Table S3). In comparison to HG TE, 9 DEG and 53 gene sets shared the same expression pattern (Supplementary Figure S6B, Tables S1 and S3). Corresponding pathways were related to transcription and translation, NF-kB signaling and regulation of cell number. A higher number of proliferating cells was detected in both HG TE and HGI TE. Decrease in HG TE, apoptosis was not altered in HGI TE. The alteration of the metabolic pathway genes and the altered expression of genes involved in epigenetic mechanisms detected in HG TE were not identified in HGI TE. Inversely, OXPHOS and ROS pathways were only overrepresented in HGI TE (Supplementary Table S3).

3.3.3. High Glucose and High Insulin Induced Common and Specific Responses in ICM and TE

Between HGI ICM and HGI TE, whereas only one DEG was shared (*ARRDC4*, log2FC = 2.09 and 2.02, respectively), half (n = 54) of the enriched gene sets were shared and exhibited the same enrichment pattern (Supplementary Tables S1 and S3). Transcription, translation, OXPHOS, ROS, and NF-kB signaling were impacted in ICM and TE in response to HGI (Supplementary Table S2). Specific responses between compartments can be noticed, such as cell commitment dysregulations occurring exclusively in the ICM.

4. Discussion

Prediabetes and the early stages of T2D are characterized by hyperglycemia and hyperinsulinemia [1,2]. Unfortunately, these first metabolic dysregulations are often asymptomatic, resulting in nearly half of people with T2D being undiagnosed and untreated [1]. In the early stages of pregnancy, including the preimplantation period, women are not yet aware of their gestational status; therefore, in undiagnosed diabetic women, pregnancies are not adequately intervened. Increased glucose and insulin concentrations are reflected in oviductal and uterine fluids [9,11]. Preimplantation embryos are responsive to glucose and insulin through the activation of signaling and metabolic pathways [14,15,41–43]. The preimplantation period corresponds to a critical window of susceptibility during which variations in the environment can have a major impact on the offspring. Here, we have established an in vitro model using the rabbit to study the effects of high glucose and/or high insulin on preimplantation embryo development.

As growth factors, glucose and insulin are key regulators of proliferation and apoptosis. In the present study, the presence of high glucose stimulated blastocyst development and growth. Observations in mice and bovine embryos mainly described a negative impact of glucose on blastocyst development, obtained with glucose concentrations above 20 mM [11,44]. Here, we have shown that high glucose exposure led to the alteration of proliferation and apoptosis in mirror patterns. Consistent with our findings, mouse and rat embryos exposed to glucose showed increased apoptosis in the ICM [45,46]. Inversely to ICM, and to our knowledge first described here, proliferation was increased, and apoptosis decreased in the TE of embryos exposed to high glucose. When embryos were exposed to high levels of insulin alone, blastocyst rate and growth were not impacted, and changes in the expression of genes involved in proliferation and apoptosis were not identified. The mitogenic actions of insulin are well known [18]; however, in preimplantation embryos, this remains controversial [47–50]. Our findings show that when high levels of insulin were added in addition to high glucose, the increased rate and growth of blastocysts observed in the presence of high glucose alone persisted. Despite changes in the proliferation and apoptosis gene expression, only the proliferation rate remained increased in the TE. These results suggest a crosstalk between glucose and insulin in mediating growth-related effects.

Glucose and insulin play a central role in regulating energy homeostasis and metabolism [18,51]. Here, embryos developed in the presence of high glucose exhibited OXPHOS signatures in the ICM and mTORC1 signaling and glycolytic and lipid metabolism signatures in the TE. Glucose, via glycolysis and OXPHOS, leads to the production of cellular energy in the form of ATP [52]. Until the morula stage, preimplantation embryos metabolize lactate and pyruvate preferentially as an energy source through OXPHOS [51]. Around the morula stage and onward, glucose is preferentially metabolized, although the metabolic pathway used may differ between the ICM and TE [51,53]. Exposure to hyperglycemia in vitro and in vivo led to hyperactivation of mTORC1 signaling in rabbit blastocysts, especially in the TE [41]. The mTORC1 and mTORC2 complexes stimulate anabolic processes such as protein, lipid, and nucleotide synthesis and regulate glucose metabolism by favoring glycolysis over OXPHOS [54]. To sustain cell growth, the mTORC1 and mTORC2 complexes stimulate anabolic processes such as protein, lipid, and nucleotide synthesis, regulate glucose metabolism by favoring glycolysis over OXPHOS, and promote cell survival and proliferation [54]. In the present study, both ICM and TE developed in an insulin-rich environment and exhibited OXPHOS gene expression signatures. The metabolic effects of insulin are well known, and insulin has been shown to stimulate the oxidative capacity of mitochondria [18,55]. In addition, the TE of embryos developed with high insulin showed ROS-related gene expression changes, suggesting insulin-mediated oxidative stress. ROS, mainly produced as a by-product of OXPHOS, plays a role in physiological cellular processes [56]. Here, in the presence of both high glucose and high insulin, OXPHOS and ROS gene expression signatures were also identified in the ICM and TE. In addition, transcriptome changes related to NF-kB and TNF-α signaling were identified in the ICM and TE. NF-kB signaling, central regulator of inflammation and immunity, also regulates multiple cellular processes, including mitochondrial respiration [57]. NF-kB is induced by environmental cues, including insulin and ROS [58,59]. Here, gene expression changes related to OXPHOS, ROS, and NF-kB suggest metabolic stress in the ICM and TE of embryos exposed to both high glucose and high insulin.

Interestingly, we identified the deregulation of a subset of genes implicated in chromatin remodeling and epigenetic regulation in the TE. Emerging research has underlined the crosslink between metabolism and chromatin dynamics and its influence on gene expression [60,61]. A clear example of this is the generation of regulators of chromatin-modifying enzymes through glucose metabolism [60,61]. Among the genes showing altered expression in the TE exposed to high glucose, we highlighted *GADD45A*, *BPTF*, *PADI2*, and *ATF7*. GADD45A mediates active DNA demethylation, facilitating transcriptional activation, and also regulates trophoblast cell migration and invasion during placentation [62,63]. BPTF encodes the largest subunit of the Nucleosome Remodeling Factor (NURF) chromatin remodeling complex and plays an essential role in extraembryonic lineage development [64,65]. As for PADI2, a catalyzer of histone citrullination, it regulates chromatin organization and transcriptional regulation of cell cycle progression, metabolism, and proliferation genes [66]. ATF7, a stress-responsive chromatin regulator that recruits histone methyltransferases to repress the transcription of metabolic genes, has been proposed to mediate paternal low protein diet-induced intergenerational programming by reducing H3K9me2 in target genes [67,68]. In the presence of both high glucose and high insulin, the number of epigenetic genes with altered expression was less than in embryos exposed to high glucose alone, whereas no gene with an epigenetic-related function showed differential expression in embryos exposed to high insulin alone. Thus, these results suggested a crosstalk between insulin and glucose in terms of epigenetic regulation, especially in the TE. Thus, the differential expression of these genes suggests alterations in the TE epigenetic landscape, alterations that could compromise trophoblast differentiation.

In addition to the altered expression of epigenetic genes in the TE, the ICM exhibited the overexpression of genes involved in the trophoblast lineage when exposed to high glucose alone or in combination with high insulin. GATA3 is a well-known transcription factor associated with TE initiation and trophoblast differentiation [69,70]. Overexpression of GATA3 was sufficient to induce trophoblast fate in mouse embryonic stem cells (ESCs) [69]. PLET1 is an epigenetically-regulated cell surface protein essential to drive the differentiation of the trophoblast lineage [71]. *PEG10*, a paternally expressed imprinted gene highly expressed in the placenta, is essential for placenta formation in early development [72]. In the mouse, it has been recently demonstrated that glucose metabolism is required for the specification of the TE lineage through the hexosamine biosynthetic pathway (HBP), the pentose phosphate pathway (PPP), and the activation of the mTOR pathway [70]. Furthermore, when human ESCs were cultured with high glucose, the differentiation of the definitive endoderm was impaired [73]. In addition, we observed the enrichment of NF-kB signatures on the ICM of embryos exposed to high glucose alone or in combination with high insulin, and the NF-kB signaling pathway is known to regulate trophoblast differentiation and function [74,75]. Moreover, the ICM of embryos exposed to high glucose alone or in combination with high insulin showed signatures of oxidative rather than glycolytic metabolism, which in the mouse has been described to be characteristic of the TE rather than of the ICM [20]. Our findings indicate a potential impairment in cell commitment in the ICM. Perturbations in ICM cell allocation could directly influence the TE lineage [46,76]. Blastocysts with different amounts of ICM cells led to limited trophoblast proliferation, suggesting the necessity for cell allocation homeostasis between these two compartments [46,76]. Moreover, the crosstalk between ICM and TE influencing TE differentiation has been previously described [17].

In conclusion, exposure to high glucose and high insulin alone or in combination during preimplantation development results in lineage-specific responses in the progenitors of the future individual and the embryonic portion of the placenta. We showed here that in the presence of high insulin, the impact of high glucose was lowered in some cases, suggesting significant crosstalk between glucose metabolism and insulin signaling in the early embryo. These results suggested that a mismatch in the glucose and insulin axis represents a risk for early embryonic development and, thus, for offspring health. Moreover, despite being present in the preimplantation maternal environment, insulin is usually absent in in vitro culture systems. Integration of insulin may be useful in improving embryo culture media.

Supplementary Materials: The following supporting information can be downloaded at: https://www.mdpi.com/article/10.3390/cells11233766/s1, Figure S1: Quantification of total cell number in vitro-developed blastocysts with CNTRL, HI, HG or HGI; Table S1: Differentially expressed genes (DEGs) in Inner Cell Mass (ICM) and Trophectoderm (TE) of embryos developed with High Insulin (HI), High Glucose (HG) or High Glucose and high Insulin (HGI) versus Control (CNTRL); Table S2: Gene Set Enrichment Analysis (GSEA); Table S3: Shared gene sets; Table S4: Functional annotation of Differentially expressed genes (DEGs)

Author Contributions: Conceptualization, S.C. and V.D.; methodology, C.A., N.D., R.V.y.R. and S.C.; software, A.F., L.J. and R.V.y.R.; validation, S.C. and R.V.y.R.; formal analysis, A.F., C.A., G.C., R.V.y.R. and S.C.; investigation, N.D., R.V.y.R., S.C. and Y.J.; resources, N.D., R.V.y.R. and S.C.; data curation, A.F., C.A., L.J. and Y.J.; writing-original draft preparation, R.V.y.R. and S.C.; writing-review and editing, S.C. and V.D.; visualization, A.F. and R.V.y.R.; supervision, S.C. and V.D., project administration, S.C., funding acquisition, S.C. and V.D. All authors have read and agreed to the published version of the manuscript.

Funding: This research was funded by the Societé Francophone du Diabète (SFD), by the INRAE with dedicated help from the PhASE department (CI 2017), and by the Agence Nationale de la Recherche REVIVE LabEx (Investissement d'Avenir, ANR-10-LABX-73). RV is the recipient of a doctoral fellowship from the Ministère de l'Enseignement supérieur, de la Recherche et de l'Innovation (MESRI), from the INRAE PhASE department and from the REVIVE LabEx.

Institutional Review Board Statement: The animal study protocol was approved by the Ethics Committee of INRAE (approved protocol code APAFIS#2180-2015112615371038 v2, approved on 16 December 2015 and approved protocol code APAFIS#26907-2020070115104375 v3, approved on 1 October 2020).

Informed Consent Statement: Not applicable.

Data Availability Statement: The data are available under accession number GSE218009GEO.

Acknowledgments: We are grateful for the technical support provided by the INRAE animal experimentation unit (UE1298–SAAJ, INRAE Jouy-en-Josas, France). We thank the @BRIDGe facility (GABI, AgroParisTech, INRAE, Université Paris-Saclay, Jouy-en-Josas, France) for their valuable technical assistance in assessing total RNA quality. This work has benefited from the facilities and expertise of the high throughput sequencing core facility of I2BC (Centre de Recherche de Gif—http://www.i2bc.paris-saclay.fr/, accessed on 4 October 2022). We thank the MIMA2 platform (Microscopie et Imagerie des Microorganismes, Animaux et Aliments, https://doi.org/10.15454/1.5572348210007727E12, accessed on 4 October 2022) and particularly Pierre Adenot for his help on ApoTome microscopy observations. We thank Laura Hua, Virginie Marcinek, and Pierre Midrouillet for their help during their short-term internships. We thank Delphine Rousseau-Ralliard, Bertrand Duvillié, and Patrice Humblot for their helpful discussions on data analysis.

Conflicts of Interest: The authors declare no conflict of interest.

References

1. IntelInternational Diabetes Federation. *IDF Diabetes Atlas*, 10th ed.; International Diabetes Federation: Brussels, Belgium, 2021; ISBN 978-2-930229-98-0.
2. Thomas, D.D.; Corkey, B.E.; Istfan, N.W.; Apovian, C.M. Hyperinsulinemia: An Early Indicator of Metabolic Dysfunction. *J. Endocr. Soc.* **2019**, *3*, 1727–1747. [CrossRef]
3. Hjort, L.; Novakovic, B.; Grunnet, L.G.; Maple-Brown, L.; Damm, P.; Desoye, G.; Saffery, R. Diabetes in Pregnancy and Epigenetic Mechanisms—How the First 9 Months from Conception Might Affect the Child's Epigenome and Later Risk of Disease. *Lancet Diabetes Endocrinol.* **2019**, *7*, 796–806. [CrossRef] [PubMed]
4. DiMeglio, L.A.; Evans-Molina, C.; Oram, R.A. Type 1 Diabetes. *Lancet* **2018**, *391*, 2449–2462. [CrossRef]
5. Francis, E.C.; Dabelea, D.; Ringham, B.M.; Sauder, K.A.; Perng, W. Maternal Blood Glucose Level and Offspring Glucose–Insulin Homeostasis: What Is the Role of Offspring Adiposity? *Diabetologia* **2020**, *64*, 83–94. [CrossRef]
6. Langley-Evans, S.C. Nutrition in Early Life and the Programming of Adult Disease: A Review. *J. Hum. Nutr. Diet.* **2015**, *28*, 1–14. [CrossRef]
7. Watkins, A.J.; Ursell, E.; Panton, R.; Papenbrock, T.; Hollis, L.; Cunningham, C.; Wilkins, A.; Perry, V.H.; Sheth, B.; Kwong, W.Y.; et al. Adaptive Responses by Mouse Early Embryos to Maternal Diet Protect Fetal Growth but Predispose to Adult Onset Disease1. *Biol. Reprod.* **2007**, *78*, 299–306. [CrossRef] [PubMed]
8. Velazquez, M.A. Impact of Maternal Malnutrition during the Periconceptional Period on Mammalian Preimplantation Embryo Development. *Domest. Anim. Endocrinol.* **2015**, *51*, 27–45. [CrossRef] [PubMed]
9. Kaye, P.L.; Gardner, H.G. Preimplantation Access to Maternal Insulin and Albumin Increases Fetal Growth Rate in Mice. *Hum. Reprod.* **1999**, *14*, 3052–3059. [CrossRef] [PubMed]
10. Acevedo, J.J.; Mendoza-Lujambio, I.; de la Vega-Beltrán, J.L.; Treviño, C.L.; Felix, R.; Darszon, A. KATP Channels in Mouse Spermatogenic Cells and Sperm, and Their Role in Capacitation. *Dev. Biol.* **2006**, *289*, 395–405. [CrossRef]
11. Fraser, R.B.; Waite, S.L.; Wood, K.A.; Martin, K.L. Impact of Hyperglycemia on Early Embryo Development and Embryopathy: In Vitro Experiments Using a Mouse Model. *Hum. Reprod.* **2007**, *22*, 3059–3068. [CrossRef] [PubMed]
12. Fleming, T.P.; Sun, C.; Denisenko, O.; Caetano, L.; Aljahdali, A.; Gould, J.M.; Khurana, P. Environmental Exposures around Conception: Developmental Pathways Leading to Lifetime Disease Risk. *Int. J. Environ. Res. Public Health* **2021**, *18*, 9380. [CrossRef] [PubMed]
13. Jungheim, E.S.; Moley, K.H. The Impact of Type 1 and Type 2 Diabetes Mellitus on the Oocyte and the Preimplantation Embryo. *Semin. Reprod. Med.* **2008**, *26*, 186–195. [CrossRef] [PubMed]
14. Ramin, N.; Thieme, R.; Fischer, S.; Schindler, M.; Schmidt, T.; Fischer, B.; Santos, A.N. Maternal Diabetes Impairs Gastrulation and Insulin and IGF-I Receptor Expression in Rabbit Blastocysts. *Endocrinology* **2010**, *151*, 4158–4167. [CrossRef]
15. Rousseau-Ralliard, D.; Couturier-Tarrade, A.; Thieme, R.; Brat, R.; Rolland, A.; Boileau, P.; Aubrière, M.-C.; Daniel, N.; Dahirel, M.; Derisoud, E.; et al. A Short Periconceptional Exposure to Maternal Type-1 Diabetes Is Sufficient to Disrupt the Feto-Placental Phenotype in a Rabbit Model. *Mol. Cell. Endocrinol.* **2019**, *480*, 42–53. [CrossRef] [PubMed]
16. Moley, K.H.; Chi, M.M.Y.M.-Y.; Mueckler, M.M. Maternal Hyperglycemia Alters Glucose Transport and Utilization in Mouse Preimplantation Embryos. *Am. J. Physiol. Metab.* **1998**, *275*, E38–E47. [CrossRef]
17. Leunda-Casi, A.; de Hertogh, R.; Pampfer, S. Decreased Expression of Fibroblast Growth Factor-4 and Associated Dysregulation of Trophoblast Differentiation in Mouse Blastocysts Exposed to High D-Glucose in Vitro. *Diabetologia* **2001**, *44*, 1318–1325. [CrossRef]
18. Boucher, J.; Kleinridders, A.; Kahn, C.R. Insulin Receptor Signaling in Normal and Insulin-Resistant States. *Cold Spring Harb. Perspect. Biol.* **2014**, *6*, a009191. [CrossRef]
19. Purcell, S.H.; Moley, K.H. Glucose Transporters in Gametes and Preimplantation Embryos. *Trends Endocrinol. Metab.* **2009**, *20*, 483–489. [CrossRef]
20. Gardner, D.K.; Harvey, A.J. Blastocyst Metabolism. *Reprod. Fertil. Dev.* **2015**, *27*, 638–654. [CrossRef]
21. Navarrete Santos, A.; Ramin, N.; Tonack, S.; Fischer, B. Cell Lineage-Specific Signaling of Insulin and Insulin-Like Growth Factor I in Rabbit Blastocysts. *Endocrinology* **2008**, *149*, 515–524. [CrossRef]
22. Canon, E.; Jouneau, L.; Blachère, T.; Peynot, N.; Daniel, N.; Boulanger, L.; Maulny, L.; Archilla, C.; Voisin, S.; Jouneau, A.; et al. Progressive Methylation of POU5F1 Regulatory Regions during Blastocyst Development. *Reproduction* **2018**, *156*, 145–161. [CrossRef] [PubMed]
23. Bouchereau, W.; Jouneau, L.; Archilla, C.; Aksoy, I.; Moulin, A.; Daniel, N.; Peynot, N.; Calderari, S.; Joly, T.; Godet, M.; et al. Major Transcriptomic, Epigenetic and Metabolic Changes Underlie the Pluripotency Continuum in Rabbit Preimplantation Embryos. *Development* **2022**, *149*, dev200538. [CrossRef] [PubMed]
24. Fleming, T.P.; Kwong, W.Y.; Porter, R.; Ursell, E.; Fesenko, I.; Wilkins, A.; Miller, D.J.; Watkins, A.J.; Eckert, J.J. The Embryo and Its Future. *Biol. Reprod.* **2004**, *71*, 1046–1054. [CrossRef] [PubMed]
25. Staud, F.; Karahoda, R. Trophoblast: The Central Unit of Fetal Growth, Protection and Programming. *Int. J. Biochem. Cell Biol.* **2018**, *105*, 35–40. [CrossRef]
26. Fischer, B.; Chavatte-Palmer, P.; Viebahn, C.; Navarrete Santos, A.; Duranthon, V. Rabbit as a Reproductive Model for Human Health. *Reproduction* **2012**, *144*, 1–10. [CrossRef]

27. Laskowski, D.; Sjunnesson, Y.; Humblot, P.; Sirard, M.A.; Andersson, G.; Gustafsson, H.; Båge, R. Insulin Exposure during in Vitro Bovine Oocyte Maturation Changes Blastocyst Gene Expression and Developmental Potential. *Reprod. Fertil. Dev.* **2017**, *29*, 876–889. [CrossRef]
28. Tarrade, A.; Rousseau-Ralliard, D.; Aubrière, M.C.; Peynot, N.; Dahirel, M.; Bertrand-Michel, J.; Aguirre-Lavin, T.; Morel, O.; Beaujean, N.; Duranthon, V.; et al. Sexual Dimorphism of the Feto-Placental Phenotype in Response to a High Fat and Control Maternal Diets in a Rabbit Model. *PLoS ONE* **2013**, *8*, e83458. [CrossRef]
29. Sanz, G.; Daniel, N.; Aubrière, M.C.; Archilla, C.; Jouneau, L.; Jaszczyszyn, Y.; Duranthon, V.; Chavatte-Palmer, P.; Jouneau, A. Differentiation of Derived Rabbit Trophoblast Stem Cells under Fluid Shear Stress to Mimic the Trophoblastic Barrier. *Biochim. Biophys. Acta-Gen. Subj.* **2019**, *1863*, 1608–1618. [CrossRef]
30. Love, M.I.; Huber, W.; Anders, S. Moderated Estimation of Fold Change and Dispersion for RNA-Seq Data with DESeq2. *Genome Biol.* **2014**, *15*, 1–21. [CrossRef]
31. Heberle, H.; Meirelles, V.G.; da Silva, F.R.; Telles, G.P.; Minghim, R. InteractiVenn: A Web-Based Tool for the Analysis of Sets through Venn Diagrams. *BMC Bioinform.* **2015**, *16*, 1–7. [CrossRef]
32. Huang, D.W.; Sherman, B.T.; Lempicki, R.A. Systematic and Integrative Analysis of Large Gene Lists Using DAVID Bioinformatics Resources. *Nat. Protoc.* **2009**, *4*, 44–57. [CrossRef]
33. Subramanian, A.; Tamayo, P.; Mootha, V.K.; Mukherjee, S.; Ebert, B.L.; Gillette, M.A.; Paulovich, A.; Pomeroy, S.L.; Golub, T.R.; Lander, E.S.; et al. Gene Set Enrichment Analysis: A Knowledge-Based Approach for Interpreting Genome-Wide Expression Profiles. *Proc. Natl. Acad. Sci. USA* **2005**, *102*, 15545–15550. [CrossRef] [PubMed]
34. Liberzon, A.; Birger, C.; Thorvaldsdóttir, H.; Ghandi, M.; Mesirov, J.P.; Tamayo, P. The Molecular Signatures Database Hallmark Gene Set Collection. *Cell Syst.* **2015**, *1*, 417–425. [CrossRef] [PubMed]
35. Jassal, B.; Matthews, L.; Viteri, G.; Gong, C.; Lorente, P.; Fabregat, A.; Sidiropoulos, K.; Cook, J.; Gillespie, M.; Haw, R.; et al. The Reactome Pathway Knowledgebase. *Nucleic Acids Res.* **2019**, *48*, D498–D503. [CrossRef] [PubMed]
36. Ashburner, M.; Ball, C.A.; Blake, J.A.; Botstein, D.; Butler, H.; Cherry, J.M.; Davis, A.P.; Dolinski, K.; Dwight, S.S.; Eppig, J.T.; et al. Gene Ontology: Tool for the Unification of Biology. *Nat. Genet.* **2000**, *25*, 25–29. [CrossRef] [PubMed]
37. Carbon, S.; Douglass, E.; Good, B.M.; Unni, D.R.; Harris, N.L.; Mungall, C.J.; Basu, S.; Chisholm, R.L.; Dodson, R.J.; Hartline, E.; et al. The Gene Ontology Resource: Enriching a GOld Mine. *Nucleic Acids Res.* **2021**, *49*, D325–D334. [CrossRef]
38. Savage, S.R.; Shi, Z.; Liao, Y.; Zhang, B. Graph Algorithms for Condensing and Consolidating Gene Set Analysis Results. *Mol. Cell. Proteomics* **2019**, *18*, S141–S152. [CrossRef] [PubMed]
39. Morris, J.H.; Apeltsin, L.; Newman, A.M.; Baumbach, J.; Wittkop, T.; Su, G.; Bader, G.D.; Ferrin, T.E. ClusterMaker: A Multi-Algorithm Clustering Plugin for Cytoscape. *BMC Bioinform.* **2011**, *12*, 1–14. [CrossRef]
40. Shannon, P. Cytoscape: A Software Environment for Integrated Models of Biomolecular Interaction Networks. *Genome Res.* **2003**, *13*, 2498–2504. [CrossRef]
41. Gürke, J.; Schindler, M.; Pendzialek, S.M.; Thieme, R.; Grybel, K.J.; Heller, R.; Spengler, K.; Fleming, T.P.; Fischer, B.; Navarrete Santos, A. Maternal Diabetes Promotes MTORC1 Downstream Signalling in Rabbit Preimplantation Embryos. *Reproduction* **2016**, *151*, 465–476. [CrossRef]
42. Shao, W.-J.; Tao, L.-Y.; Xie, J.-Y.; Gao, C.; Hu, J.-H.; Zhao, R.-Q. Exposure of Preimplantation Embryos to Insulin Alters Expression of Imprinted Genes. *Comp. Med.* **2007**, *57*, 482–486. [PubMed]
43. Schindler, M.; Pendzialek, S.M.; Grybel, K.; Seeling, T.; Santos, A.N. Metabolic Profiling in Blastocoel Fluid and Blood Plasma of Diabetic Rabbits. *Int. J. Mol. Sci.* **2020**, *21*, 919. [CrossRef] [PubMed]
44. Jiménez, A.; Madrid-Bury, N.; Fernández, R.; Pérez-Garnelo, S.; Moreira, P.; Pintado, B.; de la Fuente, J.; Gutiérrez-Adán, A. Hyperglycemia-Induced Apoptosis Affects Sex Ratio of Bovine and Murine Preimplantation Embryos. *Mol. Reprod. Dev.* **2003**, *65*, 180–187. [CrossRef] [PubMed]
45. Moley, K.H. Diabetes and Preimplantation Events of Embryogenesis. *Semin. Reprod. Endocrinol.* **1999**, *17*, 137–151. [CrossRef] [PubMed]
46. Pampfer, S. Apoptosis in Rodent Peri-Implantation Embryos: Differential Susceptibility of Inner Cell Mass and Trophectoderm Cell Lineages—A Review. *Placenta* **2000**, *21*, 3–10. [CrossRef]
47. Harvey, M.B.; Kaye, P.L. Insulin Increases the Cell Number of the Inner Cell Mass and Stimulates Morphological Development of Mouse Blastocysts in Vitro. *Development* **1990**, *110*, 963–967. [CrossRef]
48. Gardner, H.G.; Kaye, P.L. Insulin Increases Cell Numbers and Morphological Development in Mouse Pre-Implantation Embryos in Vitro. *Reprod. Fertil. Dev.* **1991**, *3*, 79–91. [CrossRef]
49. Augustin, R.; Pocar, P.; Wrenzycki, C.; Niemann, H.; Fischer, B. Mitogenic and Anti-Apoptotic Activity of Insulin on Bovine Embryos Produced in Vitro. *Reproduction* **2003**, *126*, 91–99. [CrossRef]
50. Chi, M.M.-Y.; Schlein, A.L.; Moley, K.H. High Insulin-Like Growth Factor 1 (IGF-1) and Insulin Concentrations Trigger Apoptosis in the Mouse Blastocyst via Down-Regulation of the IGF-1 Receptor. *Endocrinology* **2000**, *141*, 4784–4792. [CrossRef]
51. Kaneko, K.J. *Metabolism of Preimplantation Embryo Development: A Bystander or an Active Participant?* 1st ed.; Elsevier Inc.: Amsterdam, The Netherlands, 2016; Volume 120, ISBN 9780128014288.
52. May-Panloup, P.; Boguenet, M.; El Hachem, H.; Bouet, P.E.; Reynier, P. Embryo and Its Mitochondria. *Antioxidants* **2021**, *10*, 139. [CrossRef]

53. Leese, H.J.; Conaghan, J.; Martin, K.L.; Hardy, K. Early Human Embryo Metabolism. *BioEssays* **1993**, *15*, 259–264. [CrossRef] [PubMed]
54. Liu, G.Y.; Sabatini, D.M. MTOR at the Nexus of Nutrition, Growth, Ageing and Disease. *Nat. Rev. Mol. Cell Biol.* **2020**, *8*, 183–203. [CrossRef] [PubMed]
55. Cheng, Z.; Tseng, Y.; White, M.F. Insulin Signaling Meets Mitochondria in Metabolism. *Trends Endocrinol. Metab.* **2010**, *21*, 589–598. [CrossRef]
56. Sies, H.; Jones, D.P. Reactive Oxygen Species (ROS) as Pleiotropic Physiological Signalling Agents. *Nat. Rev. Mol. Cell Biol.* **2020**, *21*, 363–383. [CrossRef] [PubMed]
57. Albensi, B.C. What Is Nuclear Factor Kappa B (NF-KB) Doing in and to the Mitochondrion? *Front. Cell Dev. Biol.* **2019**, *7*, 154. [CrossRef] [PubMed]
58. Bertrand, F.; Philippe, C.; Antoine, P.J.; Baud, L.; Groyer, A.; Capeau, J.; Cherqui, G. Insulin Activates Nuclear Factor KB in Mammalian Cells through a Raf-1- Mediated Pathway. *J. Biol. Chem.* **1995**, *270*, 24435–24441. [CrossRef] [PubMed]
59. Morgan, M.J.; Liu, Z. Crosstalk of Reactive Oxygen Species and NF-KB Signaling. *Cell Res.* **2011**, *21*, 103–115. [CrossRef]
60. Reid, M.A.; Dai, Z.; Locasale, J.W. The Impact of Cellular Metabolism on Chromatin Dynamics and Epigenetics. *Nat. Cell Biol.* **2017**, *19*, 1298–1306. [CrossRef] [PubMed]
61. Martínez-Reyes, I.; Chandel, N.S. Mitochondrial TCA Cycle Metabolites Control Physiology and Disease. *Nat. Commun.* **2020**, *11*, 102. [CrossRef]
62. Arab, K.; Karaulanov, E.; Musheev, M.; Trnka, P.; Schäfer, A.; Grummt, I.; Niehrs, C. GADD45A Binds R-Loops and Recruits TET1 to CpG Island Promoters. *Nat. Genet.* **2019**, *51*, 217–223. [CrossRef]
63. Qian, X.; Zhang, Y. EZH2 Enhances Proliferation and Migration of Trophoblast Cell Lines by Blocking GADD45A-Mediated P38/MAPK Signaling Pathway. *Bioengineered* **2022**, *13*, 12583–12597. [CrossRef] [PubMed]
64. Goller, T.; Vauti, F.; Ramasamy, S.; Arnold, H.-H. Transcriptional Regulator BPTF/FAC1 Is Essential for Trophoblast Differentiation during Early Mouse Development. *Mol. Cell. Biol.* **2008**, *28*, 6819–6827. [CrossRef] [PubMed]
65. Landry, J.; Sharov, A.A.; Piao, Y.; Sharova, L.V.; Xiao, H.; Southon, E.; Matta, J.; Tessarollo, L.; Zhang, Y.E.; Ko, M.S.H.; et al. Essential Role of Chromatin Remodeling Protein Bptf in Early Mouse Embryos and Embryonic Stem Cells. *PLoS Genet.* **2008**, *4*, e1000241. [CrossRef]
66. Beato, M.; Sharma, P. Peptidyl Arginine Deiminase 2 (PADI2)-Mediated Arginine Citrullination Modulates Transcription in Cancer. *Int. J. Mol. Sci.* **2020**, *21*, 1351. [CrossRef]
67. Yoshida, K.; Maekawa, T.; Ly, N.H.; Fujita, S.; Muratani, M.; Ando, M.; Katou, Y.; Araki, H.; Miura, F.; Shirahige, K.; et al. ATF7-Dependent Epigenetic Changes are Required for the Intergenerational Effect of a Paternal Low-Protein Diet. *Mol. Cell* **2020**, *78*, 445–458.e6. [CrossRef] [PubMed]
68. Liu, Y.; Maekawa, T.; Yoshida, K.; Muratani, M.; Chatton, B.; Ishii, S. The Transcription Factor ATF7 Controls Adipocyte Differentiation and Thermogenic Gene Programming. *iScience* **2019**, *13*, 98–112. [CrossRef] [PubMed]
69. Ralston, A.; Cox, B.J.; Nishioka, N.; Sasaki, H.; Chea, E.; Rugg-Gunn, P.; Guo, G.; Robson, P.; Draper, J.S.; Rossant, J. Gata3 Regulates Trophoblast Development Downstream of Tead4 and in Parallel to Cdx2. *Development* **2010**, *137*, 395–403. [CrossRef]
70. Chi, F.; Sharpley, M.S.; Nagaraj, R.; Roy, S.S.; Banerjee, U. Glycolysis-Independent Glucose Metabolism Distinguishes TE from ICM Fate during Mammalian Embryogenesis. *Dev. Cell* **2020**, *53*, 9–26.e4. [CrossRef] [PubMed]
71. Murray, A.; Sienerth, A.R.; Hemberger, M. Plet1 Is an Epigenetically Regulated Cell Surface Protein That Provides Essential Cues to Direct Trophoblast Stem Cell Differentiation. *Sci. Rep.* **2016**, *6*, 25112. [CrossRef]
72. Ono, R.; Nakamura, K.; Inoue, K.; Naruse, M.; Usami, T.; Wakisaka-Saito, N.; Hino, T.; Suzuki-Migishima, R.; Ogonuki, N.; Miki, H.; et al. Deletion of Peg10, an Imprinted Gene Acquired from a Retrotransposon, Causes Early Embryonic Lethality. *Nat. Genet.* **2006**, *38*, 101–106. [CrossRef] [PubMed]
73. Chen, A.C.H.; Lee, Y.L.; Fong, S.W.; Wong, C.C.Y.; Ng, E.H.Y.; Yeung, W.S.B. Hyperglycemia Impedes Definitive Endoderm Differentiation of Human Embryonic Stem Cells by Modulating Histone Methylation Patterns. *Cell Tissue Res.* **2017**, *368*, 563–578. [CrossRef] [PubMed]
74. Armistead, B.; Kadam, L.; Drewlo, S.; Kohan-Ghadr, H.-R.R. The Role of NFκB in Healthy and Preeclamptic Placenta: Trophoblasts in the Spotlight. *Int. J. Mol. Sci.* **2020**, *21*, 1775. [CrossRef]
75. Marchand, M.; Horcajadas, J.A.; Esteban, F.J.; McElroy, S.L.; Fisher, S.J.; Giudice, L.C. Transcriptomic Signature of Trophoblast Differentiation in a Human Embryonic Stem Cell Model. *Biol. Reprod.* **2011**, *84*, 1258–1271. [CrossRef] [PubMed]
76. Ansell, J.D.; Snow, M.H.L. The Development of Trophoblast in Vitro from Blastocysts Containing Varying Amounts of Inner Cell Mass. *J. Embryol. Exp. Morphol.* **1975**, *33*, 177–185. [CrossRef] [PubMed]

MDPI

Article

The Long Terminal Repeats of ERV6 Are Activated in Pre-Implantation Embryos of Cynomolgus Monkey

Kui Duan [1,2,3,4,5,†], Chen-Yang Si [1,2,3,4,5,†], Shu-Mei Zhao [1,2,3,4,5], Zong-Yong Ai [1,2,3,4,5], Bao-Hua Niu [1,2,3,4,5], Yu Yin [1,2,3,4,5], Li-Feng Xiang [1,2,3,4,5], Hao Ding [1,2,3,4,5] and Yun Zheng [1,2,3,4,5,*]

1 Faculty of Environmental Science and Engineering, Kunming University of Science and Technology, Kunming 650500, China; duank@lpbr.cn (K.D.); sicy@lpbr.cn (C.-Y.S.); zhaosm@lpbr.cn (S.-M.Z.); aizy@lpbr.cn (Z.-Y.A.); niubh@lpbr.cn (B.-H.N.); yiny@lpbr.cn (Y.Y.); xlflotus@aliyun.com (L.-F.X.); haoding2020@163.com (H.D.)
2 Faculty of Life Science and Technology, Kunming University of Science and Technology, Kunming 650500, China
3 Yunnan Key Laboratory of Primate Biomedical Research, Kunming University of Science and Technology, Kunming 650500, China
4 State Key Laboratory of Primate Biomedical Research, Institute of Primate Translational Medicine, Kunming University of Science and Technology, Kunming 650500, China
5 Yunnan Provincial Academy of Science and Technology, Kunming 650500, China
* Correspondence: zhengyun5488@gmail.com
† These authors equally contributed to this work.

Abstract: Precise gene regulation is critical during embryo development. Long terminal repeat elements (LTRs) of endogenous retroviruses (ERVs) are dynamically expressed in blastocysts of mammalian embryos. However, the expression pattern of LTRs in monkey blastocyst is still unknown. By single-cell RNA-sequencing (seq) data of cynomolgus monkeys, we found that LTRs of several ERV families, including MacERV6, MacERV3, MacERV2, MacERVK1, and MacERVK2, were highly expressed in pre-implantation embryo cells including epiblast (EPI), trophectoderm (TrB), and primitive endoderm (PrE), but were depleted in post-implantation. We knocked down MacERV6-LTR1a in cynomolgus monkeys with a short hairpin RNA (shRNA) strategy to examine the potential function of MacERV6-LTR1a in the early development of monkey embryos. The silence of MacERV6-LTR1a mainly postpones the differentiation of TrB, EPI, and PrE cells in embryos at day 7 compared to control. Moreover, we confirmed MacERV6-LTR1a could recruit Estrogen Related Receptor Beta (ESRRB), which plays an important role in the maintenance of self-renewal and pluripotency of embryonic and trophoblast stem cells through different signaling pathways including FGF and Wnt signaling pathways. In summary, these results suggest that MacERV6-LTR1a is involved in gene regulation of the pre-implantation embryo of the cynomolgus monkeys.

Keywords: MacERV6-LTR1a; primitive endoderm; trophectoderm; epiblast; cynomolgus monkey; embryos; *ESRRB*

Citation: Duan, K.; Si, C.-Y.; Zhao, S.-M.; Ai, Z.-Y.; Niu, B.-H.; Yin, Y.; Xiang, L.-F.; Ding, H.; Zheng, Y. The Long Terminal Repeats of ERV6 Are Activated in Pre-implantation Embryos of Cynomolgus Monkeys. *Cells* **2021**, *10*, 2710. https://doi.org/10.3390/cells10102710

Academic Editor: Lon J. van Winkle

Received: 18 August 2021
Accepted: 1 October 2021
Published: 9 October 2021

1. Introduction

Mammalian embryogenesis begins with a zygote that develops into a morula, followed by the formation of a blastocyst. At this stage, blastomeres undergo first lineage segregation, giving rise to the trophectoderm (TrB) and inner cell mass (ICM) [1–4]. As the embryo implants, second lineage segregation occurs and epiblast cells (EPI) in the blastocyst develop into the embryo proper and the amnion [5–7], whereas cells of the TE and primitive endoderm (PrE) of the ICM generate the placenta [8] and yolk sac [4,9,10], respectively (Figure 1A). During pre-implantation, the pluripotent development of EPI is transited from a naive to a primed state, and hypoblast cells lose pluripotent genes to generate visceral/yolk-sac endoderm [11–13]. The precise transcriptional regulation of EPI, PrE, and TrB is crucial for embryo development of pre-implantation in human and mouse [1,14,15]. Furthermore, different mechanisms were noticed in different species. For example, fibroblast growth factor/extracellular signal-regulated kinase (FGF/ERK) signaling is essential for mouse hypoblast specification. However, previous research has shown that FGF/ERK signaling is not the key factor for human hypoblast specification [16,17].

Endogenous retroviruses (ERVs) are transposable elements that have high copy numbers in mammalian genomes [18–25]. Complete ERVs are flanked by two long terminal repeats (LTRs) that recruit transcription factors (TFs) to regulate the expression of ERVs and nearby genes [26–28]. In recent years, the role of LTRs in embryonic cell fate specification and determination has received increasing attention [29–32]. Since more than 80% of LTRs are located in open chromatin regions in early-stage embryos [33], a large number of LTR elements are dynamically transcribed in early human and mouse embryos [22,29,34–36]. For example, many transcripts in two-cell human embryos are initiated by LTR elements [29,37,38], which regulate long non-coding RNAs (lincRNAs) such as HERVH-LTR7 to maintain the naive state of human embryonic stem cells [22,39]. Furthermore, HERVH-LTR7 regions recruit pluripotency factors, including *OCT4*, *SOX2*, and *NANOG* to participate in cell type-specific gene regulatory networks [24,40,41]. Specifically, human embryonic stem cells (hESCs), similar to EPIs at post-implantation stages, may be converted to pre-implantation stage EPI-like cells when blastocyst-specific LTR7 elements are activated [34]. In contrast, the epigenetic mediator, the KRAB-ZFP/KAP1 system, has the ability to silence LTR elements activation in post-implantation embryos in human and mouse [33,34,37,42]. These studies exemplify that specific LTR elements act as crucial mediators in the early embryos of human and mouse. Although many LTR elements have been identified and dynamically expressed in early embryos [24,34], certain stage-specific LTR families have not yet been systematically investigated, especially in the implantation stage of embryos. Moreover, the molecular mechanisms underlying the temporal development of monkey pre-implantation embryos remain unclear [11].

Monkeys are important model species for studying human embryo development because of their high similarity to the human reproduction system [43–45]. Here, we examined the transcriptional dynamics of LTR elements in monkeys. Then, we knocked down MacERV6-LTR1a and observed differentiation was postponed in embryonic cells compared to control in the pre-implantation stage. We also validated that MacERV6-LTR1a could recruit ESRRB, which plays an important role in the maintenance of self-renewal and pluripotency of embryonic and trophoblast stem cells through different signaling pathways including FGF signaling pathway and Wnt signaling pathways [46–51]. These results suggest that MacERV6-LTR1a is involved in the regulations of the temporal development of early TrB, EPI, and PrE cells in monkey embryos.

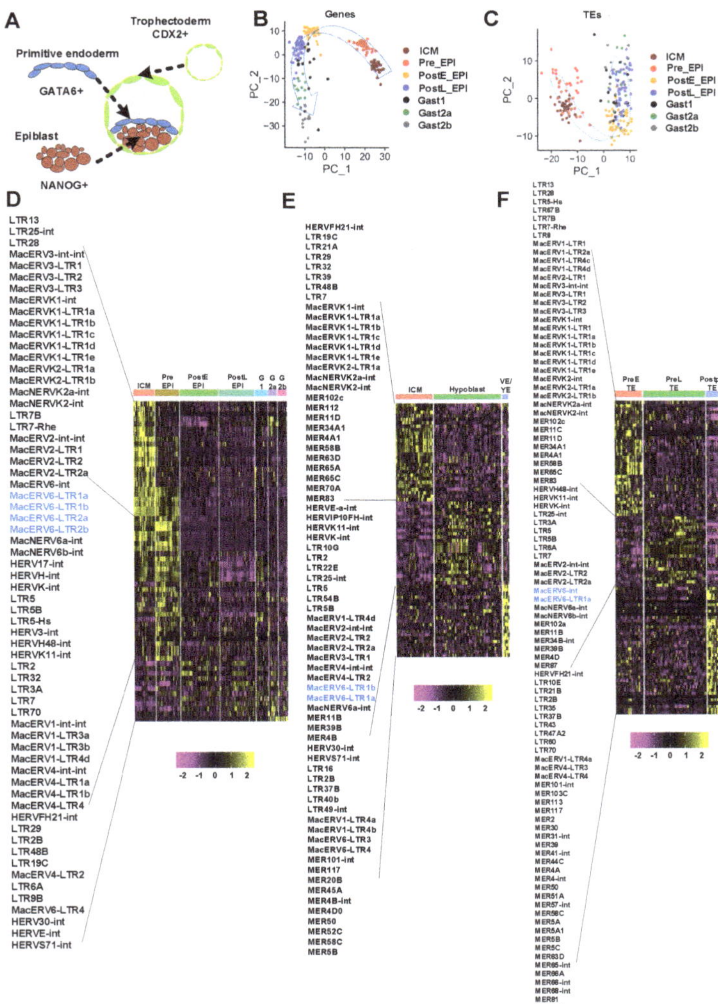

Figure 1. The dynamic expression of LTR elements during the development of *Macaca fascicularis* embryo from blastocysts to the E17 gastrulation stage. (**A**) Schematic view of three lineages of *Macaca fascicularis* blastocyst. (**B**) Principal component analysis with gene expression levels during *Macaca fascicularis* epiblast development with Seurat. (**C**) Principal component analysis with expression levels of transposable element (TEs) during *Macaca fascicularis* epiblast development. (**D**) Heatmap of LTR elements expression during epiblast development. The cell type include inner cell mass (ICM) (n = 30), pre-implantation epiblasts (PreEPI) (n = 34), post-implantation early EPI (PostEEPI) (n = 55), post-implantation late EPI (postLEPI) (n = 50), gastrulating cells (G1) (n = 18), (G2a) (n = 13), and (G2b) (n = 13) stages. (**E**) Heatmap of LTR elements exoression during primitive endoderm development. The cell type include inner cell mass (ICM) (n = 30), Hypoblast (also named PrE) (n = 54), visceral/yolk-sac endoderm (VE/YE) (n = 5). (**F**) Heatmap of LTR elements expression during trophectoderm (TrB) development. The cell type include pre-implantation early trophectoderm (PreETE) (n = 23), pre-implantation late TE (PreLTE) (n = 52), post-implantation parietal trophectoderm (PostpaTE) (n = 11).

2. Materials and Methods

2.1. Animal Ethics

Female cynomolgus monkeys (*Macaca fascicularis*), ranging from 5 to 8 years and weighing 4–6 kg were used in this study. All animals were housed at Yunnan Key Laboratory of Primate Biomedical Research and individually bred in an American standard cage in a light/dark cycle of 12 h/12 h. All animal procedures were approved in advance by the Institutional Animal Care and Use Committee, and protocols were performed in accordance with the Assessment and Accreditation of Laboratory Animal Care International for the ethical treatment of primates.

2.2. Calculation of the Expression Levels of LTR Elements and Genes

The 390 single cell RNA-sequencing (scRNA-seq) data relating to monkey pre- and post-implantation embryos were downloaded from previously published studies (GEO (Gene Expression Omnibus): GSE74767) [52]. RNA-seq reads were filtered by fast quality control (QC) procedures and adaptors cut using Cutadapt software (v1.11). After this, sequences were mapped to the *Macaca fascicularis* genome (macFas5) using HISAT2 software. Then, we quantified transposable element (TEs) and gene expression of single-cell sequencing data using the scTE software [53].

The single cells of monkey embryos were prepared according to smart-seq2 protocol, and sequenced with 2 × 150 base pair (bp) end method using the HiSeq X Ten platform (Illumina, San Diego CA, USA). Library construction and sequencing were performed by Annoroad Gene Technology (http://www.annoroad.com/, accessed on 28 September 2021). The sequencing reads were mapped against the *Macaca fascicularis* genome (macFas5) using HISAT2 (v2.2.1) software. Then, we quantified transposable element (TEs) and gene expression of single-cell sequencing data using the scTE software as before.

2.3. Oocyte Collection and In Vitro Fertilization

Superovulation, oocyte collection, and fertilization procedures were previously described [54]. In short, 10 healthy female cynomolgus monkeys aged 5–8 years with regular menstrual cycles were selected as oocyte donors for superovulation. This was performed by intramuscular injection with recombinant human follitropin-α (rhFSH) (GONAL-F, Merck Serono, Darmstadt, Germany) for 8 days. Then, recombinant human chorionic gonadotropin-α (rhCG) (OVIDREL, Merck Serono, Darmstadt, Germany) was injected on day 9. Oocytes were collected by laparoscopic follicular aspiration approximately 32–35 h after rhCG administration. Follicular contents were added to HEPES-buffered Tyrode's albumin lactate pyruvate (TALP) medium containing 0.3% bovine serum albumin (BSA) (Sigma-Aldrich, Saint-Louis, MO, USA) at 37 °C. Oocytes were stripped of cumulus cells by pipetting after a brief exposure (<1 min) to hyaluronidase (0.5 mg/mL) in TALP-HEPES to visually select nuclear-based mature metaphase II (MII; first polar body present) oocytes. The mature oocytes were immediately subjected to intracytoplasmic sperm injection and then cultured in Connaught Medical Research Laboratories (CMRL) -1066 media (Thermo Fisher Scientific, Waltham, MA, USA) containing 10% fetal bovine serum (FBS, Gibco, USA) at 37 °C in 5% CO_2. Fertilization was confirmed by the presence of a second polar body and two pronuclei. Zygotes were then cultured in chemically defined hamster embryo culture medium-9 containing 10% FBS at 37 °C in 5% CO_2 to allow embryo development. The culture medium was replaced every other day until the blastocyst stage.

2.4. Lentivirus Production and Purification

293T cells in DMEM + 10% FBS + 1% NEAA were split into 6 × 10 cm dishes (3 × 10^6 cells/dish) and incubated overnight at 37 °C in 5% CO_2. Evolutionary analyses were conducted in MEGAX [55]. The psicoR-EF1A-GFP-shRNA-scramble, psicoR-EF1A-GFP-shRNA-109, psicoR-EF1A-GFP-shRNA-149, and psicoR-EF1A-GFP-shRNA-346 plasmids were prepared according to Jacks laboratory protocols (http://web.mit.edu/jacks-lab/protocols/pSico.html, accessed on 28 September 2021) (Table S1). The psicoR-

EF1A-GFP contain a GFP gene driven by EF1A promoter. Therefore, GFP expression could be used as a signal of successful transfection of the plasmid. Per 1 × 10 cm dish, 11 μg shRNA plasmids, and two lentiviral packaging plasmids (3 μg of pMD2.g, and 8 μg of psPAX2) were used. Plasmids were mixed and added to 0.25 M $CaCl_2$ (500 μL), mixed gently in 500 μL BBS solution, incubated at room temperature for 15 min, and added to 293T cells (3 × 10^6 cells). Approximately 48 h later, the supernatants were collected, sieved through a 0.45-μm filter (Millipore, Sigma-Aldrich, Saint-Louis, MO, USA), and centrifuged (Beckman, Brea, CA, USA) at 100,000× *g* for 2 h at 4 °C. Viral pellets were then resuspended in 50 μL PBS (pH 7.2–7.4) and infection titers were determined using the gradient dilution method. Simply, a 10-fold serial dilution of the lentivirus solution was added to a 96-well plate (1 × 10^5 293T cells/well) and GFP fluorescence was observed 48 h after infection.

2.5. Isolation of Green Fluorescent Protein (GFP) Positive Single Cell and scRNA-Seq

Embryos were dissected into single cells in 0.25% trypsin (Thermo Fisher Scientific, Waltham, MA, USA), washed in Dulbecco's Phosphate-Buffered Saline (DPBS) (Thermo Fisher Scientific, Waltham, MA, USA), and aspirated using a Pasteur pipette under a dissecting microscope. GFP-positive embryonic cells were collected for gene expression analysis. The synthesis and amplification of full-length cDNAs were performed using the Smart-seq2 protocol [56]. Reverse transcription reactions and pre-amplification steps were performed using SuperScript II (Thermo Fisher Scientific, Waltham, MA, USA) and KAPA HiFi HotStart ReadyMix (KAPA Biosystems, Sigma-Aldrich, Saint-Louis, MO, USA), respectively. The cDNA quality was evaluated using the Bioanalyzer 2100 instrument. Library construction and sequencing were performed by Annoroad Gene Technology. GFP-positive cells were first screened using our previously described PCR method using primers for *GAPDH*. Then, MacERV6-LTR1a expression in GFP positive cells was determined by qRT-PCR and it was performed with iTaq™ universal SYBR Green supermix (Biorad, Hercules, CA, USA) (Table S2). In total, we used 166 single cells to perform pair-end sequencing on the Illumina HiSeq X-Ten platform.

2.6. Single-Cell Cluster Analysis

The output of single-cell gene counts was used to create objects using the Seurat package (v3.0) [57]. Principal component analysis (PCA) dimensional reduction techniques were used to visualize gene expression and perform cluster analyses based on the principal components (PCs) input. The VinPlot tool was used to show the probability distribution of marker gene expression across clusters. Differentially expressed genes (DEGs) between different cell types were detected by FindMarkers function of Seurat, test.use with DESeq2 (1.32.0) [58]. p-value < 0.05 and min.pct = 0.1 were used to denote significant differences in gene expression. Gene ontology (GO) and Kyoto Encyclopedia of Genes and Genomes (KEGG) pathway analyses of DEGs were conducted using the Metascape website (https://metascape.org/, accessed on 28 September 2021) [59]. GO terms with a p-value < 0.05 were defined as significantly enriched.

2.7. TF Binding Site Predictions and Dual Luciferase Expression Assays

We first identified motifs that potentially bound to TFs based on conserved nucleotide sequences of MacERV6-LTR1a using MEME algorithm [60]. We then predicted candidate TFs recruited by MacERV-LTR1a based on potential motifs using Tomtom algorithm [61]. We performed Dual-Glo luciferase assays (Promega, Madison WI, USA) to estimate the regulatory potential of MacERV6-LTR1a. We used 293T cells (ATCC CRL-3216) grown in DMEM (Thermo Fisher Scientific, Waltham, MA, USA) supplemented with 10% FBS (Thermo Fisher Scientific, Waltham, MA, USA) and 1% non-essential amino acids (NEAA) (Thermo Fisher Scientific, Waltham, MA, USA). The MacERV6-LTR1a_39 element sequence that contained multiple pluripotency TF-binding sites has cloned into PGL3-basic vector and driven firefly luciferase activity (named pERV6-LTR1a_39). Then, pERV6-LTR1a_39 and pRL-TK vector (expressing renilla luciferase) (Promega, Madison WI, USA) were co-transfected into 293T cells using Lipofectamine 2000 transfection reagent (Thermo Fisher Scientific, Waltham, MA, USA) (2 μL for 1 μg plasmid DNA). All transfections were performed in 24-well plates (Corning, Corning NY, USA). After 24–48 h, luciferase expression levels were processed according to the Dual-Glo reporter assay protocol (Promega, Madison WI, USA). We performed each luciferase experiment in triplicate and repeated each experiment three times. Luciferase fold change was estimated as the ratio of firefly and renilla luciferase for pERV6-LTR1a_39, and then normalized to the empty vector (pGL3 plasmid with no MacERV6-LTR1a_39 insert). We also individually cloned the coding sequences of the genes, *ESRRB*, *KLF4*, *POU5F1*, *SOX2*, *SMAD3*, and *HNF4A* into the PM2 expression vector (pEASY-Blunt M2 Expression Kit, TransGen Biotech, Beijing, CHINA) (Table S2). Furthermore, we synthesized the nucleotide sequences of MacERV6-LTR1a_39 which ESRRB binding site had mutated. Furthermore, then, cloned that sequences into the PGL3-basic vector (named pmERV6-LTR1a_39). The firefly and renilla luciferase fluorescence intensity was measured using a GENios (TECAN, Männedorf, Switzerland). Statistical parameters, including statistical analyses, statistical significance, and n values were reported in figure legends. For significance analysis, Student's *t*-test was used to indicate a * p-value < 0.05 or ** p-value < 0.01.

2.8. Immunofluorescence

Monkey embryos at day 7 were fixed in 4% paraformaldehyde and 0.1% polyvinyl pyrrolidone (PVP) in PBS (pH 7.2–7.4) for 20 min at room temperature. They were then washed three times in PBS and incubated with 0.3% Triton X-100 and 0.1% PVP for 30 min at room temperature. After washing three times in PBS, embryos were blocked for 2 h in blocking buffer (3% BSA + 10% FBS + 0.1% PVP) and separately incubated overnight at 4 °C with the following primary antibodies: chicken anti-GFP (Abcam, ab13970, 1:500), mouse anti-Oct4 (Santa Cruz, SC5279, C-10, 1:400), and goat anti-Gata6 (R&D Systems, AF1700, polyclonal, 1:400). Embryos were washed three times in PBS/0.05% Tween-20 and incubated with the following secondary antibodies for 2 h at room temperature Alexa Fluor 488 Donkey Anti-Chicken IgY† (IgG) (H + L) (Jackson Immunoresearch, 703-545-155, 1:500), Alexa Fluor 568 donkey anti-mouse IgG Thermo Fisher Scientific, Waltham, MA, USA A-10037, 1:600), and Alexa Fluor 647 donkey anti-goat IgG (Thermo Fisher Scientific, Waltham, MA, USA, A-21447, 1:600). DAPI (Roche Life Science, Basel, Switzerland, 10236276001, 1:1000) was used to stain nuclei.

3. Results

3.1. Temporal Expression of LTR Elements during Monkey Embryo Development

Implantation is an important event during embryo development. To understand the process, we analyzed the dynamic expression of transposable elements (TEs) in pre- and post-implantation embryos of cynomolgus monkeys based on publicly available scRNA-seq data for *Macaca fascicularis* (GEO accession: GSE74767) which covered embryonic EPI from blastocyst to E17 gastrulation-stages [52]. These scRNA-seq data including ICM (n = 30), pre-implantation EPI (pre-EPI, n = 34), early post-implantation EPI (postE-EPI, n = 55), late post-implantation EPI (postL-EPI, n = 50), and three gastrulation cell types (Gast1, n = 18, Gast2a, n = 13, and Gast2b, n = 13). In agreement with gene expression profiles [52], EPI cells can be separated clearly into pre-implantation EPI (ICM and pre-EPI) and post-implantation EPI (postE-EPI, postL-EPI, and gastrulation cells) by expression levels of transposable elements in PCA analysis (Figure 1B,C), showing that the expression of transposable element experienced a dramatic change during implantation. Then, we were particularly interested in LTR elements and found that many LTR elements were dynamically expressed in different stages of three lineages cell types. Firstly, there are 52 LTRs specifically activated in pre-implantation stages and others 13 LTR expressed in post-implantation (Figure 1D). We also observed that the LTR elements of several families, LTR5 (LTR5, LTR5B and LTR5-Hs), LTR7 (LTR7, LTR7B and LTR7-Rhe), and MacERV6-LTR (MacERV6-LTR1a, MacERV6-LTR1b, MacERV6-LTR2a and MacERV6-LTR2b) were specifically activated at ICM and Pre-EPI stages and found that MacERV6-LTRs were highly expressed at Pre-EPI stages, but all of LTRs above were rapidly depleted after implantation (Figure 1D). Of these LTR elements, the LTR7 (LTR7, LTR7B and LTR7-Rhe) and LTR5 (LTR5, LTR5B and LTR5-Hs) were previously reported as templates for initiating stage-specific transcripts in human embryonic cells [34].

To further evaluate whether MacERV6-LTR expression is stage-specific in other lineages, we analyzed its temporal expression in primitive endoderm development (ICM (n = 30), Hypoblast (n = 54), VE/YE (n = 5)). We noticed that MacERV6-LTR (MacERV6-LTR1a, MacERV6-LTR1b) were activated in the Hypoblast cell (Figure 1E). Meanwhile, we evaluated MacERV6-LTR expression in trophectoderm development (PreETE (n = 23), PreLTE (n = 52), PostpaTE (n = 11)) and found that MacERV6-LTR (MacERV6-LTR1a, MacERV6-int) were specifically expressed in the PreLTE stage of pre-implantation embryos (Figure 1F), suggesting that MacERV6-LTR may have specific functions during monkey embryo development. The function of the MacERV6-LTR in pre-implantation embryos is obscure. We selected MacERV6-LTR1a as a family representative for further exprimental validation, because MacERV6s have as many as eight different types of LTRs (Figure 2A).

3.2. *MacERV6-LTR1a Knockdown in Cynomolgus Monkey Embryos*

In order to further understand the conservation of MacERV6-LTRs in various species, we aligned the sequences of MacERV6-LTRs, and other two control (LTR5RM and LTR7) to the genomes of different species, and found that MacERV6-LTRs specifically exist in *Macaca fascicularis* and *Macaca mulatta* (Figure 2B). Previous research suggests that several transcripts in two-cell embryos are initiated by LTRs derived from endogenous retroviruses [29]. Then, we knocked down MacERV6-LTR1a expression to investigate the potential functions of MacERV6-LTR1a during monkey embryo development. Firstly, we designed three shRNAs (shRNA-109, shRNA-149 and shRNA-346) based on conserved regions in the MacERV6-LTR1a, and one scramble shRNA (shRNA-scramble) as the control (Figure 2B,C). Knockdown efficiency was verified using the MacERV6-LTR1a-driven luciferase assay. We observed the luciferase activity has been significantly knocked down by shRNA-149 and shRNA-346 plasmids in 293T cells (Figure 2D). Secondly, the shRNA-346 and shRNA-scramble were selected for virus packaging. To evaluate lentivirus toxicity, 50–100 pL virus (shRNA-346) suspensions expressing shRNA-346 or the shRNA-scrambled control were injected into the perivitelline space of mouse one-cell embryos. GFP fluorescence signals were observed in 90% of embryos at 3–5 days after injection (Figure 2E). No differences in development efficiency were observed in embryos, suggesting that these viral reagents exerted no toxicity on embryo development (Figure 2F).

To further assess the functions of MacERV6-LTR1a in *Macaca fascicularis* embryos, we injected shRNA-346 or the shRNA-scrambled lentivirus into fertilized one-cell cynomolgus monkey embryos (Figure 3A,B). A similar development efficiency of blastocysts were observed in embryos infected by these reagents (Figure 3C). We verified the knockdown efficiency in GFP-positive single cells of embryonic blastocysts on day 7 using qRT-PCR. MacERV6-LTR1a expression was significantly knocked down in shRNA-346-treated cells compared with the shRNA-scramble treated embryonic cells (Figure 3D). Single-cell transcriptome analyses also demonstrated the knockdown efficiency (Figure 3E). Meanwhile, we stained embryos to evaluate the phenotypes on day 7 with specific lineage markers, OCT4 for pre-EPI and GATA6 for hypoblasts to analyze cell fates. However, no significant differences were observed in OCT4 EPI and GATA6 hypoblast development at the whole embryonic level in both embryo groups (Figure 3F).

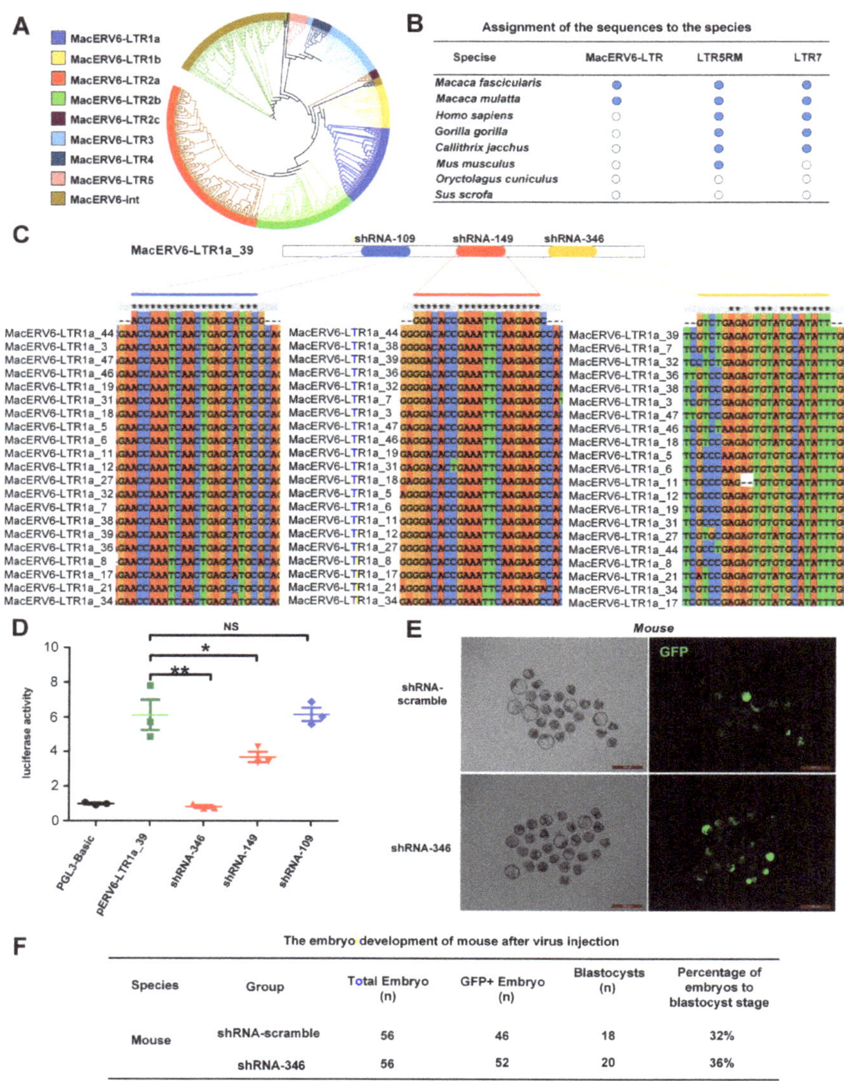

Figure 2. MacERV6-LTR1a knockdown in pre-implantation monkey embryos. (**A**) Phylogenetic tree based on the MacERV6 family sequences. (**B**) The conservation of MacERV6-LTRs, LTR5RM, and LTR7 in selected mammals. The sequences of MacERV6-LTRs, LTR5RM, and LTR7 were aligned to the genomes of different species. Blue circles represent there is the annotated sequence in the genome and white circles are reverse. (**C**) The schematic view of shRNAs, including shRNA-109 (psicoR-EF1A-GFP-shRNA-109), shRNA-149 (psicoR-EF1A-GFP-shRNA-149) and shRNA-346 (psicoR-EF1A-GFP-shRNA-346), targeting MacERV6-LTR1a_39 sites and the conserved sequences of MacERV6-LTR1a. (**D**) The knockdown efficiency of shRNAs targeting MacERV6-LTR1a. PGL3-basic, control plasmid with no promoter. pERV6-LTR1a_39, luciferase reporter was driven by the MacERV6-LTR1a_39 element. pERV6-LTR1a_39 plasmid was co-transfected with shRNA plasmids (shRNA-346, shRNA-149, shRNA-109) into 293T cells. Data were presented as the mean ± standard deviation (SD), ($n = 3$). * $p < 0.05$, ** $p < 0.01$. NS, not significant. Student's t-test. (**E**) GFP signals in mouse embryos that were treated with shRNA-scramble and shRNA-346 lentivirus, separately. Scale bar = 200 μm. (**F**) The development efficiency of mouse embryo after injection with shRNA-scrambled and shRNA-346 lentivirus separately.

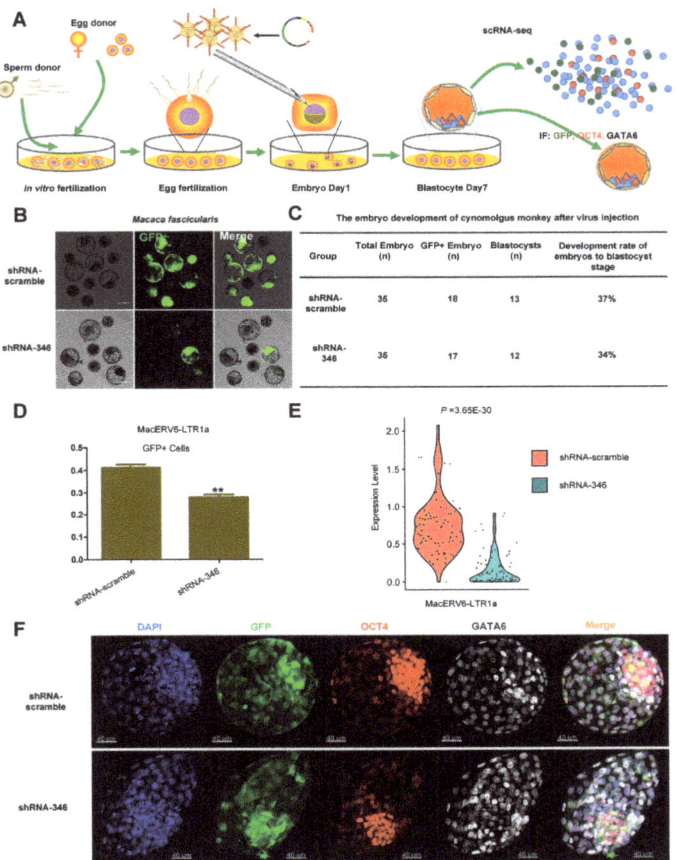

Group	Total Embryo (n)	GFP+ Embryo (n)	Blastocysts (n)	Development rate of embryos to blastocyst stage
shRNA-scramble	35	18	13	37%
shRNA-346	35	17	12	34%

Figure 3. The potential functions of MacERV6-LTR1a in preimplantation cynomolgus monkey embryos. (**A**) Schematic representation of MacERV6-LTR1a functional assays during embryonic development. The mature oocytes were immediately subjected to intracytoplasmic sperm injection and then cultured in Connaught Medical Research Laboratories (CMRL) -1066 media containing 10% fetal bovine serum at 37 °C in 5% CO_2. Fertilization was confirmed by the presence of a second polar body and two pronuclei. Zygotes were then cultured in chemically defined hamster embryo culture medium-9 containing 10% FBS at 37 °C in 5% CO_2 to allow embryo development. The lentivirus of shRNA-346 and shRNA-scramble were injected at Day 1 of monkey embryos. The culture medium was replaced every other day until the blastocyst stage at Day 7. The embryos were performed Immunofluorescence (IF) in part **F** and scRNA-seq. (**B**) The GFP signals in *Macaca fascicularis* blastocysts with injection of shRNA-scrambled or shRNA-346 lentivirus. Scale bar = 200 µm. (**C**) Blastocyst development efficiency in cynomolgus monkey embryos treated with shRNA-scrambled or shRNA-346 lentivirus. (**D**) The knockdown efficiency of MacERV6-LTR1a in embryonic GFP positive cells treated with shRNA-scramble (n = 30 cells) or shRNA-346 (n = 30 cells) lentivirus. Quantitative RT-PCR was used to evaluate MacERV6-LTR1a expression. ** $p < 0.01$, Student's t-test. (**E**) Vlnplot showing the expression of MacERV6-LTR1a in cynomolgus monkey embryonic cells at day 7. shRNA-346 (n = 86), shRNA-scramble (n = 80). p-value was evaluated with DEseq2. (**F**) Representative embryo staining images at day 7: GFP (green), OCT4 (red), GATA6 (white), and DAPI (blue). shRNA-scramble (n = 3 embryos), shRNA-346 (n = 3 embryos). Scale bar = 40 µm.

3.3. Transcriptomic Changes after MacERV6-LTR1a Knockdown

To further explore the functions of MacERV6-LTR1a in monkey embryos, GFP positive single cells were collected from blastocysts which were treated with shRNA-346- and the shRNA-scramble at day 7. We performed scRNA-seq using the Smart-seq2 protocol [56]. PCA analyses categorized these cells into three groups based on lineage marker expression (Figure 4A–C). *CDX2*, *GATA2* and *ESRRB* are highly expressed in trophoblast (TrB) populations, along with intermediate *GATA6* expression. Meanwhile, *NANOG*, *OCT4*, *PRDM14*, *DPPA3*, *FGF4*, *FGFR1* and *FGFR2* are heterogeneously expressed in EPI (Pre-implantation epiblast) population. *GATA4* and *PDGFRA* are enriched in primitive endoderm (also named hypoblast or PrE) cluster, along with intermediate *GATA6* expression (Figure 4D). The presence of TrB, PrE, and EPI populations indicated that monkey embryos at day 7 exhibit the segregations of the three lineage. We observed the shutdown state of MacERV6-LTR1a in three lineages cell types, and also found MacERV6-LTR1b, MacERV6-LTR2a and MacERV6-LTR4 were knocked down when treating with shRNA-346 (Figure 4D,E). Interestingly, we observed that the expression of *ESRRB* has significantly changed when treating with shRNA-346 (Figure 4F).

Next, we evaluated the DEGs (different expression genes) between shRNA-346 and shRNA-scramble embryonic cells. We noted that 247 and 423 genes were significantly upregulated and downregulated in shRNA-346-treated TrB cells when compared to shRNA-scrambled control (Figure 5). The upregulated genes were mainly associated with stem cell population maintenance, oxidative phosphorylation, DNA methylation or demethylation, positive regulation of apoptotic signaling pathway, and cell fate specification. The downregulated genes were mainly related to trophectodermal cell differentiation, the establishment of cell polarity, establishment or maintenance of apical/basal cell polarity, placenta development, and regulation of canonical Wnt signaling pathway (Table S3). These data suggest that MacERV6-LTR1a potentially regulates the development of TrB cells.

We also compared EPI cells treated with shRNA-346 and shRNA-scramble. We observed that 310 and 298 genes were significantly upregulated and downregulated in shRNA-346-treated EPI cells, respectively (Figure 5). Upregulated genes were mainly associated with oxidative phosphorylation, epithelial cell migration, regulation of embryonic development, insulin signaling pathway, regulation of cell adhesion, and reproductive structure development. Whereas downregulated genes were related to signaling pathways regulating pluripotency of stem cells, I-kappaB kinase/NF-kappaB signaling, epithelial cell proliferation, epithelial cell development, regulation of canonical Wnt signaling pathway, and epithelial cell differentiation (Table S3), these results suggest that MacERV6-LTR1a is required for regulation of the pluripotency of EPI cells.

We also compared PrE cells treated with shRNA-346 and shRNA-scramble. We noted that 160 and 174 genes were significantly upregulated and downregulated in shRNA-346-treated PrE cells, respectively (Figure 5). The upregulated genes were mainly enriched for insulin-like growth factor receptor signaling pathway, phosphatidylcholine biosynthetic process, DNA-templated transcription, initiation, positive regulation of gene expression, epigenetic, histone modification. Whereas downregulated genes were mainly associated with stem cell differentiation, regulation of hematopoietic stem cell differentiation, positive regulation of programmed cell death, positive regulation of collagen biosynthetic process, regulation of MAP kinase activity (Table S3). These data suggest that MacERV6-LTR1a mainly participates in the transcriptional activity and stem cell differentiation of early embryo development.

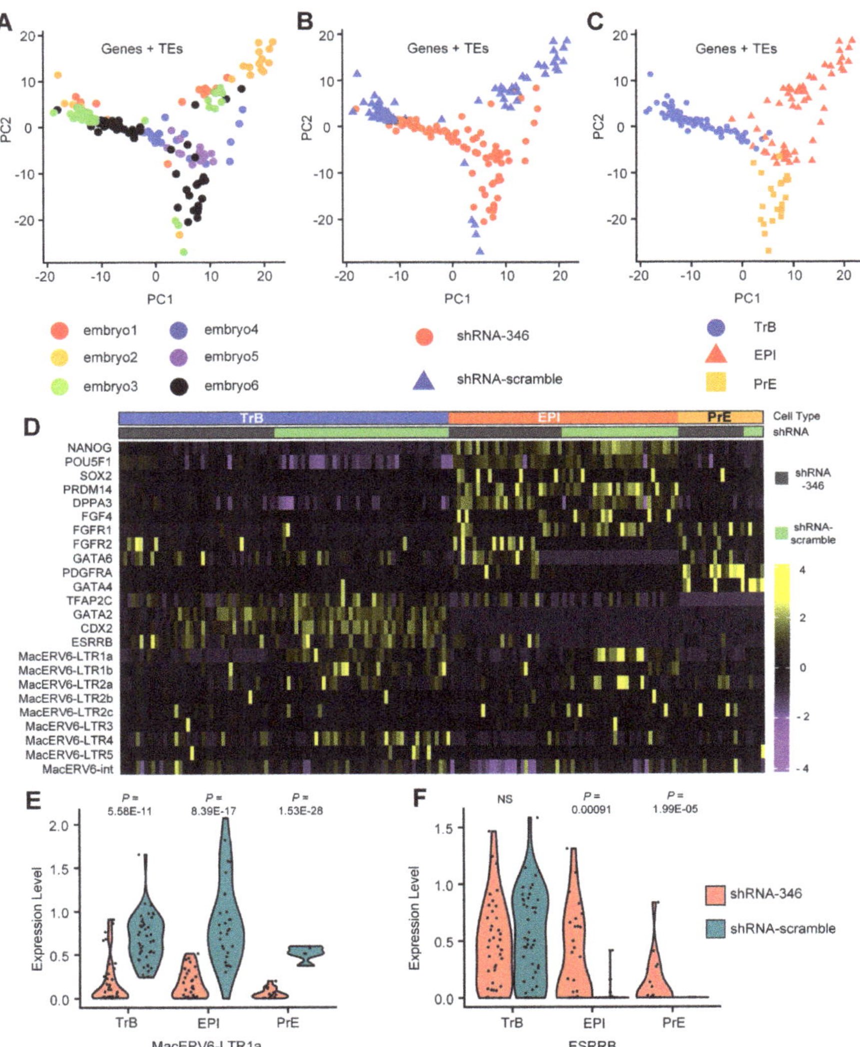

Figure 4. Transcriptional changes in shRNA-346- and shRNA-scramble-treated cynomolgus monkey embryos at day 7. (**A–C**) Principal component analysis (PCA) of single cells of embryos according to the expression of genes and TEs. (**A**) The distribution of single cells from six embryos, of which three were the control group (treated with shRNA-scramble, embryo1, embryo2, embryo3) and three were the experimental group (treated with shRNA-346, embryo4, embryo5, embryo6). (**B**) The distribution of single cells of embryos treated with shRNA-scramble (blue triangle) or shRNA-346 (red circle). shRNA-346 represents embryonic cells infected by shRNA-346 lentivirus, whereas shRNA-scramble represents embryonic cells infected by shRNA-scramble control lentivirus. (**C**) Single cells of embryos were annotated as three lineages: TrB, EPI, and PrE, according to the expression of the marker gene in part **D**. (**D**) Heatmap of selected marker genes. *NANOG*, *OCT4*, *PRDM14*, and *DPPA3* for epiblast (EPI), *GATA4* and *PDGFRA* for primitive endoderm/hypoblast (PrE), *CDX2* and *GATA2* for trophectoderm (TrB). (**E**) Vinplots of MacERV6-LTR1a expression in TrB, EPI, and PrE cells that were treated with shRNA-346 and shRNA-scramble lentivirus. (**F**) Vinplots of ESRRB expression in TrB, EPI, and PrE cells that were treated with shRNA-346 and shRNA-scramble lentivirus. NS, not significant. Furthermore, see Figure S2.

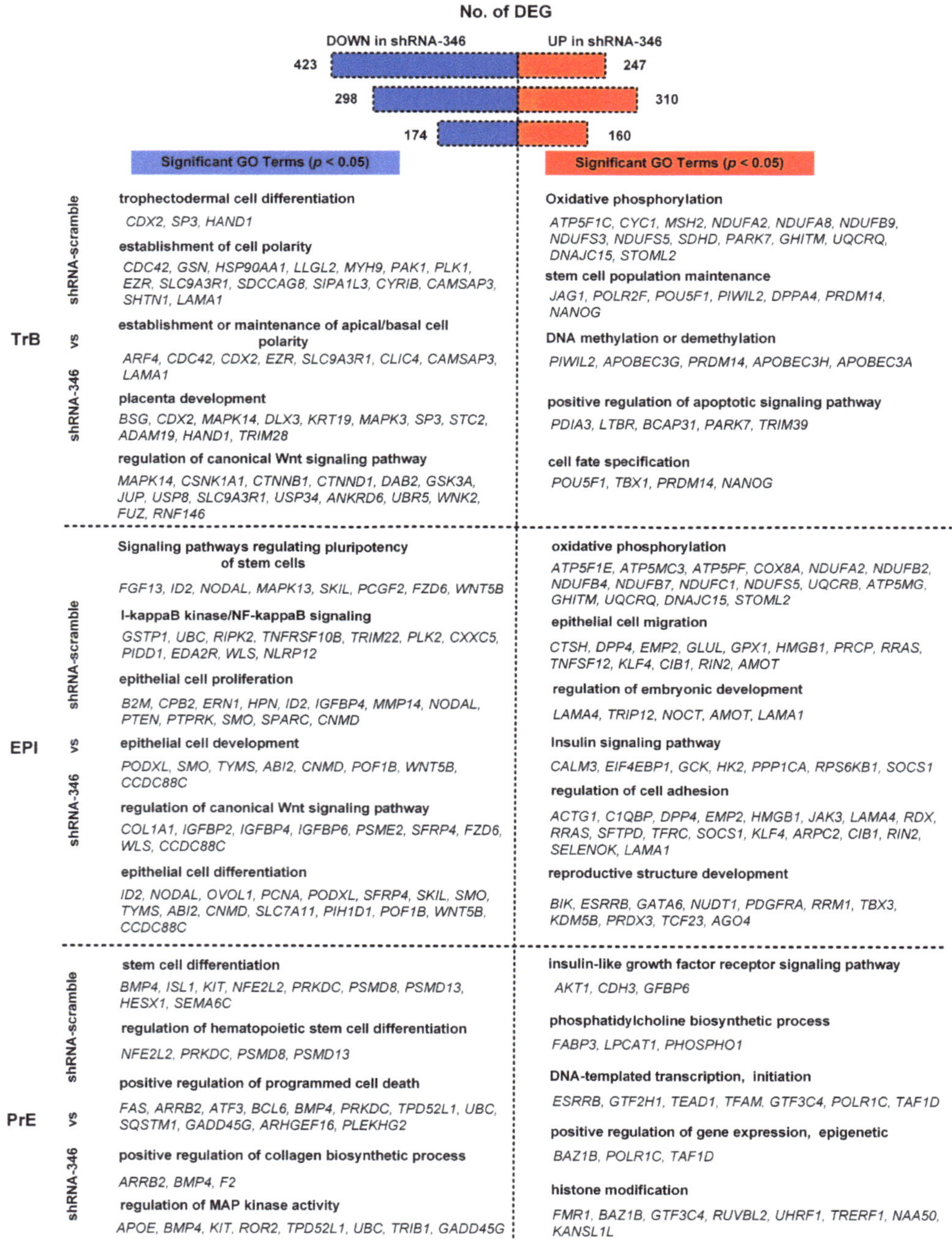

Figure 5. Differentially expressed genes (DEGs) and their corresponding gene ontology (GO) terms in TrB, EPI, and PrE from shRNA-346- or shRNA-scramble-treated embryos. DEGs were identified with function of FindMarkers in Seurat, test.use with DEseq2. p-value < 0.05, min.pct = 0.1. Significant upregulated and downregulated genes were used to identify enriched GO terms with the Metascape website (https://metascape.org/, accessed on 28 September 2021).

3.4. The Interaction between MacERV6-LTR1a and Pluripotency Factors In Vitro

Several LTR elements have been reported as tissue-specific enhancers or promoters in a variety of mammalian cell types, including early mouse embryos, placenta, pluripotent stem cells [29,35,62], and the initiation of transcriptional activity in specific cell lineages [37]. So, we hypothesized that MacERV6-LTR1a may act as a promoter or enhancer to recruit ICM, pre-EPI-specific TFs.

To address this hypothesis, we identified motifs and predicted candidate TFs recruited by MacERV6-LTR1a. By integrating these candidate TFs with TFs specific for ICM and Pre-EPI cells, we identified *ESRRB*, *KLF4*, *POU5F1*, *SOX2*, *SMAD3*, and *HNF4A*, which were highly expressed in ICM and PreEPI cells, and may be recruited by MacERV6-LTR1a (Figures 6A,B and S1A).

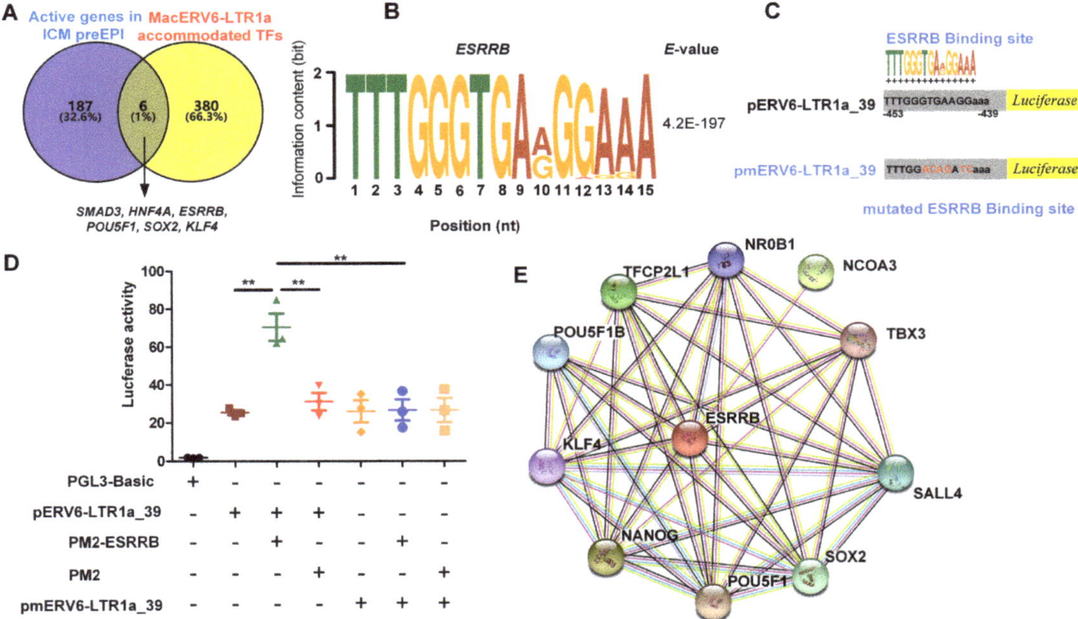

Figure 6. In vitro MacERV6-LTR1a interactions with pluripotency factors. (**A**) A Venn diagram showing overlapping transcription factors (TFs) specifically activated in ICM and pre-EPI cells and potentially recruited by MacERV6-LTR1a. Overall, 193 active genes in ICM and preEPI cells were reported in [52]. Additionally, 386 MacERV6-LTR1a accommodated TFs were predicted with MEME [60] and Tomtom [61]. (**B**) The motif of ESRRB in MacERV6-LTR1a and its E-value were predicted and calculated by MEME, respectively. The height of the nucleotide (Y-axis) is information content (in bit) and the X-axis is the position in the motif (nt). (**C**) ESRRB binding sites mutation in MacERV6-LTR1a.The MacERV6-LTR1a_39 element sequence that contained multiple pluripotency TF-binding sites have cloned into PGL3-basic vector and driven firefly luciferase activity (named pERV6-LTR1a_39). The nucleotide sequences of MacERV6-LTR1a_39, which the ESRRB binding site had mutated and cloned into the PGL3-basic vector (named pmERV6-LTR1a_39). (**D**) ESRRB interacting with MacERV6-LTR1a by luciferase assay in 293T cells. PM2-ESRRB, the PM2 vector containing coding sequence of ESRRB. pmERV6-LTR1a_39, pERV6-LTR1a_39, the plasmid contains luciferase reporter driven by ESRRB binding sites mutated MacERV6-LTR1a_39 and MacERV6-LTR1a_39, respectively. * p-value < 0.05 or ** p-value < 0.01. NS, not significant. Student's t-test. (**E**) The protein–protein interaction network of ESRRB protein and others identified with the STRING web tool (https://string-db.org/, accessed on 28 September 2021).

To further investigate the interaction between MacERV6-LTR1a and TFs, we performed luciferase assays. After transfecting the MacERV6-LTR1a-driven luciferase plasmid (pERV6-LTR1a_39) into 293T cells, we detected strong luciferase signals suggesting MacERV6-LTR1a had a strong promoter activity. Next, we co-transfected TF expressing vectors and pERV6-LTR1a_39 plasmid into 293T cells and observed that the *KLF4*, *POU5F1*, *SOX2*, *SMAD3*, and *HNF4A* co-transfections did not induce a significant increase in luciferase activity (Figure S1B–E). In contrast, the luciferase activity has significantly increased when added ESRRB, suggesting MacERV6-LTR1a recruited ESRRB. Since ESRRB induced the highest luciferase activity, we then mutated putative ESRRB binding sites in MacERV6-LTR1a to block ESRRB binding (pmERV6-LTR1a_39). We observed pmERV6-LTR1a_39 had lost their ability to enhance luciferase activity even in the presence of ESRRB (Figure 6C,D), suggesting ESRRB was recruited by MacERV6-LTR1a through the identified motif. Finally, we predicted functional associations between ESRRB and other proteins with STRING [63], and found that it is associated with self-renewal and pluripotency regulating genes, including KLF4, SOX2, NANOG, TFCP2L1, and POU5F1 (Figure 6E). These results may imply MacERV6-LTR1a recruited ESRRB to regulate gene expression, transcription activity, maintenance of self-renewal, and pluripotency of embryonic cells in an indirect way.

4. Discussion

The precise temporal development of embryonic cells is key to normal embryo development. At the beginning of the pre-implantation period, the embryo undergoes major transcriptional changes that establish a base for subsequent developmental phases. For example, FGF-4 ligands and FGF/ERK signaling are considered as the key cell-fate determination factors during this process in mice, but not in humans [64,65]. As mouse and primate embryogenesis are significantly diverse during successive stages, it is vital to investigate temporal development directly in monkey embryos.

In this study, we identified a large number of LTR elements, including HERVH, LTR7, LTR7B, LTR7-Rhe, LTR5 and LTR5-Hs, which are highly expressed in monkeys pre-implantation embryos but rapidly shutdown in post-implantation which was similar to mouse and human embryos [30,34]. A close relationship between LTR elements and embryonic development has been reported in human and mouse embryos [30,66,67]. However, it is unknown whether monkey-specific LTR elements are involved in the precise development of earlier pre-implantation embryos.

In humans, the embryo forms a blastocyst consisting of the ICM (inner cell mass), trophectoderm, and blastocoelic cavity at embryonic day 5-6 [68]. Similarly, the blastocysts of cynomolgus monkey can be observed at embryonic day 5-6 [52]. After blastocyst maturation in humans, the process of adhesion and implantation is occurring between embryonic days 7 and 8 [69], and cynomolgus monkey occurred in 7–9 days [45,52]. Lineage segregation occurred, but incompletely in human day 6, and cell fates of embryos became more fixed on day 7–9 [70]. While the cynomolgus monkeys began on day 7 [52], the lineage separation was not completed until day 10 [45]. Generally, the pregnancy period of cynomolgus monkeys is 24–30 weeks, while that of humans is 38–40 weeks. These results suggest that the early development of monkey embryos is different from that of human embryos.

In our research, we observed some monkey-specific LTRs (MacERV6-LTR, MacERVK1-LTR, MacERVK2-LTR, MacERV3-LTR, MacERV2-LTR, and MacERV1-LTR) were specifically expressed in blastocyst of pre-implantation stages and rapidly diminished in post-implantation stages. Even though blastocysts appeared morphologically identical between shRNA-scrambled and shRNA-346-treated embryos at day 7, changes in the differentiation of embryonic cells were noticed at the transcriptomic level. Importantly, we observed that knockdown of MacERV6-LTR1a can postpone the differentiation of TrB, EPI, and PrE cells.

MacERV6-LTR1a can recruit ESRRB, which plays an important role in the self-renewal, pluripotency of embryonic and trophoblast stem cells through different signaling pathways including FGF signaling pathway and Wnt signaling pathways [46–51]. In addition, ESRRB interacts with other important stem-cell TF, such as OCT4, NANOG, SOX2, KLF4, and TFCP2L1 (Figure 6E). We observed that *NANOG*, *POU5F1*, and *KLF4* were significantly highly expressed in TrB cells of shRNA-346 treated (Figure S2A–C, respectively), which may prevent TrB differentiation during embryonic development [71,72]. *TBX3* and *KLF4* are usually act as a transcriptional repressor in embryonic developmental processes, which highly expressed in EPI of shRNA-346 treated (Figure S2C,F, respectively), that means the gene relate differentiation may be inhibited in EPI cells of shRNA-346 treated embryos [73–75]. *TFCP2L1*, *SALL4*, and *NCOA3* play a key role in facilitates establishment, self-renewal, and transcriptional activities of embryonic stem cells, which have the tendency highly expressed in PrE of shRNA-346 treated (Figure S2D,E,G, respectively), suggesting that the changes of pluripotency in PrE may relate to MacERV6-LTR1a [75]. Thus, these might give us a new mechanism controlling early development in monkey embryos.

In this study, we observed that many LTRs of ERVs, including MacERV6-LTR1a, are activated in ICM and preEPI cells of monkey embryos, suggesting that other LTRs may involve in embryonic development via compensatory mechanisms. Therefore, it will be interesting to identify the function of other LTR elements during monkey embryo development in the future.

In summary, our results suggest that MacERV6-LTR1a is involved in precise gene regulation in the early developmental stages of monkey embryos. Because ERV6 is monkey specific, the activation of MacERV6-LTR in pre-implantation stages may contribute to different developmental patterns of monkey embryos.

Supplementary Materials: The following are available online at https://www.mdpi.com/article/10.3390/cells10102710/SupplementaryData. Figure S1: MacERV6-LTR1a interactions with pluripotency factors. The specific TFs, such as *SMAD3*, *SOX2*, *KLF4*, *POU5F1* or *HNF4A*, interact with pERV6-LTR1a_39 elements in 293T cells by luciferase assay. Figure S2: Vinplots and comparisons of *NANOG*, *POU5F1*, *KLF4*, *TFCP2L1*, *SALL4*, *TBX3*, and *NCOA3* expression in TrB, EPI, and PrE cells which treated with shRNA-346 and shRNA-scramble lentivirus. Table S1: Short hairpin RNA (shRNA) primers targeting MacERV6-LTR1a and shRNA scrambled control primers. Table S2: Primers used to clone coding sequences of pluripotent genes and primers for qRT-PCR. Table S3: Differentially expressed genes (DEGs) and their corresponding gene ontology (GO) terms and Kyoto Encyclopedia of Genes and Genomes (KEGG) pathways in TrB, EPI, and PrE from shRNA-346- or shRNA-scramble-treated embryos.

Author Contributions: Y.Z. and K.D. initiated the project. Y.Z. and K.D. designed the experiments, organized and supervised the entire project, and wrote the manuscript. K.D. performed lentivirus packaging, production and other molecular cell biology experiments. K.D., Y.Y. and H.D. performed single-cell isolation, data collection and data analysis. C.-Y.S., S.-M.Z., Z.-Y.A., B.-H.N. and L.-F.X. performed embryo culture and immunofluorescence staining. All authors have read and agreed with the published version of the manuscript.

Funding: This study was funded by a grant from the National Natural Science Foundation of China (NSFC) (No. 31960155) given to K.D.; a grant of NSFC (No. 31760314) and a grant of the Ministry of Science and Technology of China (No. 2018YFA0108502) given to Y.Z.

Institutional Review Board Statement: Female cynomolgus monkeys (Macaca fascicularis), ranging from 5 to 8 years and weighing 4–6 kg, were used in this study. All animals were housed at Yunnan Key Laboratory of Primate Biomedical Research and individually bred in an American standard cage in a light/dark cycle of 12 h/12 h. All animal procedures were approved in advance by the Institutional Animal Care and Use Committee, and protocols were performed in accordance with the Assessment and Accreditation of Laboratory Animal Care International for the ethical treatment of primates.

Informed Consent Statement: Not applicable.

Data Availability Statement: The scRNA-seq data generated in this study have been deposited at NCBI GEO database under accession number GSE182061.

Conflicts of Interest: The authors declare no conflict of interest.

Abbreviations

The following abbreviations are used in this manuscript:

TF	Transcription factor
TrB	Trophectoderm
TE	Transposable Element
ICM	Inner cell mass
EPI	Epiblast
preEEPI	Pre-implantation early epiblast
preETE	Pre-implantation early trophectoderm
Pre-EPI	Pre-implantation epiblast
postE-EPI	Post-implantation early epiblast
postL-EPI	Post-implantation late epiblast
PrE	Primitive endoderm
ERV	Endogenous retroviruse
LTR	Long terminal repeat
hESC	Human embryonic stem cell
KRAB-ZFP	KRAB-containing zinc finger protein
rhFSH	Recombinant human follitropin alpha
rhCG	Recombinant human chorionic gonadotropin alpha
TALP	Tyrode's albumin lactate pyruvate
BSA	Bovine serum albumin
MOR	Morula
DEGs	Differentially expressed genes
Gast1	Gastrulating cells 1
Gast2a	Gastrulating cells 2a
Gast2b	Gastrulating cells 2b

References

1. Chazaud, C.; Yamanaka, Y. Lineage specification in the mouse preimplantation embryo. *Development* **2016**, *143*, 1063–1074. [CrossRef] [PubMed]
2. Leung, C.Y.; Zhu, M.; Zernicka-Goetz, M. Polarity in Cell-Fate Acquisition in the Early Mouse Embryo. *Curr. Top. Dev. Biol.* **2016**, *120*, 203–234. [CrossRef]
3. Rossant, J. Genetic Control of Early Cell Lineages in the Mammalian Embryo. *Annu. Rev. Genet.* **2018**, *52*, 185–201. [CrossRef] [PubMed]
4. Wamaitha, S.E.; Niakan, K.K. Human Pre-gastrulation Development. *Curr. Top. Dev. Biol.* **2018**, *128*, 295–338. [CrossRef]
5. Chazaud, C.; Yamanaka, Y.; Pawson, T.; Rossant, J. Early lineage segregation between epiblast and primitive endoderm in mouse blastocysts through the Grb2-MAPK pathway. *Dev. Cell.* **2006**, *10*, 615–624. [CrossRef] [PubMed]
6. Marikawa, Y.; Alarcon, V.B. Establishment of trophectoderm and inner cell mass lineages in the mouse embryo. *Mol. Reprod. Dev.* **2009**, *76*, 1019–1032. [CrossRef] [PubMed]
7. Zheng, Y.; Xue, X.; Shao, Y.; Wang, S.; Esfahani, S.N.; Li, Z.; Muncie, J.M.; Lakins, J.N.; Weaver, V.M.; Gumucio, D.L.; et al. Controlled modelling of human epiblast and amnion development using stem cells. *Nature* **2019**, *573*, 421–425. [CrossRef]
8. Chang, C.W.; Wakeland, A.K.; Parast, M.M. Trophoblast lineage specification, differentiation and their regulation by oxygen tension. *J Endocrinol.* **2018**, *236*, R43–R56. [CrossRef]
9. Luckett, W.P. Origin and differentiation of the yolk sac and extraembryonic mesoderm in presomite human and rhesus monkey embryos. *Am. J. Anat.* **1978**, *152*, 59–97. [CrossRef]
10. Mole, M.A.; Weberling, A.; Zernicka-Goetz, M. Comparative analysis of human and mouse development: From zygote to pre-gastrulation. *Curr. Top. Dev. Biol.* **2020**, *136*, 113–138. [CrossRef]
11. Boroviak, T.; Stirparo, G.G.; Dietmann, S.; Hernando-Herraez, I.; Mohammed, H.; Reik, W.; Smith, A.; Sasaki, E.; Nichols, J.; Bertone, P. Single cell transcriptome analysis of human, marmoset and mouse embryos reveals common and divergent features of preimplantation development. *Development* **2018**, *145*, dev167833. [CrossRef] [PubMed]
12. Nichols, J.; Smith, A. Naive and primed pluripotent states. *Cell Stem Cell* **2009**, *4*, 487–492. [CrossRef] [PubMed]
13. Peng, G.; Suo, S.; Cui, G.; Yu, F.; Wang, R.; Chen, J.; Chen, S.; Liu, Z.; Chen, G.; Qian, Y.; et al. Molecular architecture of lineage allocation and tissue organization in early mouse embryo. *Nature* **2019**, *572*, 528–532. [CrossRef]
14. Frankenberg, S.; Gerbe, F.; Bessonnard, S.; Belville, C.; Pouchin, P.; Bardot, O.; Chazaud, C. Primitive endoderm differentiates via a three-step mechanism involving Nanog and RTK signaling. *Dev. Cell.* **2011**, *21*, 1005–1013. [CrossRef] [PubMed]
15. Guo, G.; Huss, M.; Tong, G.Q.; Wang, C.; Li Sun, L.; Clarke, N.D.; Robson, P. Resolution of cell fate decisions revealed by single-cell gene expression analysis from zygote to blastocyst. *Dev. Cell.* **2010**, *18*, 675–685. [CrossRef] [PubMed]

16. Kuijk, E.W.; van Tol, L.T.; Van de Velde, H.; Wubbolts, R.; Welling, M.; Geijsen, N.; Roelen, B.A. The roles of FGF and MAP kinase signaling in the segregation of the epiblast and hypoblast cell lineages in bovine and human embryos. *Development* **2012**, *139*, 871–882. [CrossRef]
17. Roode, M.; Blair, K.; Snell, P.; Elder, K.; Marchant, S.; Smith, A.; Nichols, J. Human hypoblast formation is not dependent on FGF signalling. *Dev. Biol.* **2012**, *361*, 358–363. [CrossRef]
18. Cordaux, R.; Batzer, M.A. The impact of retrotransposons on human genome evolution. *Nat. Rev. Genet.* **2009**, *10*, 691–703. [CrossRef]
19. Fort, A.; Hashimoto, K.; Yamada, D.; Salimullah, M.; Keya, C.A.; Saxena, A.; Bonetti, A.; Voineagu, I.; Bertin, N.; Kratz, A.; et al. Deep transcriptome profiling of mammalian stem cells supports a regulatory role for retrotransposons in pluripotency maintenance. *Nat. Genet.* **2014**, *46*, 558–566. [CrossRef] [PubMed]
20. Kapusta, A.; Kronenberg, Z.; Lynch, V.J.; Zhuo, X.; Ramsay, L.; Bourque, G.; Yandell, M.; Feschotte, C. Transposable elements are major contributors to the origin, diversification, and regulation of vertebrate long noncoding RNAs. *PLoS Genet.* **2013**, *9*, e1003470. [CrossRef] [PubMed]
21. Kelley, D.; Rinn, J. Transposable elements reveal a stem cell-specific class of long noncoding RNAs. *Genome Biol.* **2012**, *13*, R107. [CrossRef] [PubMed]
22. Lu, X.; Sachs, F.; Ramsay, L.; Jacques, P.E.; Goke, J.; Bourque, G.; Ng, H.H. The retrovirus HERVH is a long noncoding RNA required for human embryonic stem cell identity. *Nat. Struct. Mol. Biol.* **2014**, *21*, 423–425. [CrossRef]
23. Macia, A.; Munoz-Lopez, M.; Cortes, J.L.; Hastings, R.K.; Morell, S.; Lucena-Aguilar, G.; Marchal, J.A.; Badge, R.M.; Garcia-Perez, J.L. Epigenetic control of retrotransposon expression in human embryonic stem cells. *Mol. Cell. Biol.* **2011**, *31*, 300–316. [CrossRef]
24. Santoni, F.A.; Guerra, J.; Luban, J. HERV-H RNA is abundant in human embryonic stem cells and a precise marker for pluripotency. *Retrovirology* **2012**, *9*, 111. [CrossRef]
25. Weiss, R.A. The discovery of endogenous retroviruses. *Retrovirology* **2006**, *3*, 67. [CrossRef]
26. Cohen, C.J.; Lock, W.M.; Mager, D.L. Endogenous retroviral LTRs as promoters for human genes: a critical assessment. *Gene* **2009**, *448*, 105–114. [CrossRef] [PubMed]
27. Jern, P.; Coffin, J.M. Effects of retroviruses on host genome function. *Annu. Rev. Genet.* **2008**, *42*, 709–732. [CrossRef]
28. Maksakova, I.A.; Mager, D.L.; Reiss, D. Keeping active endogenous retroviral-like elements in check: the epigenetic perspective. *Cell. Mol. Life Sci.* **2008**, *65*, 3329–3347. [CrossRef]
29. Macfarlan, T.S.; Gifford, W.D.; Driscoll, S.; Lettieri, K.; Rowe, H.M.; Bonanomi, D.; Firth, A.; Singer, O.; Trono, D.; Pfaff, S.L. Embryonic stem cell potency fluctuates with endogenous retrovirus activity. *Nature* **2012**, *487*, 57–63. [CrossRef] [PubMed]
30. Thompson, P.J.; Macfarlan, T.S.; Lorincz, M.C. Long Terminal Repeats: From Parasitic Elements to Building Blocks of the Transcriptional Regulatory Repertoire. *Mol. Cell.* **2016**, *62*, 766–776. [CrossRef]
31. Wang, J.; Wang, L.; Feng, G.; Wang, Y.; Li, Y.; Li, X.; Liu, C.; Jiao, G.; Huang, C.; Shi, J.; et al. Asymmetric Expression of LincGET Biases Cell Fate in Two-Cell Mouse Embryos. *Cell* **2018**, *175*, 1887–1901.e18. [CrossRef]
32. Liu, D.; Liu, L.; Duan, K.; Guo, J.; Li, S.; Zhao, Z.; Zhang, X.; Zhou, N.; Zheng, Y. Transcriptional dynamics of transposable elements when converting fibroblast cells of Macaca mulatta to neuroepithelial stem cells. *BMC Genom.* **2021**, *22*, 405. [CrossRef]
33. Jacques, P.E.; Jeyakani, J.; Bourque, G. The majority of primate-specific regulatory sequences are derived from transposable elements. *PLoS Genet.* **2013**, *9*, e1003504. [CrossRef] [PubMed]
34. Goke, J.; Lu, X.; Chan, Y.S.; Ng, H.H.; Ly, L.H.; Sachs, F.; Szczerbinska, I. Dynamic transcription of distinct classes of endogenous retroviral elements marks specific populations of early human embryonic cells. *Cell Stem Cell* **2015**, *16*, 135–141. [CrossRef] [PubMed]
35. Peaston, A.E.; Evsikov, A.V.; Graber, J.H.; de Vries, W.N.; Holbrook, A.E.; Solter, D.; Knowles, B.B. Retrotransposons regulate host genes in mouse oocytes and preimplantation embryos. *Dev. Cell.* **2004**, *7*, 597–606. [CrossRef]
36. Rowe, H.M.; Trono, D. Dynamic control of endogenous retroviruses during development. *Virology* **2011**, *411*, 273–287. [CrossRef]
37. Grow, E.J.; Flynn, R.A.; Chavez, S.L.; Bayless, N.L.; Wossidlo, M.; Wesche, D.J.; Martin, L.; Ware, C.B.; Blish, C.A.; Chang, H.Y.; et al. Intrinsic retroviral reactivation in human preimplantation embryos and pluripotent cells. *Nature* **2015**, *522*, 221–225. [CrossRef] [PubMed]
38. Wang, J.; Xie, G.; Singh, M.; Ghanbarian, A.T.; Rasko, T.; Szvetnik, A.; Cai, H.; Besser, D.; Prigione, A.; Fuchs, N.V.; et al. Primate-specific endogenous retrovirus-driven transcription defines naive-like stem cells. *Nature* **2014**, *516*, 405–409. [CrossRef] [PubMed]
39. Ohnuki, M.; Tanabe, K.; Sutou, K.; Teramoto, I.; Sawamura, Y.; Narita, M.; Nakamura, M.; Tokunaga, Y.; Nakamura, M.; Watanabe, A.; et al. Dynamic regulation of human endogenous retroviruses mediates factor-induced reprogramming and differentiation potential. *Proc. Natl. Acad. Sci. USA* **2014**, *111*, 12426–12431. [CrossRef]
40. Kunarso, G.; Chia, N.Y.; Jeyakani, J.; Hwang, C.; Lu, X.; Chan, Y.S.; Ng, H.H.; Bourque, G. Transposable elements have rewired the core regulatory network of human embryonic stem cells. *Nat. Genet.* **2010**, *42*, 631–634. [CrossRef] [PubMed]
41. Sundaram, V.; Cheng, Y.; Ma, Z.; Li, D.; Xing, X.; Edge, P.; Snyder, M.P.; Wang, T. Widespread contribution of transposable elements to the innovation of gene regulatory networks. *Genome Res.* **2014**, *24*, 1963–1976. [CrossRef]

42. Ecco, G.; Cassano, M.; Kauzlaric, A.; Duc, J.; Coluccio, A.; Offner, S.; Imbeault, M.; Rowe, H.M.; Turelli, P.; Trono, D. Transposable Elements and Their KRAB-ZFP Controllers Regulate Gene Expression in Adult Tissues. *Dev. Cell* **2016**, *36*, 611–623. [CrossRef] [PubMed]
43. Chan, A.W. Progress and prospects for genetic modification of nonhuman primate models in biomedical research. *ILAR J.* **2013**, *54*, 211–223. [CrossRef] [PubMed]
44. Ma, H.; Zhai, J.; Wan, H.; Jiang, X.; Wang, X.; Wang, L.; Xiang, Y.; He, X.; Zhao, Z.A.; Zhao, B.; et al. In vitro culture of cynomolgus monkey embryos beyond early gastrulation. *Science* **2019**, *366*. [CrossRef] [PubMed]
45. Niu, Y.; Sun, N.; Li, C.; Lei, Y.; Huang, Z.; Wu, J.; Si, C.; Dai, X.; Liu, C.; Wei, J.; et al. Dissecting primate early post-implantation development using long-term in vitro embryo culture. *Science* **2019**, *366*. [CrossRef] [PubMed]
46. Festuccia, N.; Dubois, A.; Vandormael-Pournin, S.; Gallego Tejeda, E.; Mouren, A.; Bessonnard, S.; Mueller, F.; Proux, C.; Cohen-Tannoudji, M.; Navarro, P. Mitotic binding of Esrrb marks key regulatory regions of the pluripotency network. *Nat. Cell Biol.* **2016**, *18*, 1139–1148. [CrossRef] [PubMed]
47. Zhang, X.; Zhang, J.; Wang, T.; Esteban, M.A.; Pei, D. Esrrb activates Oct4 transcription and sustains self-renewal and pluripotency in embryonic stem cells. *J. Biol. Chem.* **2008**, *283*, 35825–35833. [CrossRef]
48. Festuccia, N.; Osorno, R.; Halbritter, F.; Karwacki-Neisius, V.; Navarro, P.; Colby, D.; Wong, F.; Yates, A.; Tomlinson, S.R.; Chambers, I. Esrrb is a direct Nanog target gene that can substitute for Nanog function in pluripotent cells. *Cell Stem Cell* **2012**, *11*, 477–490. [CrossRef]
49. Festuccia, N.; Osorno, R.; Wilson, V.; Chambers, I. The role of pluripotency gene regulatory network components in mediating transitions between pluripotent cell states. *Curr. Opin. Genet. Dev.* **2013**, *23*, 504–511. [CrossRef] [PubMed]
50. Festuccia, N.; Owens, N.; Navarro, P. Esrrb, an estrogen-related receptor involved in early development, pluripotency, and reprogramming. *FEBS Lett.* **2018**, *592*, 852–877. [CrossRef]
51. Martello, G.; Sugimoto, T.; Diamanti, E.; Joshi, A.; Hannah, R.; Ohtsuka, S.; Gottgens, B.; Niwa, H.; Smith, A. Esrrb is a pivotal target of the Gsk3/Tcf3 axis regulating embryonic stem cell self-renewal. *Cell Stem Cell* **2012**, *11*, 491–504. [CrossRef] [PubMed]
52. Nakamura, T.; Okamoto, I.; Sasaki, K.; Yabuta, Y.; Iwatani, C.; Tsuchiya, H.; Seita, Y.; Nakamura, S.; Yamamoto, T.; Saitou, M. A developmental coordinate of pluripotency among mice, monkeys and humans. *Nature* **2016**, *537*, 57–62. [CrossRef] [PubMed]
53. He, J.; Babarinde, I.A.; Sun, L.; Xu, S.; Chen, R.; Shi, J.; Wei, Y.; Li, Y.; Ma, G.; Zhuang, Q.; et al. Identifying transposable element expression dynamics and heterogeneity during development at the single-cell level with a processing pipeline scTE. *Nat. Commun.* **2021**, *12*, 1456. [CrossRef]
54. Niu, Y.; Yu, Y.; Bernat, A.; Yang, S.; He, X.; Guo, X.; Chen, D.; Chen, Y.; Ji, S.; Si, W.; et al. Transgenic rhesus monkeys produced by gene transfer into early-cleavage-stage embryos using a simian immunodeficiency virus-based vector. *Proc. Natl. Acad. Sci. USA* **2010**, *107*, 17663–17667. [CrossRef] [PubMed]
55. Kumar, S.; Stecher, G.; Li, M.; Knyaz, C.; Tamura, K. MEGA X: Molecular Evolutionary Genetics Analysis across Computing Platforms. *Mol. Biol. Evol.* **2018**, *35*, 1547–1549. [CrossRef]
56. Picelli, S.; Faridani, O.R.; Bjorklund, A.K.; Winberg, G.; Sagasser, S.; Sandberg, R. Full-length RNA-seq from single cells using Smart-seq2. *Nat. Protoc.* **2014**, *9*, 171–181. [CrossRef] [PubMed]
57. Stuart, T.; Butler, A.; Hoffman, P.; Hafemeister, C.; Papalexi, E.; Mauck, W.M., III; Hao, Y.; Stoeckius, M.; Smibert, P.; Satija, R. Comprehensive Integration of Single-Cell Data. *Cell* **2019**, *177*, 1888–1902.e21. [CrossRef]
58. Love, M.I.; Huber, W.; Anders, S. Moderated estimation of fold change and dispersion for RNA-seq data with DESeq2. *Genome Biol.* **2014**, *15*, 550. [CrossRef]
59. Zhou, Y.; Zhou, B.; Pache, L.; Chang, M.; Khodabakhshi, A.H.; Tanaseichuk, O.; Benner, C.; Chanda, S.K. Metascape provides a biologist-oriented resource for the analysis of systems-level datasets. *Nat. Commun.* **2019**, *10*, 1523. [CrossRef]
60. Bailey, T.L.; Elkan, C. Fitting a mixture model by expectation maximization to discover motifs in biopolymers. *Proc. Int. Conf. Intell. Syst. Mol. Biol.* **1994**, *2*, 28–36.
61. Gupta, S.; Stamatoyannopoulos, J.A.; Bailey, T.L.; Noble, W.S. Quantifying similarity between motifs. *Genome Biol.* **2007**, *8*, R24. [CrossRef]
62. Wolf, G.; Greenberg, D.; Macfarlan, T.S. Spotting the enemy within: Targeted silencing of foreign DNA in mammalian genomes by the Kruppel-associated box zinc finger protein family. *Mob. DNA* **2015**, *6*, 17. [CrossRef]
63. Szklarczyk, D.; Gable, A.L.; Lyon, D.; Junge, A.; Wyder, S.; Huerta-Cepas, J.; Simonovic, M.; Doncheva, N.T.; Morris, J.H.; Bork, P.; et al. STRING v11: protein-protein association networks with increased coverage, supporting functional discovery in genome-wide experimental datasets. *Nucleic Acids Res.* **2019**, *47*, D607–D613. [CrossRef] [PubMed]
64. Morris, S.A.; Teo, R.T.; Li, H.; Robson, P.; Glover, D.M.; Zernicka-Goetz, M. Origin and formation of the first two distinct cell types of the inner cell mass in the mouse embryo. *Proc. Natl. Acad. Sci. USA* **2010**, *107*, 6364–6369. [CrossRef]
65. Yamanaka, Y.; Lanner, F.; Rossant, J. FGF signal-dependent segregation of primitive endoderm and epiblast in the mouse blastocyst. *Development* **2010**, *137*, 715–724. [CrossRef]
66. Feschotte, C.; Gilbert, C. Endogenous viruses: insights into viral evolution and impact on host biology. *Nat. Rev. Genet.* **2012**, *13*, 283–296. [CrossRef]
67. Rossant, J.; Tam, P.P.L. New Insights into Early Human Development: Lessons for Stem Cell Derivation and Differentiation. *Cell Stem Cell* **2017**, *20*, 18–28. [CrossRef]

68. Richter, K.S.; Harris, D.C.; Daneshmand, S.T.; Shapiro, B.S. Quantitative grading of a human blastocyst: optimal inner cell mass size and shape. *Fertil. Steril.* **2001**, *76*, 1157–1167. [CrossRef]
69. Lindenberg, S.; Hyttel, P.; Sjogren, A.; Greve, T. A comparative study of attachment of human, bovine and mouse blastocysts to uterine epithelial monolayer. *Hum. Reprod.* **1989**, *4*, 446–456. [CrossRef] [PubMed]
70. Xiang, L.; Yin, Y.; Zheng, Y.; Ma, Y.; Li, Y.; Zhao, Z.; Guo, J.; Ai, Z.; Niu, Y.; Duan, K.; et al. A developmental landscape of 3D-cultured human pre-gastrulation embryos. *Nature* **2020**, *577*, 537–542. [CrossRef] [PubMed]
71. Chambers, I.; Tomlinson, S.R. The transcriptional foundation of pluripotency. *Development* **2009**, *136*, 2311–2322. [CrossRef] [PubMed]
72. Shi, G.; Jin, Y. Role of Oct4 in maintaining and regaining stem cell pluripotency. *Stem Cell Res. Ther.* **2010**, *1*, 39. [CrossRef]
73. Chia, N.Y.; Chan, Y.S.; Feng, B.; Lu, X.; Orlov, Y.L.; Moreau, D.; Kumar, P.; Yang, L.; Jiang, J.; Lau, M.S.; et al. A genome-wide RNAi screen reveals determinants of human embryonic stem cell identity. *Nature* **2010**, *468*, 316–320. [CrossRef] [PubMed]
74. Takashima, Y.; Guo, G.; Loos, R.; Nichols, J.; Ficz, G.; Krueger, F.; Oxley, D.; Santos, F.; Clarke, J.; Mansfield, W.; et al. Resetting transcription factor control circuitry toward ground-state pluripotency in human. *Cell* **2014**, *158*, 1254–1269. [CrossRef]
75. Takahashi, K.; Yamanaka, S. A decade of transcription factor-mediated reprogramming to pluripotency. *Nat. Rev. Mol. Cell. Biol.* **2016**, *17*, 183–193. [CrossRef] [PubMed]

 cells

MDPI

Article

Hypoblast Formation in Bovine Embryos Does Not Depend on NANOG

Claudia Springer [1,2], Valeri Zakhartchenko [1,2], Eckhard Wolf [1,2,3] and Kilian Simmet [1,2,*]

1 Institute of Molecular Animal Breeding and Biotechnology, Gene Center and Department of Veterinary Sciences, Ludwig-Maximilians-Universität München, 85764 Oberschleissheim, Germany; c.springer@gen.vetmed.uni-muenchen.de (C.S.); v.zakhartchenko@gen.vetmed.uni-muenchen.de (V.Z.); ewolf@genzentrum.lmu.de (E.W.)

2 Center for Innovative Medical Models (CiMM), Ludwig-Maximilians-Universität München, 85764 Oberschleissheim, Germany

3 Laboratory for Functional Genome Analysis (LAFUGA), Gene Center, Ludwig-Maximilians-Universität München, 81377 Munich, Germany

* Correspondence: k.simmet@gen.vetmed.uni-muenchen.de

Abstract: The role of the pluripotency factor NANOG during the second embryonic lineage differentiation has been studied extensively in mouse, although species-specific differences exist. To elucidate the role of NANOG in an alternative model organism, we knocked out *NANOG* in fibroblast cells and produced bovine *NANOG*-knockout (KO) embryos via somatic cell nuclear transfer (SCNT). At day 8, *NANOG*-KO blastocysts showed a decreased total cell number when compared to controls from SCNT (NT Ctrl). The pluripotency factors OCT4 and SOX2 as well as the hypoblast (HB) marker GATA6 were co-expressed in all cells of the inner cell mass (ICM) and, in contrast to mouse *Nanog*-KO, expression of the late HB marker SOX17 was still present. We blocked the MEK-pathway with a MEK 1/2 inhibitor, and control embryos showed an increase in NANOG positive cells, but SOX17 expressing HB precursor cells were still present. *NANOG*-KO together with MEK-inhibition was lethal before blastocyst stage, similarly to findings in mouse. Supplementation of exogenous FGF4 to *NANOG*-KO embryos did not change SOX17 expression in the ICM, unlike mouse *Nanog*-KO embryos, where missing SOX17 expression was completely rescued by FGF4. We conclude that NANOG mediated FGF/MEK signaling is not required for HB formation in the bovine embryo and that another—so far unknown—pathway regulates HB differentiation.

Keywords: NANOG; SOX17; bovine preimplantation development; MEK; second lineage differentiation

Citation: Springer, C.; Zakhartchenko, V.; Wolf, E.; Simmet, K. Hypoblast Formation in Bovine Embryos Does Not Depend on NANOG. *Cells* **2021**, *10*, 2232. https://doi.org/10.3390/cells10092232

Academic Editor: Lon J. van Winkle

Received: 15 July 2021
Accepted: 25 August 2021
Published: 28 August 2021

Publisher's Note: MDPI stays neutral with regard to jurisdictional claims in published maps and institutional affiliations.

1. Introduction

Before implantation, mammalian embryos undergo two consecutive lineage specifications. First, outer and inner cells in the morula form the surrounding CDX2-expressing trophectoderm (TE) and the inner cell mass (ICM), respectively, during blastocyst development. Second, within the ICM, NANOG-expressing cells form the pluripotent epiblast (EPI) and exhibit a mutually exclusive expression pattern with differentiated cells expressing GATA6 and SOX17 from the primitive endoderm (PE) or hypoblast (HB) in bovine and human embryos. Consequently, three distinct cell lineages arise: the EPI, which will give rise to the embryo proper, the PE/HB, which will form the yolk sac, and the TE, responsible for extraembryonic tissues and implantation [1–8], reviewed in [9]. While these landmarks of preimplantation embryonic development are conserved between mammalian species, fundamental differences exist regarding the regulation of the second lineage segregation. Human and bovine *OCT4*-knockout (KO) embryos lose NANOG and maintain GATA6 expression, whereas mouse *Oct4*-KO embryos still express NANOG and fail to develop a PE [10–13]. Additionally, FGF4 signaling via the mitogen-activated protein kinase (MAPK) pathway, also called MEK pathway, has different roles in the regulation of the second lineage differentiation between species. It is known that in mouse, EPI precursor cells express FGF4, which via the FGF receptors 1/2 (FGFR) and the MEK-pathway induces and regulates the formation of the PE (reviewed in [14]). Inhibition of the MEK pathway or the FGFR in mouse embryos results in an ICM only expressing NANOG, while GATA6 expression is completely lost [15–17]. In both human and bovine embryos, inhibition upstream of MEK via FGFR inhibitors has no effect on EPI or HB formation [18]. In bovine embryos, MEK inhibition increases NANOG expression and reduces HB markers, but HB marker expression is still present. In human embryos, MEK inhibition has no effect, suggesting that bovine and human HB formation is partly or completely independent of this pathway, and that these species regulate the second lineage differentiation differently [18–21].

Recently, a dosage-dependent effect of the MEK-inhibitory compound PD0325901 (PD032) was found in bovine embryos. A concentration of 2.5 μM eliminated expression of the later HB-marker SOX17 completely [22], while previous studies observed maintenance of the early marker GATA6 at concentrations of 0.5 and 1 μM [18,20], challenging the hypothesis of a partly MEK-independent HB formation in bovine.

Supplementing exogenous FGF4 from morula to blastocyst stage leads to ubiquitous GATA6 expression in mouse, bovine, pig and rabbit embryos (reviewed in [23]), revealing a non-cell-autonomous role for FGF4.

NANOG is a member of the homeobox family of DNA binding transcription factors that is known to maintain the pluripotency of embryonic stem cells (ESCs) together with OCT4 and SOX2 [6,24]. In mouse, *Nanog*-KO does not affect the formation of the blastocyst, but during the second lineage differentiation, the EPI lineage fails. Thereby, ubiquitous expression of the early PE marker GATA6 within the ICM was reported, whereas the late PE markers SOX17 and GATA4 were lost [1,2,25]. SOX17 expression in mouse *Nanog*-KO embryos is rescued by supplementing exogenous FGF4, confirming the crucial role of FGF4 expressed by EPI cells for PE differentiation [2]. Mouse *Nanog*-KO embryos and ESCs lose viability in the presence of MEK inhibitors, resulting in early cell death [2,26]. So far, the phenotype of *NANOG*-KO in human embryos or ESCs has not been investigated, but a *NANOG*-knockdown experiment in human ESCs showed that, while NANOG represses embryonic ectoderm differentiation, it does not influence the expression of OCT4 or SOX2 [27]. In bovine embryos, NANOG is first expressed at the morula stage and becomes specific to EPI precursor cells in the ICM at day 7 [18,28,29]. Only recently, a *NANOG*-KO via zygote injection was first described in bovine, displaying pan-ICM GATA6 expression and reduced transcript levels for the pluripotency factors SOX2 and H2AFZ [30].

In the present study, we addressed the role of NANOG in bovine preimplantation embryos using a reverse genetics approach. After induction of a *NANOG* frameshift mutation in fibroblasts and production of embryos via somatic cell nuclear transfer (SCNT), we characterized the *NANOG*-KO phenotype by immunofluorescence staining of day 8 blastocysts for markers of EPI and HB precursor cells. We further addressed the roles of FGF4 and the MEK pathway by treating *NANOG*-KO embryos with exogenous FGF4 and with an inhibitor of the MEK pathway, respectively, revealing new insights into the second lineage differentiation in bovine embryos.

2. Materials and Methods

2.1. CRISPR/Cas9-Mediated KO of NANOG in Adult Fibroblasts

We induced a frameshift mutation in a non-homologous end joining approach with an sgRNA (5′-CTCTCCTCTTCCCTCCTCCA-3′) designed by Synthego software (V2.0) using ENSBTAT00000027863 (*NANOG*) as reference gene (design.synthego.com, accessed on 15 July 2021). The sgRNA targeting exon 2 of *NANOG* was cloned into pSpCas9(BB)-2A-Puro (PX459) V2.0, a gift from Feng Zhang [31]. All experiments are based on a cell line with origin in bovine adult ear fibroblast cells that were isolated in the laboratory in the Chair for Molecular Animal Breeding and Biotechnology, Ludwig-Maximilians-Universität München, 85764 Oberschleissheim, Germany. Bovine fibroblasts were transfected with the Nucleofector device (Lonza; Basel, Switzerland) according to the manufacturer's instructions. After selection with 2 μg/mL puromycin (Sigma-Aldrich; St. Louis, MO, USA) for 48 h, we produced single-cell clones as described previously [32]. After PCR amplification with primers 5′-GGAAGGGATTCCTGAAATGAG-3′ (forward) and 5′-GTGGGATCTTAGTTGCGACAT-3′ (reverse), gene editing-induced modifications in the *NANOG* alleles and naturally occurring SNPs were examined by Sanger sequencing using the primers 5′-AAGGTCTGGGTTGCAATAGG-3′ (forward) and 5′-CCACCAGGGAAATCCCTTATTT-3′ (reverse). All primers were synthesized by Biomers.net (Ulm, Germany; accessed on 15 July 2021).

2.2. Production and Analysis of SCNT and IVP Embryos

SCNT and in vitro production (IVP) procedures were performed as described previously [33]. Briefly, bovine ovaries were collected at a slaughterhouse, and retrieved cumulus-oocyte-complexes were matured in vitro. After SCNT or fertilization of the oocytes (day 0), fused complexes and presumptive zygotes were cultivated in synthetic oviductal fluid including Basal Medium Eagle's amino acids solution (BME, Sigma-Aldrich), Minimum Essential Medium (MEM, Sigma-Aldrich) and 5% estrous cow serum (OCS) from day 0 or day 1 up to day 8 for SCNT or IVP zygotes, respectively. After 8 days of culture, the zona pellucida was removed enzymatically using Pronase (Merck Millipore; Burlington, MA, USA), and embryos were fixed in a solution containing 2% paraformaldehyde [34].

2.3. Modulation of Signaling Pathways

Growth factors or inhibitors were supplemented from day 5 morula stage until day 8 blastocyst stage. The MEK1 and MEK2 inhibitor PD032 (Tocris; Bristol, UK) was used at 0.5 or 2.5 μM, and controls were cultured in equal amounts of DMSO (Sigma-Aldrich). 1 μg/mL human recombinant FGF4 (R&D Systems; Minneapolis, MN, USA) was added to synthetic oviductal fluid with 1 μg/mL heparin (Sigma-Aldrich).

2.4. Immunofluorescence Staining and Confocal Laser Scanning Microscopy

Before staining, embryos were incubated for 1 h at room temperature in a blocking solution containing 0.5% Triton X-100 (Sigma-Aldrich) and 5% donkey serum (Jackson ImmunoResearch; West Grove, PA, USA) or fetal calf serum (Thermo Fisher Scientific; Waltham, MA, USA) or both sera sequentially, depending on the species origin of the secondary antibodies. Double staining for either NANOG/GATA6, NANOG/SOX17, SOX17/SOX2, OCT4/SOX2 or GATA6/CDX2 was achieved by incubation overnight at 4 °C in primary antibody solution and transfer to secondary antibody solution at 37 °C for 1 h after washing 3 times. The antibodies used and the applied dilutions are presented in Supplementary Table S1. Labeled embryos were mounted in Vectashield mounting medium containing 4′,6-diamidino-2-phenylindole (DAPI, Vector Laboratories; Burlingame, CA, USA) in a manner that conserved the 3D structure of the specimen [35]. Z-stacks of optical sections with an interval of 1.2 μm were recorded using an LSM710 Axio Observer confocal laser scanning microscope (CLSM; Zeiss, Jena, Germany) with a 25× water immersion objective (LD LCI Plan-Apochromat 25×/0.8 Imm Korr DIC M27) or a Leica SP8 CLSM (Leica; Wetzlar, Germany) with a 40× water immersion objective (Leica; 1.1NA), respectively. DAPI, Alexa Fluor 488, 555, and 647 were excited with laser lines of 405 nm, 499 nm, 553 nm, and 653 nm (LSM710), respectively, or with a white light laser (SP8).

2.5. Statistical Analysis

Statistical analysis was performed using Graphpad Prism 5.04. After checking normal distribution of data with a Kolmogorov–Smirnov test, we performed nonparametric tests. For pairwise comparisons, a two-tailed Mann–Whitney U test was performed, whereas for three experimental groups, a Kruskal–Wallis test and subsequent Dunn's multiple comparisons test as post hoc test was applied. The level of significance was set to $p < 0.05$. Data are presented as mean ± standard deviation (SD). For quantitative image analysis, the ImageJ (V 1.53c) cell counter plugin was used [36]; control embryos with less than 8 cells in the ICM were excluded from the analysis.

3. Results

3.1. NANOG-KO Has No Effect on Blastocyst Rate but Results in Reduced Total Cell Number

After selection with puromycin, we generated 57 single-cell clones and achieved a mutation rate of 54.4% including 8 homozygous mutations (14.0%). Two different cell clones with an identical homozygous insertion of a single nucleotide, which induces a frameshift mutation, were used for SCNT to produce embryos without NANOG (*NANOG*-KO). SCNT embryos from two different cell clones with no mutation from the same transfection experiment (NT Ctrl) and embryos produced by in vitro fertilization (IVP Ctrl) served as controls. There were no differences regarding blastocyst rates between embryos from all four cell clones. SCNT embryos from both *NANOG*-KO cell clones showed consistent alterations as described below. The group of NT Ctrl embryos derived from the two *NANOG*-intact cell clones was phenotypically homogeneous. To verify the absence of NANOG protein, *NANOG*-KO blastocysts ($n = 9$) were stained for NANOG using two different antibodies, and no positive cells were observed (Supplementary Figure S1A). There was no significant difference between *NANOG*-KO and Ctrl embryos regarding blastocyst rates. *NANOG*-KO embryos were able to expand but appeared to be smaller than NT Ctrl blastocysts (Supplementary Figure S1B), with significantly decreased diameters of day 8 *NANOG*-KO compared to NT Ctrl blastocysts (Supplementary Figure S1C). We analyzed the total cell number and the number of ICM and TE cells using SOX2 and CDX2 as markers, respectively. We found a significant reduction in both lineages in *NANOG*-KO embryos, while no significant difference in the ratio of ICM to total cell number was seen, showing a proportionally normal distribution of cells to ICM and TE during the first lineage differentiation in the absence of NANOG (Table 1).

Table 1. Developmental rates and cell numbers of day 8 *NANOG*-knockout (*NANOG*-KO), nuclear transfer control (NT Ctrl), and in vitro-produced control (IVP Ctrl) embryos. Data are presented as mean ± standard deviation. Different superscript letters (a, b) within a row indicate significant differences ($p < 0.05$). Data were analyzed by Kruskal–Wallis test with Dunn's multiple comparisons test as post hoc test.

Experimental Group	*NANOG*-KO	NT Ctrl	IVP Ctrl
No. of experiments	8	5	17
Blastocysts/zygotes [%]	29.0 ± 12.9	36.1 ± 9.9	28.3 ± 4.2
No. of analyzed embryos	25	32	51
Total cell number	66.4 ± 27.3 ^{a}	148.6 ± 65.6 ^{b}	177.4 ± 52.2 ^{b}
ICM/total cell number [%]	23.8 ± 11.3	29.2 ± 11.0	29.0 ± 7.3
ICM number	15.8 ± 10.5 ^{a}	43.5 ± 24.5 ^{b}	51.3 ± 20.0 ^{b}
TE cell number	50.6 ± 22.6 ^{a}	105.1 ± 47.2 ^{b}	126.1 ± 40.5 ^{b}

3.2. NANOG Is Dispensable for Expression of Pluripotency Factors and Hypoblast Markers

We stained *NANOG*-KO, NT Ctrl, and IVP Ctrl day 8 blastocysts for GATA6/CDX2 (Figure 1A) and SOX17/SOX2 (Figure 1B). In both control groups, embryos showed consistent co-expression of CDX2 with GATA6, and a subset of the CDX2 negative ICM cells was also GATA6 negative. Staining of NT Ctrl ($n = 9$) and IVP Ctrl blastocysts ($n = 18$) for NANOG and GATA6 (Supplementary Figure S2A) confirmed that GATA6 negative cells express NANOG, resulting in the previously reported mutually exclusive expression of these lineage markers [16,18]. SOX2 was expressed in the entire ICM, and a subset of cells already expressed the late HB marker SOX17. In *NANOG*-KO day 8 blastocysts, an ICM was clearly discernible by CDX2 negative cells, while GATA6 was expressed ubiquitously with no negative cells in the ICM or TE.

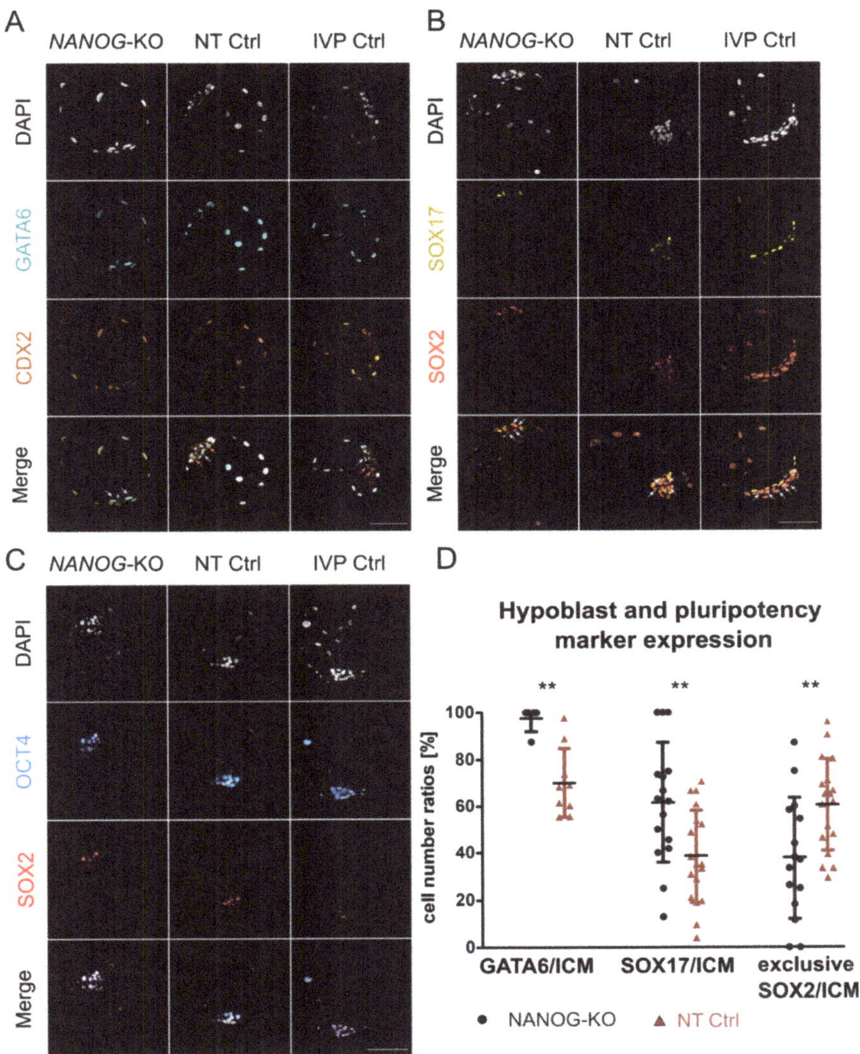

Figure 1. Expression of hypoblast and pluripotency markers in *NANOG*-KO and control groups. Representative confocal planes of day 8 blastocysts stained for GATA6/CDX2 (**A**) and SOX17/SOX2 (**B**). Sample sizes of GATA6/CDX2 were $n = 5$, 9, 14 and of SOX17/SOX2 $n = 16$, 18, 10 for *NANOG*-KO, NT Ctrl, and IVP Ctrl, respectively. White arrows indicate ICM cells with GATA6 expression (**A**) or exclusive SOX2 expression (**B**), red arrows indicate GATA6/CDX2 double negative cells (**A**) in the ICM. (**C**) Expression of pluripotency factors OCT4 and SOX2. Sample sizes were $n = 4$, 9, 4 for *NANOG*-KO, NT Ctrl, and IVP Ctrl, respectively. Color codes were: Grey (DAPI), cyan (GATA6), orange hot (CDX2), yellow (SOX17), red (SOX2), and cyan hot (OCT4). (**D**) The ratio of GATA6, SOX17, and SOX2 positive cells within the ICM of *NANOG*-KO (black) and NT Ctrl (red) embryos. SOX2 served as ICM marker in the quantification of SOX17. SOX2 exclusive expression represents cells positive for SOX2 while negative for SOX17. Data were analyzed using a two-tailed Mann–Whitney U test and are presented as mean (%) ± standard deviation. Asterisks (**) indicate significant differences between groups ($p < 0.01$). Sample sizes of GATA6 were $n = 5$, 9; of SOX17 $n = 16$, 19; and of SOX2 exclusive $n = 15$, 18 for *NANOG*-KO and NT Ctrl embryos, respectively. Scale bars indicate 100 μm.

The ratio of SOX17 positive cells within the ICM increased significantly compared to NT Ctrl (61.6% ± 25.7% vs. 38.6% ± 19.6%, respectively) but not IVP Ctrl (56.1% ± 13.5%), while cells with exclusive SOX2 expression were still present, albeit at reduced numbers (Figure 1D). We conclude that NANOG is required for the repression of GATA6 in the ICM. In contrast to mouse *Nanog*-KO embryos that show complete loss of SOX17 [2], we still found SOX17 positive cells in the ICM. However, absence of NANOG and a ubiquitous GATA6 expression is not sufficient to induce a pan-ICM expression of SOX17 in bovine blastocysts.

Staining for OCT4 and SOX2 showed that in NT Ctrl and IVP Ctrl embryos, both factors are co-expressed throughout the entire ICM and that in the absence of NANOG, this pattern is maintained (Figure 1C). None of the embryos showed OCT4 expression in the TE at day 8.

3.3. Inhibition of MEK Induces Cell Death in NANOG-KO Embryos

In the next step, we aimed to investigate the effect of *NANOG*-KO while inhibiting the MEK signaling pathway. Because previous reports on the effect of the MEK 1/2 inhibitor PD032 in bovine embryos are in conflict [18,22], we first set out to test the effect of different dosages on the expression of NANOG and GATA6 in IVP Ctrl embryos. There was no difference between the DMSO control (n = 11) and the dosages 0.5 (n = 4) and 2.5 µM (n = 10) PD032 regarding the blastocyst per morula rate (45.6% ± 12.5%, 46.8% ± 6.1%, 50.0% ± 16.3%, respectively) and the ratio of ICM to total cell number (30.4% ± 5.8%, 29.2% ± 7.8%, 32.5% ± 8.2%, respectively). The number of ICM cells was determined without a specific staining on the basis of the embryos' morphology. In agreement with Kuijk, et al. [18], the proportion of NANOG positive cells was markedly increased, while the expression of GATA6 was reduced but not completely switched off at both concentrations (Figure 2A). Similarly to GATA6, SOX17 was significantly reduced but still present at a concentration of 2.5 µM (Figure 2B). At 2.5 µM, both HB markers were always co-expressed with NANOG and thus failed to establish a mutually exclusive expression pattern.

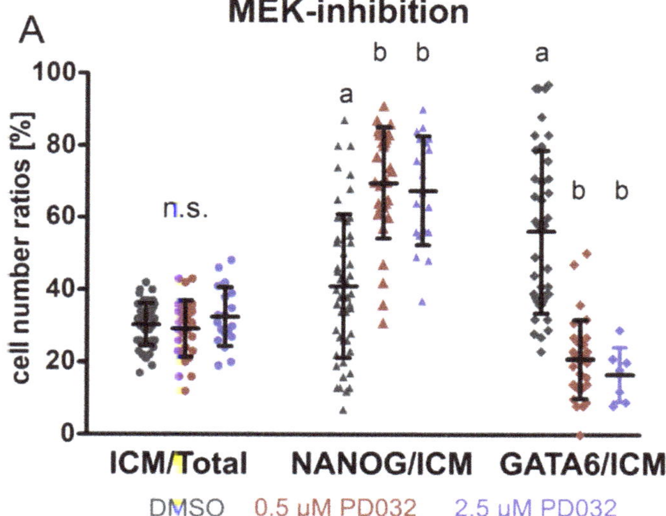

Figure 2. *Cont.*

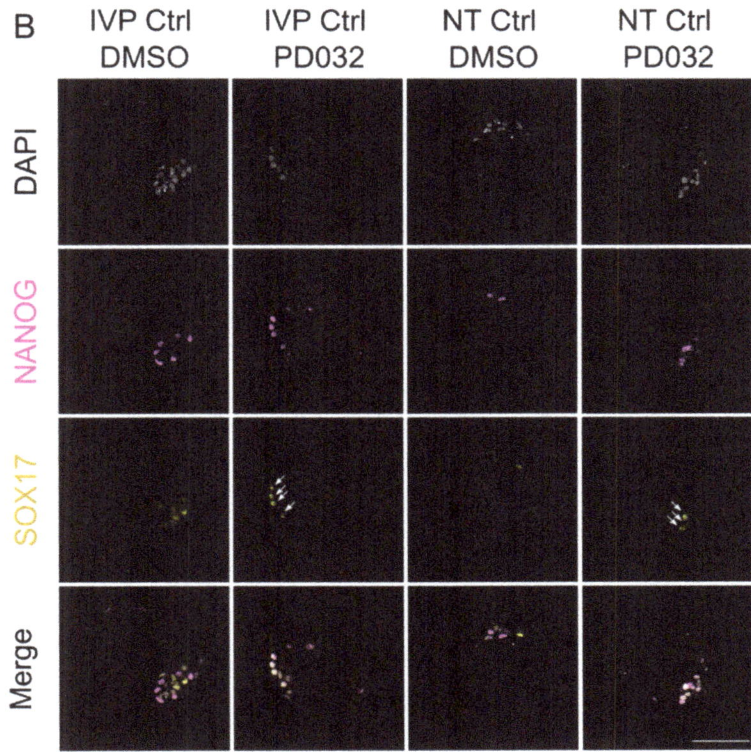

Figure 2. The effect of different dosages of MEK-inhibitor PD0325901 (PD032) on the expression of NANOG, GATA6, and SOX17. (**A**) The ratio of ICM/Total, NANOG/ICM, and GATA6/ICM in the presence of 0.5 and 2.5 µM PD032 in IVP Ctrl embryos. The proportion of ICM to total cell number (Total) is shown, and the number of NANOG and GATA6 expressing cells was set in relation to the number of ICM cells. Embryos were cultured from day 5 morula to day 8 blastocyst in the presence of 0.5 and 2.5 µM PD032. Data were analyzed by Kruskal–Wallis with Dunn's multiple comparisons test as post hoc test and are presented as mean ± standard deviation. Different superscripts (a, b) indicate significant differences between groups ($p < 0.0001$), n.s. = not significant. Sample sizes of ICM/Total and NANOG/ICM were $n = 49, 30, 19$ and of GATA6/ICM $n = 40, 30, 7$ for DMSO (grey), 0.5 µM (red) and 2.5 µM (blue) PD032, respectively. (**B**) Representative confocal planes of NANOG and SOX17 expression in IVP Ctrl and NT Ctrl blastocysts cultured with DMSO or 2.5 µM PD032. Sample sizes of embryos stained for NANOG (magenta) and SOX17 (yellow) were $n = 49$ for IVP Ctrl DMSO, $n = 19$ for IVP Ctrl PD032, $n = 8$ for NT Ctrl DMSO, and $n = 9$ for NT Ctrl PD032. DAPI = grey; arrows indicate SOX17 expression in the presence of 2.5 µM PD032; scale bar indicates 100 µm.

As a higher dosage (2.5 µM) of PD032 did not affect blastocyst development or cell numbers, we performed inhibition of the MEK pathway in SCNT embryos using this concentration. We found similar blastocyst per morula rates ($p > 0.05$) of NT Ctrl in DMSO ($n = 3$; 52.4% ± 16.7%) and PD032 ($n = 4$; 42.5% ± 14.8%). The expression pattern of NT Ctrl embryos incubated with the MEK inhibitor was comparable to that of IVP Ctrl embryos that underwent the same treatment, as HB markers were still present (Figure 2B). Although treatment of *NANOG*-KO embryos with DMSO did not affect the blastocyst per morula rate ($n = 3$, 57.8% ± 16.2%) when compared to NT Ctrl treated with DMSO, incubating *NANOG*-KO embryos in the presence of PD032 ($n = 5$) resulted in severely compromised viability, and all embryos died. This agrees with findings in mouse embryos and mouse ESCs, where loss of NANOG and inhibition of MEK also result in cell death [2,26].

3.4. FGF4 in NANOG-KO Embryos Does Not Convert the Entire ICM to Hypoblast Precursor Cells

Subsequently, we investigated whether exogenous FGF4 can induce full SOX17 expression in *NANOG*-KO bovine embryos. In IVP Ctrl and NT Ctrl embryos, treatment with FGF4 completely switched off the expression of NANOG (Figure 3D), and most ICM cells expressed SOX17 (Figure 3E, Supplementary Figure S2B). FGF4 had no effect regarding blastocyst per morula rate and total cell number, while the ratio of SOX2 positive cells, i.e., the ICM, to total cell number was significantly reduced in both groups (Figure 3A–C). Treatment of *NANOG*-KO embryos with FGF4 did not affect the blastocyst per morula rate or the ICM to total cell number ratio, but the total cell number increased significantly with *NANOG*-KO embryos (117.6 ± 48.7 vs. 66.4 ± 27.3 without FGF4 treatment) reaching a total cell number similar to that of untreated NT Ctrl (148.6 ± 65.6, Table 1). In all embryos treated with FGF4, the ubiquitous expression of SOX2 in the ICM was maintained (Figure 4). As the SOX17 expression increased in FGF4 treated Ctrl groups, the exclusive expression of SOX2 was significantly reduced, whereas in mutant embryos, SOX2 exclusive expression remained unchanged (Figure 3F, Figure 4). The percentage of SOX17 positive cells in the ICM did not increase in *NANOG*-KO embryos (Figure 3E), which is in contrast to mouse *Nanog*-KO embryos, where exogenous FGF4 induces SOX17 expression in most of the ICM cells [2]. We conclude that in bovine, NANOG is required for FGF4 mediated expression of SOX17, as FGF4 alone was not sufficient to convert all ICM cells to SOX17 expressing HB precursor cells.

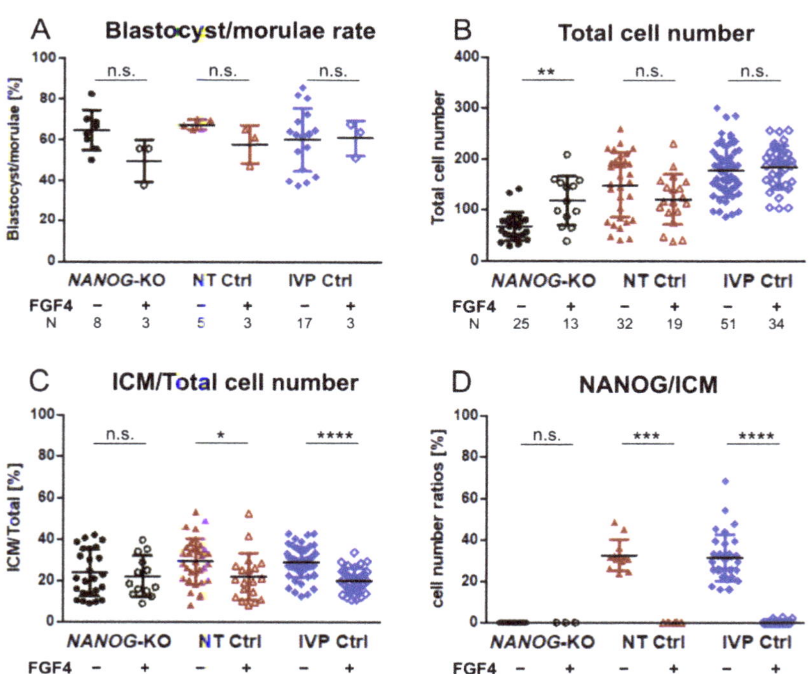

Figure 3. *Cont.*

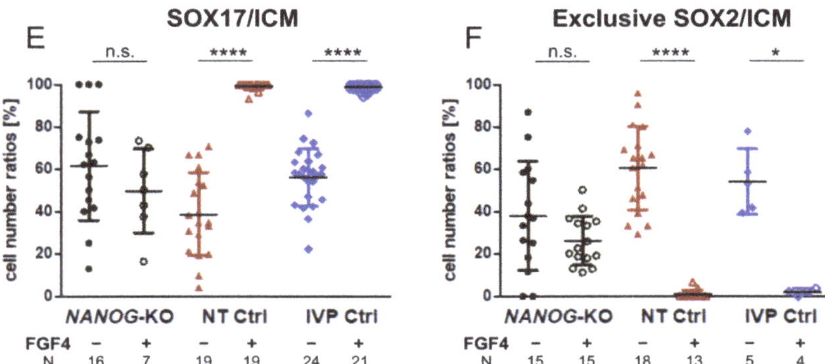

Figure 3. Developmental rates and cell number ratios of NANOG, SOX17, and exclusive SOX2 in *NANOG*-KO, NT Ctrl, and IVP Ctrl day 8 embryos treated with exogenous FGF4 and heparin. (**A**) Blastocyst per morula rate, (**B**) total cell number, (**C**) proportion of ICM to total cell number, (**D**) ratio of NANOG-positive cells in the ICM, (**E**) ratio of SOX17-positive cells in the ICM, and (**F**) ratio of cells exclusively expressing SOX2 in the ICM of *NANOG*-KO (black), NT Ctrl (red), and IVP Ctrl (blue) embryos without (−) and with (+) FGF4 and heparin are presented. ICM cells were determined by staining of SOX2. Data were analyzed by two-tailed Mann–Whitney U test and are presented as mean (%) ± standard deviation. Asterisks indicate significant effects of FGF4 treatment within embryo group. N = number of analyzed embryos, * $p < 0.05$; ** $p < 0.01$; *** $p < 0.001$; **** $p < 0.0001$; n.s. = not significant.

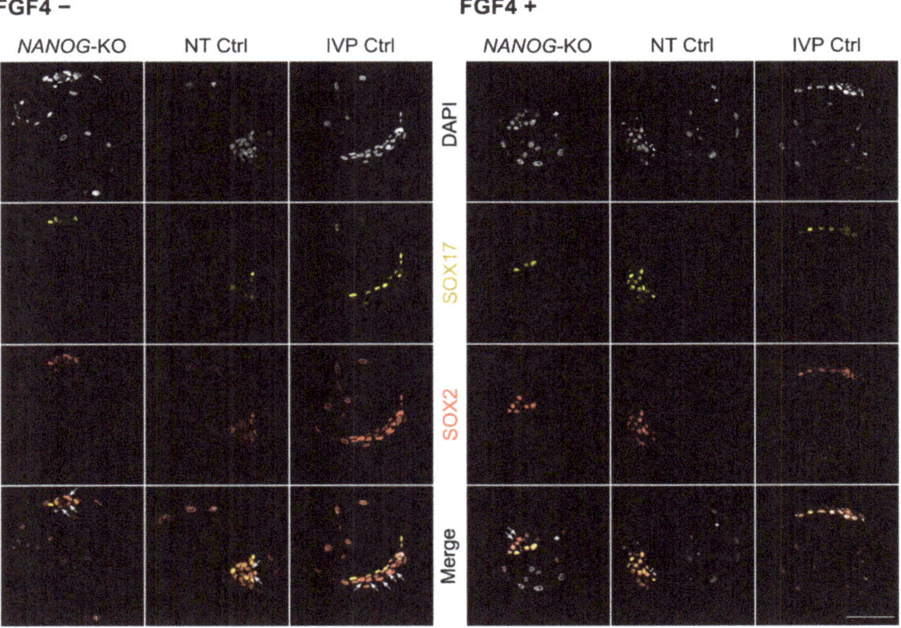

Figure 4. Expression of pluripotency and late hypoblast markers in *NANOG*-KO, NT Ctrl, and IVP Ctrl day 8 embryos treated with exogenous FGF4 and heparin. Representative confocal planes of day 8 blastocysts stained for SOX17/SOX2. Embryos without (−) and with (+) FGF4 and heparin treatment are presented. Arrows indicate ICM cells with exclusive SOX2 expression (SOX17 negative). Color codes are: Grey (DAPI), yellow (SOX17) and red (SOX2). Scale bar indicates 100 μm.

4. Discussion

To investigate the regulation of differentiation and maintenance of pluripotency during mammalian preimplantation development, it is vital to examine models other than mouse, as species-specific differences exist. The bovine embryo is a very suitable alternative, as IVP procedures are highly developed and similarities to human embryo development have been reported (reviewed in [23]).

In this study, we focused on NANOG, because it is not clear whether the role of this pluripotency factor during the second lineage segregation is conserved between mammals. In order to achieve uniform modification of all cells of the embryo, we used SCNT to produce *NANOG*-KO embryos instead of zygote injection, where mosaicism may hamper analyses. To exclude the effects of the SCNT procedure on the phenotype, we implemented two control groups: SCNT embryos generated from transfected cells that maintained the wildtype genotype (NT Ctrl) and embryos from in vitro fertilization (IVP Ctrl). NT Ctrl did not vary from IVP Ctrl embryos in any of the examined parameters, except for the proportion of SOX17 cells in the ICM, which was decreased in NT Ctrl embryos. We set the number of cells expressing lineage marker proteins in relation to the number of ICM cells in order to account for variations due to the different sizes of the embryos, especially the reduced size of *NANOG*-KO embryos. Staining embryos with markers for TE and ICM, i.e., CDX2 and SOX2, respectively, enabled us to quantify reliably the cell numbers in each lineage after the first differentiation. We found that the ratio of ICM to total cell number did not change in *NANOG*-KO embryos, showing that NANOG is not required for proper segregation of TE and ICM, as reported in mouse. On the other hand, we found a significant reduction in total cell numbers, which is in contrast to mouse *Nanog*-KO embryos, where the loss of NANOG does not impede cell proliferation until the E3.5 blastocyst stage [1]. Interestingly, the reduced total cell number in *NANOG*-KO embryos reached normal levels when the embryos were cultivated with exogenous FGF4, where the proliferative impact of FGF4 [17,37–39] evidently alleviated the reduction of total cell numbers in the absence of NANOG. We hypothesize that in *NANOG*-KO embryos, the absence of FGF4 expressing EPI precursor cells causes the reduced cell number. This suggests that EPI cells express FGF4, which to our knowledge has not been shown yet in bovine but is known in mouse [1].

Although Ortega, et al. [30] found a reduction of *SOX2* transcripts in bovine *NANOG*-KO embryos, we detected SOX2 and OCT4 expression in the absence of NANOG on the protein level. To our knowledge, this is the first report on SOX2 expression in the absence of NANOG in a mammalian embryo. We were not able to detect OCT4 in the TE of day 8 blastocysts using a monoclonal antibody, which is in contrast to Berg, et al. [40] and Simmet, et al. [11], who detected OCT4 in the TE of ex vivo day 11 or in vitro day 7 blastocysts using a different polyclonal antibody, respectively.

In bovine *NANOG*-KO embryos, the ICM ubiquitously expresses the early HB marker GATA6, which agrees with previous reports on bovine and mouse NANOG deficient embryos [2,30]. Interestingly, in bovine *NANOG*-KO embryos, expression of the later HB marker SOX17 was still present, but the absence of NANOG and a ubiquitous GATA6 expression was not sufficient to induce a pan-ICM expression of SOX17, as some cells in the ICM still showed exclusive SOX2 expression, making the regulation of SOX17 in the bovine embryo partly independent of NANOG and GATA6.

This is in sharp contrast to the mouse *Nanog*-KO, where expression of the late PE markers GATA4 and SOX17 completely fails but can be rescued in a chimeric complementation assay or fully induced by exogenous FGF4 [1,2,25].

We further investigated the second lineage segregation in bovine blastocysts by inhibiting the MEK pathway with PD032. In line with previous reports, MEK-inhibition did not completely ablate GATA6 positive cells [18,20], and also SOX17 was still expressed in the ICM. Canizo, et al. [22] found a dosage-dependent effect of PD032 with the concentration also applied in this study abolishing all SOX17. The reasons for these contrasting results remain unclear, and we can only speculate that the different embryo culture media have an effect on SOX17 expression in the presence of PD032. Bovine embryos were cultured in PD032 concentrations of up to 100 µM, and reduction of *SOX17* transcripts was already achieved at 10 µM, while higher dosages did not further decrease transcript abundance [19]. Treating bovine embryos with a broad-spectrum inhibitor of receptor tyrosine kinases (RTKs) including MEK (BI-BF1120) increased the abundance of *SOX17* transcripts, suggesting that SOX17 does not depend on direct activation via the MEK pathway [41]. We further found that HB markers were generally co-expressed with NANOG when the MEK pathway was blocked. Therefore, we suggest that NANOG mediated repression of HB markers is dependent on MEK signaling. Our data and previous reports indicate that, in bovine embryos, GATA6 and SOX17 are partly independent of the MEK pathway and that a so far unknown factor plays an important role in the regulation of HB differentiation.

When combining inhibition of MEK with loss of NANOG, we found that the viability of those embryos was severely compromised, resulting in cell death. Similar reports exist in NANOG deficient mouse embryos and ESCs, where cell death is observed after adding inhibitors of the MEK-pathway [2,26]. We conclude that HB formation, i.e., expression of GATA6 and SOX17, in the absence of both NANOG and a functioning MEK pathway is associated with cell death. We speculate that apoptosis is induced during the cell sorting process to eliminate cells that do not commit to either EPI or HB, as selective apoptosis was described for appropriate segregation of PE and EPI in mouse blastocysts [5,42]. Nevertheless, our hypothesis cannot explain why the TE is also affected by cell death.

In the mouse, the loss of FGF4 expressing EPI precursor cells leads to complete ablation of late PE marker expressing cells that can be rescued with exogenous FGF4 [2,43]—evidence that FGF4 alone is sufficient to induce PE differentiation. The regulation of SOX17 expression in bovine embryos appears to be different, as FGF4 alone without functional NANOG was not sufficient to convert all ICM cells to SOX17 expressing HB precursor cells. Thus, we conclude that NANOG is required for FGF4 mediated expression of SOX17.

Our results show that in the bovine embryo, the establishment of HB precursor cells is independent of EPI-cell mediated FGF/MEK signaling. This is in sharp contrast to mouse but similar to human, where the FGF/MEK pathway does not regulate the second lineage differentiation [18,21]. An unknown factor induces HB differentiation, and it is of utmost interest to further investigate this pathway and whether it also exists in human embryos as well.

Supplementary Materials: The following are available online at https://www.mdpi.com/article/10.3390/cells10092232/s1, Figure S1: Producing bovine *NANOG*-KO embryos via somatic cell nuclear transfer (SCNT); Figure S2: Expression of epiblast and hypoblast markers in NT Ctrl and IVP Ctrl day 8 embryos; Table S1: Targets, antibodies, suppliers, and applied dilutions for immunofluorescence staining.

Author Contributions: Conceptualization—K.S.; writing—original draft preparation, C.S. and K.S.; writing—review and editing, C.S., V.Z., E.W. and K.S.; funding acquisition, E.W. and K.S. All authors have read and agreed to the published version of the manuscript.

Funding: Our projects involving the investigation of bovine and porcine preimplantation development and nuclear transfer are funded by the Deutsche Forschungsgemeinschaft (DFG, German Research Foundation) under grant 405453332; by the Bayerische Forschungsstiftung under grant AZ-1300-17; and by the German Center for Diabetes Research (DZD) under grant 82DZD00802.

Institutional Review Board Statement: Not applicable.

Informed Consent Statement: Not applicable.

Data Availability Statement: Not applicable.

Acknowledgments: We thank Eva-Maria Hinrichs, Maximilian Moraw, and Tuna Güngör for their excellent technical assistance and Christophe Jung for instructions and access to confocal laser scanning microscopy in the Center for Advanced Light Microscopy (CALM), LMU Munich.

Conflicts of Interest: The authors declare no conflict of interest. The funders had no role in the design of the study; in the collection, analyses, or interpretation of data; in the writing of the manuscript, or in the decision to publish the results.

Abbreviations

BME	Basal Medium Eagle's amino acids solution
CLSM	Confocal laser scanning microscope
EPI	Epiblast
ESCs	Embryonic stem cells
FGFR	FGF-receptor
HB	Hypoblast
ICM	inner cell mass
IVP	In vitro production
KO	Knockout
MAPK	Mitogen-activated protein kinase
MEM	Minimum Essential Medium
OCS	Estrous cow serum
PD032	PD0325901
PE	Primitive endoderm
RTK	Receptor tyrosine kinase
SCNT	Somatic cell nuclear transfer
SD	Standard deviation
TE	Trophectoderm

References

1. Messerschmidt, D.M.; Kemler, R. Nanog is required for primitive endoderm formation through a non-cell autonomous mechanism. *Dev. Biol.* **2010**, *344*, 129–137. [CrossRef]
2. Frankenberg, S.; Gerbe, F.; Bessonnard, S.; Belville, C.; Pouchin, P.; Bardot, O.; Chazaud, C. Primitive endoderm differentiates via a three-step mechanism involving Nanog and RTK signaling. *Dev. Cell* **2011**, *21*, 1005–1013. [CrossRef]
3. Cai, K.Q.; Capo-Chichi, D.C.; Rula, M.E.; Yang, D.-H.; Xu, X.-X. Dynamic GATA6 expression in primitive endoderm formation and maturation in early mouse embryogenesis. *Dev. Dyn.* **2008**, *237*, 2820–2829. [CrossRef]
4. Morris, S.A.; Teo, R.T.Y.; Li, H.; Robson, P.; Glover, D.M.; Zernicka-Goetz, M. Origin and formation of the first two distinct cell types of the inner cell mass in the mouse embryo. *Proc. Natl. Acad. Sci. USA* **2010**, *107*, 6364–6369. [CrossRef]
5. Plusa, B.; Piliszek, A.; Frankenberg, S.; Artus, J.; Hadjantonakis, A.-K. Distinct sequential cell behaviours direct primitive endoderm formation in the mouse blastocyst. *Development* **2008**, *135*, 3081–3091. [CrossRef] [PubMed]
6. Mitsui, K.; Tokuzawa, Y.; Itoh, H.; Segawa, K.; Murakami, M.; Takahashi, K.; Maruyama, M.; Maeda, M.; Yamanaka, S. The homeoprotein Nanog is required for maintenance of pluripotency in mouse epiblast and ES cells. *Cell* **2003**, *113*, 631–642. [CrossRef]
7. Chambers, I.; Silva, J.C.R.; Colby, D.; Nichols, J.; Nijmeijer-Winter, B.; Robertson, M.; Vrana, J.; Jones, K.; Grotewold, L.; Smith, A. Nanog safeguards pluripotency and mediates germline development. *Nature* **2007**, *450*, 1230–1234. [CrossRef]
8. Chambers, I.; Colby, D.; Robertson, M.; Nichols, J.; Lee, S.; Tweedie, S.; Smith, A. Functional expression cloning of Nanog, a pluripotency sustaining factor in embryonic stem cells. *Cell* **2003**, *113*, 643–655. [CrossRef]
9. Piliszek, A.; Madeja, Z.E. Pre-implantation development of domestic animals. *Curr. Top. Dev. Biol.* **2018**, *128*, 267–294. [CrossRef]
10. Fogarty, N.M.E.; McCarthy, A.; Snijders, K.E.; Powell, B.E.; Kubikova, N.; Blakeley, P.; Lea, R.; Elder, K.; Wamaitha, S.; Kim, D.; et al. Genome editing reveals a role for OCT4 in human embryogenesis. *Nature* **2017**, *550*, 67–73. [CrossRef] [PubMed]
11. Simmet, K.; Zakhartchenko, V.; Philippou-Massier, J.; Blum, H.; Klymiuk, N.; Wolf, E. OCT4/POU5F1 is required for NANOG expression in bovine blastocysts. *Proc. Natl. Acad. Sci. USA* **2018**, *115*, 2770–2775. [CrossRef] [PubMed]
12. Frum, T.; Halbisen, M.A.; Wang, C.; Amiri, H.; Robson, P.; Ralston, A. Oct4 cell-autonomously promotes primitive endoderm development in the mouse blastocyst. *Dev. Cell* **2013**, *25*, 610–622. [CrossRef]
13. Le Bin, G.C.; Muñoz-Descalzo, S.; Kurowski, A.; Leitch, H.; Lou, X.; Mansfield, W.; Etienne-Dumeau, C.; Grabole, N.; Mulas, C.; Niwa, H.; et al. Oct4 is required for lineage priming in the developing inner cell mass of the mouse blastocyst. *Development* **2014**, *141*, 1001–1010. [CrossRef]
14. Simmet, K.; Zakhartchenko, V.; Wolf, E. Comparative aspects of early lineage specification events in mammalian embryos—Insights from reverse genetics studies. *Cell Cycle* **2018**, *17*, 1688–1695. [CrossRef] [PubMed]
15. Nichols, J.; Silva, J.; Roode, M.; Smith, A. Suppression of Erk signalling promotes ground state pluripotency in the mouse embryo. *Development* **2009**, *136*, 3215–3222. [CrossRef] [PubMed]
16. Chazaud, C.; Yamanaka, Y.; Pawson, T.; Rossant, J. Early lineage segregation between epiblast and primitive endoderm in mouse blastocysts through the Grb2-MAPK pathway. *Dev. Cell* **2006**, *10*, 615–624. [CrossRef] [PubMed]
17. Yamanaka, Y.; Lanner, F.; Rossant, J. FGF signal-dependent segregation of primitive endoderm and epiblast in the mouse blastocyst. *Development* **2010**, *137*, 715–724. [CrossRef]

18. Kuijk, E.W.; van Tol, L.T.A.; Van de Velde, H.; Wubbolts, R.; Welling, M.; Geijsen, N.; Roelen, B.A.J. The roles of FGF and MAP kinase signaling in the segregation of the epiblast and hypoblast cell lineages in bovine and human embryos. *Development* **2012**, *139*, 871–882. [CrossRef]
19. McLean, Z.; Meng, F.; Henderson, H.; Turner, P.; Oback, B. Increased MAP kinase inhibition enhances epiblast-specific gene expression in bovine blastocysts. *Biol. Reprod.* **2014**, *91*, 1–10. [CrossRef]
20. Warzych, E.; Pawlak, P.; Lechniak, D.; Madeja, Z.E. WNT signalling supported by MEK/ERK inhibition is essential to maintain pluripotency in bovine preimplantation embryo. *Dev. Biol.* **2020**, *463*, 63–76. [CrossRef]
21. Roode, M.; Blair, K.; Snell, P.; Elder, K.; Marchant, S.; Smith, A.; Nichols, J. Human hypoblast formation is not dependent on FGF signalling. *Dev. Biol.* **2012**, *361*, 358–363. [CrossRef]
22. Canizo, J.R.; Rivolta, A.E.Y.; Echegaray, C.V.; Suvá, M.; Alberio, V.; Aller, J.F.; Guberman, A.S.; Salamone, D.F.; Alberio, R.H.; Alberio, R. A dose-dependent response to MEK inhibition determines hypoblast fate in bovine embryos. *BMC Dev. Biol.* **2019**, *19*, 13. [CrossRef] [PubMed]
23. Springer, C.; Wolf, E.; Simmet, K. A new toolbox in experimental embryology—Alternative model organisms for studying preimplantation development. *J. Dev. Biol.* **2021**, *9*, 15. [CrossRef] [PubMed]
24. Wu, G.; Schöler, H.R. Role of Oct4 in the early embryo development. *Cell Regen.* **2014**, *3*, 7. [CrossRef] [PubMed]
25. Silva, J.; Nichols, J.; Theunissen, T.W.; Guo, G.; van Oosten, A.L.; Barrandon, O.; Wray, J.; Yamanaka, S.; Chambers, I.; Smith, A. Nanog is the gateway to the pluripotent ground state. *Cell* **2009**, *138*, 722–737. [CrossRef] [PubMed]
26. Hastreiter, S.; Skylaki, S.; Loeffler, D.; Reimann, A.; Hilsenbeck, O.; Hoppe, P.S.; Coutu, D.L.; Kokkaliaris, K.D.; Schwarzfischer, M.; Anastassiadis, K.; et al. Inductive and selective effects of GSK3 and MEK inhibition on Nanog heterogeneity in embryonic stem cells. *Stem Cell Rep.* **2018**, *11*, 58–69. [CrossRef] [PubMed]
27. Wang, Z.; Oron, E.; Nelson, B.; Razis, S.; Ivanova, N. Distinct lineage specification roles for NANOG, OCT4, and SOX2 in human embryonic stem cells. *Cell Stem Cell* **2012**, *10*, 440–454. [CrossRef] [PubMed]
28. Madeja, Z.E.; Sosnowski, J.; Hryniewicz, K.; Warzych, E.; Pawlak, P.; Rozwadowska, N.; Plusa, B.; Lechniak, D. Changes in sub-cellular localisation of trophoblast and inner cell mass specific transcription factors during bovine preimplantation development. *BMC Dev. Biol.* **2013**, *13*, 32. [CrossRef]
29. Kuijk, E.W.; Du Puy, L.; Van Tol, H.T.A.; Oei, C.H.Y.; Haagsman, H.P.; Colenbrander, B.; Roelen, B.A.J. Differences in early lineage segregation between mammals. *Dev. Dyn.* **2008**, *237*, 918–927. [CrossRef] [PubMed]
30. Ortega, M.S.; Kelleher, A.M.; O'Neil, E.; Benne, J.; Cecil, R.; Spencer, T.E. NANOG is required to form the epiblast and maintain pluripotency in the bovine embryo. *Mol. Reprod. Dev.* **2020**, *87*, 152–160. [CrossRef]
31. Ran, F.A.; Hsu, P.D.; Wright, J.; Agarwala, V.; Scott, D.A.; Zhang, F. Genome engineering using the CRISPR-Cas9 system. *Nat. Protoc.* **2013**, *8*, 2281–2308. [CrossRef] [PubMed]
32. Vochozkova, P.; Simmet, K.; Jemiller, E.-M.; Wünsch, A.; Klymiuk, N. Gene Editing in Primary Cells of Cattle and Pig. *Breast Cancer* **2019**, *1961*, 271–289. [CrossRef]
33. Bauersachs, S.; Ulbrich, S.E.; Zakhartchenko, V.; Minten, M.; Reichenbach, M.; Reichenbach, H.-D.; Blum, H.; Spencer, T.; Wolf, E. The endometrium responds differently to cloned versus fertilized embryos. *Proc. Natl. Acad. Sci. USA* **2009**, *106*, 5681–5686. [CrossRef]
34. Messinger, S.M.; Albertini, D.F. Centrosome and microtubule dynamics during meiotic progression in the mouse oocyte. *J. Cell Sci.* **1991**, *100*, 289–298. [CrossRef] [PubMed]
35. Wuensch, A.; Habermann, F.A.; Kurosaka, S.; Klose, R.; Zakhartchenko, V.; Reichenbach, H.-D.; Sinowatz, F.; McLaughlin, K.J.; Wolf, E. Quantitative Monitoring of Pluripotency Gene Activation after Somatic Cloning in Cattle1. *Biol. Reprod.* **2007**, *76*, 983–991. [CrossRef] [PubMed]
36. O'Brien, J.; Hayder, H.; Peng, C. Automated quantification and analysis of cell counting procedures using ImageJ plugins. *J. Vis. Exp.* **2016**, *117*, e54719. [CrossRef] [PubMed]
37. Tanaka, S.; Kunath, T.; Hadjantonakis, A.-K.; Nagy, A.; Rossant, J. Promotion of trophoblast stem cell proliferation by FGF4. *Science* **1998**, *282*, 2072–2075. [CrossRef]
38. Nichols, J.; Zevnik, B.; Anastassiadis, K.; Niwa, H.; Klewe-Nebenius, D.; Chambers, I.; Schöler, H.; Smith, A. Formation of pluripotent stem cells in the mammalian embryo depends on the POU transcription factor Oct4. *Cell* **1998**, *95*, 379–391. [CrossRef]
39. Molè, M.A.; Coorens, T.H.H.; Shahbazi, M.N.; Weberling, A.; Weatherbee, B.A.T.; Gantner, C.W.; Sancho-Serra, C.; Richardson, L.; Drinkwater, A.; Syed, N.; et al. A single cell characterisation of human embryogenesis identifies pluripotency transitions and putative anterior hypoblast centre. *Nat. Commun.* **2021**, *12*, 3679. [CrossRef] [PubMed]
40. Berg, D.K.; Smith, C.S.; Pearton, D.J.; Wells, D.N.; Broadhurst, R.; Donnison, M.; Pfeffer, P.L. Trophectoderm lineage determination in cattle. *Dev. Cell* **2011**, *20*, 244–255. [CrossRef]
41. Meng, F.; Forrester-Gauntlett, B.; Turner, P.; Henderson, H.; Oback, B. Signal inhibition reveals JAK/STAT3 pathway as critical for bovine inner cell mass development. *Biol. Reprod.* **2015**, *93*, 131–139. [CrossRef] [PubMed]
42. Meilhac, S.M.; Adams, R.J.; Morris, S.A.; Danckaert, A.; Le Garrec, J.-F.; Zernicka-Goetz, M. Active cell movements coupled to positional induction are involved in lineage segregation in the mouse blastocyst. *Dev. Biol.* **2009**, *331*, 210–221. [CrossRef] [PubMed]
43. Kang, M.; Piliszek, A.; Artus, J.; Hadjantonakis, A.-K. FGF4 is required for lineage restriction and salt-and-pepper distribution of primitive endoderm factors but not their initial expression in the mouse. *Development* **2013**, *140*, 267–279. [CrossRef] [PubMed]

 cells

MDPI

Article

Progressive Domain Segregation in Early Embryonic Development and Underlying Correlation to Genetic and Epigenetic Changes

Hui Quan [1,†], Hao Tian [1,†], Sirui Liu [1,†], Yue Xue [1], Yu Zhang [2,3], Wei Xie [2,3] and Yi Qin Gao [1,4,5,*]

1 Beijing National Laboratory for Molecular Sciences, College of Chemistry and Molecular Engineering, Peking University, Beijing 100871, China; quanhui@pku.edu.cn (H.Q.); ccme_th@pku.edu.cn (H.T.); liusirui@pku.edu.cn (S.L.); xueyccme@pku.edu.cn (Y.X.)
2 Center for Stem Cell Biology and Regenerative Medicine, MOE Key Laboratory of Bioinformatics, School of Life Sciences, Tsinghua University, Beijing 100871, China; zhangyu-11@mails.tsinghua.edu.cn (Y.Z.); xiewei121@tsinghua.edu.cn (W.X.)
3 THU-PKU Center for Life Sciences, Tsinghua University, Beijing 100871, China
4 Biomedical Pioneering Innovation Center (BIOPIC), Peking University, Beijing 100871, China
5 Beijing Advanced Innovation Center for Genomics, Peking University, Beijing 100871, China
* Correspondence: gaoyq@pku.edu.cn
† These authors contributed equally.

Abstract: Chromatin undergoes drastic structural organization and epigenetic reprogramming during embryonic development. We present here a consistent view of the chromatin structural change, epigenetic reprogramming, and the corresponding sequence-dependence in both mouse and human embryo development. The two types of domains, identified earlier as forests (CGI-rich domains) and prairies (CGI-poor domains) based on the uneven distribution of CGI in the genome, become spatially segregated during embryonic development, with the exception of zygotic genome activation (ZGA) and implantation, at which point significant domain mixing occurs. Structural segregation largely coincides with DNA methylation and gene expression changes. Genes located in mixed prairie domains show proliferation and ectoderm differentiation-related function in ZGA and implantation, respectively. The chromatin of the ectoderm shows the weakest and the endoderm the strongest domain segregation in germ layers. This chromatin structure difference between different germ layers generally enlarges upon further differentiation. The systematic chromatin structure establishment and its sequence-based segregation strongly suggest the DNA sequence as a possible driving force for the establishment of chromatin 3D structures that profoundly affect the expression profile. Other possible factors correlated with or influencing chromatin structures, including transcription, the germ layers, and the cell cycle, are discussed for an understanding of concerted chromatin structure and epigenetic changes in development.

Keywords: sequence; domain segregation; epigenetic modification; ZGA and implantation

Citation: Quan, H.; Tian, H.; Liu, S.; Xue, Y.; Zhang, Y.; Xie, W.; Gao, Y.Q. Progressive Domain Segregation in Early Embryonic Development and Underlying Correlation to Genetic and Epigenetic Changes. *Cells* **2021**, *10*, 2521. https://doi.org/10.3390/cells10102521

Academic Editor: Lon J. van Winkle

Received: 20 August 2021
Accepted: 17 September 2021
Published: 23 September 2021

1. Introduction

In mammals, chromatin undergoes drastic organizational and epigenetic reprogramming after fertilization [1,2]. These processes are essential for gene regulation, either globally or specifically, by generating a chromatin environment that is permissive or repressive to gene expression [3]. The chromatin of mouse zygotes and two-cell embryos has obscure high-order structures, existing in markedly relaxed states, which undergo the consolidation of TADs (topologically associating domains, that is, self-interacting domains in the 3D space) and segregation of chromatin compartments through development [4,5]. The TAD structure and compartmentalization are also gradually established following human fertilization [6]. Along with the remodeling of the 3D chromatin architecture, genome-wide epigenetic reprogramming also takes place during embryonic development [2,7–12]. The global hypomethylation of the genome occurs, and histone modifications are globally reset, changing from a non-canonical distribution to a canonical one [13–15] during early embryonic development, leading to the specification of the germ layers and cell differentiation.

The progression of the mammalian embryo from fertilization to germ-layer formation, concurrent with transcriptional changes and cell fate transitions [3], involves an ordered series of hierarchical lineage determinations that ensures the establishment of a blueprint for the whole animal body. One of the most notable transcriptional changes is the zygotic genome activation (ZGA), during which the embryo changes from a state where there is little transcription to another state in which thousands of genes are transcribed [16]. The ZGA is mechanistically coordinated with changes in the chromatin state and cell cycle [17]. Mammalian ZGA may primarily prepare for differentiation toward the inner cell mass (ICM) and the trophectoderm (TE), which begins at the morula stage, then the TE can further develop into the extraembryonic tissue necessary for embryo implantation and receiving nutrients. After implantation, the ICM then gives rise to three germ layers (ectoderm, mesoderm and endoderm) through gastrulation, generating founder tissues for subsequent somatic development [18,19].

Thanks to recent developments in low-input chromatin analysis technologies, the chromatin structural, epigenetic and transcriptional properties have been roundly explored in the early embryonic development process [20]. Assuming that chromatin structural properties at the mouse pre-implantation stages have been investigated, analysis on the latest Hi-C data describing chromatin structural properties at post-implantation stages in mice and embryonic development in humans ought to be able to provide a fairly complete view of the structural change during development. On the other hand, the relationship between global genome structural changes, epigenetic reprogramming, and the DNA sequence remains largely unknown and, therefore, needs to be investigated further.

In our previous study, we analyzed the DNA sequence dependence in the formation of 3D chromatin structures for different cell types [21]. Based on CpG island (CGI) densities, the genome was divided into alternative forest (high CGI density, F) and prairie (low CGI density, P) domains with average lengths of 1 to 3 million bases (forest and prairie domains are therefore cell type-invariant in one species). CGI forests and prairies effectively separate the linear DNA sequence into domains with distinctly different genetic, epigenetic and structural properties. 78.5% of human genes (and 91.3% of human housekeeping genes) reside in forest domains. Besides, compartments A and B are mainly composed of forest and prairie regions, and the segregation of the two domains tends to intensify during embryonic development, cell differentiation and senescence as a result of sequence-based thermodynamic stabilization. However, the segregation degree of some somatic cells is less than that of ICM, implying that the chromatin structure changes in a non-monotonic way from zygote to differentiated somatic cells. The way that chromatin conformation gradually changes to establish cell identity during development is thus an interesting open question.

In the present study, we conducted an analysis of global chromosome structural changes during early embryonic development. Two specific stages, ZGA and implantation, during which gene activation and lineage specification occur, respectively, were both found to involve the mixing of the two types of genomic domains, with the latter showing a more significant change than the former. The segregation level positively correlates with the proportion of prairie domains residing in compartment B, the larger value of which suggests a more segregated chromatin structure. We also found that, compared to prairie regions constantly residing in compartment B, genes in switchable prairie regions tend to be tissue-specific. The DNA methylation distribution in early embryonic development also correlates with the trend of domain segregation in this process, which indicates that the chromatins of the endoderm and mesoderm are more segregated than that of the ectoderm. Moreover, the domain segregation levels of the earliest fate-committed germ layers correlate with those of the differentiated cells. We further investigated in detail the chromatin structure changes at different scales during ZGA and found a correlation between LAD mixing and domain mixing in ZGA, which was related to the functions of cell proliferation. The detailed functions of genes residing in more mixed domains during implantation were then analyzed, all of which are ectoderm-related. Finally, we present a consistent view of the structural change of chromatin from birth to senescence and discuss possible factors influencing global chromatin structure, such as transcription, germ layers, and cell division.

2. Materials and Methods

2.1. Overall Segregation Ratio

Based on the Hi-C contact matrix, the inter-domain contact ratio between the same domain types was calculated as:

$$R_f^s = \frac{\sum_{i,\, j\in F,\, i\neq j} D_{ij}}{\sum_{i\in F,\, j\in A} D_{ij}} \text{ and } R_p^s = \frac{\sum_{i,\, j\in P,\, i\neq j} D_{ij}}{\sum_{i\in P,\, j\in A} D_{ij}}$$

for forests and prairies, respectively. In the above equations, D_{ij} represents the sum of contacts between the two domains i and j, which is further divided by the product of domain length of i and j. F is the collection for all forest domains, P is the collection for all prairie domains, and A is the union of sets F and P.

The inter-domain contact between different types was calculated similarly as:

$$R_f^d = \frac{\sum_{i\in F,\, j\in P} D_{ij}}{\sum_{i\in F,\, j\in A} D_{ij}} \text{ and } R_p^d = \frac{\sum_{i\in P,\, j\in F} D_{ij}}{\sum_{i\in P,\, j\in A} D_{ij}}$$

The overall segregation ratio OR_s was then defined as the ratio of inter-domain contacts between the same types and different types:

$$OR_s = \frac{R_f^s}{R_f^d} \text{ or } OR_s = \frac{R_p^s}{R_p^d}$$

for forests and prairies, respectively. All Hi-C data in this calculation were normalized by ICE normalization [22].

2.2. 3D Chromatin Structure Modeling

Detailed information can be found in our previous work [23]. Briefly, we first coarse-grained a chromosome as a string of beads. The equilibrium distance between two beads was obtained by converting the contact frequency to the spatial distance. A randomly generated initial structure was then used for further structure optimization using MD (Molecular Dynamics) simulations until the RMSD (root-mean-square deviation) of the modeled structure became convergent.

2.3. Domain Segregation Ratio Calculation

For each forest/prairie domain i, the domain-segregation ratio DR_s was defined as the ratio between its inter-domain contacts with the same types and its inter-domain contacts with different types:

$$DR_s^{i\in F} = \frac{\sum_{j\in F, j\neq i} D_{ij}}{\sum_{j\in P} D_{ij}} \text{ and } DR_s^{i\in P} = \frac{\sum_{j\in P, j\neq i} D_{ij}}{\sum_{j\in F} D_{ij}}$$

In the above equations, D_{ij} represents the sum of contacts between the two domains i and j, F is the collection of all forest domains, and P is the collection of all prairie domains.

2.4. Distance-Dependent Segregation Ratio Calculation

The segregation ratio at each distance d was calculated as the ratio of contacts with that distance apart between the same types and different types:

$$R_s^f(d) = \frac{\sum_{i\in F, j=i\pm d\in F} C_{ij}}{\sum_{i\in F, j=i\pm d\in P} C_{ij}} \text{ and } R_s^p(d) = \frac{\sum_{i\in P, j=i\pm d\in P} C_{ij}}{\sum_{i\in P, j=i\pm d\in F} C_{ij}}$$

for forests and prairies, respectively. In the above equations, C_{ij} is the normalized Hi-C contact probability between bins i and j, F is the collection for all forest domains, and P is the collection for all prairie domains.

2.5. Open-Sea Methylation Difference Index (MDI) between Forest and Prairie

We quantified the methylation difference between adjacent forests and prairies by:

$$MDI_i = \left(q_i - \frac{q_{i-1} + q_{i+1}}{2}\right) / \left(\frac{q_{i-1} + q_i + q_{i+1}}{3}\right)$$

where q_i, q_{i-1} and q_{i+1} are the average open-sea methylation levels for the i th domain and its two flanking domains.

2.6. Overall Relative Segregation Ratios

Overall relative segregation ratios (R_{or}) between one tissue and another were identified as:

$$R_{or} = ln\left(\frac{N_{ij}^o}{N_{ji}^o}\right)$$

where N_{ij}^o is the number of domains possessing higher DR_s value in tissue i, compared to tissue j. A positive value of R_{or} indicates that the chromatin of the former tissue is more segregated than that of the latter. Thus, the parameter R_{or} is used to generally reflect domain-segregation behavior differences between two tissues.

2.7. Significantly More Segregated or More Mixed Domains

To identify forest or prairie domains that are significantly more (or less) segregated in one tissue when compared to another, we first identified the threshold values to distinguish more (or less) segregated domains at certain stages. We calculated the logarithm ratio of DR_s values between the two ICM replicates from the same laboratory. Significantly more strongly segregated domains are taken as those where the variation of DR_s values are in the top 2.5-percent tier of all domains, and more mixed domains in the bottom 2.5 percent. The corresponding threshold values of DR_s variation $T_{DR_s}^{log}$ for segregation, and mix are then 0.1469 and −0.1469, respectively.

We calculated the logarithm ratio of DR_s between early and late 2-cell embryos for ZGA, ICM and E6.5 epiblast for implantation, respectively. Significantly more strongly segregated domains were identified as those with a logarithm ratio of DR_s that is higher than 0.1469, and significantly more mixed domains were defined as those with a logarithm ratio of DR_s that is lower than −0.1469. For convenience, we denoted more strongly segregated and more mixed domains in forest and prairie as F_{seg}, P_{seg}, F_{mix}, and P_{mix}, respectively.

2.8. Compartment Index Calculation

To quantify the degree of compartmentalization, we defined a compartment index (C-index) I_i for 200-kb bin i as the logarithm ratio of the average contact between this bin and all A compartments, over that between this bin and all B compartments:

$$I_i = ln\frac{\left(\sum_j C_{ij}\delta_j\right)/N_A}{\left(\sum_j C_{ij}(1-\delta_j)\right)/N_B} \delta_j = \begin{cases} 1, & \text{if bin } j \text{ is in compartment A} \\ 0, & \text{if bin } j \text{ is in compartment B} \end{cases}$$

where C_{ij} is the distance-normalized Hi-C contact probability (normalized by dividing each contact by the average contact probability at that genomic distance) between bins i and j. N_A and N_B are the total number of bins belonging to compartments A and B, respectively. For each region, a higher value of I_i indicates a more compartment A-like environment, and a lower one, a compartment B-like environment.

2.9. Gene Function Analysis

We analyzed the functional enrichment of genes located in various selected regions using ClusterProfiler and DAVID (https://david.ncifcrf.gov, accessed on 16 March 2021), which yielded similar results. The results were demonstrated via GO terms with p-values. The functional annotation clustering was analyzed using DAVID, and the results were shown via GO terms with enrichment scores.

2.10. Identification of Lineage-Specific Genes

To identify lineage-specific genes, we used a Shannon-entropy-based method [24]. Genes with an entropy score of less than 1.7 were selected as candidates for stage-specific genes. Among them, we selected E6.5 epiblast-specific genes that satisfied the following conditions: the gene is highly expressed in E6.5 epiblast (FPKM $\geq$ 1); its relative expression is higher than 1/7; and its expression level in epiblast is higher than that in ICM. These genes were then reported in the final lineage-specific gene lists.

2.11. Comparison of Segregation Extent between SINE and CpG Density

For each 40 kb bin (the resolution corresponds to Hi-C matrix resolution), we calculated its CpG and SINE density. Bins where the CpG (SINE) density was larger than the median value of CpG (SINE) density were regarded as high-CpG (SINE) bins, and the remaining bins were low-CpG (SINE) bins. For the two groups of bins, such as hChS and hClS, the segregation extent was calculated as the average contact probability between hChS and hChS, divided by hChS and hClS, and the average contact probability between hClS and hClS, divided by hChS and hClS, respectively.

3. Results

3.1. Domain Segregation in Early Embryonic Development

Previous studies on chromatin structures during early embryonic development were mainly about changes to structural elements, such as compartments and TADs [4,5]. Here we present a structural analysis based on two types of genomic domains, to show domain segregation behaviors in different scales. We analyzed chromatin structural changes by calculating the overall segregation ratio OR_s (one parameter used to globally reflect chromatin segregation behavior), the domain segregation ratio DR_s (which was used to reflect segregation behavior of individual forest/prairie domains), the segregation ratio regarding genomic distances $R_s(d)$ and the F–F (forest–forest)/F–P (forest–prairie)/P–P (prairie–prairie) spatial interaction ratio at each stage.

In our previous study [21], based on the division of the mammalian genomes into forest and prairie domains, OR_s was defined as the inter-domain contact ratio between domains of the same types and different types, regarding each domain as one unit (see "Methods", for instance, OR_s for forest is calculated as forest–forest interactions divided by forest–prairie interactions). A higher OR_s for a sample indicates a stronger segregation. Here, we calculated OR_s using Hi-C data [4,6,20] of 21 mouse cells and 5 human cells (see Additional File 2: Data Sources), which allowed us to investigate the chromatin structure changes following early embryonic development. The calculated OR_s for each stage of mouse and human embryo development is given in 1A and 1B, respectively. We also constructed the modeled 3D chromatin structures following our previous work [23] to show visually the degree of segregation at each stage (see "Methods").

As seen from Figure 1A, OR_s increases during the development of early mouse embryonic cells, except for two dips. Such a trend suggests the increased segregation of forest and prairie domains from each other, in line with our previous observation that more inter-compartment interactions occur at early stages than late stages [4]. The two dips observed along the OR_s curve correspond to the early to late 2-cell, and ICM to E6.5 epiblast transitions, respectively. The decrease during the former transition is seen in both alleles (Figure S1A). For the latter transition, the modeled chromatin structure clearly changes from a domain-separated state to a forest–prairie (F–P) mixed state (Figure 1A). Interestingly, these two processes (that is, early to late 2-cell, and ICM to E6.5 epiblast) exactly coincide with ZGA [25] and implantation [26], respectively. The domain-level F–P chromatin mixing suggested by the decrease in OR_s is possibly related to LAD (lamina-associated domains) mixing at the ZGA stage, as well as germ layer differentiation following implantation, which is analyzed below.

The overall chromatin segregation follows a similar trend in human embryonic development (Figure 1B). The OR_s for prairies generally increases except for the blastocyst-to-6-week transition, in accordance with the observed trend of global segregation and domain-level mixing after the ICM stage in the mouse sample. These changes of domain segregation are also reflected by the 3D structures reconstructed via Hi-C data (Figure 1B). Again, these results suggest that forest and prairie domains tend to segregate from each other before implantation, and domain mixing is observed at the post-implantation stage, showing an increased interaction of the prairie with forest domains.

Besides the overall structure parameter OR_s, we also calculated the domain segregation ratio DR_s on each forest/prairie domain during embryonic development (see "Methods"; a higher value indicates that the corresponding domain is more segregated), and the variation trend is similar to OR_s (Figure 1C (mouse) and Figure 1D (human)). Interestingly, for the mouse embryo, the change in DR_s is more significant for the ICM to E6.5 epiblast transition $P = 3.7 \times 10^{-96}$ by Welch's unequal variance test) than the early to late 2-cell transition $P = 1.9 \times 10^{-16}$ by Welch's unequal variance test) for prairie domains, while for forest domains the former one is less pronounced ($P : 2.4 \times 10^{-5}$ VS 2.1×10^{-54}, Welch's unequal variance test).

Furthermore, to investigate chromatin segregation at different genomic distances, we calculated the segregation ratio as a function of genomic distance $R_s(d)$ (Figures 1E, S1B–D and S2, see "Methods"). It can be seen from Figure 1E that in early mouse embryo development, prairies are characterized by elevated segregation at genomic distances of several million bases (Mb). The segregation becomes more pronounced at long distances (>1 Mb) after the 2-cell stage for both forests and prairies, as is again consistent with an enhancement in domain segregation. Besides, Figure S1B shows a slight decrease of $R_s(d)$ for forests at the early to late 2-cell stage and a conspicuous decrease of $R_s(d)$ for prairies at the ICM to E6.5 epiblast stage for mouse embryos; the underlying biological significances behind these phenomena will be discussed below. Enhanced segregation at a several-Mb scale also occurs during human embryonic development (Figure S1C), and an obvious drop in $R_s(d)$ for prairies at the blastocyst-to-6-week transition in human embryos was also observed (Figure S1D).

Finally, the top 10% of contact probabilities under the genomic distance d (that is, the top 10% of elements of one diagonal of the Hi-C matrix), $TC_{10}(d)$, were extracted, within which the proportions of F–F, F–P and P–P interactions were calculated (notably, the premise of such a calculation is that all elements in the $TC_{10}(d)$ should be non-zero, otherwise the proportions of F–F, F–P and P–P are assigned zero). The genomic distance, after which the proportions of F–F/F–P/P–P remain at zero, is named the critical genomic distance. During the development of mouse embryos, the critical genomic distance gradually increases (Figure 1F), showing the gradual establishment of long-range chromatin contact. At the same time, the F-P ratio descends, while the P–P ratio increases from the PN3 to ICM stages, again hinting that forest and prairie become more spatially segregated, along with the development (Figure 1G). These phenomena were also observed in human embryo development (Figure S1E).

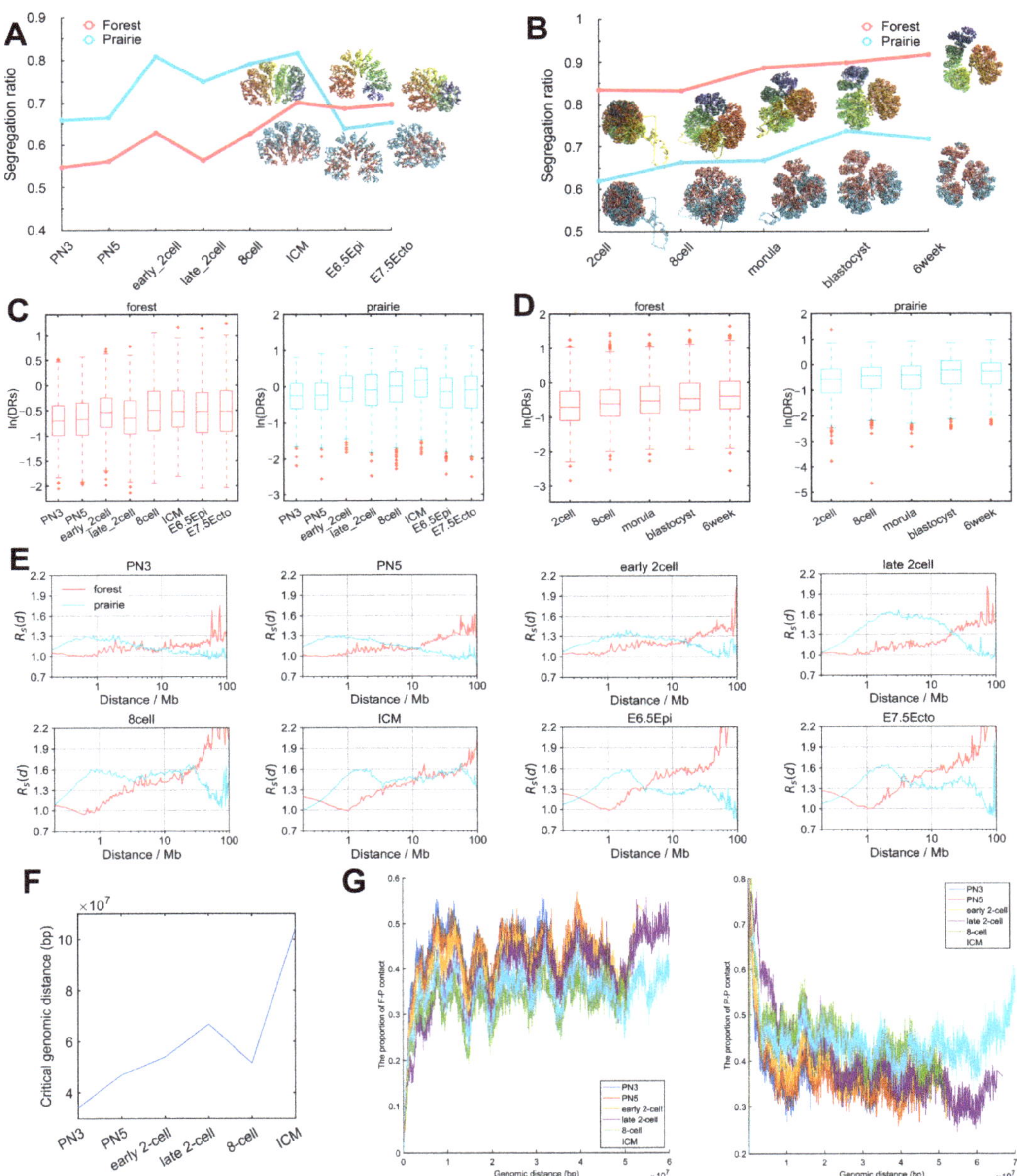

Figure 1. Domain segregation in early embryonic development. (**A**,**B**) The chromatin overall segregation ratio OR_s of forest and prairie domains at each stage in mouse (**A**) and human (**B**) embryonic development. Modeled chromatin structures are also shown at corresponding stages. The upper structures are colored by the sequence from blue to red, and the lower structures are colored by forest/prairie domains. (**C**,**D**) The domain segregation ratio DR_s of forest and prairie domains at each stage in mouse (**C**) and human (**D**) embryonic development. (**E**) The distance-dependent segregation ratio $R_s(d)$ of forest and prairie at each stage in mouse embryonic development. (**F**) The change of critical genomic distance during mouse embryonic development. (**G**) The proportions of F–P and P–P spatial interactions within the top 10% contact probabilities under one certain genomic distance, at each stage in mouse embryonic development.

3.2. Compartment Changes Related to Domain Segregation

By means of the Hi-C measurement, chromatin can be partitioned into two spatial compartments (Figure 2A,B) [27]. It was also found that forest domains reside mainly in compartment A and prairie domains mainly in compartment B in different cells, manifesting that DNA with similar sequence properties tends to spatially interact [21]. Based on these findings, we calculated the proportion for the two sequential domains (forests and prairies) as distributed in the two spatial compartments A and B during development.

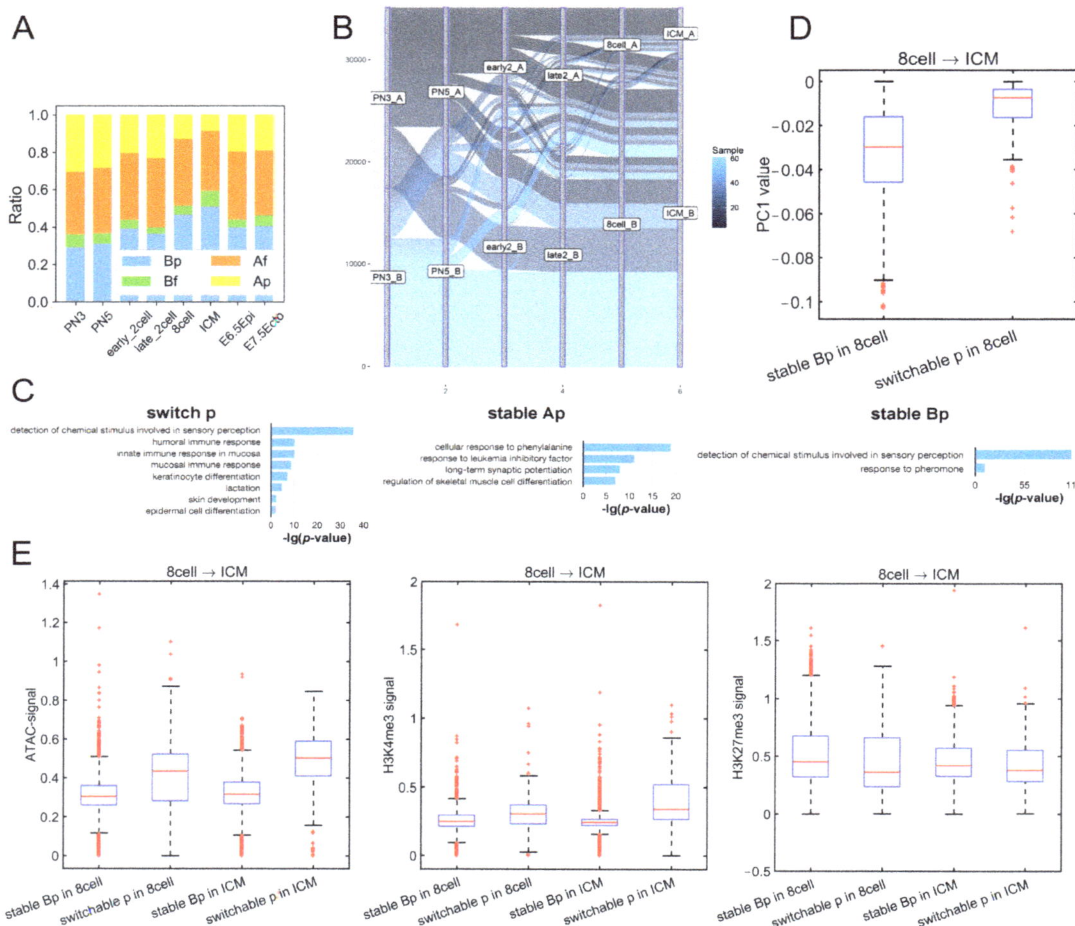

Figure 2. Compartment changes related to domain segregation. (**A**) The proportions of four sequence components, Af, Bf, Ap, Bp (calculated according to the positioning of forests and prairies in compartments **A** and **B**), for each stage in mouse embryonic development. (**B**) The dynamic partition of prairie into compartments **A** and **B** in mouse embryonic development. (**C**) Function annotation clustering of genes located in switch p, stable Ap and stable Bp regions, respectively. (**D**) The PC1 values of stable Bp (i.e., prairie regions constantly remaining in compartment B in both 8-cell and ICM) and switchable p (i.e., prairie regions located in compartment B in 8-cell while switching to compartment A in ICM) regions in 8-cell. (**E**) From 8-cell to ICM, the ATAC-seq, H3K4me3 and H3K27me3 signals of stable Bp and switchable p regions (see notes for Figure 2D) in 8-cell and ICM.

The corresponding four types of DNA sequences are named as Af, Bf, Ap, and Bp, respectively, with the first letter denoting the compartments and the second, the forest (f) or prairie (p) domain. One can see from Figure 2A that, during mouse development, the proportion of prairies in compartment B (Bp) changes in the same trend as the overall segregation ratio, which increases from the PN3 to ICM stage, except for the early to late 2-cell stage, then decreases from ICM to E6.5 epiblast, and slightly increases when the E6.5 epiblast develops into the E7.5 ectoderm. The ratio of Ap changes in a direction opposite to that of Bp. A similar phenomenon was also observed in human embryos, where the changes of OR_S are in accordance with the size of Bp (Figure S3A). A recent study also found that the attractions between heterochromatic regions are crucial for the compartmentalization and domain segregation of the active and inactive genome [28]. Similarly, we found here that, during embryo development, the aggregation behavior of prairie domains (low CpG density region) strengthens, and more and more prairie regions belong to compartment B.

Since the genome compartmentalization is weak at the early stages of development [4], for the robustness of the compartment partition, we calculated the distribution of the compartment index (C-index, a parameter to quantify the degree of compartmentalization, a larger value of which corresponds to a more compartment-A-like environment; see "Methods") at each stage of mouse embryo development for compartments A and B, respectively. The gradually increased discrepancy of the C-index between compartment sA and B supports the global structure segregation from zygote to ICM (Figure S3B). Then, we identified compartment-A regions with a positive C-index as strict compartment A (sA), B regions with a negative C-index as strict compartment B (sB), and calculated the length of the genome located in sAf (forest regions in sA), sAp, sBf and sBp, respectively. The result (Figure S3C) showed a trend similar to Figure 2A; that is, the proportion of prairies in strict compartment B (sBp) changes in the same trend as the chromatin segregation level, which rules out the possibility that the growth of Bp is an artifact due to compartment definition.

In comparison with stable Bp regions, the prairie domains that switch the compartment type are particularly interesting, as they may be regarded as mediators of nuclear architecture establishment during development. The analysis of compartment transformation during mouse preimplantation development shows that 7.7% of prairies belong to compartment A (stable Ap), 26.3% of prairies always reside in compartment B (stable Bp), and the remaining 66.0% switch compartment type at least once (switchable p) (Figure 2B). Genes in the stable Ap and switchable p regions are enriched in immune- and ectoderm-related functions, while those in the stable Bp regions are not (Figure 2C). Previous analyses showed that forest and prairie domains tend to spatially segregate (and mainly correspond to compartment A and B, respectively), but to a different extent in different cells [21]. Moreover, forest–prairie spatial intermingling is cell-type specific, which is thought to be associated with prairie tissue-specific gene activation and the establishment of cell identity. Therefore, compartment B (heterochromatin) provides a silent environment for prairie genes, which wait to be activated through spatial interactions with forest/compartment A in the following differentiation stages (as seen in their spatial contact and expression properties in differentiated cells [29]). The function of genes in the stable Ap and switchable p regions thus supports that the gene expression in prairie domains plays an important role in cell fate determination. Indeed, our earlier analysis shows that prairie genes in compartment A are highly tissue-specific and, by examining the functions of the related genes, one can deduce the tissue type of the associated sample [21]. Besides, compared to forest genes, the expression of genes in prairie domains is more likely to be highly correlated with the compartment environment in which they reside [29].

Next, to further investigate the differences between switchable prairie and stable prairie regions, we compared the chromatin features of these two kinds of regions between two adjacent stages. The cells we analyzed here include PN5, early 2-cell, late 2-cell, 8-cell and ICM, due to the availability of mouse Hi-C data. For every two adjacent stages (i.e., PN5 vs early 2-cell, early 2-cell vs late 2-cell, late 2-cell vs 8-cell, 8-cell vs ICM), we identified stable Bp regions and switchable p regions (p regions located in compartment B in the earlier stage, then switched to compartment A in the later stage). Although these two kinds of prairie regions are both located in compartment B in the cell of the earlier stage, the PC1 value of switchable p regions was significantly more positive than stable Bp regions (Figure 2D and Figure S3D); accordingly, switchable p regions are significantly closer to A–B compartment boundaries (Figure S3E), indicating that Bp regions near the A–B boundary are more likely to switch. The analyses on ATAC-seq signal, H3K4me3 and H3K27me3 also support this observation (Figure 2E). For example, from 8-cell to ICM, the ATAC-seq and H3K4me3 signals of switchable p regions are significantly higher than stable Bp regions in both the 8-cell and ICM, while the H3K27me3 signal of switchable p regions is weaker than stable Bp regions.

3.3. The Association between Domain Segregation and DNA Methylation

In the earlier study, we showed that differences in the methylation levels between forests and prairies correlate well with chromatin spatial packing [21]. In the following, we analyzed methylation data for early mouse embryonic cells obtained by four different research groups (see Additional File 2: Data Sources). These data all show that the open-sea (defined as the genomic regions excluding CGIs, CGI shores and CGI shelves [30]) methylation differences between forest and prairie domains in different cell types correlate well to their corresponding chromatin structural segregation behaviors (Figures 3A and S4A–C). During mouse embryonic development, the absolute value of MDI (F–P open-sea methylation difference index, see "Methods", where a domain possessing a positive (negative) value indicates the methylation level of this domains is generally higher (lower) than its two flanking domains; therefore, a higher absolute value indicates the larger methylation level difference) increases from 2-cell to ICM stages, decreases at the ICM to E6.5 epiblast stage, and increases again in the further development to the E7.5 stage (especially for the E7.5 endoderm stage, Figure 3A). The variation of DNA methylation difference resembles that of the chromatin structural segregation. The correspondence between methylation difference and segregation degree during mouse embryonic development further supports a connection between this epigenetic mark and the chromatin structure. In fact, such a correlation might have a simple explanation: since forests contribute dominantly to the more accessible chromatin regions, they are presumably more prone to both DNA demethylation and methylation than prairies.

Following such an observation, the extent of domain segregation of three germ layers can be inferred from the DNA methylation difference between the forest and prairie domains. As can be seen from Figure 3A, the calculated absolute value of MDI for the endoderm is greater than that for the mesoderm (the corresponding value of the ectoderm is slightly less than that of the mesoderm). This result suggests that the forest–prairie domains of the endoderm are more segregated than those of the mesoderm and ectoderm, which is indeed in accordance with the analysis on mouse Hi-C data (unpublished results). Interestingly, the domain segregation levels of the different tissues can be seen to reflect the germ layers from which they originate, as shown below from the analyses of both DNA methylation and the structural data.

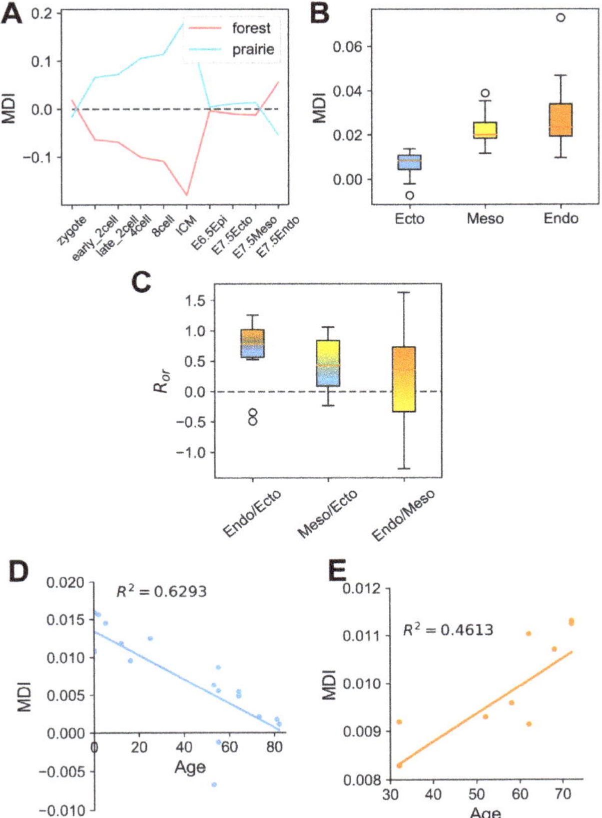

Figure 3. The association between domain segregation and DNA methylation. (**A**) The average forest–prairie open-sea methylation level difference (MDI) at different stages during mouse embryonic development. (**B**) The box plots of the average MDI for human differentiated cells originating from endoderm, mesoderm and ectoderm, respectively, for forest domains. (**C**) The box plots of the overall relative segregation ratio R_{or} (see "Methods") of one tissue over another for forests in human embryos. The three categories are tissues derived from endoderm and mesoderm over those derived from ectoderm, and endoderm over mesoderm. The parameter R_{or} was used to evaluate the differences in domain segregation behaviors between two tissues. For example, as for tissues 1 and 2, N_{12} and N_{21} represent the number of domains showing a higher DR_s in tissues 1 and 2, respectively. The overall relative segregation ratio for tissues 1 and 2 is then defined as $\ln(N_{12}) - \ln(N_{21})$, a positive value of which thus indicates that the chromatin structure of tissue 1 is more segregated, compared to tissue 2. (**D**,**E**) The scatter-plots of age and methylation difference for brain cells (**D**) and mature neutrophil cells (**E**) for forest domains in human embryos.

It was found, based on the Hi-C data, that ectoderm-derived cortex chromatin is less segregated than that of the endoderm-derived liver [21]. Since the above analysis suggests chromatin in the ectoderm to be less segregated than endoderm (Figure 3A), we decided to examine whether chromatin structures of differentiated tissues also show such a trend, therefore exhibiting a germ-layer dependence. We analyzed the average methylation difference levels of 138 mouse differentiated cells and 45 human differentiated cells (38, 34, 66 cells originate from endoderm, mesoderm and ectoderm, respectively, for mouse embryos and 13, 19, 13 for human embryos, see Additional File 2: Data Sources) [31–33].

As shown in Figure 3B, the absolute values of MDI for forest domains for human ectoderm-derived tissues are significantly smaller than those for mesoderm- ($P = 4.938 \times 10^{-7}$ by two-sample t-test) and endoderm-derived $P = 5.696 \times 10^{-4}$ by two-sample t-test) tissues, in the same as the segregation degree of the three embryonic germ layers. The data on the prairie domains of human tissues show the same trend, as do those on both the forest and prairie domains of mouse tissues (Figure S4D–F). To further validate this finding, we analyzed the Hi-C data of 14 human tissues [34] (see Additional File 2: Data Sources) and compared the segregation ratio of these tissues, derived from different germ layers. To quantify the difference between two tissues, we defined and calculated the overall relative segregated ratios R_{or}) of one tissue over another, a positive value of which represents that chromatin in the former is more segregated than that in the latter (see "Methods"). For human tissues derived from the endoderm and mesoderm, their overall relative segregated ratios over those originating from the ectoderm (cortex and hippocampus) are generally positive (Figure 3C), indicating that the chromatin of the former indeed tends to be more segregated than that of the latter. Similarly, the overall relative segregation ratios of tissues derived from the endoderm over those derived from mesoderm also tend to be positive (Figures 3C and S4G). Together, these results are all consistent with the notion that the segregation level of a differentiated cell shows the corresponding germ layer signature, that is, the order of segregation degrees for the three germ layers is in accordance with that of differentiated tissues.

More interestingly, we found that the absolute values of MDI for the brain decrease with aging, and the trend is opposite for mature neutrophil cells in humans, as shown in Figure 3D,E. Earlier studies have found that DNA hypomethylation, which is more likely to occur in prairie domains, correlates with the cell cycles. Partially methylated domains (PMDs) in tumor cells are mainly composed of prairies [35], and PMD hypomethylation increases with age, which appears to track the accumulation of cell divisions [36]. Xuan Ming et al. also found that solo-WCGW sites display aging- and cancer-associated hypomethylation, which exhibits low maintenance efficiency during the cell cycle [37]. For chromatin structural changes, we compared the segregation ratio in the G1 stage to that in the late S~G2 stage, and found that forests and prairies tend to become more separated in the late S~G2 stage [35]. The enhanced segregation supports the hypothesis that mitosis is conducive to a more segregated chromatin structure. Therefore, we speculate that the different methylation patterns associated with aging among different tissues may reflect their different cell division patterns: the ectoderm-originating brain cells hardly divide, while liver cells constantly undergo cell cycles. The observed MDI differences between these cells are consistent with their different dividing patterns in life. Such consistency makes an understanding of the mechanistic connection between methylation and cell-division patterns highly desirable.

3.4. ZGA and Associated 3D Genome Architecture Change

To further investigate how transcription is associated with chromatin structure, we analyzed the chromatin structure changes from mouse early 2-cell to late 2-cell at different genomic distance scales, since ZGA typically occurs at this period. At small genomic distances (<500 kb), the F–P ratio for the late 2-cell is larger than that for early 2-cell, while the P–P ratio is smaller for the former (Figure 4A, upper two figures). Our previous work has revealed that prairie–forest intermingling was associated with prairie gene activation [21]; therefore, we speculate that the increase of F–P ratio may be associated with ZGA. At large distances (500 kb~20 Mb), the P–P ratio (F–P ratio) of the late 2-cell was larger (smaller) than the early 2-cell (Figure 4A, upper two figures), indicating that at such a scale, the forest–prairie separation is enhanced, in accordance with the increase of compartmentalization degree [4] (given that compartment A and B are mainly composed of forest and prairie regions, respectively). Furthermore, compared to the early 2-cell, the chromatin of a late 2-cell establishes more spatial interactions over longer distances (>50 Mb, Figure 4A, down two figures). However, the compartmentalization is weak at this scale (compared to the F–P/P–P ratio of ICM in the same range (i.e., >50 Mb) and the late 2-cell itself in a small range (i.e., <50 Mb), Figure 4A, down two figures).

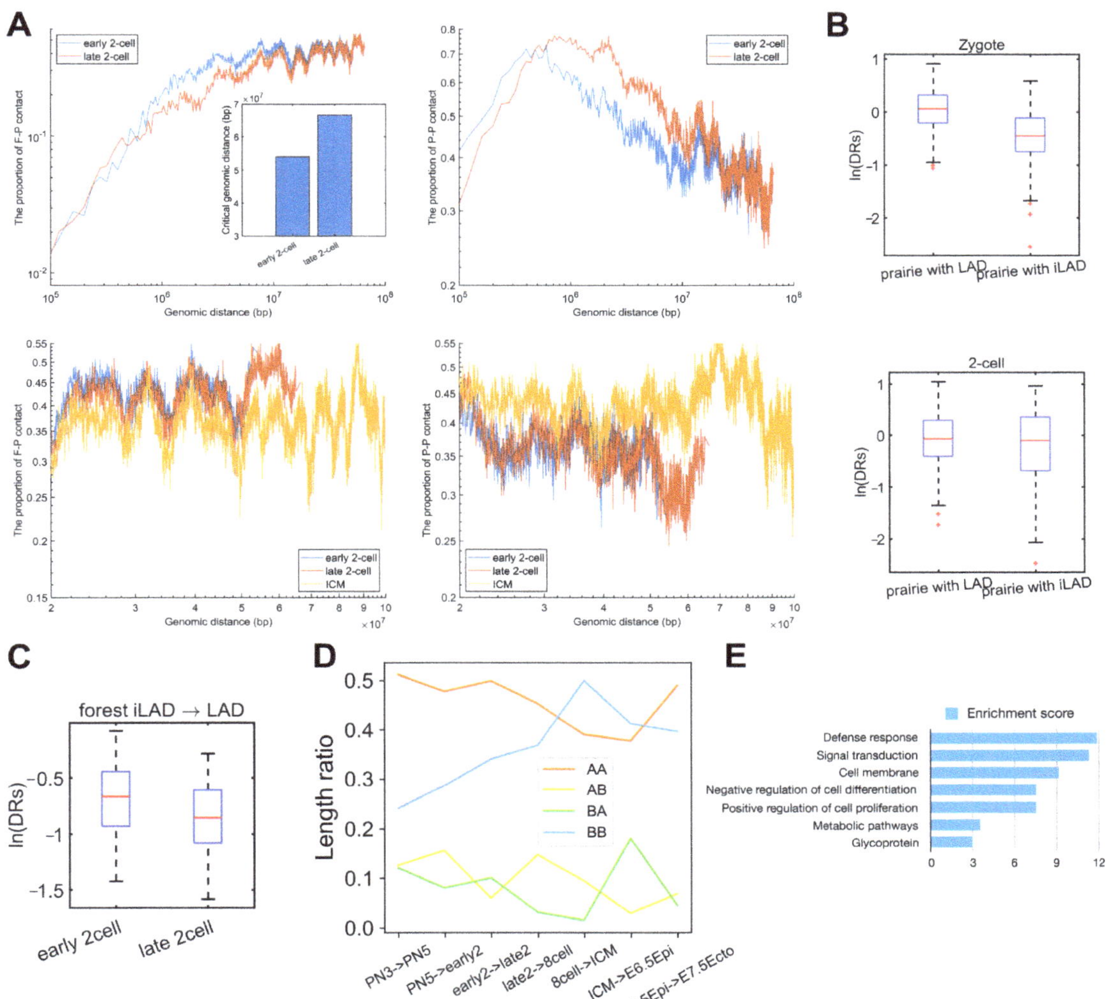

Figure 4. ZGA and associated 3D genome architecture change. (**A**) The upper two figures represent the proportion of F–P and P–P spatial interactions within the top 10% contact probabilities under one certain genomic distance, for normal mouse early 2-cell, late 2-cell. The lower two figures were the amplifications of the partial regions of two upper figures and ICM was also included for comparison. The inner plot represents the critical genomic distance of early 2-cell and late 2-cell. (**B**) The domain segregation ratio (DR_s) of prairie domains that are/are not in LAD regions in mouse embryos. (**C**) The domain segregation ratio of forest domains switching from iLAD to LAD during mouse ZGA, in early 2-cell and late 2-cell. (**D**) The proportion of four components (AA: A→A, AB: A→B, BA: B→A, BB: B→B) in each transition of mouse embryos (e.g., from early 2-cell to late 2-cell). The length ratios of, e.g., AA for PN3 -> PN5 indicates the ratio between the genome length of stable A regions (between PN3 and PN5) and the entire genome length. (**E**) Function annotation clustering of prairie genes switching from B to A during mouse ZGA.

LAD domains, which resemble compartment B, are established quickly after fertilization. More than 75% of LAD domains are prairies across preimplantation development in mouse embryos, and the domain segregation ratio (DR_s) of prairie domains overlapping with LAD is significantly larger than those that are not (Figure 4B, $P = 2.9 \times 10^{-34}$ and 0.01 for mouse zygote and 2-cell, respectively, Welch's unequal variance test). We further analyzed the correlation between LAD mixing and prairie/forest mixing at the ZGA stage in mouse embryos. LAD/iLAD (inter-LAD) regions in the zygote and 2-cell stages were obtained from previous studies [38]. We then identified forest/prairie domains that become significantly more intermingled or segregated (i.e., $F_{mix}F_{seg}/P_{mix}/P_{seg}$, see "Methods") during mouse ZGA. The number of forest domains switching from an iLAD to LAD state is 139, half of which (70 domains) belong to F_{mix}. In contrast, the proportion of F_{mix} domains within forest domains remaining in the iLAD state is only 24.8% ($P = 1 \times 10^{-7}$ by Fisher's exact test). Accordingly, forests that change from iLAD to LAD regions show a significant decrease in domain segregation ratios (Figure 4C, $P = 6.7 \times 10^{-5}$ by two-sample t-test). Such a result indicates that the forests changing from iLAD to LAD gain contacts with prairies, and a correlation does exist between LAD-mixing and forest-mixing during ZGA.

We next examined how a compartment switch is associated with ZGA. Firstly, we examined the compartment change during mouse embryonic development. The overall length of genomic domains changing from compartment A to B is shorter than those switching from compartment B to A in ZGA (Figure 4D). Such a phenomenon was also observed in implantation (from ICM to E6.5Epi). In addition, the prairie genes that move from compartment B to A in ZGA were found to be rich in the functional annotation of "defense response", "signal transduction", "cell membrane", "negative regulation of cell differentiation" and "positive regulation of cell proliferation," identified using DAVID [39,40] (see "Methods", Figure 4E). These results indicate that the domain mixing during ZGA is heavily related to the functions of cell proliferation.

3.5. Implantation-Related Domain Mixing in Differentiation

Since it was shown that the forest–prairie inter-domain interaction coincides with tissue-specific gene activation in differentiation [21], one wonders how the achievements of tissue-specific functions are reflected by domain mixing in implantation. We again calculated the domain segregation ratio DR_s for each forest/prairie domain at related mouse embryonic developmental stages (i.e., ICM and E6.5Epi), and observed that DR_s decreases for 89.0% of prairie domains at implantation (Figure 5A). To be more specific, we identified those DNA domains that undergo significant changes of segregation, and denoted more strongly segregated and more mixed domains in forest and prairie as F_{seg}, P_{seg}, F_{mix}, and P_{mix}, respectively (see "Methods"). In the implantation stage, 71.0% of prairie domains belong to P_{mix}, while 21.3% of forest domains become more mixed (Figure 5B).

As for P_{mix} regions following mouse implantation, we first examined their structural changes by calculating the compartment index (C-index) changes, a larger value of which corresponds to a more compartment A-like environment (see "Methods"). Along with the ICM to E6.5 epiblast transition, 86.9% of P_{mix} domains have an increase in the C-index (entering a more compartment A-like environment), implying that genes in these regions move to a more active environment (Figure 5C). Next, we analyzed the functional enrichment for genes located in the P_{mix} domains using ClusterProfiler [41] and DAVID [39,40] (see "Methods"). The associated P_{mix} genes are characterized by "sensory perception of taste", "regulation of lactation", "keratinocyte differentiation", "epidermal cell differentiation", "epidermis development", "skin development", "mammary gland development" and "long-term synaptic potentiation" (Figure 5D). Interestingly, these terms are related to mammary glands, the epidermis and nervous system, all of which are differentiated from the ectoderm [42]. In fact, chromatin structures of E6.5 epiblast and E7.5 ectoderm are also similar, as seen by their similar DR_s values (Figure 5E). In addition, during the mouse implantation process, the expression level of 309 prairie genes increased, while 162 prairie genes showed a decreased expression level. Here, we further found that prairie domains harboring genes with elevated expression levels tend to become more forest–prairie-mixed from ICM to E6.5Epi, compared to genes with lowered expression (Figure S5A, $P = 4.163 \times 10^{-4}$ by two-sample t-test), as is consistent with our previous finding on the vital role of forest–prairie spatial interactions in prairie gene activation [21]. Moreover, we found that P_{mix} genes with increased expression tend to be enriched in neuro-related functions (Figure S5B), indicating that the neuroectoderm may differentiate ahead of the epidermal ectoderm. Finally, to confirm the roles of domain segregation changes on biological functions, we analyzed the structural changes of domains containing lineage-specific genes. We identified 552 lineage-specific genes for the E6.5 epiblast stage (see "Methods", i.e., genes specifically and highly expressed in the E6.5 epiblast stage), among which 474 genes locate in forest domains and 78 genes in prairie domains. Ninety-seven percent of the prairie lineage-specific genes were located in P_{mix}, suggesting that the mixing of prairies into forests does associate with lineage specification during implantation. Meanwhile, the prairie genes that move from compartment B to A during implantation are also functionally enriched in ectoderm differentiation and in utero embryonic development (Figure S5C), in accordance with the above analysis on P_{mix} regions.

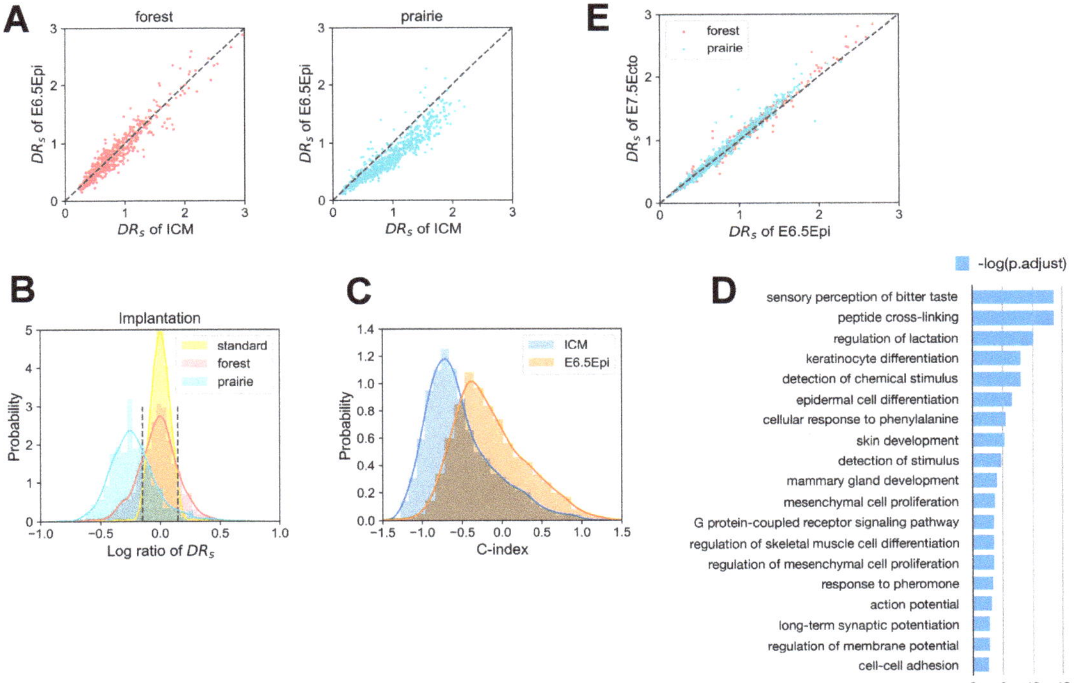

Figure 5. Implantation-related domain mixing in differentiation. (**A**) The scatter diagrams of the domain segregation ratio DR_s for forests and prairies during mouse implantation. (**B**) The distribution of the logarithm ratio of DR_s between mouse E6.5Epi and ICM for forests and prairies. The plot, the legend of which is "standard", represents the logarithm ratio of DR_s between two ICM replicates from the same lab. Significantly more strongly segregated domains are taken, as those whose logarithm ratio of DR_s values exceed the 97.5th percentile of the "standard" distribution; accordingly, domains where the logarithm ratio of DR_s is smaller than the 2.5th percentile of the "standard" distribution are regarded as significantly more mixed domains. The corresponding threshold values are 0.1469 and –0.1469, respectively, which was labeled using two black dotted lines. (**C**) The change of compartment index (C-index) for less-segregated prairies, P_{mix}, from the mouse ICM to E6.5 epiblast stage. (**D**) Functional assignment of genes in the P_{mix} during mouse implantation. (**E**) The scatter plot of DR_s for forests and prairies between the mouse E6.5 epiblast and E7.5 ectoderm.

4. Discussion

4.1. Sequence-Based Chromatin Domain Segregation

In the present study, we described the chromatin structural changes during embryonic development from the zygote to the post-implantation stages and, in particular, how the DNA sequence is associated with the chromatin structural changes. We also investigated the relationship between DNA domain segregation and genetic/epigenetic properties. Earlier studies have shown a gradual compartmentalization in embryonic development [4]. Here, we connect the 3D structure change to DNA sequence properties and perform a functional analysis of these structural changes. The sequence-structure-function analysis provides a common ground for the understanding of different biological processes.

In particular, two types of genomic domains as defined earlier, forests and prairies, are shown to generally undergo an enhanced spatial separation in early embryonic development in both mouse and human embryos, which is also reflected in the DNA methylation and gene expression difference between the two domain types. In fact, it is possible that the different behaviors of DNA with different sequence properties could be a result of the different epigenetic events and different proteins differentially enriched at forest/prairie domains, which are DNA-sequence dependent. It was reported that L1 and B1 became gradually segregated during early embryonic development [43]. Herein, we divided the human genome into four parts: high-CpG-high-SINE (hChS), high-CpG-low-SINE (hClS), low-CpG-high-SINE (lChS) and low-CpG-low-SINE (lClS) (see "Methods") and compared the segregation behaviors among hChS and hClS, lChS and lClS, hChS and lChS, hClS and lClS (see "Methods"), aiming to identify the fundamental sequence factors in chromatin segregation. The results revealed that the segregation extent of the latter two groups (that is, hChS and lChS (1.238 and 1.115), hClS and lClS (1.257 and 1.224)) was generally higher than the former groups (hChS and hClS (1.164 and 0.939), lChS and lClS (1.127 and 0.983)), indicating that compared to interspersed elements, CpG density variation may play a fundamental role in chromatin compartmentalization.

Interestingly, noticeable domain mixing does occur at two important stages, ZGA, when zygotic genes begin to express, and implantation, during which differentiation starts to form germ layers. During the mouse ZGA process, short-range (<500 kb) forest–prairie spatial interactions increase, which event was thought to be associated with gene activation since our previous work has revealed that prairie genes tend to move to a more forest environment for activation [29]. Besides, we also investigated in detail the chromatin structural changes during the ZGA process at different scales. Notably, from human 2-cell to 8-cell, we did not observe domain mixing (Figure 1B). It is believed that human ZGA begins at ~4-8-cell stage; unfortunately, the Hi-C data of the human 4-cell is not available. Structural analyses between the 2-cell and 4-cell, as well as between 4-cell and 8-cell, may help us to gain more insights into the relationship between the human ZGA and chromatin structure. During implantation, we observed a conspicuous decrease in the segregation ratio $R_s(d)$ for prairies at large genomic distances (>3 Mb). Intriguingly, almost all mouse E6.5 epiblast-specific prairie genes resided in P_{mix} domains, again emphasizing the intimate link between domain mixing and gene activation (lineage specification). Genes in the prairie domains that become more mixed in implantation are prominently associated with ectoderm differentiation.

An analysis of Hi-C data in differentiated and senescent cells [44] showed a consistently enhanced segregation of chromatin structures, with a gradual establishment of long-range DNA contacts from zygote to senescence. At the zygotic stage, few high-order structural features exist, and the chromatin is organized similarly to a random coil (Figure 6A). As the embryo develops, local structures (loops and TADs) become more prominent and the two different types of domains segregate from each other to form compartments. Such a trend continues through ICM (Figure 6B). After implantation and as differentiation starts, a subset of prairie domains tends to mix into the active environment, activating associated genes (Figure 6C). During senescence, prairie domains congregate further, some of which detach from nuclear membranes and cluster inside the nucleus, permitting long-range contact establishment [45] (Figure 6D). Our previous analysis showed that chromatin domain segregation continues in cell differentiation and senescence [44]. It appears that the overall trend of chromatin structure change follows the establishment of long-range contacts from birth to senescence, as seen from the segregation level change (Figure S6A–C). In this process, intra-domain contacts are established first, followed by long-range inter-domain contact formation, and then the establishment of lineage-specific inter-domain contacts, which relates to cell differentiation (Figure S6D–G).

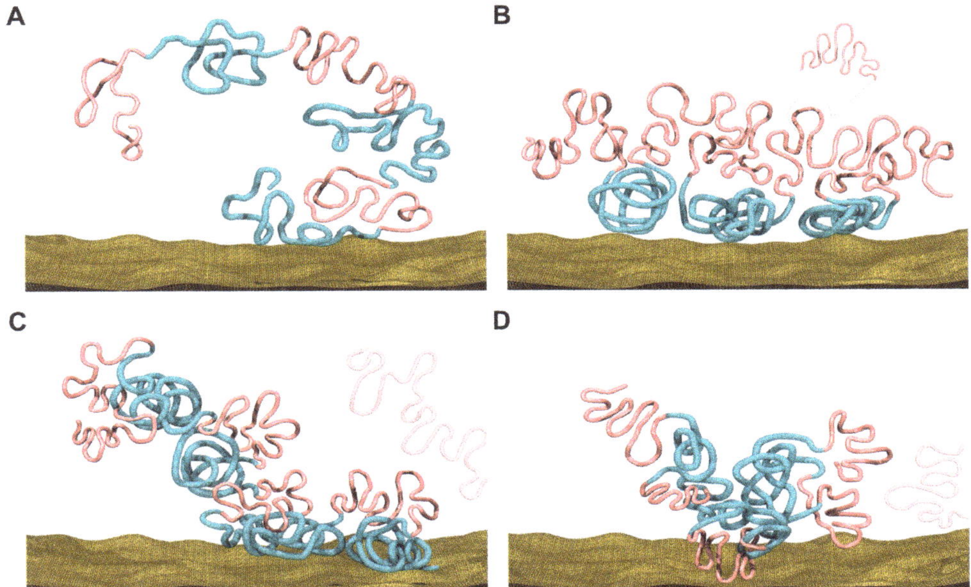

Figure 6. Schematic pictures of chromatin structural patterns at different stages. (**A**) In the beginning zygotic stage, chromatin has few long-range structural features. Here, the red lines represent CGI forest domains, and the blue ones represent CGI prairie domains. (**B**) In ICM stage, the two different types of domains segregate from each other to form compartments. (**C**) In differentiated cells, prairies containing tissue-specific genes tend to form contacts with forests. (**D**) During senescence, prairies congregate further (with an increased probability to detach from the nuclear membrane).

4.2. The Association between Transcription Inhibition and Genome Architecture

In this study, we also investigated the association between transcription and chromatin structure. The increase in the F–P ratio at small genomic distances (<500 kb) tends to be associated with ZGA. The effect of transcription on domain mixing during ZGA can also be tested using available mouse α-amanitin-treated Hi-C data. To investigate how transcription inhibition affects the chromatin architecture, we compared the F–P and P–P ratios within the top 10% contact probabilities under one certain genomic distance between the PN3, normal late 2-cell, and α-amanitin-treated-20 h cell (α-amanitin-treated late 2-cell). The critical genomic distance of α-amanitin-treated-20 h cell is significantly smaller than that of the normal late 2-cell, and the F–P and P–P ratios of α-amanitin-treated-20 h cell are between PN3 and normal late 2-cell (Figure S7A), indicating that transcription inhibition can slow down the establishment of long-range contact and forest–prairie separation. We further compared the F–P and P–P ratios between the α-amanitin-treated-20 h cell and α-amanitin-treated-45 h cell, and found that the critical genomic distance and P–P ratio of the latter are larger than the former, the F–P ratio of the latter is smaller than the former (Figure S7B), hinting that when transcription is inhibited, long-range chromatin contacts can still be established, and forest–prairie separation also progresses, although at a reduced rate. This observation is in keeping with previous findings that the maturation of higher-order chromatin organization can at least partially proceed in the absence of zygotic transcription [4].

4.3. Other Factors That May Relate to Chromatin Structure Change

Sequence-based chromatin-domain segregation suggests that factors affecting domain interactions could lead to functional chromatin structure changes. The cell cycle is an important factor that could influence chromatin domain segregation. During embryonic development, the establishment of a TAD structure was reported as requiring DNA replication but not zygotic transcription [4,5], implying the critical role of cell division in 3D structural formation. In fact, our analysis shows that the different cell division behaviors of different tissues, including those derived from different germ layers, further increase their discrepancy in structural segregation as well as in methylation pattern. The current and earlier analyses indicate that DNA methylation is in fact associated closely with chromatin conformation [21]. The age-related hypomethylation appears to closely track the accumulation of cell divisions [36]. We speculate that, along with hypomethylation (leading to the formation of partially methylated domains), prairie domains tend to segregate more stably in the less active spatial domains of heterochromatin [46].

During mouse embryonic development (from zygote to blastocyst), the nucleocytoplasmic ratio gradually increases, accompanied by a decrease in nuclear size [47]. The change of nucleocytoplasmic ratio, in principle, could have an effect on the decay of contact probability by affecting the crosslinking efficiency in the Hi-C experiment, although in the procedure of sisHi-C, excessive reagent was used to ensure a sufficient and quick fixation. In addition, nuclear lamina is known to play an indispensable role in chromatin structure formation and maintenance. LADs contribute to priming B compartments and remain as the most stable chromatin domains, playing an important role in global chromatin segregation. One previous study [48] also unveiled a connection between nuclear size and nuclear lamina, which may also contribute to a change to the landscape of chromatin domain segregation. These issues remain to be fully addressed by experimental studies.

Since precise gene expression profiles are essential for normal embryo development, factors influencing gene regulation were suggested as impacting the development process [49]. For instance, OCT4, one of the four Yamanaka factors, was thought to play a vital role in development. Accurate epigenetic patterns (such as DNA methylation, histone modifications), as well as high order genome organization, are needed to ensure the normal regulation of OCT4. Therefore, it is worthwhile to understand the regulatory mechanism behind specific gene regulation during development. These include but are not limited to how DNA methylation, active/repressive histone mark loading, and chromatin domain segregation vary before and after gene activation/repression. It is interesting to uncover the interplay between different epigenetic makers and the high-order chromatin structure, to gain more insight into the roles that these various factors play in embryo development.

5. Conclusions

In this study, we delineated chromatin structural and epigenetic reprogramming in early embryonic development, based on the sequence-based domain segregation model, and correlated the overall chromatin structural changes with functional implementation during ZGA and implantation. The two types of DNA domains gradually separate from each other during embryonic development, but they show a tendency to mix when transcription and implantation happen. Increased interactions between the two types of domains indicate the start of transcription or set the stage for further cell differentiation, leading to lineage commitment. Genes of changed segregation states show lineage-relevant functions and important roles in the corresponding processes; thus, more detailed analyses on gene expression and the regulation of transcription factors are needed for understanding the related molecular mechanisms and, more importantly, to predict biological functions at the molecular level from the perspective of domain segregation and mixing-related chromatin structure formation.

Supplementary Materials: The following are available online at https://www.mdpi.com/article/10.3390/cells10102521/s1, Additional File 1: Supplemental information.docx (Supplementary text: Further investigation on the relationship among sequence, compartmentalization and expression, Figure S1: Domain segregation in early embryonic development, Figure S2: Comparisons between maternal and paternal genomes, Figure S3: Compartment changes related to domain segregation, Figure S4: Domain segregation and differentiated DNA methylation associated with germ layer formation, Figure S5: Implantation-related domain mixing in differentiation, Figure S6: Chromatin structural patterns at different stages, Figure S7: The association between transcription inhibition and genome architecture), Additional file 2: Data_Sources.xlsx.

Author Contributions: Conceptualization, Y.Q.G., H.Q., H.T. and S.L.; methodology, Y.Q.G., H.Q., H.T. and S.L.; software, H.Q., H.T., S.L. and Y.X.; validation, Y.Z. and W.X.; formal analysis, H.Q., H.T., S.L. and Y.X.; investigation, Y.Z. and W.X.; writing—original draft preparation, H.Q. and H.T.; writing—review and editing, Y.Q.G., H.Q., H.T., S.L. and Y.X.; visualization, H.Q., H.T. and Y.X.; supervision, Y.Q.G.; funding acquisition, Y.Q.G. All authors have read and agreed to the published version of the manuscript.

Funding: This research was funded by National Natural Science Foundation of China [92053202,22050003].

Institutional Review Board Statement: The data used in this study is downloaded from TCGA, GEO and ENCODE, which are public databases that have obtained ethical approval.

Informed Consent Statement: The study does not require informed consent as it only used information from TCGA, GEO and ENCODE.

Data Availability Statement: All data analyzed during this study are publicly available. The detailed data accession can be found in Additional file 2: Data_Sources.xlsx.

Acknowledgments: We would like to thank Xiaoliang S. Xie for helpful discussions, Zhenhai Du for the initial processing of Hi-C data, and Yupeng Huang for helping us to draw Figure 6.

Conflicts of Interest: The authors declare no conflict of interest.

References

1. Zheng, H.; Xie, W. The role of 3D genome organization in development and cell differentiation. *Nat. Rev. Mol. Cell Biol.* **2019**, *20*, 535–550. [CrossRef]
2. Xu, Q.; Xie, W. Epigenome in early mammalian development: Inheritance, reprogramming and establishment. *Trends Cell Biol.* **2018**, *28*, 237–253. [CrossRef]
3. Burton, A.; Torres-Padilla, M.-E. Chromatin dynamics in the regulation of cell fate allocation during early embryogenesis. *Nat. Rev. Mol. Cell Biol.* **2014**, *15*, 723–735. [CrossRef]
4. Du, Z.; Zheng, H.; Huang, B.; Ma, R.; Wu, J.; Zhang, X.; He, J.; Xiang, Y.; Wang, Q.; Li, Y.; et al. Allelic reprogramming of 3D chromatin architecture during early mammalian development. *Nature* **2017**, *547*, 232–235. [CrossRef]
5. Ke, Y.; Xu, Y.; Chen, X.; Feng, S.; Liu, Z.; Sun, Y.; Yao, X.; Li, F.; Zhu, W.; Gao, L.; et al. 3D chromatin structures of mature gametes and structural reprogramming during mammalian embryogenesis. *Cell* **2017**, *170*, 367–381.e20. [CrossRef]
6. Chen, X.; Ke, Y.; Wu, K.; Zhao, H.; Sun, Y.; Gao, L.; Liu, Z.; Zhang, J.; Tao, W.; Hou, Z.; et al. Key role for CTCF in establishing chromatin structure in human embryos. *Nature* **2019**, *576*, 306–310. [CrossRef]
7. Morgan, H.D.; Santos, F.; Green, K.; Dean, W.; Reik, W. Epigenetic reprogramming in mammals. *Hum. Mol. Genet.* **2005**, *14*, R47–R58. [CrossRef]
8. Burton, A.; Torres-Padilla, M.-E. Epigenetic reprogramming and development: A unique heterochromatin organization in the preimplantation mouse embryo. *Brief. Funct. Genom.* **2010**, *9*, 444–454. [CrossRef] [PubMed]
9. Wang, C.; Liu, X.; Gao, Y.; Yang, L.; Li, C.; Liu, W.; Chen, C.; Kou, X.; Zhao, Y.; Chen, J.; et al. Reprogramming of H3K9me3-dependent heterochromatin during mammalian embryo development. *Nat. Cell Biol.* **2018**, *20*, 620–631. [CrossRef]
10. Liu, X.; Wang, C.; Liu, W.; Li, J.; Li, C.; Kou, X.; Chen, J.; Zhao, Y.; Gao, H.; Wang, H.; et al. Distinct features of H3K4me3 and H3K27me3 chromatin domains in pre-implantation embryos. *Nature* **2016**, *537*, 558–562. [CrossRef]
11. Dahl, J.A.; Jung, I.; Aanes, H.; Greggains, G.D.; Manaf, A.; Lerdrup, M.; Li, G.; Kuan, S.; Li, B.; Lee, A.Y.; et al. Broad histone H3K4me3 domains in mouse oocytes modulate maternal-to-zygotic transition. *Nature* **2016**, *537*, 548–552. [CrossRef] [PubMed]
12. Zhang, B.; Zheng, H.; Huang, B.; Li, W.; Xiang, Y.; Peng, X.; Ming, J.; Wu, X.; Zhang, Y.; Xu, Q.; et al. Allelic reprogramming of the histone modification H3K4me3 in early mammalian development. *Nature* **2016**, *537*, 553–557. [CrossRef] [PubMed]

13. Zheng, H.; Huang, B.; Zhang, B.; Xiang, Y.; Du, Z.; Xu, Q.; Li, Y.; Wang, Q.; Ma, J.; Peng, X.; et al. Resetting epigenetic memory by reprogramming of histone modifications in mammals. *Mol. Cell* **2016**, *63*, 1066–1079. [CrossRef] [PubMed]
14. Smith, Z.D.; Meissner, A. DNA methylation: Roles in mammalian development. *Nat. Rev. Genet.* **2013**, *14*, 204–220. [CrossRef]
15. Lee, H.J.; Hore, T.A.; Reik, W. Reprogramming the methylome: Erasing memory and creating diversity. *Cell Stem Cell* **2014**, *14*, 710–719. [CrossRef] [PubMed]
16. Lu, F.; Liu, Y.; Inoue, A.; Suzuki, T.; Zhao, K.; Zhang, Y. Establishing chromatin regulatory landscape during mouse preimplantation development. *Cell* **2016**, *165*, 1375–1388. [CrossRef]
17. Jukam, D.; Shariati, S.A.M.; Skotheim, J.M. Zygotic genome activation in vertebrates. *Dev. Cell* **2017**, *42*, 316–332. [CrossRef]
18. Arnold, S.J.; Robertson, E.J. Making a commitment: Cell lineage allocation and axis patterning in the early mouse embryo. *Nat. Rev. Mol. Cell Biol.* **2009**, *10*, 91–103. [CrossRef]
19. Lawson, K.A.; Meneses, J.J.; Pedersen, R.A. Clonal analysis of epiblast fate during germ layer formation in the mouse embryo. *Development* **1991**, *113*, 891–911. [CrossRef] [PubMed]
20. Zhang, Y.; Xiang, Y.; Yin, Q.; Du, Z.; Peng, X.; Wang, Q.; Fidalgo, M.; Xia, W.; Li, Y.; Zhao, Z.; et al. Dynamic epigenomic landscapes during early lineage specification in mouse embryos. *Nat. Genet.* **2018**, *50*, 96–105. [CrossRef]
21. Liu, S.; Zhang, L.; Quan, H.; Tian, H.; Meng, L.; Yang, L.; Feng, H.; Gao, Y.Q. From 1D sequence to 3D chromatin dynamics and cellular functions: A phase separation perspective. *Nucleic Acids Res.* **2018**, *46*, 9367–9383. [CrossRef]
22. Servant, N.; Varoquaux, N.; Lajoie, B.R.; Viara, E.; Chen, C.-J.; Vert, J.-P.; Heard, E.; Dekker, J.; Barillot, E. HiC-Pro: An optimized and flexible pipeline for Hi-C data processing. *Genome Biol.* **2015**, *16*, 259. [CrossRef]
23. Xie, W.J.; Meng, L.; Liu, S.; Zhang, L.; Cai, X.; Gao, Y.Q. Structural modeling of chromatin integrates genome features and reveals chromosome folding principle. *Sci. Rep.* **2017**, *7*, 2818. [CrossRef]
24. Xie, W.; Schultz, M.D.; Lister, R.; Hou, Z.; Rajagopal, N.; Ray, P.; Whitaker, J.W.; Tian, S.; Hawkins, R.D.; Leung, D.; et al. Epigenomic analysis of multilineage differentiation of human embryonic stem cells. *Cell* **2013**, *153*, 1134–1148. [CrossRef]
25. Svoboda, P.; Franke, V.; Schultz, R.M. Sculpting the transcriptome during the oocyte-to-embryo transition in mouse. *Curr. Top. Dev. Biol.* **2015**, *113*, 305–349. [CrossRef]
26. Bedzhov, I.; Zernicka-Goetz, M. Self-organizing properties of mouse pluripotent cells initiate morphogenesis upon implantation. *Cell* **2014**, *156*, 1032–1044. [CrossRef]
27. Lieberman-Aiden, E.; van Berkum, N.L.; Williams, L.; Imakaev, M.; Ragoczy, T.; Telling, A.; Amit, I.; Lajoie, B.R.; Sabo, P.J.; Dorschner, M.O.; et al. Comprehensive mapping of long range interactions reveals folding principles of the human genome. *Science* **2009**, *326*, 289–293. [CrossRef]
28. Falk, M.; Feodorova, Y.; Naumova, N.; Imakaev, M.; Lajoie, B.R.; Leonhardt, H.; Joffe, B.; Dekker, J.; Fudenberg, G.; Solovei, I.; et al. Heterochromatin drives compartmentalization of inverted and conventional nuclei. *Nature* **2019**, *570*, 395–399. [CrossRef]
29. Tian, H.; Yang, Y.; Liu, S.; Quan, H.; Gao, Y.Q. Toward an understanding of the relation between gene regulation and 3D genome organization. *Quant. Biol.* **2020**, *8*, 295–311. [CrossRef]
30. Sandoval, J.; Heyn, H.; Moran, S.; Serra-Musach, J.; Pujana, M.A.; Bibikova, M.; Esteller, M. Validation of a DNA methylation microarray for 450,000 CpG sites in the human genome. *Epigenetics* **2011**, *6*, 692–702. [CrossRef]
31. Schultz, M.D.; He, Y.; Whitaker, J.W.; Hariharan, M.; Mukamel, E.A.; Leung, D.; Rajagopal, N.; Nery, J.R.; Urich, M.A.; Chen, H.; et al. Human body epigenome maps reveal noncanonical DNA methylation variation. *Nature* **2015**, *523*, 212–216. [CrossRef]
32. Lister, R.; Mukamel, E.A.; Nery, J.R.; Urich, M.; Puddifoot, C.A.; Johnson, N.D.; Lucero, J.; Huang, Y.; Dwork, A.J.; Schultz, M.D.; et al. Global epigenomic reconfiguration during mammalian brain development. *Science* **2013**, *341*, 1237905. [CrossRef]
33. The ENCODE Project Consortium, An integrated encyclopedia of DNA elements in the human genome. *Nature* **2012**, *489*, 57–74. [CrossRef]
34. Schmitt, A.D.; Hu, M.; Jung, I.; Xu, Z.; Qiu, Y.; Tan, C.L.; Li, Y.; Lin, S.; Lin, Y.; Barr, C.L.; et al. A compendium of chromatin contact maps reveals spatially active regions in the human genome. *Cell Rep.* **2016**, *17*, 2042–2059. [CrossRef]
35. Xue, Y.; Yang, Y.; Tian, H.; Quan, H.; Liu, S.; Zhang, L.; Gao, Y.Q. Domain segregated 3D chromatin structure and segmented DNA methylation in carcinogenesis. *bioRvix* **2020**. [CrossRef]
36. Zhou, W.; Dinh, H.Q.; Ramjan, Z.; Weisenberger, D.J.; Nicolet, C.M.; Shen, H.; Laird, P.W.; Berman, B.P. DNA methylation loss in late-replicating domains is linked to mitotic cell division. *Nat. Genet.* **2018**, *50*, 591–602. [CrossRef]
37. Ming, X.; Zhang, Z.; Zou, Z.; Lv, C.; Dong, Q.; He, Q.; Yi, Y.; Li, Y.; Wang, H.; Zhu, B. Kinetics and mechanisms of mitotic inheritance of DNA methylation and their roles in aging-associated methylome deterioration. *Cell Res.* **2020**, *30*, 980–996. [CrossRef] [PubMed]
38. Borsos, M.; Perricone, S.M.; Schauer, T.; Pontabry, J.; de Luca, K.L.; de Vries, S.S.; Ruiz-Morales, E.R.; Torres-Padilla, M.-E.; Kind, J. Genome–lamina interactions are established de novo in the early mouse embryo. *Nature* **2019**, *569*, 729–733. [CrossRef]
39. Huang, D.W.; Sherman, B.T.; Lempicki, R.A. Systematic and integrative analysis of large gene lists using DAVID bioinformatics resources. *Nat. Protoc.* **2008**, *4*, 44–57. [CrossRef]
40. Huang, D.W.; Sherman, B.T.; Lempicki, R.A. Bioinformatics enrichment tools: Paths toward the comprehensive functional analysis of large gene lists. *Nucleic Acids Res.* **2009**, *37*, 1–13. [CrossRef]
41. Yu, G.; Wang, L.-G.; Han, Y.; He, Q.-Y. ClusterProfiler: An R package for comparing biological themes among gene clusters. *OMICS* **2012**, *16*, 284–287. [CrossRef] [PubMed]

42. Li, L.; Liu, C.; Biechele, S.; Zhu, Q.; Song, L.; Lanner, F.; Jing, N.; Rossant, J. Location of transient ectodermal progenitor potential in mouse development. *Development* **2013**, *140*, 4533–4543. [CrossRef] [PubMed]
43. Lu, J.Y.; Chang, L.; Li, T.; Wang, T.; Yin, Y.; Zhan, G.; Han, X.; Zhang, K.; Tao, Y.; Percharde, M.; et al. Homotypic clustering of L1 and B1/Alu repeats compartmentalizes the 3D genome. *Cell Res.* **2021**, *31*, 613–630. [CrossRef] [PubMed]
44. Quan, H.; Yang, Y.; Liu, S.; Tian, H.; Xue, Y.; Gao, Y.Q. Chromatin structure changes during various processes from a DNA sequence view. *Curr. Opin. Struct. Biol.* **2020**, *62*, 1–8. [CrossRef] [PubMed]
45. Chandra, T.; Ewels, P.A.; Schoenfelder, S.; Furlan-Magaril, M.; Wingett, S.W.; Kirschner, K.; Thuret, J.-Y.; Andrews, S.; Fraser, P.; Reik, W. Global reorganization of the nuclear landscape in senescent cells. *Cell Rep.* **2015**, *10*, 471–483. [CrossRef] [PubMed]
46. DEEP Consortium; Salhab, A.; Nordström, K.; Gasparoni, G.; Kattler, K.; Ebert, P.; Ramirez, F.; Arrigoni, L.; Müller, F.; Polansky, J.K.; et al. A comprehensive analysis of 195 DNA methylomes reveals shared and cell-specific features of partially methylated domains. *Genome Biol.* **2018**, *19*, 150. [CrossRef] [PubMed]
47. Tsichlaki, E.; FitzHarris, G. Nucleus downscaling in mouse embryos is regulated by cooperative developmental and geometric programs. *Sci. Rep.* **2016**, *6*, 28040. [CrossRef]
48. Mukherjee, R.N.; Chen, P.; Levy, D.L. Recent advances in understanding nuclear size and shape. *Nucleus* **2016**, *7*, 167–186. [CrossRef]
49. Anifandis, G.; Messini, C.I.; Dafopoulos, K.; Messinis, I.E. Genes and conditions controlling mammalian pre- and post-implantation embryo development. *Curr. Genomics* **2015**, *16*, 32–46. [CrossRef]

 *cells*

MDPI

Article

Intrachromosomal Looping and Histone K27 Methylation Coordinately Regulates the lncRNA *H19*-Fetal Mitogen *IGF2* Imprinting Cluster in the Decidual Microenvironment of Early Pregnancy

Xue Wen [1], Qi Zhang [1], Lei Zhou [1], Zhaozhi Li [1], Xue Wei [1], Wang Yang [1], Jiaomei Zhang [1], Hui Li [1], Zijun Xu [1], Xueling Cui [2], Songling Zhang [1], Yufeng Wang [1], Wei Li [1], Andrew R. Hoffman [3], Zhonghui Liu [2,*], Ji-Fan Hu [1,3,*] and Jiuwei Cui [1,*]

1 Key Laboratory of Organ Regeneration and Transplantation of Ministry of Education, Cancer Center, First Hospital of Jilin University, Changchun 130061, China
2 Department of Immunology, College of Basic Medical Sciences, Jilin University, Changchun 130021, China
3 Stanford University Medical School, VA Palo Alto Health Care System, Palo Alto, CA 94304, USA
* Correspondence: liuzh@jlu.edu.cn (Z.L.); jifan@stanford.edu or hujifan@jlu.edu.cn (J.-F.H.); cuijw@jlu.edu.cn (J.C.); Tel.: +86-431-8561-9476 (Z.L.); +86-650-852-3275 (J.-F.H.); +86-431-8878-2178 (J.C.); Fax: +86-650-849-1213 (J.-F.H.)

Citation: Wen, X.; Zhang, Q.; Zhou, L.; Li, Z.; Wei, X.; Yang, W.; Zhang, J.; Li, H.; Xu, Z.; Cui, X.; et al. Intrachromosomal Looping and Histone K27 Methylation Coordinately Regulates the lncRNA *H19*-Fetal Mitogen *IGF2* Imprinting Cluster in the Decidual Microenvironment of Early Pregnancy. *Cells* **2022**, *11*, 3130. https://doi.org/10.3390/cells11193130

Academic Editors: Lon J. van Winkle and Silvia Garagna

Received: 28 July 2022
Accepted: 22 September 2022
Published: 5 October 2022

Abstract: Recurrent spontaneous abortion (RSA) is a highly heterogeneous complication of pregnancy with the underlying mechanisms remaining uncharacterized. Dysregulated decidualization is a critical contributor to the phenotypic alterations related to pregnancy complications. To understand the molecular factors underlying RSA, we explored the role of longnoncoding RNAs (lncRNAs) in the decidual microenvironment where the crosstalk at the fetal–maternal interface occurs. By exploring RNA-seq data from RSA patients, we identified *H19*, a noncoding RNA that exhibits maternal monoallelic expression, as one of the most upregulated lncRNAs associated with RSA. The paternally expressed fetal mitogen *IGF2*, which is reciprocally coregulated with *H19* within the same imprinting cluster, was also upregulated. Notably, both genes underwent loss of imprinting, as *H19* and *IGF2* were actively transcribed from both parental alleles in some decidual tissues. This loss of imprinting in decidual tissues was associated with the loss of the H3K27m3 repressive histone marker in the *IGF2* promoter, CpG hypomethylation at the central CTCF binding site in the imprinting control center (ICR), and the loss of CTCF-mediated intrachromosomal looping. These data suggest that dysregulation of the *H19/IGF2* imprinting pathway may be an important epigenetic factor in the decidual microenvironment related to poor decidualization.

Keywords: decidualization; recurrent spontaneous abortion; long noncoding RNA; epigenetics; H3K27 methylation

1. Introduction

Spontaneous abortion is the most common complication of pregnancy, affecting >20% of recognized pregnancies [1,2]. Most spontaneous abortions are sporadic and occur prior to the second trimester [3,4]. A subset of women suffer from recurrent spontaneous abortion (RSA), defined as three or more consecutive spontaneous abortions before 20 weeks of gestation. This common gynecological emergency poses significant challenges to future fertility and general psychological health.

A successful pregnancy depends upon complex crosstalk between the developmentally competent embryo and the receptive maternal endometrium [5,6]. Upon implantation, embryos elicit a complex response in the decidua, characterized by transformation of stromal fibroblasts into secretory, epithelioid-like decidual cells, accompanied by the influx of specialized uterine immune cells and vascular remodeling. Decidual cells produce growth factors and cytokines [7,8], including insulin-like growth factor binding protein 1 (*IGFBP1*) and prolactin (*PRL*), which can be used as biomarkers for decidualized cells. Abnormal endometrial receptivity is a key factor leading to implantation failure. However, the molecular factors that regulate this crosstalk in decidualization reactions remains largely uncharacterized.

Longnoncoding RNAs (lncRNAs) act as prominent epigenetic factors in normal development and numerous diseases, often by interacting with chromatin remodeling complexes [9–11]. Differential expression and risk analyses have identified multiple lncRNAs that are associated with recurrent miscarriage [12]. However, little is known about the specific mechanisms of these lncRNAs. Decidualization of the endometrium plays an essential role for the establishment of a successful pregnancy. In order to identify key RNA molecules that mediate the crosstalk at the fetal–maternal interface, we explored RNA transcriptome sequencing datasets from RSA patients. We found that *H19*, an imprinted lncRNA that is expressed from the maternal allele [13,14], and its reciprocally coregulated *IGF2*, a fetal mitogen gene that is expressed from the paternal allele [15,16], were highly upregulated in decidual tissues.

Genomic imprinting of the *H19*/*IGF2* cluster is regulated by the methylation status of CpG islands in the imprinting control region (ICR) located upstream of the *H19* gene. The ICR contains seven CTCF binding sites. The sixth CTCF binding site is differentially methylated [17] and serves as a CTCF "boundary insulator" [18]. Specific binding of CTCF to the unmethylated maternal allele orchestrates the formation of an intrachromosomal loop that links the *IGF2* promoters. CTCF recruits polycomb repressive complex 2 (PCR2) via the docking factor SUZ12, leading to allelic histone 3 lysine 27 (H3K27) methylation that silences the maternal *IGF2* allele. On the other hand, paternal-specific methylation of the ICR prevents CTCF binding and permits expression of *IGF2* while silencing *H19* from the paternal allele. As a result, differential methylation at the CTCF site serves as an "imprint" to ensure the reciprocal imprinting of these two neighboring genes [19]. Importantly, imprinting is dynamically regulated in gametes and in early development. Imprinting defects, including those at the *H19*/*IGF2* locus, are associated with increased risk of developmental disorders [20,21]. Aberrant DNA methylation of the CTCF binding sites in the ICR is associated with an increased risk for abortion [22] and for male infertility [23]. Furthermore, imprinting is frequently dysregulated in IVF embryos [24,25].

Given the critical role of *H19* in in vitro fertilization (IVF) [24] and male infertility [26], we examined the imprinting status of the *H19*/*IGF2* cluster in decidual tissues. We show that there is loss of imprinting of both *H19* and *IGF2* in some decidual tissues. Using human primary endometrial stromal cells as an in vitro model, we studied the epigenetic mechanisms underlying abnormal *H19/IGF2* imprinting in decidualization.

2. Results

2.1. Identification of H19 as a Recurrent Spontaneous Abortion-Associated lncRNA

To search for key factors that might be involved in fetal–maternal regulatory crosstalk in RSA, we explored the differentially expressed lncRNAs in GSE178535, which contained the RNA-seq data of decidual tissues from three RSA patients and three healthy control subjects. The Kyoto Encyclopedia of Genes and Genomes (KEGG) pathway analysis showed associations with cytokine–cytokine receptor interaction, ECM–receptor interaction, hematopoietic cell lineage, chemokine signaling pathway, *PI3K-Akt* signaling pathway, as well as signaling pathways in the regulation of stem cell pluripotency (Figure S1, Table S2).

We focused on the role of the imprinted lncRNA *H19* (Figure 1A, Table S3). In normal tissues, *H19* is expressed only from the maternal allele, while the paternal allele is silenced. Aberrant imprinting of the *H19* gene occurs frequently in tumors [19]. Using an in vitro fertilization model, we previously showed that *H19* imprinting was frequently lost in IVF embryos [24]. We were therefore interested in examining if aberrant regulation of lncRNA *H19* in decidual tissues played a role in the fetal–maternal regulatory crosstalk in RSA.

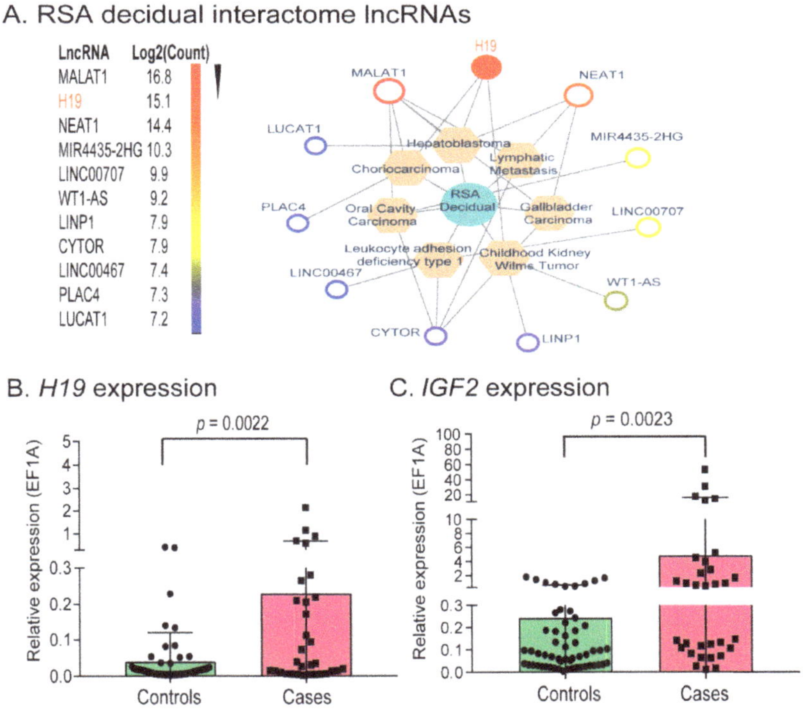

Figure panel labels: A. RSA decidual interactome lncRNAs; B. *H19* expression; C. *IGF2* expression

LncRNA	Log2(Count)
MALAT1	16.8
H19	15.1
NEAT1	14.4
MIR4435-2HG	10.3
LINC00707	9.9
WT1-AS	9.2
LINP1	7.9
CYTOR	7.9
LINC00467	7.4
PLAC4	7.3
LUCAT1	7.2

Figure 1. Identification of RSA-associated lncRNAs by integrating RNA-seq data from RSA patients and senescent decidualized human endometrial stromal cells. (**A**) Identification of RSA-associated lncRNAs. Differentially expressed lncRNAs were analyzed using the RNA transcriptome sequencing dataset GSE178535. The top 11 differentially expressed lncRNAs are ranked based on the RNA expression-fold from high (red) to low (blue) between the RSA patients and the controls. (**B**) Up-regulated *H19* in decidual tissues from RSA cases. Thirty-two decidual tissues with unexplained RSA were collected as the case group. As the control, 57 decidual tissues samples were obtained from healthy adult women who were diagnosed with early pregnancy and were undergoing legal elective termination. Gene expression was measured by qPCR and standardized over the value of the *EEF1A1* control. All data shown are mean ± SD. Error bars represent the SD of the average of three independent PCR reactions. $p = 0.0022$ as compared with the CTL control. (**C**) Reciprocal upregulation of *IGF2* in decidual tissues of RSA cases. $p = 0.0023$ as compared with the CTL control.

We quantitated the expression of *H19* in decidual tissues from 32 patients with RSA. For comparison, decidual tissues were also collected from 57 healthy adult women at 7–10 weeks of gestation who were undergoing early pregnancy termination (Table S4). Using EF1A (*EEF1A1*) as the RT-qPCR control, we found that the expression of *H19* was significantly higher in decidual tissues from the patients with RSA than in decidua from healthy subjects (Figure 1B, $p < 0.01$).

The *H19* gene is located in an imprinting cluster on human chromosome 11 and is coregulated with the adjacent gene *IGF2*, a gene that encodes a mitogen that is required for normal fetal growth. The hierarchical cluster heat map analysis showed that *IGF2* was among the top six of the differentially expressed genes in the analysis (Figure S1B), despite the variability among the subjects (Figure S1C). Therefore, we also quantitated the mRNA abundance of *IGF2* in decidual tissues using quantitative PCR and found that, like *H19*, *IGF2* was significantly upregulated in decidual tissues derived from patients who had

RSA (Figure 1C, $p < 0.01$). Similar data were also obtained by using β-Actin (*ACTB*) as the RT-qPCR control (Figure S1D,E).

2.2. Loss of Genomic Imprinting in Decidual Tissues

To examine the status of *H19* and *IGF2* imprinting in decidual tissues, we genotyped genomic DNA using single nucleotide polymorphisms (SNPs) in *H19* and *IGF2*. Heterozygous SNPs were used to distinguish between the two parental alleles, and the imprinting status was examined in those tissues that were SNP-informative. Twenty-one of the decidual tissues derived from patients who had RSA were informative for *H19* heterozygosity and 20 were informative for *IGF2* heterozygosity. We found that *H19/IGF2* imprinting was lost in 39% (11/28) of *H19/IGF2* informative decidual tissues from the RSA cases (Figure 2A, left panel). Among them, 2 out of 21 samples (9.5%) showed loss of *H19* imprinting, and 7 out of 20 samples (35%) exhibited *IGF2* LOI. Two samples (#22 and #U20) showed loss of imprinting of both *H19* and *IGF2* (Table 1). Imprinting was also lost in some decidual tissues collected from the controls (Figure 2A, right panel).

Table 1. Loss of *H19* and *IGF2* imprinting in RSA decidua.

Cases (ID)		*H19* Genotype	*H19* cDNA	*IGF2* Genotype	*IGF2* cDNA
		Loss of imprinting of *H19* (9.5%) *			
1	U18	A/C	a/c	C/T	t
2	U21	A/C	a/c	T/T	-
		Loss of imprinting of *IGF2* (35%) **			
1	8	A/C	a	C/T	c/t
2	E1	A/C	c	C/T	c/t
3	E3	A/A	-	C/T	c/t
4	E5	A/C	c	C/T	c/t
5	U11	A/A	-	C/T	c/t
6	U14	A/A	-	C/T	c/t
7	U17	A/A	-	C/T	c/t
		Loss of imprinting of *H19* and *IGF2* ***			
1	M22	A/C	a/c	C/T	c/t
2	U20	A/C	a/c	C/T	c/t

* After genotyping, 21 informative samples were used for *H19* allelic analysis. ** 20 *IGF2*-informative samples were used to examine *IGF2* imprinting. *** Informative for both *H19* and *IGF2*. - Tissues that are not informative for allelic analysis of either *H19* or *IGF2*.

As an example, the decidual tissue from Control #13 showed normal imprinting of *H19* (maintenance of imprinting) (Figure 2B, left top panel). The genomic DNA carried both the "A" and "C" alleles, but the cDNA showed the exclusive expression of the "A" allele. The "C" allele was silenced. The decidual tissues from two cases (#U18 and #M22) were also informative for the SNP (Figure 2B, left panels 2–3). However, both the "A" and "C" alleles were detected in their cDNA samples, demonstrating loss of imprinting (LOI).

Similarly, the genotyping of an SNP at the 3′-UTR of *IGF2* showed the presence of the "C/T" alleles. In normal informative decidual tissue #4, only the "T" allele was expressed (Figure 2C, top right panel). However, in two cases of RSA (U11, M22), both the "C" and "T" alleles were expressed in decidual tissues (LOI) (Figure 2C, right panels 2–3). In case U29T, however, both *H19* and *IGF2* maintained normal imprinting.

Loss of *IGF2/H19* imprinting is an early oncogenic event that is detected in tumor-paired adjacent normal tissues [19]. Therefore, we examined the allelic expression of *IGF2/H19* in decidual samples of control subjects. We also detected the presence of *IGF2/H19* LOI in the decidua of several control subjects (Table S5), suggesting epigenetic vulnerability in the decidual microenvironment of early embryo development. The chi-squared analyses showed more LOI cases in the RSA case group for *IGF2* ($p < 0.05$, $\chi^2 = 6.93$), but not for *H19* ($p = 0.721$, $\chi^2 = 0.407$) (Figure S2). The quantitative expression data of *IGF2* and *H19* in LOI and maintenance of imprinting subgroups are presented in Figure S3. Polymorphic imprinting has been observed in placenta [27,28]. Thus, imprinting erosion as observed in both RSA and normal decidual tissues here may represent a decidua-specific polymorphic imprinting trait.

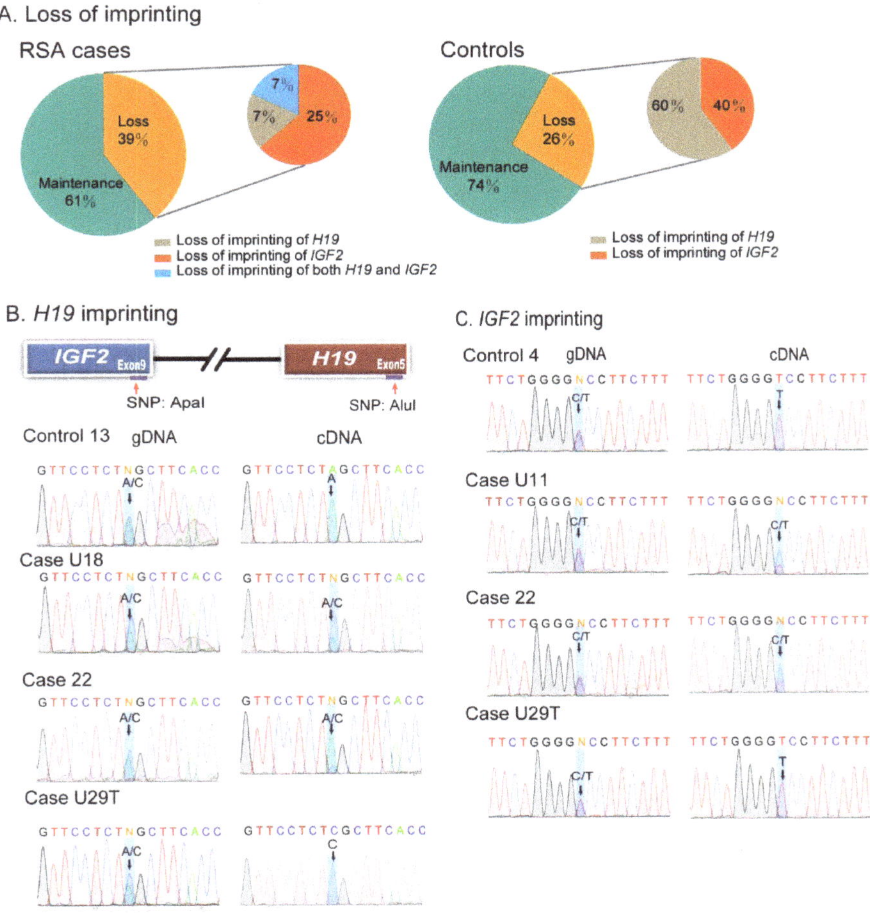

Figure 2. Loss of *H19*/*IGF2* imprinting in decidual tissues in RSA. (**A**) Percentage of abnormal *H19*/*IGF2* imprinting. Among the decidual tissues that are Apa1-informative, 39% cases in RSA cases and 26% in control subjects show the loss of either *H19* or/and *IGF2* imprinting. (**B**) Example of aberrant *H19* allelic expression in RSA cases. Genomic DNAs (gDNA) of both cases U18 and M22 are Apa1 SNP informative (A/C alleles). In the cDNA samples, both parental alleles are expressed in decidual tissues. (**C**) Loss of *IGF2* imprinting in RSA. In control 4, only the T allele is expressed. However, in cases U11 and M22, decidual cDNAs show biallelic expression of *IGF2*.

2.3. The Role of Altered Epigenotypes in the In Vitro Induced Decidualization Model

In vitro cell-induced decidualization is a good model for studying the complex process of implantation [29,30]. We thus cultured two human primary endometrial stromal cell lines (U29T and N45T) (Figure S4A). N45T cells were cultured from the decidual tissues collected from a normal control subject. U29T cells were derived from an RSA case who had suffered four spontaneous abortions (Figure S4B). Genotyping of genomic DNA showed that U29T cells were informative for both *H19* and *IGF2*. N45T cells, however, were only informative for *H19*. No informative SNPs were available for *IGF2* in N45T cells to distinguish the two parental alleles.

We examined the role of altered epigenotypes in this *in vitro* decidualization model. We pretreated U29T and N45T cells with the histone deacetylase inhibitor valproic acid (VPA) (Figure 3A), which is known to modify epigenotypes and alter allelic expression [31]. Following VPA treatment, cells were induced for decidualization. We found that this VPA treatment upregulated *IGF2* and *H19*, particularly in cells with induced decidualization (Figure 3B). However, two decidualization markers (*PRL* and *IGFBP1*) were significantly lower in VPA-treated decidualized cells (Figure 3C), suggesting an impaired decidualization process in VPA-induced cells.

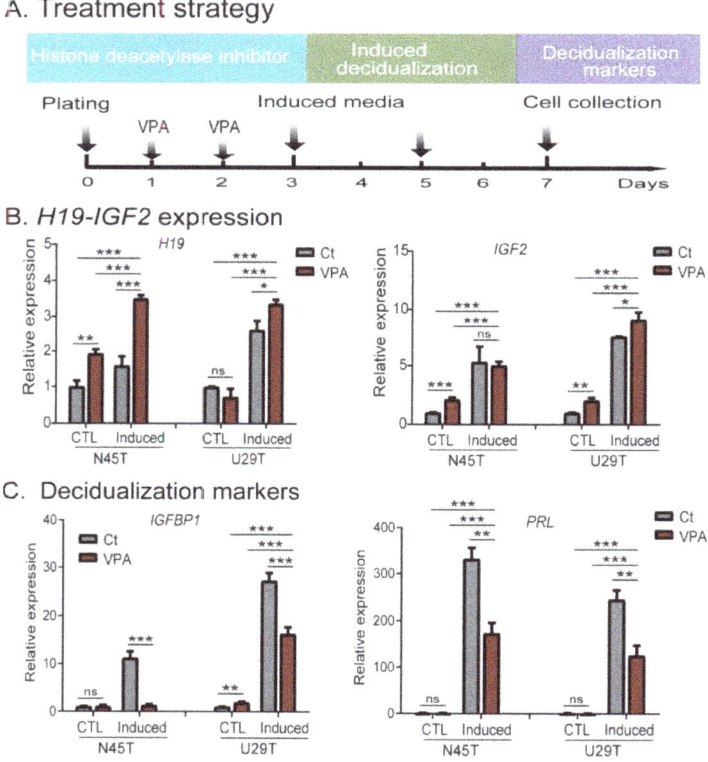

Figure 3. The role of disturbed epigenetics in in vitro decidualization. (**A**) Strategy of inducing epigenetic disturbance by the histone deacetylase inhibitor VPA in primary cultures of endometrial cells. Cells were pretreated with VPA and then were induced for in vitro decidualization. (**B**) Expression of *H19* and *IGF2* in decidualized endometrial cells. VPA: cells treated with the histone deacetylase inhibitor valproic acid. Ct: cells treated with PBS control. Induced: in vitro induction of decidualization. CTL: PBS-treated control cells. (**C**) Expression of two decidualization markers *IGFBP1* and *PRL* in decidualized endometrial cells. The data are the mean ± SD from three independent experiments. * $p < 0.05$, ** $p < 0.01$, and *** $p < 0.001$ as compared with the vector lentivirus control group (CTL). ns: not statistically significant.

2.4. Histone Deacetylase Inhibitor Alters Imprinting in Decidualized Cells

We then used informative SNPs to examine the allelic expression in decidualized cells (Figure S5A). Both U29T and N45T cells were informative for *H19* through gDNA genotyping and both maintained normal *H19-IGF2* imprinting after being placed in culture.

Maintenance of *H19* imprinting was also observed after induced decidualization, with only the "C" allele expressed in U29T cells and the "A" allele expressed in N45T cells (Figure S5B, CTL-Induced cDNA). However, VPA pretreatment induced biallelic expression of *H19* in both decidualized cell lines (Figure S5B, VPA-Induced cDNA). These data suggest that pretreatment with a histone deacetylase inhibitor predisposed endometrial stromal cells to lose imprinting control during decidualization.

By using informative SNP rs680, we also examined the imprinting status of *IGF2* in U29T cells (Figure S5C). The untreated cells maintained normal imprinting, with only the "T" allele expressed (Figure S5D, CTL cDNA). However, *IGF2* imprinting was lost, with both parental alleles (C/T) expressed in the decidualized cells (both CTL-induced and VPA-induced cDNA). *IGF2* and *H19* expression are normally tightly coordinated and reciprocally controlled by an "enhancer competition" mechanism [32]. The data from these treated primary endometrial stromal cells, however, suggest that the control of *IGF2* and *H19* imprinting can be uncoupled.

2.5. Loss of Imprinting Is Associated with Aberrant Histone H3 Lysine 27 Methylation

We then examined the epigenetic mechanisms underlying the loss of imprinting in these two decidualized cell lines. The expression of *IGF2* is driven by four promoters, including an upstream nonimprinted P1 promoter and three downstream imprinted promoters (P2–P4). While they are rich in CpG islands, promoters P2–P4 are not regulated by DNA methylation. Instead, gene silencing of the maternal *IGF2* allele is mediated by polycomb repressive complex 2 (PCR2) component SUZ12-catalyzed H3K27 methylation [19]. We thus focused on the status of H3K27 methylation in the three imprinted *IGF2* promoters (Figure 4A) [32].

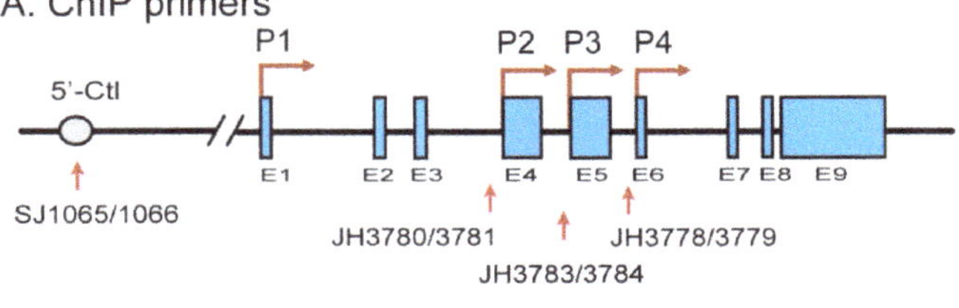

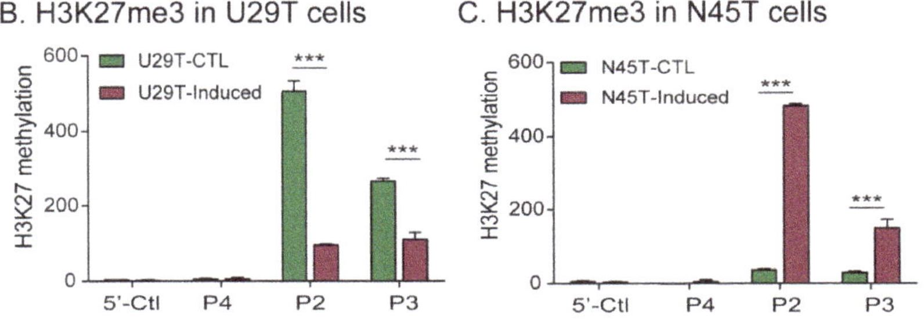

Figure 4. H3K27 methylation in the promoter of *IGF2*. (**A**) Location of PCR primers used for H3K27 methylation. Two primer sets JH3780/JH3781 and JH3783/JH3784 are used to quantitate H3K27 methylation in promoters 2 and 3 of *IGF2*. The primer set (SJ1065/SJ1066) for the P1 promoter upstream site (5′-Ctl) is used as the negative control. (**B**,**C**) Histone methylation in the *IGF2* promoter of U29T cells (**B**) and N45T (**C**) cell lines. U29T cells exhibited *IGF2* LOI after decidualization, while N45T cells maintained normal imprinting. Histone modifications in the *IGF2* promoter were measured by ChIP assay using antibodies specific for H3K27me3. Normal rabbit IgG was used as a negative control and was used for normalization. The data are the mean ± SD from three independent experiments. *** $p < 0.001$as compared with control cells (CTL). Note the reduced H3K27me3 level in decidualized N29T cells that demonstrate *IGF2* LOI.

Using antibodies specific for H3K27me3, we examined H3K27 methylation in *IGF2* promoters in U29T cells that exhibited *IGF2* LOI. We found that H3K27 methylation in the first two *IGF2*-imprinted promoters (P2, P3) was significantly reduced in decidualized cells (Figure 4B). As a control, the 5′-Ctl site upstream of the nonimprinted P1 promoter showed no significant change in the H3K27me3 mark during decidualization. In N45T cells that kept normal imprinting after in vitro decidualization, however, the ChIP signal for H3K27me3 was increased following in vitro decidualization (Figure 4C).

It is known that the key decidual marker gene *IGFBP1* in decidualization is controlled by H3K27 methylation [33]. It was therefore used as the positive control in the ChIP assay. We confirmed the reduction of H3K27 methylation in the *IGFBP1* promoter in both N45T and U29T cells following decidualization (Figure S6A,B). As expected, decidualization did not alter the status of H3K27 methylation in the negative control gene *GPD1* (Figure S6C,D).

2.6. Aberrant Imprinting Is Accompanied by the Loss of Intrachromosomal Looping

The status of histone 3 lysine 27 (H3K27) in the *IGF2* promoters is determined by CTCF-orchestrated intrachromosomal looping [34,35]. CTCF binds to unmethylated DNA motifs in ICR located between the *H19* and IGF2 genes and orchestrates the formation of an intrachromosomal loop, where polycomb repressive complex 2 (PCR2) is recruited via the docking factor SUZ12, leading to allelic H3K27 methylation which then silences the imprinted allele [36].

We used chromosome conformation capture (3C) methodology to examine the chromatin three-dimensional (3D) structure surrounding the *IGF2/H19* locus, with the focus on the CTCF-binding site in the ICR [37]. Using the β-Globin gene (*HBB*) as a positive control, we detected intrachromosomal looping between the LCR (locus control region) and the 3′-enhancer in two decidualized cell lines (Figure S7). In the same 3C samples, we detected an intrachromosomal loop structure between the ICR-enhancers and ICR-*IGF2* promoters in untreated U29T primary decidual cells (Figure 5A). The 3C products were purified, and DNA sequencing confirmed the loop joint separated by the Bgl2/BamH1, Bgl2/Bgl2, and BamH1/BamH1 ligation sites (Figure 5B). However, after induced decidualization in vitro, all three intrachromosomal loops were abolished (Figure 5C) in parallel with the loss of *IGF2* imprinting.

The intrachromosomal looping, however, was not significantly affected in decidualized N45T cells that maintained normal imprinting (Figure 5D). Thus, as was previously reported in cancer cells with LOI [34], CTCF-orchestrated intrachromosomal looping may be essential for maintaining normal imprinting of *IGF2* in decidual tissues.

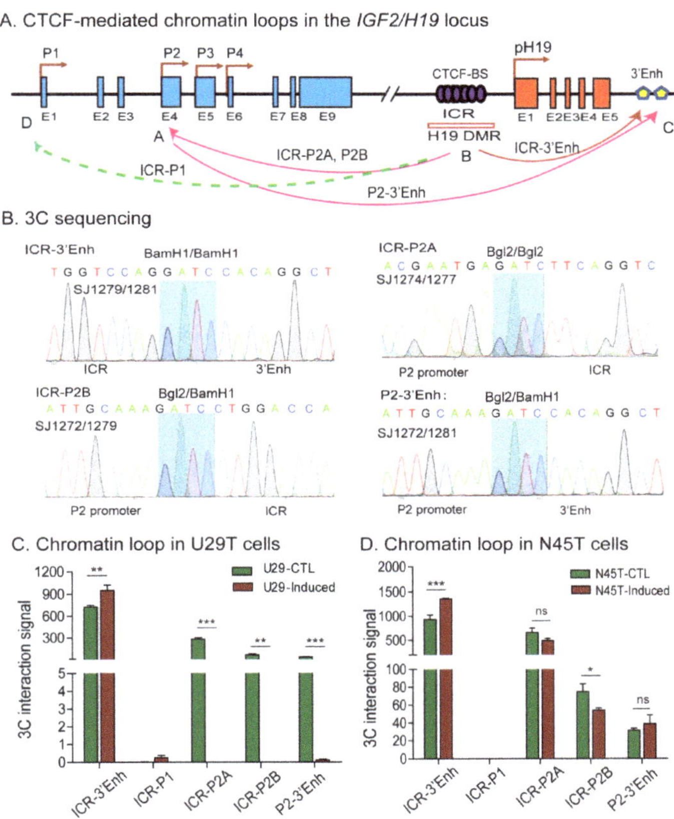

Figure 5. Intrachromosomal loop interactions in the *H19/IGF2* imprinting locus. (**A**) Location of 3C primers used to detect the interaction between the *IGF2* promoter, CTCT6, and *H19* enhancer. P1–P4: *IGF2* promoters. E1–E9: *IGF2* exons 1–9. E1–E5: *H19* exons 1–5. Enh: enhancers. Arrows: intrachromosomal interactions. (**B**) Sequencing of the *IGF2/H19* intrachromosomal loop 3C products. Blue background on the sequence: the 3C ligation product between the BamH1 and/or Bgl2 restriction sites. (**C**,**D**) Quantitation of 3C intrachromosomal interaction signals. The 3C interaction was quantitated by qPCR. U29T cells showed normal imprinting but exhibited *IGF2* LOI after decidualization. Note the lack of intrachromosomal interaction in decidualized N29T cells. Decidualized N45T cells with normal imprinting were used as the control. The data represent the mean ± SD from three independent experiments. * $p < 0.05$, ** $p < 0.01$, and *** $p < 0.001$ as compared with the vector lentivirus control group (CTL). ns: Not statistically significant.

2.7. Loss of Imprinting Is Associated with De Novo DNA Methylation in the Imprinting Control Region

Allelic expression of *IGF2* is regulated by the methylation status of CpG islands in the ICR. We examined allele-specific DNA methylation in the ICR for decidual tissues that were informative for two SNPs in the ICR and one SNP in the *H19* promoter (Figure 6A). The status of CpG DNA methylation was examined using sodium bisulfite sequencing. After converting the unmethylated cytosines into uracils by sodium bisulfite, the ICR and *H19* promoter regions were amplified with DNA methylation-specific primers and cloned into a pJet vector for DNA sequencing. As expected, a typical semimethylated pattern was observed in control #Z4 that had normal monoallelic expression of *H19* and *IGF2* (Figure S8). Case #M22, derived from a patient with RSA, was homozygous for the two

SNPs, and therefore, we were not be able to distinguish the two parental alleles. However, we detected hyper-methylation in the ICR and the *H19* promoter (Figure 6B, top panel). Case U11, which was heterozygous for the ICR SNP, had a hyper-methylated "AA" allele and increased DNA methylation in the "AG" allele (36.5%) (left top panel).

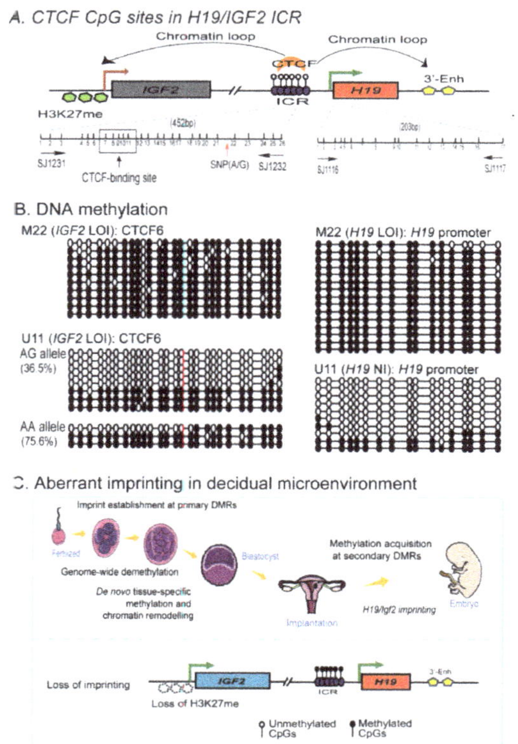

Figure 6. Abnormal DNA methylation at the CTCF6 site in the imprinting control region. (**A**) Schematic diagram of CpGs in the ICR. Locations of PCR primers are indicated by numbered arrows. Vertical lines: location of CpGs. Red arrows: single nucleotide polymorphisms allowing for discrimination of the two parental alleles in case U11. CTCF site 6 carrying CpG 7–11 is boxed. CTCF, a tethering protein, binds to the unmethylated ICR and serves as a molecular glue to secure intrachromosomal interactions. The CTCF-mediated looping brings the ICR and the *IGF2* promoters into close contact, where the polycomb repressive complex 2 (PCR2) is recruited via SUZ12, inducing allelic H3K27 methylation and epigenetic imprinting. (**B**) Alteration of DNA methylation at the CTCF6 site and the *H19* promoter in decidual tissues of cases with RSA (M22, U11). LOI: loss of imprinting. NI: not informative. Numbers in parenthesis: percentage of methylated CpGs. Note the hypermethylation status in case M22 and biallelic DNA methylation in case U11 at the CTCF6 site. (**C**) A model of aberrant imprinting in decidual tissues. After fertilization, genome-wide demethylation occurs, except for DMRs. Following embryo implantation, a global de novo methylation occurs in response to organ development. Parental-specific DMR imprints determine tissue-specific allelic expression of imprinting genes, including *H19/IGF2*. Loss of *H19/IGF2* imprinting occurs in the decidual microenvironment due to the aberrant control of the ICR epigenotype, intrachromosomal looping, and H3K27m3 repressive histone marks. Dysregulation of *H19/IGF2* in preimplantation development and postimplantation stages may represent an epigenetic risk factor contributing to abnormal decidual microenvironment, in addition to locally secreted cytokines and growth factors.

We also observed increased CpG DNA methylation at the ICR CTCF6 site (AA allele, 19.2%) in decidualized U29T cells that exhibited *IGF2* LOI, as compared with the control cells (AA allele, 4.6%) (Figure S9). These data suggest that aberrant imprinting of *H19/IGF2* may be associated with CpG DNA epimutations in the ICR region.

3. Discussion

The molecular mechanisms underlying the spontaneous loss of a pregnancy are unknown [38]. Decidualization plays a critical role in the implantation of the embryo through a regulatory network that coordinates trophoblast invasion of the maternal decidua-myometrium and remodeling of maternal uterine spiral arteries [39,40]. Many factors, including locally secreted cytokines and growth factors, are involved in this complicated network. We have identified the lncRNA *H19* as one of the most upregulated RNA molecules in decidual tissue, where the molecular crosstalk at the fetal–maternal interface occurs. *H19* is also significantly upregulated in the decidua derived from patients with RSA. *IGF2*, a gene that encodes an important fetal mitogen, is located at the adjacent chromosomal locus. *IGF2* is also increased in the decidua in patients who have suffered an RSA. In most normal tissues, the *H19/IGF2* locus is imprinted. Notably, we demonstrate that there is loss of *H19* and *IGF2* imprinting in decidual tissues of some RSA patients. Loss of imprinting also occurs following induced decidualization in primary endometrial stromal cells. Mechanistically, we show that this aberrant imprinting in decidual tissues was associated with the loss of the H3K27m3 repressive histone mark as well as with the loss of intrachromosomal looping and CpG demethylation in the imprinting control center. Pretreatment with histone deacetylase inhibitor VPA predisposed primary endometrial stromal cells to develop abnormal in vitro decidualization. Collectively, these studies suggest that the disturbance of *H19/IGF2* epigenetic regulation, in addition to the locally secreted cytokines and growth factors, may be an epigenetic risk factor for poor decidualization (Figure 6C).

Both the maternal and paternal genomes are necessary for normal embryogenesis and fetal development [41,42]. *H19* is a maternally expressed imprinted gene, and its transcription gives rise to a fetal lncRNA that also functions as a precursor to microRNA miR675 [43], which negatively affects cell proliferation and tumor metastasis [44]. *H19* is abundantly expressed prior to implantation or shortly thereafter, and its expression is specifically confined to progenitor cells of the placenta and extraembryonic tissues [45,46]. *H19* is expressed coordinately with its neighboring gene *Igf2*, a gene that plays a key role in regulating fetal–placental development [47,48]. Genomic deletion of *Igf2* causes placental and fetal growth restriction. In contrast, overexpression of *Igf2* induces placental and fetal overgrowth via paracrine and/or autocrine IGF pathways. The serum levels of IGF-II have been positively linked to infant birth weight. *H19* and *Igf2* regulate embryonic development [49,50]. The allelic expression of *IGF2/H19* is coordinately controlled by a differentially methylated imprinting control region upstream of the *H19* promoter [19,51]. In this study, we demonstrate that both *H19* and *IGF2* are upregulated in decidual tissues of RSA patients as compared with the control cohorts. Moreover, there is loss of imprinting of both genes in many decidual tissues. Major epigenetic events take place in the embryo both in preimplantation development and in postimplantation stages, including the genome-wide resetting of imprints in the PGCs [52,53]. Aberrant methylation of imprinted genes correlates with the risk of abortion [22]. Specifically, CpG hypomethylation in the ICR is correlated with recurrent pregnancy loss [54]. As a result, the periconceptional stage is very sensitive to environmental stressors, leading to epigenetic disturbances.

Loss-of-imprinting has been linked to a number of diseases characterized by abnormal growth phenotypes and behavioral disorders, including Beckwith–Wiedemann syndrome, Silver–Russell syndrome, Angelman syndrome, and Prader–Willi syndrome [55,56], as well as multiple malignancies [57]. Placental-specific imprinting plays a critical role in coordinating the crosstalk between nutrient acquisition and fetal development. Human placentas exhibit widespread placental-specific imprints inherited from the oocyte, including maternally biased DNA methylation DMRs and histone modifications [45,50]. In particular, *H19* shows a unique placenta epigenotype, with the paternal allele-specific DNA methylation covering the core ICR to the gene body [58]. In this study, we also observed more loss of imprinting of *H19/IGF2* in RSA decidual tissues. Pretreatment of two human primary endometrial stromal cells with a histone deacetylase inhibitor induced loss of imprinting and reduced in vitro decidualization. Loss of imprinting in the placenta is associated with intrauterine growth restriction [27,59]. Future studies are needed to elucidate whether dysregulated imprinting plays a role in regulating fetal growth as well as other pregnancy-related pathologies.

It is noteworthy that the mouse and human genome contain a subset of genes that undergo polymorphic imprinting, including *IGF2*, *IGF2R*, *WT1*, *SLC22A2*, and *HTR2A*, with the imprinting status varying among individuals and tissues. For example, human *WT1* is biallelically expressed in kidney, but is monoallelically expressed in brain. In the placenta, *WT1* is maternally expressed in ~60% of the population. The human nc886 gene, encoding a tumor-suppressing ncRNA at chromosome 5q31 is another typical example of nonplacental polymorphic imprinting, with allele-specific methylation predominantly found on the maternal allele in many tissues [60]. Moreover, profiling of placental-specific imprinted DMRs shows that human placenta preferentially maintains maternal germline-derived imprint marks and appears to be highly polymorphic in the population [28]. Thus, the biallelic expression of *H19* and *IGF2* as observed in the present study may be associated with a decidua-specific polymorphic imprinting trait.

It should be noted that this study also has several weaknesses. First, two primary endometrial stromal cells yielded some discrepancies in in vitro decidualization. U29T cells, derived from the decidua of an RSA patient, were more vulnerable to hormone induction and exhibited loss of *IGF2* imprinting following in vitro decidualization. N45T cells derived from a normal subject, on the other hand, maintained normal imprinting unless they were also pretreated with histone deacetylase inhibitor. Although this discrepancy may be related to the polymorphic imprinting trait in primary endometrial stromal cells, we still do not know the specific mechanisms by which these differences arise. Second, several other lncRNAs are also upregulated in the decidual samples of RSA cases. For instance, *MALAT1* was the most upregulated lncRNA on the list. *NEAT1* was also upregulated in RSA decidual tissues. *Neat1* knockout mice stochastically show decreased fertility due to corpus luteum dysfunction and concomitant low progesterone [61], suggesting a critical role of *Neat1* in the establishment of pregnancy. Thus, future studies are needed to address if these lncRNAs are also involved in the dysregulated decidualization related to RSA.

In summary, this study reveals the first evidence that the imprinting status of *H19/IGF2* is dysregulated in decidual tissues. Using primary endometrial stromal cells as a model, we demonstrate that the in vitro decidualization process is affected by altered epigenotypes induced by a histone deacetylase inhibitor. The loss of imprinting in decidual tissues was associated with a dysregulated H3K27m3 histone marker and altered CTCF-mediated intrachromosomal looping. Altered levels of *H19* lncRNA and/or IGF-II protein in fetal decidua may alter normal fetal–placental development. It would be interesting to explore whether epigenetic targeting of the *H19*/*IGF2* epimutation [19] may provide an alternative strategy to prevent the poor decidualization seen in some pregnancy-related disorders.

4. Materials and Methods

4.1. Identification of RSA-Associated lncRNAs Using RNA-Seq Data

To identify RSA-associated lncRNAs, we downloaded the RSA dataset GSE178535 from the NIH GEO database website. The dataset contained the RNA-seq data of decidual tissues from three RSA patients and three healthy control subjects [62] (Next Generation Sequencing Facilitates Quantitative Analysis of healthy controls and RSA patients Transcriptomes. Available online: https://www.ncbi.nlm.nih.gov/geo/query/acc.cgi?acc=GSE178535, accessed on 22 June 2021).

The in vitro decidualization of embryonic stem cells (ESCs) was induced using differentiation media containing 0.5 mM dibutyryl cAMP, 1 μM medroxyprogesterone 17-acetate, and 10 nM β-estradiol. Decidualized cells were used for RNA-seq [63].

Differentially expressed RNAs were calculated as the log2-transformed gene expression values (Fold Change). The Kyoto Encyclopedia of Genes and Genomes (KEGG) pathway analysis (KEGG_PATHWAY) was carried out using DAVID Bioinformatics Resources 6.8 (https://david.ncifcrf.gov, accessed on 21 September 2022) [64,65]. Hierarchical Cluster Heatmap was generated using HIPLOT (https://hiplot.com.cn, accessed on 21 September 2022) [66]. LncRNAs with the fold-change >2 and $p < 0.001$ were chosen for further functional characterization.

4.2. Human Decidual Samples

Decidual tissue samples were collected at The First Hospital of Jilin University between 2017–2022. Ethical approval for this study was provided by the Research Ethics Board of the First Hospital of Jilin University, and written informed consent was obtained from all patients prior to sample collection.

A total of 32 decidual tissues were collected from women with unexplained RSA. The inclusion criteria for this group were women aged under 40 years with a history of > three consecutive pregnancy losses. Clinical examination showed that they had normal uterine cavity shape and size; normal follicle-stimulating hormone (FSH), estradiol (E2), prolactin (PRL), luteinizing hormone (LH), and thyroid-stimulating hormone (TSH) levels at menstrual day 2–3; no mutations detected in Factor V (Leiden) and prothrombin gene analysis; normal antithrombin III, protein C, and S activity; negative results for lupus anticoagulant evaluation; cardiolipin antibody; beta2-glycoprotein antibody; and normal karyotype. Their partners have normal semen analyses and normal karyotype. None of the patients had received a prior infertility treatment.

In addition, 57 decidual samples were obtained as the control group from healthy adult women at 7–10 weeks of gestation undergoing legal elective termination. The inclusion criteria were women aged under 40 years with regular menstrual cycles, at least one live birth, no previous miscarriages, no history of infertility/treatment, and no associated gynecologic (endometriosis, fibroids, active or history of pelvic inflammatory disease) or other medical comorbidities (e.g., hyperprolactinemia, thyroid disease). The male partners of control subjects had normal semen analysis results and karyotypes. The characteristics of RSA patients and controls are listed in Table S5.

All the decidual samples were collected by the same pathology lab technician at Jilin University First Hospital. The placenta was rinsed with saline to remove blood. Decidual tissues were collected by carefully dissecting the maternal basal plate of the placenta. Collected tissues were rinsed with 1× PBS, frozen with liquid nitrogen, and saved in −80 °C freezer for analysis.

4.3. Culture of Human Primary Endometrial Stromal Cells

Primary endometrial stromal cells were cultured from U29T and N45T decidual tissues that were *H19-IGF2* informative and that maintained normal imprinting. N45T cells were cultured from the decidual tissues collected from a normal control subject. U29T cells were derived from an RSA case who had suffered four spontaneous abortions (Figure S4).

After curettage, the tissues were immediately collected under sterile conditions into prechilled PBS and divided into decidua and villi. Two or three pieces of decidual tissues were collected and washed 2–3 times again with prechilled PBS to exclude villous contamination. Fresh tissues were cut into approximately 2 mm^3 fragments, washed in DMEM (high glucose; Sigma, MO, USA), and directly cultured at 37 °C in 5% CO_2 by attaching to the substratum in a 10 cm dish with complete medium consisting of DMEM medium (Sigma, St. Louis, MO, USA) supplemented with 10% (*v*/*v*) fetal bovine serum (Sigma, St. Louis, MO, USA), 100 U/mL of penicillin sodium, and 100 μg/mL of streptomycin sulfate (Invitrogen, Carlsbad, CA, USA). After approximately 12 days in culture, cells migrated out from the edges. Migrating cells were collected with 0.1% trypsin and 0.25 mM EDTA and passaged for allelic study and in vitro decidualization assays (Figure S4). After culturing, cells were aliquoted and stored in liquid nitrogen for further studies.

4.4. In Vitro Decidualization

In vitro artificially induced decidualization was performed following the method as described in [29]. Briefly, U29T and N45T primary endometrial stromal cells were cultured in complete medium containing 10 nM E2, 1 μM P4, and 0.5 mM 8-Br-cAMP. Culture medium was changed every 2 days. Cells were harvested for subsequent experiments 96 h after the treatment.

4.5. The Role of Histone Deacetylase Inhibitor VPA in Decidualization

To examine the role of aberrant epigenotypes in in vitro decidualization, we pretreated primary endometrial stromal cells with the histone deacetylase inhibitor valproic acid (VPA), which is known to modify epigenotypes and alter allelic expression [31]. U29T and N45T cells were treated with 2 mM VPA. Cells treated with equal volume of PBS were used as the control (Ct). Culture medium was changed daily. Forty-eight hours after VPA treatment, cells were used for in vitro decidualization experiments. After 96-h treatment, cells were collected for imprinting assays.

4.6. RT-PCR Quantitation

Decidual tissues and cells were collected and total RNA was extracted by TRIzol reagent (Sigma, St. Louis, MO, USA) and stored at −80 °C. cDNA was synthesized using RNA reverse transcriptase (Invitrogen, CA, USA), and target amplification was performed with a Bio-Rad Thermol Cycler. PCR of 1 cycle at 95 °C for 2 min; 32 cycles at 95 °C for 15 s, 60 °C for 15 s, and 72 °C for 15 s; and 1 cycle at 72 °C for 10 min. EF1A (*EEF1A1*) and β-Actin (*ACTB*) were used as the internal controls. Quantitative real-time PCR was performed using SYBR GREEN PCR Master (Applied Biosystems, Foster City, CA, USA); the threshold cycle (Ct) values of target genes were assessed by quantitative PCR in triplicate using a sequence detector (ABI Prism 7900HT; Applied Biosystems, Foster City, CA, USA) and were normalized over the Ct of the EF1A or β-Actin controls. Primers used for PCR quantitation are listed in Table S1.

4.7. Allelic Expression of IGF2 and H19

Genomic DNA and total RNA extraction from decidual tissues and cDNA synthesis were performed as previously described. Decidual tissues were first genotyped for heterozygosity of SNPs in *IGF2* exon 9 and *H19* exon 5 (Figure 2A). Target amplification was performed with a Bio-Rad Thermol Cycler. PCR of 1 cycle at 95 °C for 2 min; 32 cycles at 95 °C for 15 s, 60 °C for 15 s, and 72 °C for 15 s; and 1 cycle at 72 °C for 10 min using primers specific for two polymorphic restriction enzymes (ApaI, AluI) in the last exon of human *IGF2* and *H19* exon 5. To determine the status of *IGF2* imprinting, the amplified products were sequenced by Comate Bioscience Co, Ltd. (Changchun, China). Decidual tissues that maintain normal imprinting expressed a single parental allele, while the LOI showed biallelic expression of *IGF2* and *H19*. PCR primers used for *IGF2* imprinting are listed in Supplementary Table S1.

4.8. DNA Methylation Analysis

Genomic DNA was collected from tissues or cells using dBIOZOL Genomic DNA Extraction Reagent (BioFlux, BSC16M1, Hangzhou, China) following the manufacturer's instructions. DNA was treated with EZ DNA Methylation-Gold^{TM} Kit (ZYMO RESEARCH, D5005, Irvine, CA, USA), and PCR was performed using DNA methylation-specific primers designed for the promoter of *H19* and CTCF binding sites (Table S1). To examine the status of DNA methylation in every CpG site, the amplified PCR DNAs were cloned into pJET1.2/blunt cloning vector (Thermo, K1231, Waltham, MA, USA) and transformed into

TOP10. Plasmid DNA was collected by Wizard® Plasmid DNA Purification kit (Promega, A1223, MO, USA) and sequenced.

4.9. Chromosome Conformation Capture (3C)

Furthermore, 3C assays were performed to determine long-range intrachromosomal interactions as previously described [35,67–69]. Briefly, 1.0×10^7 cells were cross-linked with 2% formaldehyde and lysed with cell lysis buffer (10 mM Tris (pH 8.0), 10 mM NaCl, 0.2% NP-40, supplemented with protease inhibitors). Nuclei were collected and suspended in 1× restriction enzyme buffer. An aliquot of nuclei (2×10^6) was digested with 800 U of restriction enzyme BamH1/Bgl2 at 37 °C overnight. After stopping the reaction by adding 1.6% SDS and incubating the mixture at 65 °C for 20 min, chromatin DNA was diluted with NEB ligation reaction buffer, and 2 μg DNA was ligated with 4000 U of T4 DNA ligase (New England BioLabs, Irvine, CA, USA) at 16 °C for 4 h (final DNA concentration, 2.5 μg/mL). After treatment with 10 mg/mL proteinase K at 65 °C for 4 h to reverse cross-links and with 0.4 μg/mL RNase A for 30 min at 37 °C, DNA was extracted with phenol-chloroform, ethanol precipitated, and detected by PCR amplification of the ligated DNA products. Furthermore, 3C PCR products were cloned and sequenced to validate the intrachromosomal interactions by assessing for the presence of the BamH I/Bgl II ligation site. The 3C interaction was quantitated by qPCR and was standardized over the 3C ligation control. For comparison, the relative 3C interaction was calculated by setting the control as 1.

As the 3C quality control, human β-Globin (*HBB*) gene was used as a positive control. Unlike *IGF2* promoters P2-P4, *IGF2* promoter 1 (P1) is not imprinted and is biallelically expressed in all tissues. Thus, we chose a Bgl2 site upstream of P1 promoter as the 3C negative control. Human Primers used for 3C assay are listed in Supplementary Table S1.

4.10. Histone Methylation by Chromatin Immunoprecipitation (ChIP) Assay

As previously described, a ChIP assay was used to quantitate the status of histone modifications following the manufacturer's protocol (Upstate Biotechnology, Lake Placid, NY, USA). Briefly, 1.0×10^7 cells were fixed with 1% formaldehyde and then sonicated for 180 s (10 s on and 10 s off) on ice with a sonicator with a 2mm microtip at 40% output control and 90% duty cycle settings. The sonicated chromatin was collected by centrifugation, aliquoted, and stored at −80 °C. Protein A/G Magnetic Beads and a specific anti-trimethyl-histone H3 (Lys27) antibody (Merck Millipore, Darmstadt, Germany) were incubated with rotation for 30 min at room temperature. The sonication supernatant and beads were incubated with antibody at 4 °C on a rotating rack for 4–16 h or overnight. To reduce the ChIP background, we modified the manufacturer's protocol by adding two more washing steps following immunoprecipitation. As previously reported [35], anti-IgG was used as the ChIP control in parallel with testing samples. Precipitated DNA was subjected to qPCR and expressed as fold-enrichment compared to the IgG chromatin input.

For the ChIP assay, *IGFBP1*, a key decidual marker gene controlled by H3K27 methylation in decidualization, was used as the positive control. The housekeeping gene *GPD1* (G3PDH) was used as the negative control in the assay.

4.11. Statistical Analysis

All the experimental data are presented as mean ± standard deviation (SD) and were derived from at least three biological replicates. Statistical analyses were performed using GraphPad Prism v7.0 (GraphPad Software, San Diego, CA, USA). Unpaired two-tailed Student's *t*-tests were used for comparison between two groups. One-way ANOVA with Bonferroni's multiple comparison test was used to compare statistical differences for variables among three or more groups. Chi-squared tests were used to examine the association between the *H19/IGF2* imprinting status (loss of imprinting and maintenance groups) and the risk of RSA occurrence (RSA and control groups). The level of significance was indicated as * $p < 0.05$, ** $p < 0.01$, and *** $p < 0.001$, unless stated otherwise.

Supplementary Materials: The following supporting information can be downloaded at: https://www.mdpi.com/article/10.3390/cells11193130/s1, Figure S1. KEGG pathway analysis in RSA patients by RNA-seq; Figure S2. Aberrant imprinting between the RSA and control groups; Figure S3. Abundance of *H19/IGF2* between the loss and maintenance subgroups; Figure S4. Culturing of two primary endometrial stromal cells; Figure S5. Imprinting after decidualization in VPA-pretreated cells; Figure S6. H3K27 methylation positive (*IGFBP1*) and negative (*GPD1*) controls; Figure S7. 3C positive control in the human β-Globin (*HBB*) locus; Figure S8. Allelic DNA methylation in the ICR of Z4 decidual tissues; Figure S9. Allelic DNA methylation in the ICR of decidualized U29T cells; Table S1. Oligonucleotide primers used for PCR; Table S2. KEGG pathways that are associated with decidualization; Table S3. Top 11 associated lncRNAs of decidual tissues by RNA sequencing from 3 RSA patients (Log2FoldChange > 7); Table S4. Basal characteristics of controls and RSA patients; Table S5. Genotype and allelic expression of *IGF2* and *H19*.

Author Contributions: J.-F.H., J.C. and Z.L. (Zhonghui Liu) conceived and designed the study; X.W. (Xue Wen) and Q.Z. performed most of the experiments and organized the data; L.Z., Z.L. (Zhaozhi Li), X.W. (Xue Wei), H.L. and Z.X. conducted cell assays, cell culture and transduction; W.Y. omics analysis; J.Z. collecting clinical samples; J.-F.H., J.C., Z.L., W.L., S.Z., X.C., A.R.H. and Y.W. supervised experiments and reviewed the data; J.-F.H. and A.R.H. wrote and edited the manuscript. All authors have read and agreed to the published version of the manuscript.

Funding: This work was supported by the National Key R&D Program of China (2018YFA0106902, 2020YFA0707704); the Innovative Program of National Natural Science Foundation of China (82050003); National Basic Research Program of China (973 Program) (2015CB943303); National Natural Science Foundation of China (31871297, 81874052, 32000431); Fund of Jilin Provincial Science and Technology Department (20190303146SF, 20200602032ZP and 20200201390JC); Natural Science Fund of Jilin Provincial Finance Department (JLSWSRCZX2020-023, JLSWSRCZX2020-100); Youth Fund of Jilin Provincial Health Commission (2016Q035); China Guanghua Fund (2020-CXM-01 and JDYYGH2019004); Youth Fund of First Hospital of Jilin University (JDYY 102019002, JDYY 102019043); California Institute of Regenerative Medicine (CIRM) (RT2-01942), and Department of Veterans Affairs (BX002905).

Institutional Review Board Statement: Ethical approval for this study was provided by the Research Ethics Board of the First Hospital of Jilin University, and written informed consent was obtained from all patients before sample collection.

Informed Consent Statement: Not applicable.

Conflicts of Interest: The authors declare no conflict of interest.

Abbreviations

RSA	Recurrent spontaneous abortion;
lncRNAs	longnoncoding RNAs;
ICR	imprinting control center;
H3K27m3	trimethylation of lysine 27 on histone H3;
IGFBP1	insulin-like growth factor binding protein1;
PRL	prolactin;
RSA	recurrent spontaneous abortion;
SNPs	single nucleotide polymorphisms;
LOI	loss of imprinting;
MOI	maintenance of imprinting;
KEGG	The Kyoto Encyclopedia of Genes and Genomes;
E2	β-Estradiol;
P4	Progesterone;
8-Br-cAMP	8-Bromoadenosine 3′:5′-Cyclic Monophosphate;
PCR2	polycomb repressive complex 2;
3C	chromosome conformation capture;
ChIP	chromatin immunoprecipitation

References

1. How, J.; Leiva, O.; Bogue, T.; Fell, G.G.; Bustoros, M.W.; Connell, N.T. Pregnancy outcomes, risk factors, and cell count trends in pregnant women with essential thrombocythemia. *Leuk. Res.* **2020**, *98*, 106459. [CrossRef] [PubMed]
2. Garrido-Gimenez, C.; Alijotas-Reig, J. Recurrent miscarriage: Causes, evaluation and management. *Postgrad. Med. J.* **2015**, *91*, 151–162. [CrossRef] [PubMed]
3. Pinar, M.H.; Gibbins, K.; He, M.; Kostadinov, S.; Silver, R. Early Pregnancy Losses: Review of Nomenclature, Histopathology, and Possible Etiologies. *Fetal. Pediatr. Pathol.* **2018**, *37*, 191–209. [CrossRef]
4. McPherson, E. Recurrence of stillbirth and second trimester pregnancy loss. *Am. J. Med. Genet. A* **2016**, *170A*, 1174–1180. [CrossRef]
5. Ticconi, C.; Pietropolli, A.; Di Simone, N.; Piccione, E.; Fazleabas, A. Endometrial Immune Dysfunction in Recurrent Pregnancy Loss. *Int. J. Mol. Sci.* **2019**, *20*, 5332. [CrossRef] [PubMed]
6. Okada, H.; Tsuzuki, T.; Murata, H. Decidualization of the human endometrium. *Reprod. Med. Biol.* **2018**, *17*, 220–227. [CrossRef] [PubMed]
7. Durairaj, R.R.P.; Aberkane, A.; Polanski, L.; Maruyama, Y.; Baumgarten, M.; Lucas, E.S.; Quenby, S.; Chan, J.K.Y.; Raine-Fenning, N.; Brosens, J.J.; et al. Deregulation of the endometrial stromal cell secretome precedes embryo implantation failure. *Mol. Hum. Reprod.* **2017**, *23*, 478–487. [CrossRef] [PubMed]
8. Gibson, D.A.; Simitsidellis, I.; Cousins, F.L.; Critchley, H.O.; Saunders, P.T. Intracrine Androgens Enhance Decidualization and Modulate Expression of Human Endometrial Receptivity Genes. *Sci. Rep.* **2016**, *6*, 19970. [CrossRef] [PubMed]
9. Chen, J.; Wang, Y.; Wang, C.; Hu, J.F.; Li, W. LncRNA Functions as a New Emerging Epigenetic Factor in Determining the Fate of Stem Cells. *Front. Genet.* **2020**, *11*, 277. [CrossRef]
10. Patty, B.J.; Hainer, S.J. Non-Coding RNAs and Nucleosome Remodeling Complexes: An Intricate Regulatory Relationship. *Biology* **2020**, *9*, 213. [CrossRef]
11. Huang, H.; Sun, J.; Sun, Y.; Wang, C.; Gao, S.; Li, W.; Hu, J.F. Long noncoding RNAs and their epigenetic function in hematological diseases. *Hematol. Oncol.* **2019**, *37*, 15–21. [CrossRef] [PubMed]
12. Zhang, Y.; Wang, S. The possible role of long non-coding RNAs in recurrent miscarriage. *Mol. Biol. Rep.* **2022**, *49*, 9687–9697. [CrossRef] [PubMed]
13. Pope, C.; Mishra, S.; Russell, J.; Zhou, Q.; Zhong, X.B. Targeting H19, an Imprinted Long Non-Coding RNA, in Hepatic Functions and Liver Diseases. *Diseases* **2017**, *5*, 11. [CrossRef] [PubMed]
14. MacDonald, W.A.; Mann, M.R.W. Long noncoding RNA functionality in imprinted domain regulation. *PLoS Genet.* **2020**, *16*, e1008930. [CrossRef]
15. Marasek, P.; Dzijak, R.; Studenyak, I.; Fiserova, J.; Ulicna, L.; Novak, P.; Hozak, P. Paxillin-dependent regulation of IGF2 and H19 gene cluster expression. *J. Cell Sci.* **2015**, *128*, 3106–3116. [CrossRef]
16. Kasprzak, A.; Adamek, A. Insulin-Like Growth Factor 2 (IGF2) Signaling in Colorectal Cancer-From Basic Research to Potential Clinical Applications. *Int. J. Mol. Sci.* **2019**, *20*, 4915. [CrossRef]
17. Bartolomei, M.S.; Webber, A.L.; Brunkow, M.E.; Tilghman, S.M. Epigenetic mechanisms underlying the imprinting of the mouse H19 gene. *Genes Dev.* **1993**, *7*, 1663–1673. [CrossRef]
18. Lewis, A.; Murrell, A. Genomic imprinting: CTCF protects the boundaries. *Curr. Biol.* **2004**, *14*, R28426–R28428. [CrossRef]
19. Hu, J.F.; Hoffman, A.R. Targeting aberrant IGF2 epigenetics as a novel anti-tumor approach. In *DNA Methylation: Patterns, Functions and Roles in Disease*; Holland, K., Ed.; Nova Science Publishers: Hauppauge, NY, USA, 2016; pp. 91–110.
20. MacDonald, W.A.; Mann, M.R. Epigenetic regulation of genomic imprinting from germ line to preimplantation. *Mol. Reprod. Dev.* **2014**, *81*, 126–140. [CrossRef]
21. Li, E. Chromatin modification and epigenetic reprogramming in mammalian development. *Nat. Rev. Genet.* **2002**, *3*, 662–673. [CrossRef]
22. Cannarella, R.; Crafa, A.; Condorelli, R.A.; Mongioi, L.M.; La Vignera, S.; Calogero, A.E. Relevance of sperm imprinted gene methylation on assisted reproductive technique outcomes and pregnancy loss: A systematic review. *Syst. Biol. Reprod. Med.* **2021**, *67*, 251–259. [CrossRef] [PubMed]
23. Rotondo, J.C.; Lanzillotti, C.; Mazziotta, C.; Tognon, M.; Martini, F. Epigenetics of Male Infertility: The Role of DNA Methylation. *Front. Cell Dev. Biol.* **2021**, *9*, 689624. [CrossRef] [PubMed]
24. Li, T.; Vu, T.H.; Ulaner, G.A.; Littman, E.; Ling, J.Q.; Chen, H.L.; Hu, J.F.; Behr, B.; Giudice, L.; Hoffman, A.R. IVF results in *de novo* DNA methylation and histone methylation at an Igf2-H19 imprinting epigenetic switch. *Mol. Hum. Reprod.* **2005**, *11*, 631–640. [CrossRef] [PubMed]
25. Barberet, J.; Ducreux, B.; Guilleman, M.; Simon, E.; Bruno, C.; Fauque, P. DNA methylation profiles after ART during human lifespan: A systematic review and meta-analysis. *Hum. Reprod. Update* **2022**, *144*, 393. [CrossRef] [PubMed]
26. Rotondo, J.C.; Selvatici, R.; Di Domenico, M.; Marci, R.; Vesce, F.; Tognon, M.; Martini, F. Methylation loss at H19 imprinted gene correlates with methylenetetrahydrofolate reductase gene promoter hypermethylation in semen samples from infertile males. *Epigenetics* **2013**, *8*, 990–997. [CrossRef] [PubMed]
27. Monteagudo-Sanchez, A.; Sanchez-Delgado, M.; Mora, J.R.H.; Santamaria, N.T.; Gratacos, E.; Esteller, M.; de Heredia, M.L.; Nunes, V.; Choux, C.; Fauque, P.; et al. Differences in expression rather than methylation at placenta-specific imprinted loci is associated with intrauterine growth restriction. *Clin. Epigenetics* **2019**, *11*, 35. [CrossRef]

28. Hanna, C.W.; Penaherrera, M.S.; Saadeh, H.; Andrews, S.; McFadden, D.E.; Kelsey, G.; Robinson, W.P. Pervasive polymorphic imprinted methylation in the human placenta. *Genome Res.* **2016**, *26*, 756–767. [CrossRef]
29. Marquardt, R.M.; Lee, K.; Kim, T.H.; Lee, B.; DeMayo, F.J.; Jeong, J.W. Interleukin-13 receptor subunit alpha-2 is a target of progesterone receptor and steroid receptor coactivator-1 in the mouse uterusdagger. *Biol. Reprod.* **2020**, *103*, 760–768. [CrossRef]
30. Rytkonen, K.T.; Erkenbrack, E.M.; Poutanen, M.; Elo, L.L.; Pavlicev, M.; Wagner, G.P. Decidualization of Human Endometrial Stromal Fibroblasts is a Multiphasic Process Involving Distinct Transcriptional Programs. *Reprod. Sci.* **2019**, *26*, 323–336. [CrossRef]
31. Hu, J.F.; Hoffman, A.R. Examining histone acetlylation at specific genomic regions. *Methods Mol. Biol.* **2001**, *181*, 285–296.
32. Li, T.; Hu, J.F.; Qiu, X.; Ling, J.; Chen, H.; Wang, S.; Hou, A.; Vu, T.H.; Hoffman, A.R. CTCF regulates allelic expression of Igf2 by orchestrating a promoter-polycomb repressive complex-2 intrachromosomal loop. *Mol. Cell. Biol.* **2008**, *28*, 6473–6482. [CrossRef] [PubMed]
33. Grimaldi, G.; Christian, M.; Steel, J.H.; Henriet, P.; Poutanen, M.; Brosens, J.J. Down-regulation of the histone methyltransferase EZH2 contributes to the epigenetic programming of decidualizing human endometrial stromal cells. *Mol. Endocrinol.* **2011**, *25*, 1892–1903. [CrossRef] [PubMed]
34. Li, T.; Chen, H.; Li, W.; Cui, J.; Wang, G.; Hu, X.; Hoffman, A.R.; Hu, J. Promoter histone H3K27 methylation in the control of IGF2 imprinting in human tumor cell lines. *Hum. Mol. Genet.* **2014**, *23*, 117–128. [CrossRef] [PubMed]
35. Zhao, X.; Liu, X.; Wang, G.; Wen, X.; Zhang, X.; Hoffman, A.R.; Li, W.; Hu, J.F.; Cui, J. Loss of insulin-like growth factor II imprinting is a hallmark associated with enhanced chemo/radiotherapy resistance in cancer stem cells. *Oncotarget* **2016**, *7*, 51349–51364. [CrossRef] [PubMed]
36. Hu, J.F.; Hoffman, A.R. Chromatin looping is needed for iPSC induction. *Cell Cycle* **2014**, *13*, 2975–2982. [CrossRef]
37. Ulaner, G.A.; Yang, Y.; Hu, J.F.; Li, T.; Vu, T.H.; Hoffman, A.R. CTCF binding at the insulin-like growth factor-II (IGF2)/H19 imprinting control region is insufficient to regulate IGF2/H19 expression in human tissues. *Endocrinology* **2003**, *144*, 4420–4426. [CrossRef]
38. O'Connor, B.B.; Pope, B.D.; Peters, M.M.; Ris-Stalpers, C.; Parker, K.K. The role of extracellular matrix in normal and pathological pregnancy: Future applications of microphysiological systems in reproductive medicine. *Exp. Biol. Med.* **2020**, *245*, 1163–1174. [CrossRef]
39. Conrad, K.P.; Rabaglino, M.B.; Post Uiterweer, E.D. Emerging role for dysregulated decidualization in the genesis of preeclampsia. *Placenta* **2017**, *60*, 119–129. [CrossRef]
40. Vinketova, K.; Mourdjeva, M.; Oreshkova, T. Human Decidual Stromal Cells as a Component of the Implantation Niche and a Modulator of Maternal Immunity. *J. Pregnancy* **2016**, *2016*, 8689436. [CrossRef]
41. Crespi, B.J. Why and How Imprinted Genes Drive Fetal Programming. *Front. Endocrinol.* **2019**, *10*, 940. [CrossRef]
42. Zhang, K.; Smith, G.W. Maternal control of early embryogenesis in mammals. *Reprod. Fertil. Dev.* **2015**, *27*, 880–896. [CrossRef] [PubMed]
43. Cai, X.; Cullen, B.R. The imprinted H19 noncoding RNA is a primary microRNA precursor. *RNA* **2007**, *13*, 313–316. [CrossRef] [PubMed]
44. Matouk, I.J.; Halle, D.; Raveh, E.; Gilon, M.; Sorin, V.; Hochberg, A. The role of the oncofetal H19 lncRNA in tumor metastasis: Orchestrating the EMT-MET decision. *Oncotarget* **2016**, *7*, 3748–3765. [CrossRef] [PubMed]
45. Hanna, C.W. Placental imprinting: Emerging mechanisms and functions. *PLoS Genet.* **2020**, *16*, e1008709. [CrossRef]
46. Nordin, M.; Bergman, D.; Halje, M.; Engstrom, W.; Ward, A. Epigenetic regulation of the Igf2/H19 gene cluster. *Cell Prolif.* **2014**, *47*, 189–199. [CrossRef]
47. Blyth, A.J.; Kirk, N.S.; Forbes, B.E. Understanding IGF-II Action through Insights into Receptor Binding and Activation. *Cells* **2020**, *9*, 2276. [CrossRef]
48. Fowden, A.L. The insulin-like growth factors and feto-placental growth. *Placenta* **2003**, *24*, 803–812. [CrossRef]
49. Ratajczak, M.Z. Igf2-H19, an imprinted tandem gene, is an important regulator of embryonic development, a guardian of proliferation of adult pluripotent stem cells, a regulator of longevity, and a 'passkey' to cancerogenesis. *Folia Histochem. Cytobiol.* **2012**, *50*, 171–179. [CrossRef]
50. Argyraki, M.; Damdimopoulou, P.; Chatzimeletiou, K.; Grimbizis, G.F.; Tarlatzis, B.C.; Syrrou, M.; Lambropoulos, A. In-utero stress and mode of conception: Impact on regulation of imprinted genes, fetal development and future health. *Hum. Reprod. Update* **2019**, *25*, 777–801. [CrossRef]
51. Matouk, I.J.; Halle, D.; Gilon, M.; Hochberg, A. The non-coding RNAs of the H19-IGF2 imprinted loci: A focus on biological roles and therapeutic potential in Lung Cancer. *J. Transl. Med.* **2015**, *13*, 113. [CrossRef]
52. Ivanova, E.; Canovas, S.; Garcia-Martinez, S.; Romar, R.; Lopes, J.S.; Rizos, D.; Sanchez-Calabuig, M.J.; Krueger, F.; Andrews, S.; Perez-Sanz, F.; et al. DNA methylation changes during preimplantation development reveal inter-species differences and reprogramming events at imprinted genes. *Clin. Epigenetics* **2020**, *12*, 64. [CrossRef] [PubMed]
53. Marcho, C.; Cui, W.; Mager, J. Epigenetic dynamics during preimplantation development. *Reproduction* **2015**, *150*, R109–R120. [CrossRef] [PubMed]
54. Ankolkar, M.; Patil, A.; Warke, H.; Salvi, V.; Mokashi, N.K.; Pathak, S.; Balasinor, N.H. Methylation analysis of idiopathic recurrent spontaneous miscarriage cases reveals aberrant imprinting at H19 ICR in normozoospermic individuals. *Fertil. Steril.* **2012**, *98*, 1186–1192. [CrossRef] [PubMed]

55. Aygun, D.; Bjornsson, H.T. Clinical epigenetics: A primer for the practitioner. *Dev. Med. Child. Neurol.* **2020**, *62*, 192–200. [CrossRef] [PubMed]
56. Pianka, M.A.; McIntosh, A.T.; Patel, S.D.; Bakhshi, P.R.; Jung, M. Close yet so far away: A look into the management strategies of genetic imprinting disorders. *Am. J. Stem. Cells* **2018**, *7*, 72–81. [PubMed]
57. Jelinic, P.; Shaw, P. Loss of imprinting and cancer. *J. Pathol.* **2007**, *211*, 261–268. [CrossRef]
58. Court, F.; Tayama, C.; Romanelli, V.; Martin-Trujillo, A.; Iglesias-Platas, I.; Okamura, K.; Sugahara, N.; Simon, C.; Moore, H.; Harness, J.V.; et al. Genome-wide parent-of-origin DNA methylation analysis reveals the intricacies of human imprinting and suggests a germline methylation-independent mechanism of establishment. *Genome Res.* **2014**, *24*, 554–569. [CrossRef]
59. McMinn, J.; Wei, M.; Schupf, N.; Cusmai, J.; Johnson, E.B.; Smith, A.C.; Weksberg, R.; Thaker, H.M.; Tycko, B. Unbalanced placental expression of imprinted genes in human intrauterine growth restriction. *Placenta* **2006**, *27*, 540–549. [CrossRef]
60. Romanelli, V.; Nakabayashi, K.; Vizoso, M.; Moran, S.; Iglesias-Platas, I.; Sugahara, N.; Simon, C.; Hata, K.; Esteller, M.; Court, F.; et al. Variable maternal methylation overlapping the nc886/vtRNA2-1 locus is locked between hypermethylated repeats and is frequently altered in cancer. *Epigenetics* **2014**, *9*, 783–790. [CrossRef]
61. Nakagawa, S.; Shimada, M.; Yanaka, K.; Mito, M.; Arai, T.; Takahashi, E.; Fujita, Y.; Fujimori, T.; Standaert, L.; Marine, J.C.; et al. The lncRNA Neat1 is required for corpus luteum formation and the establishment of pregnancy in a subpopulation of mice. *Development* **2014**, *141*, 4618–4627. [CrossRef]
62. Yu, L.; Zhang, Y.; Xiong, J.; Liu, J.; Zha, Y.; Kang, Q.; Zhi, P.; Wang, Q.; Wang, H.; Zeng, W.; et al. Activated gammadelta T Cells With Higher CD107a Expression and Inflammatory Potential During Early Pregnancy in Patients With Recurrent Spontaneous Abortion. *Front. Immunol.* **2021**, *12*, 724662. [CrossRef] [PubMed]
63. Deryabin, P.; Domnina, A.; Gorelova, I.; Rulev, M.; Petrosyan, M.; Nikolsky, N.; Borodkina, A. "All-In-One" Genetic Tool Assessing Endometrial Receptivity for Personalized Screening of Female Sex Steroid Hormones. *Front. Cell Dev. Biol.* **2021**, *9*, 624053. [CrossRef] [PubMed]
64. Sherman, B.T.; Hao, M.; Qiu, J.; Jiao, X.; Baseler, M.W.; Lane, H.C.; Imamichi, T.; Chang, W. DAVID: A web server for functional enrichment analysis and functional annotation of gene lists (2021 update). *Nucleic Acids Res* **2022**, *10*. [CrossRef] [PubMed]
65. Huang, D.W.; Sherman, B.T.; Lempicki, R.A. Systematic and integrative analysis of large gene lists using DAVID bioinformatics resources. *Nat. Protoc.* **2009**, *4*, 44–57. [CrossRef]
66. Li, J.; Miao, B.; Wang, S.; Dong, W.; Xu, H.; Si, C.; Wang, W.; Duan, S.; Lou, J.; Bao, Z.; et al. Hiplot: A comprehensive and easy-to-use web service for boosting publication-ready biomedical data visualization. *Brief Bioinform.* **2022**, *23*. [CrossRef]
67. Zhang, Y.; Hu, J.F.; Wang, H.; Cui, J.; Gao, S.; Hoffman, A.R.; Li, W. CRISPR Cas9-guided chromatin immunoprecipitation identifies miR483 as an epigenetic modulator of IGF2 imprinting in tumors. *Oncotarget* **2017**, *8*, 34177–34190. [CrossRef]
68. Chen, N.; Yan, X.; Zhao, G.; Lv, Z.; Yin, H.; Zhang, S.; Song, W.; Li, X.; Li, L.; Du, Z.; et al. A novel FLI1 exonic circular RNA promotes metastasis in breast cancer by coordinately regulating TET1 and DNMT1. *Genome Biol.* **2018**, *19*, 218. [CrossRef]
69. Pian, L.; Wen, X.; Kang, L.; Li, Z.; Nie, Y.; Du, Z.; Yu, D.; Zhou, L.; Jia, L.; Chen, N.; et al. Targeting the IGF1R Pathway in Breast Cancer Using Antisense lncRNA-Mediated Promoter cis Competition. *Mol. Ther. Nucleic Acids* **2018**, *12*, 105–117. [CrossRef]

Article

The Effects of a Preconception Lifestyle Intervention on Childhood Cardiometabolic Health—Follow-Up of a Randomized Controlled Trial

Stijn Mintjens [1,2,3,*], Mireille N. M. van Poppel [4,5], Henk Groen [6], Annemieke Hoek [7], Ben Willem Mol [8], Rebecca C. Painter [2], Reinoud J. B. J. Gemke [1] and Tessa J. Roseboom [2,4]

1 Department of Pediatrics, Emma Children's Hospital, Amsterdam UMC, 1105 AZ Amsterdam, The Netherlands; rjbj.gemke@amsterdamumc.nl
2 Department of Obstetrics and Gynecology, Amsterdam UMC, 1105 AZ Amsterdam, The Netherlands; r.c.painter@amsterdamumc.nl (R.C.P.); t.j.roseboom@amc.uva.nl (T.J.R.)
3 Department of Pediatrics, NYC Health & Hospitals/Lincoln, New York, NY 10451, USA
4 Amsterdam Public Health Research Institute, Amsterdam UMC, 1105 AZ Amsterdam, The Netherlands; mireille.van-poppel@uni-graz.at
5 Institute of Sport Science, University of Graz, 8010 Graz, Austria
6 Department of Epidemiology, University of Groningen, University Medical Center Groningen, 9700 RB Groningen, The Netherlands; h.groen01@umcg.nl
7 Department of Obstetrics and Gynecology, University of Groningen, University Medical Center Groningen, 9700 RB Groningen, The Netherlands; a.hoek@umcg.nl
8 Department of Obstetrics and Gynecology, Monash University, Monash Medical Center, Clayton, VIC 3800, Australia; ben.mol@monash.edu
* Correspondence: mintjens@nychhc.org

Citation: Mintjens, S.; van Poppel, M.N.M.; Groen, H.; Hoek, A.; Mol, B.W.; Painter, R.C.; Gemke, R.J.B.J.; Roseboom, T.J. The Effects of a Preconception Lifestyle Intervention on Childhood Cardiometabolic Health—Follow-Up of a Randomized Controlled Trial. *Cells* **2022**, *11*, 41. https://doi.org/10.3390/cells11010041

Academic Editors: Lon J. van Winkle and Anthony Ashton

Received: 19 October 2021
Accepted: 15 December 2021
Published: 24 December 2021

Abstract: Maternal obesity is associated with adverse metabolic outcomes in her offspring, from the earliest stages of development leading to obesity and poorer cardiometabolic health in her offspring. We investigated whether an effective preconception lifestyle intervention in obese women affected cardiometabolic health of their offspring. We randomly allocated 577 infertile women with obesity to a 6-month lifestyle intervention, or to prompt infertility management. Of the 305 eligible children, despite intensive efforts, 17 in the intervention and 29 in the control group were available for follow-up at age 3–6 years. We compared the child's Body Mass Index (BMI) Z score, waist and hip circumference, body-fat percentage, blood pressure Z scores, pulse wave velocity and serum lipids, glucose and insulin concentrations. Between the intervention and control groups, the mean (±SD) offspring BMI Z score (0.69 (±1.17) vs. 0.62 (±1.04)) and systolic and diastolic blood pressure Z scores (0.45 (±0.65) vs. 0.54 (±0.57); 0.91 (±0.66) vs. 0.96 (±0.57)) were similar, although elevated compared to the norm population. We also did not detect any differences between the groups in the other outcomes. In this study, we could not detect effects of a preconception lifestyle intervention in obese infertile women on the cardiometabolic health of their offspring. Low follow-up rates, perhaps due to the children's age or the subject matter, combined with selection bias abating contrast in periconceptional weight between participating mothers, hampered the detection of potential effects. Future studies that account for these factors are needed to confirm whether a preconception lifestyle intervention may improve the cardiometabolic health of children of obese mothers.

Keywords: maternal obesity; childhood obesity; lifestyle intervention; cardiometabolic health; programming; follow-up

1. Introduction

About 25% of children worldwide are overweight or obese [1,2]. Early life adiposity impairs cardiovascular and metabolic functioning during childhood and adolescence itself, and increases risks of cardiovascular disease (CVD) later in life [3]. There is emerging evidence that an elevated Body Mass Index (BMI) in mothers before conception and during pregnancy affects early embryonic development. Obese women have oocytes with altered mitochondrial function, leading to increased redox states, which are suggestive of oxidative stress in the zygote [4]. Large studies in assisted reproduction in women with obesity showed that using autologous embryos have lower success rates of liveborn than using donor embryos of normal weight women, suggesting obesity directly affects the embryo itself [5]. These early alterations in embryo development may increase the offspring's risks of childhood obesity, increased fat mass, elevated blood pressure and disturbances in lipid and glucose concentrations. Additionally, this likely leads to increased adult cardiometabolic morbidity and all-cause mortality [6–8]. Hence, the World Health Organization (WHO) and a recent Lancet series have called to action to evaluate possible interventions to optimize preconception health [9,10]. The number of women of reproductive age with overweight and obesity is rising rapidly and is estimated to afflict more than a third of pregnancies [1,11]. Indicating the urgency to assess whether a preconception lifestyle change might be able to reduce the intergenerational development of obesity and improve the offspring's long-term cardiometabolic health [12,13].

Animal studies have shown that improving diet in pregnancy has positive effects on the offspring's adiposity and metabolic health [14]. Additionally, increasing preconception exercise improved the offspring's adiposity and lipid levels [15]. Follow-ups of human randomized controlled trials (RCT) aimed to improve maternal lifestyle in women with obesity before pregnancy are currently lacking. Studies of lifestyle interventions during pregnancy have shown limited effects on offspring health; one study found improved infant skinfolds at 6 months of age but no change in the BMI [16], and another showed improved weight-for-age Z scores at 1 year without an effect on weight-for-length Z scores [17]. Six other studies found no effects on adiposity or other CVD risk factors at infancy up to 7 years of age [18]. Here, we present, for the first time, the effects of a preconception lifestyle intervention trial, which compared a 6-month lifestyle intervention program targeting physical activity, diet and behavior modification prior to infertility management to a control group receiving infertility management as usual [19]. The intervention in this trial indeed improved maternal lifestyle, through changes in diet and an increase in physical activity and resulted in an approximately 4-kg weight loss and halving the odds of metabolic syndrome after 6 months, and in those women who successfully lost weight, beneficial effects were seen up to 6 years later [20–22]. As beneficial intervention effects were found in all the women participating in the original RCT, we found it imperative to test our hypothesis that a healthier lifestyle before conception in infertile women with obesity improves the child's cardiometabolic health at age 3–6 years.

2. Materials and Methods

2.1. Lifestyle Intervention

This follow-up study is based on the LIFE*style* study (NTR1530), a multicenter RCT performed between 2011 and 2014 in 23 hospitals in the Netherlands [19]. Both the original study and this follow-up study were approved by the medical ethics committee of the University Medical Centre Groningen, the Netherlands (NL24478.042.08). Both parents gave written informed consent. Reporting in this manuscript adheres to the CONSORT 2010 guidelines.

The original randomized trial's design has been described in detail previously [19]. In summary, infertile women (unsuccessfully tried to conceive for at least 12 months) with a BMI $\geq$ 29 kg/m^2 were randomly allocated to a 6-month lifestyle intervention program preceding infertility management or to the control arm with prompt infertility management, both for a maximum of 24 months. Infertility management was individually assigned according to Dutch guidelines.

The intervention comprised of individualized motivational counseling by trained study nurses and dieticians to improve dietary intake and increase physical activity. Women were counseled to reduce their caloric intake by 600 kcal per day, and to adhere to Dutch guidelines for a balanced diet. Women wore a step-counter and aimed to achieve 10,000 steps a day and at least 30 min of moderate-intensity exercise two or three times per week. The intervention included six face-to-face and four phone consultations over 24 weeks and aimed at a weight loss of 5–10% of body weight or achieving a BMI < 29 kg/m^2. The intervention stopped when conception occurred or was followed by up to 18 months of standard infertility management.

2.2. Eligibility for Follow-Up

The protocol of this follow-up study has previously been published [23]. In summary, all babies from a singleton pregnancy conceived within 24 months after their mother was randomized and who were known to be alive were eligible to participate in the follow-up assessments (n = 305 children of n = 577 randomized women) [19]. Children were 3–6 years old and living in the Netherlands (n = 300; n = 3 intervention group and n = 2 control group with no known address) when they were approached by mail (lay-man oriented information leaflets), and if possible, by repeated phone calls, for inclusion to the follow-up study. Parents were previously asked to fill out questionnaires about their child, during which parents could opt-out of being contacted by phone [23].

2.3. Follow-Up Assessment

Children were assessed in 2016 and 2017. We used a mobile research vehicle enabling us to conduct all assessments at/near the participant's home. One parent was present during measurements, which were performed by two assessors from a pool of six experienced assessors with appropriate training. The assessors remained blinded to the lifestyle intervention allocation of the mother of the child undergoing assessment.

Maternal- and pregnancy-related characteristics were collected during the initial trial. Child's birth weight was calculated as a gestational age and gender adjusted Z score based on Dutch reference curves with the LMS (lambda-mu-sigma) methodology [24,25].

At follow-up, we measured height using a SECA® (Hamburg, Germany) 206 wall attached measuring tape to the nearest 0.1 cm. Weight was measured using a SECA® 877 digital scale to the nearest 0.1 kg. We calculated BMI as weight in kg divided by height in meters squared. We calculated an age and sex adjusted Z-score based on the WHO reference values [26]. We measured waist and hip circumference with a SECA® 201 measuring tape to the nearest 0.1 cm. We assessed body composition by bioelectrical impedance analysis (BIA) using the Bodystat 1500 mdd® (Douglas, UK). Children were asked to empty their bladder and refrain from drinking at least 90 min prior to the BIA measurements. We used a validated, age-appropriate equation to calculate body-fat percentage (BF%) [27]. Measurements were taken in duplicate, in case there was considerable discrepancy between measurements, e.g., more than 0.5 kg for weight, a third measurement was obtained. All measurements were averaged.

We measured blood pressure (BP) seated in triplicate on the non-dominant arm, after 5 min of rest. A validated oscillometric device with age-appropriate cuff (Omron HBP-1300®, Kyoto, Japan) was used to measure BP [28]. We averaged the systolic blood pressure (SBP) and diastolic blood pressure (DBP), respectively, followed by a calculation of age and sex adjusted Z scores based on National Institute of Health (NIH) reference values [29]. We assessed arterial stiffness by Complior® (ALAM Medical, St. Quentin Fallavier, France) to measure pulse wave velocity (PWV). This technique calculates the time between standardized measurement of the carotid and femoral pulses. By dividing the time by the distance between these two reference points, the PWV was calculated.

With parental consent, a venous blood sample after an overnight fast was taken from their child during a separate appointment. We measured triglycerides (TG), total cholesterol (TC), high density lipoprotein (HDL) cholesterol, low density lipoprotein (LDL) cholesterol, insulin and glucose. We calculated the homeostatic model assessment of insulin resistance (HOMA-IR) as insulin (μIU/mL) times glucose (mmol/L) divided by 22.5.

2.4. Statistical Analyses

The sample size of the initial trial was set at 285 women per group based on the primary outcome (a healthy livebirth) [19]. No formal sample size calculation was performed for the current analysis, but the sample size of the initial trial was deemed sufficient to detect potentially relevant differences in the offspring, provided that participation rates were good.

To assess sample bias, we compared maternal- and pregnancy-related characteristics of the participating children to those that were eligible but did not participate. Similarly, of those children that participated, we compared maternal- and pregnancy-related characteristics from mothers in the intervention and control group. Differences in characteristics and outcomes were examined using Student's *t*-test or Fisher's exact test, as appropriate (Table 1).

Table 1. Maternal baseline and pregnancy related characteristics.

	n	Intervention	*n*	Control	*n*	Non-Participants	*p*-Value #
Maternal baseline characteristics:							
Age, years—mean (SD)	17	29.9 (3.4)	29	29.3 (4.1)	259	29.1 (4.3)	0.51
Caucasian—no (%)	17	16 (94.1)	29	28 (96.6)	259	227 (87.6)	0.13
Education—no (%)	17	-	29	-	246	-	0.36
Primary school	-	0 (0.0)	-	0 (0.0)	-	10 (4.1)	-
Secondary education	-	2 (11.8)	-	8 (27.6)	-	59 (24.0)	-
Intermediate vocation education	-	11 (64.7)	-	15 (51.7)	-	115 (46.7)	-
Higher education	-	2 (13.3)	-	6 (20.7)	-	62 (25.2)	-
Smoker—no. (%)	17	3 (17.6)	29	5 (17.2)	255	56 (22.0)	0.56
BMI (kg/m^2)—mean (SD)	17	36.0 (2.7)	29	35.6 (3.0)	259	35.9 (3.5)	0.80
Pregnancy related characteristics:							
Maternal age at time of pregnancy (years)—mean (SD)	17	30.5 (3.4)	29	29.8 (4.3)	254	29.8 (4.4)	0.69
Nulliparous—no. (%)	17	13 (76.5)	29	21 (72.4)	258	207 (80.2)	0.43
Delta baseline BMI and periconceptional BMI—mean (SD)	15	−0.7 (2.8)	24	−0.9 (1.5)	103	−1.0 (2.7)	0.24
Gestational weight gain (kg)—mean (SD)	10	11.8 (6.1)	18	11.3 (5.8)	195	9.9 (6.3)	0.64
Gestational diabetes—no. (%)	17	4 (23.5)	29	7 (24.1)	252	44 (17.5)	0.31
Gestational age at birth (weeks)—mean (SD)	17	39.0 (1.7)	29	39.2 (1.7)	254	39.0 (2.1)	0.79
Birth weight (grams)—mean (SD)	17	3234 (497) *	29	3652 (454) *	253	3391 (585)	0.25
Conception mode—no (%)	17	-	29	-	255	-	0.78
Natural	-	10 (58.8)	-	8 (27.6)	-	97 (38.0)	-
Ovulation Induction	-	5 (29.4)	-	11 (37.9)	-	78 (30.6)	-
IUI	-	2 (11.8)	-	5 (17.2)	-	37 (14.5)	-
IVF/ICSI/CRYO	-	0 (0.0)	-	5 (17.2)	-	43 (16.9)	-
Breastfeeding +—no (%)	17	4 (23.5)	29	9 (31.0)	259	70 (27.0)	0.86

Comparison between participants versus non-participants. * $p < 0.05$ between intervention and control. + Exclusive breastfeeding for three or more months. BMI = Body mass index, kg = kilogram, IUI = Intra-uterine insemination, IVF = In Vitro fertilization, ICSI = Intracytoplasmic sperm injection, CRYO = Cryotherapy.

As our primary analysis, we compared outcome measures (BMI Z score, BF%, BP Z score, PWV, serum lipids, glucose and insulin concentrations and HOMA-IR) of children from mothers in the intervention and control groups by means of Student's *t*-test or Fisher's exact test, as appropriate (Table 2). We adhered to the randomized trial design in our analyses. We tested in multivariable linear regression analyses whether an adjustment for the maternal- or pregnancy-related characteristics that were found to be different between group characteristics (i.e., birth weight), would alter the outcomes.

Table 2. Cardiometabolic outcome values of children of mothers from the intervention and control group.

		Anthropometry			
	n	**Intervention**	***n***	**Control**	**95% CI**
BMI (Z-score)—mean (SD)	16	0.69 (1.17)	28	0.62 (1.04)	−0.62–0.76
Waist circumference (cm)—mean (SD)	17	53.4 (4.3)	29	53.4 (5.3)	−3.04–3.10
Hip circumference (cm)—mean (SD)	17	58.3 (4.4)	29	58.4 (6.9)	−3.90–3.66
Body-fat (%)—mean (SD)	16	20.7 (7.8)	26	21.2 (9.4)	−6.16–5.16
		Cardiovascular			
	n	Intervention	*n*	Control	95% CI
SBP (Z-score)—mean (SD)	16	0.46 (0.65)	27	0.54 (0.57)	−0.46–0.30
DBP (Z-score)—mean (SD)	16	0.91 (0.66)	27	0.96 (0.57)	−0.44–0.33
PWV (m/sec)—mean (SD)	12	4.51 (0.83)	22	4.50 (1.14)	−0.75–0.77
		Metabolic			
	n	Intervention	*n*	Control	95% CI
Triglycerides (mmol/L)—mean (SD)	7	0.71 (0.63)	17	0.53 (0.17)	−0.39–0.76
Total cholesterol (mmol/L)—mean (SD)	7	4.26 (0.79)	17	4.07 (0.54)	−0.39–0.77
LDL cholesterol (mmol/L)—mean (SD)	7	2.46 (0.65)	17	2.36 (0.40)	−0.35–0.54
HDL cholesterol (mmol/L)—mean (SD)	7	1.48 (0.20)	17	1.48 (0.26)	−0.22–0.24
Insulin (μIU/mL)—mean (SD)	7	5.52 (3.12)	12	4.21 (2.87)	−1.66–4.29
Glucose (mmol/L)—mean (SD)	7	4.70 (0.33)	17	4.47 (0.42)	−0.13–0.60
HOMA-IR—mean (SD)	7	1.19 (0.75)	12	0.87 (0.64)	−0.37–1.00

BMI = Body mass index, SBP = Systolic blood pressure, DBP = Diastolic blood pressure, LDL = Low-density lipoprotein, HDL = High-density lipoprotein, HOMA-IR = Homeostatic model of insulin resistance.

As exploratory analyses for potential offspring sex differences in effects of maternal obesity, we compared outcome measures for boys and girls separately (Supplementary Table S1). Furthermore, we assessed whether children of women who successfully lost weight (i.e., 5–10% weight reduction or achieving a BMI < 29 kg/m^2) differed in outcomes from children of mothers who were unsuccessful, independent of randomization (Supplementary Table S2).

Values are presented as means and standard deviations (±SD) for continuous data and as frequency distributions for categorical data. We considered *p*-values of less than 0.05 statistically significant.

3. Results

Figure 1 shows the flowchart of the included participants. Out of the 163 women who conceived within 24 months in the invention group, 15 children were from twin/triplet pregnancies and three children were deceased, leaving 145 eligible children. From the 178 conceiving women in the control group, 14 children were from twin pairs and four children were deceased, leaving 160 children eligible. Despite intensive efforts, many parents did not respond or declined participation. A total of 51 parents provided informed consent; however, not all were able/willing to undergo measurements. Thus, we report on 17 children whose mother was randomized to the intervention group and 29 from the control group (15%).

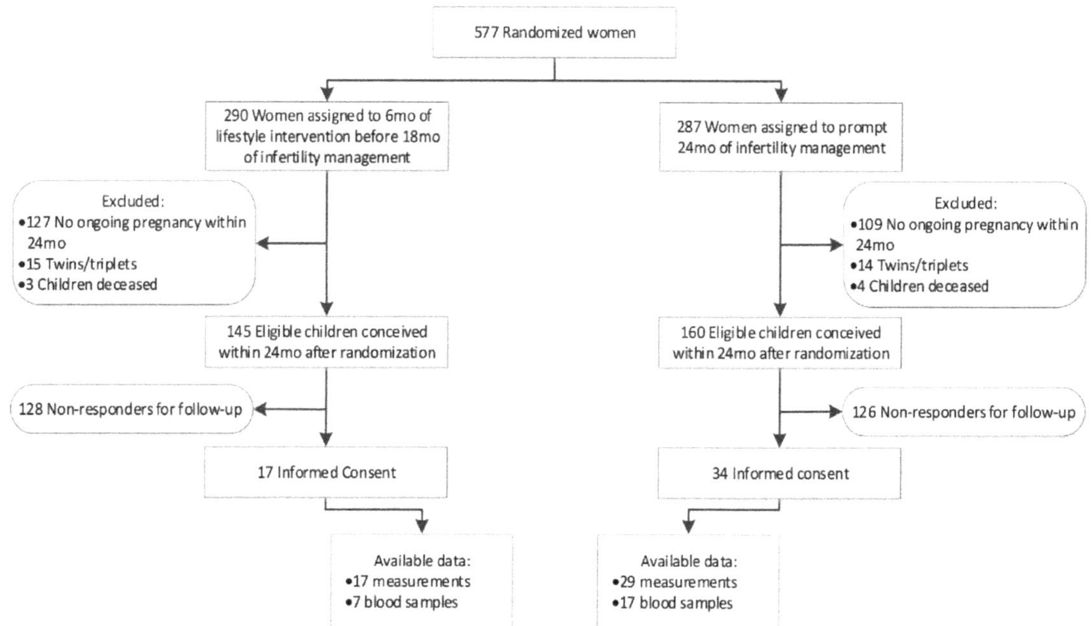

Figure 1. Flowchart of included participants. Mo = Months.

As shown in Table 1, there were no differences between the maternal baseline characteristics and the pregnancy-related characteristics between the mothers of participating children and eligible children that did not participate. Additionally, the baseline and pregnancy-related characteristics of the mothers of participating children did not differ between the intervention and control groups. In the original trial, analyzing women who conceived and who did not conceive, the women in the intervention group had improved their lifestyle, lost weight and improved metabolic indices over 6 months [21]. Up to 6 years after the intervention, the differences abated between the intervention and control groups; however, beneficial effects were still present in the women who achieved the lifestyle intervention goals (5–10% weight loss or a BMI < 29 kg/m^2) [22]. In our selected sample, all the women reduced their BMI slightly, but between the intervention and control groups there were no significant differences in the change in BMI between baseline and time of conception. Furthermore, maternal weight gain during pregnancy was equally high in both groups.

In the children, the mean birth weight of those who participated in the follow-up was lower in the intervention group compared to the control group (3234 g vs. 3652 g, $p < 0.05$; Table 1), while there was no statistically significant difference in the original trial (3312 g vs. 3341 g), and there were no differences in gestational age [19].

The mean (±SD) age of the participating children was 4.6 (±1.0) years (range 3.2–6.5 years). There were 22 boys (48%). Overall, the BMI Z score was 0.65 (±1.26), and the SBP and DBP Z scores were 0.51 (±0.59) and 0.94 (±0.60), respectively. Table 2 shows the cardiometabolic health indices of children in the intervention and control groups. We found no differences between the children of mothers from the intervention group compared to the children of mothers from the control group in childhood outcome measures (BMI Z score, BF%, BP Z score, PWV, serum lipids, glucose and insulin concentrations and

HOMA-IR). Adjusting for confounders such as the child's birth weight did not alter the effect estimates in multivariate analyses.

In exploratory analyses, cardiometabolic outcome values of boys and girls did not differ according to the maternal allocation to lifestyle intervention (Supplementary Table S1). Furthermore, there were no differences in childhood cardiometabolic outcomes between children of mothers who successfully lost weight (n = 8) compared to those whose mothers did not (Supplementary Table S2). The latter analysis was independent of the assigned groups by randomization. In both exploratory analyses, the groups had very small numbers prohibiting conclusions to be drawn, and too few lab values were available for the children of mothers who successfully lost weight.

4. Discussion

We could not detect differences in offspring cardiometabolic health at age 3–6 years in the follow-up of a preconception maternal lifestyle intervention trial. Despite numerous efforts to enhance participation (e.g., measurements near the participant's home, multiple phone calls, layman information leaflets), our study was hampered by high attrition rates, which reduced the statistical power substantially. In our sample, the maternal BMI at conception and gestational weight gain (GWG) throughout pregnancy were not different between the intervention and control groups. We were unable to examine maternal cardiovascular and metabolic factors at the time of conception in our selected sample. However, we do know from previous publications by our group, that in all the women in the original trial, regardless of their conception status, the intervention increased physical activity, reduced snacking and sugary drinks intake, led to weight loss (approximately 4 kg) and halved the odds of metabolic syndrome after 6 months [20,21]. Furthermore, up to six years later, the intervention led to decreased caloric intake, and those women that were deemed successful in the primary trial showed improved BMI and cardiometabolic indices [20,22].

A range of experimental animal studies and observational human studies have shown that during embryonic developmental, even small environmental changes will have lasting effects (Table 3) [13]. In animals, changing the maternal lifestyle before and during pregnancy, and, in turn, comparing between in utero exposure to maternal obesity or reduced weight, was associated with improved adiposity and lipid levels in offspring [14,15]. In humans, environmental factors and the lifestyle of mothers impacted the fetal and placental metabolism, oxidative stress and interactions of these, inflicting epigenetic changes that are suggested to have lasting effects [30,31]. Furthermore, in assisted reproduction variations in embryo culture conditions have led to altered metabolic and epigenetic regulation, resulting in altered growth and cardiometabolic profiles of offspring [32,33]. In a large cohort study, children born after assisted reproduction had different growth patterns in their first few years, but ended up at a grossly similar height and weight in adolescence compared to their naturally conceived peers [34]. Although these findings seem reassuring, such alterations in growth during early life are linked to a predisposition of poor cardiometabolic health later in life, suggested by early life echocardiograms alterations in cardiac shape and function in assisted reproduction offspring [35].

This follow-up was based on the first randomized controlled trial in obese women examining the effects of a preconception lifestyle intervention. Due to the randomized design, confounding factors related to maternal infertility and/or obesity were equally divided between groups. Hence, we consider this population of infertile women valid to explore the effects of maternal preconception lifestyle change and weight loss on offspring. Despite the fact that most maternal- and pregnancy-related factors were similar between the groups, there was selective participation in our follow-up sample. This was indicated by a lower birth weight in children in the intervention group, a difference that was not present in the original trial [19]. Since our study had a null result and adjusting our analyses for birth weight did not change our results, we consider it unlikely that this selection has led to bias.

Table 3. Summary of selected current (pre-)pregnancy lifestyle intervention studies in animal and human settings and effects on offspring's health.

Study Identifier	Animal (A) or Human (H)	Intervention	Results
Gallou-Kabani et al., (2007) [36]	A	Dietary at time of conception/pregnancy and lacation	Female, not male offspring, had a higher proportion that remained lean on postnatal high fat diet and improved glycemic indices and lipids.
Zambrano et al., (2010) [14]	A	Dietary (30 days prior pregancy)	(Partial) normalization of fat mass, triglycerides, leptin and insulin
Dennison et al., (2013) [37]	A	Dietary (low fat high fiber) and/or sitagliptin (8 weeks prior pregnancy)	No changes in offspring body weight. Diet had no significant effects on energy intake, leptin, fasting glucose, however some microbiome change were seen. Sitagliptin alone had largest reduction in glucemic control.
Vega et al., (2015) [15]	A	Exercise (30 days prior pregnancy)	Reduced leptin, triglycerides, glucose
Xu et al., (2018) [38]	A	Dietary (up to 9 weeks prior preganncy)	Longer maternal diet intervention showed normalization of offspring's glucose and lipid metabolism
Mustilla et al., (2012) [39]	H	Lifestyle intervention on diet and physical activity durng pregnancy	At 24–48 months, the offpsring in the intervention group had slower gains in BMI z score. Over the 0–48 months there was no differences in BMI z score gain between groups. (Follow-up rate, 72%)
Tanvig et al., (2014) [40]	H	Diet, exercise and coaching during pregnancy (RCT)	At 2.8 years follow-up, there were no differences between groups in BMI z-scores, nor in skinfold, anthropometrics, total fat mass, lean mass or fat percentage. (Follow-up rate 29%)
Rauh et al., (2015) [41]	H	Lifestyle intervention including dietary and physical acivity counseling twice during pregnancy	At 10–12 months after birth, there were no significant differences in offspring's weight. (Follow-up rate, 85%)
Horan et al., (2016) [42]	H	Dietary intervention during pregnancy in women with previous LGA infant	No effects on offspring at 6 months, at 2 years improvement of anthropometrics indices with heatlhier dietary intake during pregancy. (Follow-up rate, 35%)
Kolu et al., (2016) [43]	H	Lifestyle intervention of diet and physical activity during 5 antenatel visits during pregnancy	No differences in child's BMI up to 7 years. Children of mothers who adhered to all lifestyle aims had signifcantly lower BMIs. (Follow-up rate, 43%)
Vesco et al., (2016) [17]	H	Weekly weight management intervention focused on diet and exercise during pregnancy	At 1 year of age there was significant reduction in weight-for-age z scores in children in the intervention group, but no differences in weight-for-height z score between groups.
Ronnberg et al., (2017) [44]	H	Lifestyle intervention on diet and physical activity during pregnancy, focus on healthy gestational weight gain	Follow-up of children's BMI until 5 years of age showed no differences between groups in child's BMI z score. (Follow-up rate, 80%)
Dalrymple et al., (2021) [45]	H	Diet and physical activity intervention of 8 weeks during pregnancy (RCT)	6 months lower skinfold measures in interventions, at 3 year follow-up no significant differences in BMI or skinfold between groups. Significan lower pulse rate in offspring of intervention (Follow-up rate, 33%)

More importantly, in contrast to the original trial, participating women from both groups had similar weight loss between randomization and periconception. Additionally, the GWG was above the recommended levels for obese women (5–9 kg) in both groups [46]. This resulted in a very limited contrast in maternal body weight at time of conception between the groups, which may have contributed to our null finding. While the intervention induced effects on maternal parameters other than body weight, i.e., glucose metabolism or nutritional quality [20,22], these changes may not have been able to mitigate the detrimental effects of maternal obesity and/or excessive GWG on the offspring's cardiometabolic health [47]. Since studies assessing children born after their mothers had bariatric surgery showed improved cardiometabolic health [48,49], more substantial maternal weight loss may be needed to elicit changes in childhood health outcomes. On the other hand, observational evidence suggests a graded 'dose response' association between maternal BMI and offspring's cardiometabolic health [47,50], indicating even modest weight changes could carry positive effects, but we were not able to detect any effect.

To provide more reliable conclusions about the potential effects of maternal lifestyle change before conception on children, future studies should aim to maximize follow-up rates and power calculations based on childhood outcomes, which should account for high attrition. Compared to our study, a higher participation rate of 52% was present 3 years after a different lifestyle intervention during pregnancy [51]. Still, attrition in follow-up studies is generally high [52], and reasons why often remain unknown. Although parents were not required to state their reasons for declining participation, those that did indicated that they refused due to time restraints, did not want to burden their child or that they mainly wanted to become pregnant and were not interested in a further follow-up. Some stated that they were aware of the negative association of maternal obesity with childhood health, while many others were not previously aware; many women indicated that they did not want to contribute to further evidence of that association. While we attempted to involve participants in the planning of the study, only a few provided limited information on topics of interest to them that corresponded to the outcomes we were considering. The future investigation of offspring health-related themes considered relevant by obese women and their partners may provide guidance into a strategy that achieves lower attrition rates in follow-up studies such as our own.

Individual interventions in obese adults have been marginally successful [52]; community-based interventions and policies could thus be better suited to optimize the health of women prior to pregnancy. While the WHO has made an important step in global obesity prevention by formulating nine voluntary targets to prevent non-communicable diseases [53], and policy change has shown some local improvements [54], these have not yet been able to counteract the overall worldwide burden of obesity. Since pregnant women with obesity indicated they were mostly unaware of the effects of obesity on their (future) child [55], there is an urgent need to improve awareness in the general public of the consequences of obesity before and during pregnancy. As health behaviors are not solely individually determined and depend on environmental factors [56], and knowledge of healthy habits do not directly translate to changes in behavior [57], policies focusing on improving nutrition, physical activity and sports involving the home, school/work environment and the community are needed to curb the intergenerational cycle of obesity.

5. Conclusions

We could not detect any effect of a preconception lifestyle intervention—that did improve lifestyle, induced weight loss and improved cardiometabolic health in women 6 months after randomization—on the offspring's cardiometabolic health at age 3–6 years. Our study was hampered by limited statistical power, perhaps due to the children's age or the subject matter, as well as the minimal difference in the maternal periconception weight between the groups of participants in contrast to the original trial. Future studies should account for these factors to maximize follow-up rates to be able to draw conclusions about the potential of preconception lifestyle interventions to affect offspring cardiometabolic health.

Whether there is no effect of lifestyle interventions in women with obesity prior to conception on their offspring's cardiometabolic health needs to be confirmed in larger studies.

Supplementary Materials: The following are available online at https://www.mdpi.com/article/10.3390/cells11010041/s1, Supplementary Table S1: Cardiometabolic outcome values of boys and girls of mothers from the intervention and control group, Supplementary Table S2: Cardiometabolic outcome values of children of mothers who successfully* lost weight compared to children of mothers who did not successfully lose weight, pooled data independent of the randomization group.

Author Contributions: A.H. is the principal investigator (PI) of the Lifestyle trial, and T.J.R. is the PI of the follow-up studies. S.M., M.N.M.v.P., H.G., A.H., B.W.M., R.C.P., R.J.B.J.G. and T.J.R. planned the study procedures. S.M. collected and analyzed the data. S.M., M.N.M.v.P., R.J.B.J.G. and T.J.R. interpreted the data. S.M. wrote the manuscript. M.N.M.v.P., H.G., A.H., B.W.M., R.C.P., R.J.B.J.G. and T.J.R. revised the manuscript. All authors have read and agreed to the published version of the manuscript.

Funding: This study was supported by a grant of the Dutch Heart Foundation (2013T085) and by a DynaHealth H2020 grant (agreement No 633595). The funders had no role in the design, conduct and interpretation of the study.

Institutional Review Board Statement: The authors assert that all procedures contributing to this work comply with the ethical standards of the relevant national guidelines on human experimentation (CCMO), and with the Helsinki declaration of 1975, as revised in 2008, and has been approved by the Medical Ethical Committee of University Medical Centre Groningen, the Netherlands (NL24478.042.08). Parental consent was attained for all participants.

Informed Consent Statement: Informed consent was obtained from all subjects involved in the study.

Data Availability Statement: The datasets used and/or analyzed during the current study are available from the corresponding author on reasonable request. This manuscript has been presented as a preprint in the University of Amsterdam repository for Ph.D. thesis.

Acknowledgments: This manuscript has been presented as a preprint in the University of Amsterdam repository for Ph.D. thesis. We thank all the women and children who participated in this study. We are grateful to all members of the WOMB-project who contributed to the follow-up study, with special thanks to our colleague Ph.D. students, post-docs, research assistants and students. This follow-up study would not have been possible without the original LIFEstyle study; therefore, we thank all participating hospitals for their contribution to this study, and the lifestyle coaches, research nurses, research midwives and office members of the Dutch Consortium (www.studies-obsgyn.nl, accessed on 18 October 2021) for their hard work and dedication.

Conflicts of Interest: S.M., M.N.M.v.P., H.G., A.H., B.W.M., R.C.P., R.J.B.J.G. and T.J.R. declare no conflict of interest relevant to this study. Outside the work submitted, A.H. declares consultancy fees and grants by Ferring Pharmaceutical BV. B.W.M is supported by a NHMRC practitioner Fellowship and reports consultancy for ObsEva, Merck, Merck KGaA and Guerbet. R.P. reports board membership for the Dutch Gezondheidsraad NVVH guideline committee, and travel expenses reimbursement for ESHRE meeting Barcelona 2018 (June).

References

1. Ng, M.; Fleming, T.; Robinson, M.; Thomson, B.; Graetz, N.; Margono, C.; Mullany, E.C.; Biryukov, S.; Abbafati, C.; Abera, S.F.; et al. Global, regional, and national prevalence of overweight and obesity in children and adults during 1980–2013: A systematic analysis for the Global Burden of Disease Study 2013. *Lancet* **2014**, *384*, 766–781. [CrossRef]
2. Bentham, J.; Di Cesare, M.; Bilano, V.; Bixby, H.; Zhou, B.; Stevens, G.A.; Riley, L.M.; Taddei, C.; Hajifathalian, K.; Lu, Y.; et al. Worldwide trends in body-mass index, underweight, overweight, and obesity from 1975 to 2016: A pooled analysis of 2416 population-based measurement studies in 1289 million children, adolescents, and adults. *Lancet* **2017**, *390*, 2627–2642. [CrossRef]
3. Ayer, J.; Charakida, M.; Deanfield, J.E.; Celermajer, D.S. Childhood obesity and cardiovascular risk. *Eur. Heart J.* **2015**, *36*, 1371–1376. [CrossRef]
4. Igosheva, N.; Abramov, A.Y.; Poston, L.; Eckert, J.J.; Fleming, T.P.; Duchen, M.R.; McConnell, J. Maternal diet-induced obesity alters mitochondrial activity and redox status in mouse oocytes and zygotes. *PLoS ONE* **2010**, *5*, e10074. [CrossRef]
5. Luke, B.; Brown, M.B.; Stern, J.E.; Missmer, S.A.; Fujimoto, V.Y.; Leach, R. Female obesity adversely affects assisted reproductive technology (ART) pregnancy and live birth rates. *Hum. Reprod.* **2011**, *26*, 245–252. [CrossRef]

6. Woo Baidal, J.A.; Locks, L.M.; Cheng, E.R.; Blake-Lamb, T.L.; Perkins, M.E.; Taveras, E.M. Risk Factors for Childhood Obesity in the First 1000 Days: A Systematic Review. *Am. J. Prev. Med.* **2016**, *50*, 761–779. [CrossRef]
7. Yu, Z.; Han, S.; Zhu, J.; Sun, X.; Ji, C.; Guo, X. Pre-Pregnancy Body Mass Index in Relation to Infant Birth Weight and Offspring Overweight/Obesity: A Systematic Review and Meta-Analysis. *PLoS ONE* **2013**, *8*, e61627. [CrossRef]
8. Reynolds, R.; Allan, K.; Raja, E.; Bhattacharya, S.; McNeill, G.; Hannaford, P.C.; Sarwar, N.; Lee, A.J.; Bhattacharya, S.; Norman, J.E. Maternal obesity during pregnancy and premature mortality from cardiovascular event in adult offspring: Follow-up of 1,323,275 person years. *BMJ* **2013**, *347*, f4539. [CrossRef]
9. Stephenson, J.; Heslehurst, N.; Hall, J.; Schoenaker, D.A.; Hutchinson, J.; Cade, J.E.; Poston, L.; Barrett, G.; Crozier, S.R.; Barker, M.; et al. Before the beginning: Nutrition and lifestyle in the preconception period and its importance for future health. *Lancet* **2018**, *391*, 1830–1841. [CrossRef]
10. World Health Organization. *Ending Childhood Obesity Report of the Commission ON*; WHO Press: Geneva, Switzerland, 2016.
11. Heslehurst, N.; Rankin, J.; Wilkinson, J.R.; Summerbell, C.D. A nationally representative study of maternal obesity in England, UK: Trends in incidence and demographic inequalities in 619,323 births. *Int. J. Obes.* **2010**, *34*, 420–428. [CrossRef]
12. Hanson, M.; Barker, M.; Dodd, J.M.; Adelaide, N.; Australia, S.; Kumanyika, S.; Norris, S.; Steegers, E.; Stephenson, J. Interventions to prevent maternal obesity prior to conception, during pregnancy, and postpartum. *Lancet Diabetes Endocrinol.* **2017**, *5*, 65–67. [CrossRef]
13. Gluckman, P.D.; Hanson, M.A.; Mitchell, M.D. Developmental origins of health and disease: Reducing the burden of chronic disease in the next generation. *Genome Med.* **2010**, *2*, 14. [CrossRef]
14. Zambrano, E.; Martínez-Samayoa, P.M.; Rodríguez-González, G.L.; Nathanielsz, P.W. RAPID REPORT: Dietary intervention prior to pregnancy reverses metabolic programming in male offspring of obese rats. *J. Physiol.* **2010**, *588*, 1791–1799. [CrossRef]
15. Vega, C.; Reyes-Castro, L.A.; Bautista, C.J.; Larrea, F.; Nathanielsz, P.W.; Zambrano, E. Exercise in obese female rats has beneficial effects on maternal and male and female offspring metabolism. *Int. J. Obes.* **2015**, *39*, 712–719. [CrossRef]
16. Patel, N.; Godfrey, K.; Pasupathy, D.; Levin, J.; Flynn, A.C.; Hayes, L.; Briley, A.L.; Bell, R.; Lawlor, D.A.; Oteng-Ntim, E.; et al. Infant adiposity following a randomised controlled trial of a behavioural intervention in obese pregnancy. *Int. J. Obes.* **2017**, *41*, 1018–1026. [CrossRef]
17. Vesco, K.K.; Leo, M.C.; Karanja, N.; Gillman, M.W.; McEvoy, C.T.; King, J.C.; Eckhardt, C.L.; Smith, K.S.; Perrin, N.; Stevens, V.J. One-year postpartum outcomes following a weight management intervention in pregnant women with obesity. *Obesity* **2016**, *24*, 2042–2049. [CrossRef]
18. Dalrymple, K.V.; Martyni-Orenowicz, J.; Flynn, A.C.; Poston, L.; O'Keeffe, M. Can antenatal diet and lifestyle interventions influence childhood obesity? A systematic review. *Matern. Child Nutr.* **2018**, *14*, e12628. [CrossRef]
19. Mutsaerts, M.A.Q.; van Oers, A.M.; Groen, H.; Burggraaff, J.M.; Kuchenbecker, W.K.H.; Perquin, D.A.M.; Koks, C.A.M.; van Golde, R.; Kaaijk, E.M.; Schierbeek, J.M.; et al. Randomized Trial of a Lifestyle Program in Obese Infertile Women. *N. Engl. J. Med.* **2016**, *374*, 1942–1953. [CrossRef]
20. Van Elten, T.M.; Karsten, M.D.A.; Geelen, A.; Gemke, R.J.B.J.; Groen, H.; Hoek, A.; Van Poppel, M.N.M.; Roseboom, T.J. Preconception lifestyle intervention reduces long term energy intake in women with obesity and infertility: A randomised controlled trial. *Int. J. Behav. Nutr. Phys. Act.* **2019**, *16*, 3. [CrossRef]
21. Van Dammen, L.; Wekker, V.; van Oers, A.M.; Mutsaerts, M.A.Q.; Painter, R.C.; Zwinderman, A.H.; Groen, H.; van de Beek, C.; Muller Kobold, A.C.; Kuchenbecker, W.K.H.; et al. Effect of a lifestyle intervention in obese infertile women on cardiometabolic health and quality of life: A randomized controlled trial. *PLoS ONE* **2018**, *13*, e0190662. [CrossRef]
22. Wekker, V.; Huvinen, E.; Van Dammen, L.; Rono, K.; Painter, R.C.; Zwinderman, A.H.; Van De Beek, C.; Sarkola, T.; Mol, B.W.J.; Groen, H.; et al. Long-term effects of a preconception lifestyle intervention on cardiometabolic health of overweight and obese women. *Eur. J. Public Health* **2019**, *29*, 308–314. [CrossRef]
23. Van De Beek, C.; Hoek, A.; Painter, R.C.; Gemke, R.J.B.J.; Van Poppel, M.N.M.; Geelen, A.; Groen, H.; Willem Mol, B.; Roseboom, T.J. Women, their Offspring and iMproving lifestyle for Better cardiovascular health of both (WOMB project): A protocol of the follow-up of a multicentre randomised controlled trial. *BMJ Open* **2018**, *8*, e016579. [CrossRef]
24. Hoftiezer, L.; Hof, M.; Dijs-Elsinga, J.; Hogeveen, M.; Hukkelhoven, C.W.; Van Lingen, R.A. From population reference to national standard: New and improved birthweight charts. *Am. J. Obstet. Gynecol.* **2019**, *220*, 383.e1–383.e17. [CrossRef]
25. Cole, T.J. The LMS method for constructing normalized growth standards. *Eur. J. Clin. Nutr.* **1990**, *44*, 45–60. [PubMed]
26. World Health Organization. *WHO Child Growth Standards: Length/Height-for-Age, Weight-for-Age, Weight-for-Length, Weight-for-Height and Body Mass Index-for-Age: Methods and Development*; WHO Press: Geneva, Switzerland, 2006.
27. De Beer, M.; Timmers, T.; Weijs, P.J.; Gemke, R.J. Validation of total body water analysis by bioelectrical impedance analysis with deuterium dilution in (pre) school children. *e-SPEN Eur. e-J. Clin. Nutr. Metab.* **2011**, *6*, e223–e226. [CrossRef]
28. Meng, L.; Zhao, D.; Pan, Y.; Ding, W.; Wei, Q.; Li, H.; Gao, P.; Mi, J. Validation of Omron HBP-1300 professional blood pressure monitor based on auscultation in children and adults. *BMC Cardiovasc. Disord.* **2016**, *16*, 9. [CrossRef]
29. The National High Blood Pressure Education Program Coordinating Committee Member Organizations. *The Fourth Report on the Diagnosis, Evaluation, and Treatment of High Blood Pressure in Children and Adolescents*; NIH publications: Bethesda, MA, USA, 2005.
30. Hochberg, Z.; Feil, R.; Constancia, M.; Fraga, M.; Junien, C.; Carel, J.C.; Boileau, P.; Le Bouc, Y.; Deal, C.L.; Lillycrop, K.; et al. Child health, developmental plasticity, and epigenetic programming. *Endocr. Rev.* **2011**, *32*, 159–224. [CrossRef] [PubMed]

31. Malti, N.; Merzouk, H.; Merzouk, S.A.; Loukidi, B.; Karaouzene, N.; Malti, A.; Narce, M. Oxidative stress and maternal obesity: Feto-placental unit interaction. *Placenta* **2014**, *35*, 411–416. [CrossRef]
32. Rexhaj, E.; Paoloni-Giacobino, A.; Rimoldi, S.F.; Fuster, D.G.; Anderegg, M.; Somm, E.; Bouillet, E.; Allemann, Y.; Sartori, C.; Scherrer, U. Mice generated by in vitro fertilization exhibit vascular dysfunction and shortened life span. *J. Clin. Investig.* **2013**, *123*, 5052–5060. [CrossRef]
33. Kleijkers, S.H.M.; Van Montfoort, A.P.A.; Smits, L.J.M.; Viechtbauer, W.; Roseboom, T.J.; Nelissen, E.C.M.; Coonen, E.; Derhaag, J.G.; Bastings, L.; Schreurs, I.E.L.; et al. IVF culture medium affects post-natal weight in humans during the first 2 years of life. *Hum. Reprod.* **2014**, *29*, 661–669. [CrossRef]
34. Magnus, M.C.; Wilcox, A.J.; Fadum, E.A.; Gjessing, H.K.; Opdahl, S.; Juliusson, P.B.; Romundstad, L.B.; Håberg, S.E. Growth in children conceived by ART. *Hum. Reprod.* **2021**, *36*, 1074–1082. [CrossRef] [PubMed]
35. Valenzuela-Alcaraz, B.; Serafini, A.; Sepulveda-Martínez, A.; Casals, G.; Rodríguez-López, M.; Garcia-Otero, L.; Cruz-Lemini, M.; Bijnens, B.; Sitges, M.; Balasch, J.; et al. Postnatal persistence of fetal cardiovascular remodelling associated with assisted reproductive technologies: A cohort study. *BJOG An Int. J. Obstet. Gynaecol.* **2019**, *126*, 291–298. [CrossRef] [PubMed]
36. Gallou-Kabani, C.; Vigé, A.; Gross, M.-S.; Boileau, C.; Rabes, J.-P.; Fruchart-Najib, J.; Jais, J.-P.; Junien, C. Resistance to high-fat diet in the female progeny of obese mice fed a control diet during the periconceptual, gestation, and lactation periods. *Am. J. Physiol. Metab.* **2007**, *292*, E1095–E1100. [CrossRef] [PubMed]
37. Dennison, C. Effects of a Maternal Pre-pregnancy Dietary and Pharmacological Intervention on Maternal Fecundity and Offspring Health in Sprague-Dawley Rats. Available online: http://hdl.handle.net/11023/956 (accessed on 18 October 2021).
38. Xu, H.; Fu, Q.; Zhou, Y.; Xue, C.; Olson, P.; Lynch, E.C.; Zhang, K.K.; Wu, C.; Murano, P.; Zhang, L.; et al. A long-term maternal diet intervention is necessary to avoid the obesogenic effect of maternal high-fat diet in the offspring. *J. Nutr. Biochem.* **2018**, *62*, 210–220. [CrossRef]
39. Mustila, T.; Raitanen, J.; Keskinen, P.; Saari, A.; Luoto, R. Lifestyle counselling targeting infant's mother during the child's first year and offspring weight development until 4 years of age: A follow-up study of a cluster RCT. *BMJ Open* **2012**, *2*, e000624. [CrossRef]
40. Tanvig, M.; Vinter, C.A.; Jørgensen, J.S.; Wehberg, S.; Ovesen, P.G.; Lamont, R.F.; Beck-Nielsen, H.; Christesen, H.T.; Jensen, D.M. Anthropometrics and Body Composition by Dual Energy X-Ray in Children of Obese Women: A Follow-Up of a Randomized Controlled Trial (the Lifestyle in Pregnancy and Offspring [LiPO] Study). *PLoS ONE* **2014**, *9*, e89590. [CrossRef]
41. Rauh, K.; Günther, J.; Kunath, J.; Stecher, L.; Hauner, H. Lifestyle intervention to prevent excessive maternal weight gain: Mother and infant follow-up at 12 months postpartum. *BMC Pregnancy Childbirth* **2015**, *15*, 265. [CrossRef]
42. Horan, M.K.; Donnelly, J.M.; McGowan, C.A.; Gibney, E.R.; McAuliffe, F.M. The association between maternal nutrition and lifestyle during pregnancy and 2-year-old offspring adiposity: Analysis from the ROLO study. *J. Public Health* **2016**, *24*, 427–436. [CrossRef]
43. Kolu, P.; Raitanen, J.; Puhkala, J.; Tuominen, P.; Husu, P.; Luoto, R. Effectiveness and Cost-Effectiveness of a Cluster-Randomized Prenatal Lifestyle Counseling Trial: A Seven-Year Follow-Up. *PLoS ONE* **2016**, *11*, e0167759. [CrossRef]
44. Ronnberg, A.-K.; Hanson, U.; Nilsson, K. Effects of an antenatal lifestyle intervention on offspring obesity—A 5-year follow-up of a randomized controlled trial. *Acta Obstet. Gynecol. Scand.* **2017**, *96*, 1093–1099. [CrossRef]
45. Dalrymple, K.V.; Tydeman, F.A.S.; Taylor, P.D.; Flynn, A.C.; O'Keeffe, M.; Briley, A.L.; Santosh, P.; Hayes, L.; Robson, S.C.; Nelson, S.M.; et al. Adiposity and cardiovascular outcomes in three-year-old children of participants in UPBEAT, an RCT of a complex intervention in pregnant women with obesity. *Pediatr. Obes.* **2021**, *16*, e12725. [CrossRef]
46. Rasmussen, K.M.; Yaktine, A.L. *Weight Gain during Pregnancy: Reexamining the Guidelines Committee to Reexamine IOM Pregnancy Weight Guidelines Food and Nutrition Board and Board on Children, Youth, and Families*; National Academies Press: Washington, DC, USA, 2009.
47. Hochner, H.; Friedlander, Y.; Calderon-Margalit, R.; Meiner, V.; Sagy, Y.; Avgil-Tsadok, M.; Burger, A.; Savitsky, B.; Siscovick, D.S.; Manor, O. Associations of maternal prepregnancy body mass index and gestational weight gain with adult offspring cardiometabolic risk factors: The jerusalem perinatal family follow-up study. *Circulation* **2012**, *125*, 1381–1389. [CrossRef]
48. Smith, J.; Cianflone, K.; Biron, S.; Hould, F.S.; Lebel, S.; Marceau, S.; Lescelleur, O.; Biertho, L.; Simard, S.; Kral, J.G.; et al. Effects of maternal surgical weight loss in mothers on intergenerational transmission of obesity. *J. Clin. Endocrinol. Metab.* **2009**, *94*, 4275–4283. [CrossRef]
49. Barisione, M.; Carlini, F.; Gradaschi, R.; Camerini, G.; Adami, G.F. Body weight at developmental age in siblings born to mothers before and after surgically induced weight loss. *Surg. Obes. Relat. Dis.* **2012**, *8*, 387–391. [CrossRef]
50. Gaillard, R.; Welten, M.; Oddy, W.H.; Beilin, L.J.; Mori, T.A.; Jaddoe, V.W.V.; Huang, R.C. Associations of maternal prepregnancy body mass index and gestational weight gain with cardio-metabolic risk factors in adolescent offspring: A prospective cohort study. *BJOG An Int. J. Obstet. Gynaecol.* **2016**, *123*, 207–216. [CrossRef] [PubMed]
51. Tanvig, M.; Vinter, C.A.; Jørgensen, J.S.; Wehberg, S.; Ovesen, P.G.; Beck-Nielsen, H.; Christesen, H.T.; Jensen, D.M. Effects of lifestyle intervention in pregnancy and anthropometrics at birth on offspring metabolic profile at 2.8 years: Results from the lifestyle in pregnancy and offspring (LiPO) study. *J. Clin. Endocrinol. Metab.* **2015**, *100*, 175–183. [CrossRef] [PubMed]
52. Dombrowski, S.U.; Knittle, K.; Avenell, A.; Araújo-Soares, V.; Sniehotta, F.F. Long term maintenance of weight loss with non-surgical interventions in obese adults: Systematic review and meta-analyses of randomised controlled trials. *BMJ* **2014**, *348*, g2646. [CrossRef]

53. Menezes, M.M.; Lopes, C.T.; de Souza Nogueira, L. Impact of educational interventions in reducing diabetic complications: A systematic review. *Rev. Bras. Enferm.* **2016**, *69*, 726–737.
54. Roberto, C.A.; Swinburn, B.; Hawkes, C.; Huang, T.T.K.; Costa, S.A.; Ashe, M.; Zwicker, L.; Cawley, J.H.; Brownell, K.D. Patchy progress on obesity prevention: Emerging examples, entrenched barriers, and new thinking. *Lancet* **2015**, *385*, 2400–2409. [CrossRef]
55. Furness, P.J.; McSeveny, K.; Arden, M.A.; Garland, C.; Dearden, A.M.; Soltani, H. Maternal obesity support services: A qualitative study of the perspectives of women and midwives. *BMC Pregnancy Childbirth* **2011**, *11*, 69. [CrossRef]
56. Canoy, D.; Buchan, I. Challenges in Obesity Epidemiology. *Obes. Rev.* **2007**, *8* (Suppl. 1), 1–11. [CrossRef] [PubMed]
57. Kelly, M.; Barker, M. Why is changing health-related behaviour so difficult? *Public Health* **2016**, *136*, 109–116. [CrossRef] [PubMed]

MDPI
St. Alban-Anlage 66
4052 Basel
Switzerland
Tel. +41 61 683 77 34
Fax +41 61 302 89 18
www.mdpi.com

Cells Editorial Office
E-mail: cells@mdpi.com
www.mdpi.com/journal/cells

www.ingramcontent.com/pod-product-compliance
Lightning Source LLC
LaVergne TN
LVHW070124170726
843515LV00007B/1747
* 9 7 8 3 0 3 6 5 7 4 9 3 6 *